Process
Geomorphology

Frontispiece. G. K. Gilbert standing by perched granite boulder in Yosemite National Park, 1908. (Photo by U.S. Geological Survey)

Process
Geomorphology
Dale F. Ritter

Southern Illinois University at Carbondale

wcb
WM. C. BROWN
COMPANY PUBLISHERS
Dubuque, Iowa

Copyright © Fall 1978 by Wm. C. Brown Company Publishers

Library of Congress Catalog Card Number 77-80412

ISBN 0-697-05035-1

Sixth Printing, 1983   2-05035-03

Printed in the United States of America

To my wife, Esta, and my children, Duane, Darryl, Glen, and Lisa,
for patience, encouragement, and understanding
when the task seemed endless.

Contents

Preface

Like most branches of geology, geomorphology has undergone a drastic change in scope and philosophy during the last thirty years. Prior to World War II the discipline was mainly descriptive and primarily concerned with fitting landscapes into some evolutionary stage of development. More recently, as life became more complex and civilizations spread into all Earth environments, geomorphologists have emphasized the applied rather than the historical aspect of the field. It is increasingly clear that this change in philosophy has placed geomorphology at an interface with many other disciplines. Today's geomorphologist must relate to problems that face hydrologists, engineers, pedologists, foresters, and many other types of earth scientists. The bond that unites geology and geomorphology with all of these apparently diverse disciplines is the common need to understand the processes operating within the Earth's surficial systems. Thus, although the historical aspect of landscapes remains important, it is absolutely essential for earth scientists to have a basic understanding of surface mechanics, and in addition, of how those process mechanics are reflected in the landforms they create. This book is an attempt to satisfy those needs. However, because it is intended for undergraduate students rather than for professionals, it must be viewed as only a first step in that direction.

In writing the text I have attempted to assimilate ideas from diverse sources, some of which will be unfamiliar to most undergraduates. A lengthy bibliography has been prepared to make a nucleus of source material available to any student who wishes to pursue a given topic in greater depth. In addition, each chapter concludes with a short list of pertinent references suggested for further reading about the chapter topic. The metric system is utilized throughout the book, although English unit equivalents may be given in parentheses, and a conversion table is included to facilitate transfer from one system to another. Occasionally English units are employed in figures or equations where conversion to the metric system would have been overly cumbersome.

Any attempt to synthesize such a vast body of knowledge is doomed to failure without the counsel and criticism of experts in the field. I was especially fortunate to have the entire text reviewed by Victor R. Baker, Donald F. Eschman, Stanley M. Totten, and Sherwood D. Tuttle, and portions of the manuscript by William B. Bull, Ronald C. Flemal, and Antony R. Orme. I am deeply indebted to these professionals for their constructive advice and conscientious efforts. Errors of omission or interpretation are, of course, mine, as are any other shortcomings that the reader perceives in the book. I am also grateful for the encouragement given to me by my colleagues at Southern Illinois University and to our students, especially R. C. Kochel and J. Ward, for their help. Russell R. Dutcher deserves special thanks for eliminating many of the logistical problems that arose during preparation of the manuscript and for facilitating my work in many other significant ways.

Finally, to those readers who by persuasion or training are not students of geology, let me suggest that we all examine the Earth's surface together. Perhaps the approach in this book will convince us that we must seek each other's expertise if we ever hope to understand and solve the problems associated with Earth's surficial environments.

D.F.R.

One of the remarkable aspects of planet Earth is the infinite variety of its surface forms. It is probably safe to assume that as humans became aware of their physical environment, landscape was the first geologic characteristic they noted. Familiar surface features guided their travels and established their territorial boundaries. As time passed, people learned how best to utilize regional characteristics for different purposes, such as agriculture, trade, and military adventure. They also learned that some landforms possess certain peculiarities that somehow, almost imperceptibly, set them apart from others. Gradually these isolated observations grew into an organized collection of knowledge, and a separate branch of science was born.

Geomorphology is best and most simply defined as the study of landforms. Like most simplistic definitions, this does not do justice to a discipline that can be exciting to even the uninspired and challenging to anyone who enjoys science. Historically, landforms have been analyzed in a variety of ways because different students seek from the landscape different information and different truths. For example, since people live on landforms, geographers may justifiably be concerned with how landscapes affect human events. Engineers, on the other hand, examine surface forms to select the best construction sites or to control the physical environment in the most advantageous manner. While engineers and geographers may look at the same landscape, they probably never ask the same questions about it.

The data accumulated about landforms come from widely divergent disciplines. Synthesizing the facts into a cohesive picture of the Earth's surface, therefore, becomes a monumental task. The diverse nature of the data may explain the appearance of subdisciplines such as dynamic geomorphology or climatic

The What, Why, and How of Geomorphology

1

geomorphology (Büdel 1968) as well as the difficulties geomorphology has always had in finding a definite academic home. Today in the United States, geomorphology is still taught in both geology and geography departments, and the subject matter becomes the responsibility of anyone who will properly adopt it. This stepchild existence between geology and geography has created in the minds of some the undeserved image that geomorphology is not clearly defined as a science nor based on scientific facts.

It is true that traditional geomorphology has been excessively descriptive. Much emphasis in the past was given to placing landforms, both regional and local, into some evolutionary model, so that the field was concerned primarily with historical interpretations. In recent years, however, the discipline has become more quantitative, and research has shifted to studies with a more practical value. Modern geomorphologists often deal with problems that link them directly to other professionals working at the Earth's surface. Obviously, geomorphology is more than any definition can adequately express. Although it has identity, its boundaries are ill-defined and most certainly ephemeral.

More important than a precise definition is the fact that geomorphology is and probably always will be a field-oriented science. Map and photo analyses are necessary first steps to good geomorphic work, and laboratory data support interpretations. But the real test of geomorphic validity is outdoors, where all the evidence must be pieced together into a lucid picture showing why landforms are the way we find them and why they are located where they are. A prime requisite for a geomorphologist is to be a careful observer of relevant field relationships. This trait cannot be easily taught, and truly outstanding geomorphologists usually develop it by learning from their own mistakes. Geomorphic processes are remarkably subtle, and minor changes of basic controls can result in an infinite array of landforms. Invariably, the person with the greatest experience under varied conditions will make the most viable geomorphic interpretations. Thus a geomorphologist, like any other scientist, must learn the trade. There are no short cuts that produce geomorphic insight. It must be acquired gradually through long field experience.

The Rise of Modern Geomorphology

Although the science of geomorphology was centuries in the making, we are concerned with the views of its current practitioners. Those interested in an extensive discussion of the history of geomorphology will find excellent scholarly works already available (Chorley et al. 1964, 1973). However, to put the chapters that follow into perspective, we will briefly review some of the recent developments in geomorphic thinking.

Although every segment of geomorphology has made remarkable strides, the discussion that follows deals mainly with the historical development of fluvial geomorphology. This is not meant to denigrate other subdisciplines of geomorphology such as glacial or coastal geomorphology, nor to imply that the other fields have been stagnant over the years. Our fluvial emphasis here is simply a vehicle to demonstrate how the philosophical approach to the study of geomorphology has changed during the twentieth century.

The Western Expeditions

We begin with the late nineteenth-century explorations of the western United States, which laid the foundation for much of modern geomorphology. That vast area opened a new geomorphic world for geologists who were accustomed to the humid settings of Europe and the eastern United States. Its surface was not veiled by dense vegetation but stood bare of foliage as if meticulously prepared for a thorough geologic examination. The excellent preservation of critical field relationships in the West allowed the early observers to formulate concepts about surface form that had been previously understood in only the vaguest terms. Revelations, in fact, seemed to be everywhere. In response to the dogged determination of J. W. Powell and the genius of G. K. Gilbert, the enigmatic relationship between form and process began to unravel. One by one, observations and partially understood facts were pieced together into a basic model that stands as the framework of today's geomorphology.

The scientific discoveries of John Wesley Powell (1834–1902) were presented in his reports on the exploration of the Colorado River (Powell 1875) and the geology of the Uinta Mountains (Powell 1876). He was impressed by the magnitude of the erosive processes that he saw and suggested that landscapes theoretically might be denuded to some ultimate level. Powell termed this the *base level,* an idea now accepted as fundamental to geologic processes. Many recognize this concept as his greatest achievement. Although a form of the concept had been expressed earlier (Chorley et al. 1964), no one before Powell had visualized the extent to which erosion might be carried. His feeling that mountain systems are ephemeral under the erosive attack of rivers became a major cornerstone in the later Davisian school of denudational chronology (Chorley et al. 1973).

Powell clearly was impressed also by the effects of uplift on landforms. His genetic classification of river valleys as *antecedent, consequent,* and *superimposed* stemmed from his recognition that vertical erosion was generated by crustal movements. As he pondered the origin of deep canyons, he realized that some landscapes had a history—that certain deeply incised valleys inherited their position and character from some time earlier than the uplift that initiated their development. This insight surely set the stage for the later concepts of polycyclic landscapes and erosion cycles.

Powell's interests in geomorphology were not all large-scale. His descriptions of weathering in arid climates are excellent. He also recognized the influence of vegetation on process, suggesting that sediment yield would not increase proportionately with precipitation because of a damping effect of vegetation.

While Powell gained fame for his conquest of the Grand Canyon and his leadership of the infant U.S. Geological Survey, he also left his mark on geomorphology. Perhaps his greatest contribution, however, was in creating the scientific climate of the time and organizing the team of extraordinary geologists who worked with him in the western surveys. Powell was the driving force in these early expeditions and must be credited with their success.

One of Powell's colleagues, Grove Karl Gilbert (1843–1918) can justly be recognized as the father of modern American geomorphology. His report on the Henry Mountains of Utah (Gilbert 1877) stands today as a classic example of geomorphic thinking and certainly represents the first real attempt by a geologist to relate processes to the origin of landforms.

Many geomorphologists consider Gilbert's major contribution to be his concept of the *graded river*: A stream is in a graded condition when all its available energy is needed to transport the sediment provided to it from the drainage basin. Although some confusion exists as to Gilbert's precise meaning, it seems reasonable to say that in 1877 he believed *grade* to be an equilibrium condition between load and transporting power which was maintained by adjustments in channel slope. He later altered his view about the total predominance of slope in perpetuating the balanced state, but it was his original perception of grade that was utilized by most geomorphologists.

Gilbert's preoccupation with equilibrium was integrated into much of his thinking about landforms. He postulated a series of "laws" that became the basic framework for his geomorphic analyses. For example, his laws of structure (geology), divides, and declivities (slopes) came from his belief that many factors play an active role in the development of a landscape. To Gilbert, topography represented the result of a continuous interaction between these various controls, as demonstrated in figure 1.1; terms like "dynamic adjustment" and "balanced condition" permeate his writing. This recurring theme that landforms reflect some unique accommodation between process and geology is, in my opinion, his greatest contribution to modern geomorphology. The theme manifests itself in passages like the following:

> Erosion is most rapid where the resistance is least; and hence as the soft rocks are worn away the hard are left prominent. The differentiation continues until an equilibrium is reached through the law of declivities. When the ratio of erosive action as dependent on declivities becomes equal to the ratio of resistances as dependent on rock character, there is equality of action. (Gilbert 1877, pp. 115-16)

Figure 1.1. Interpretation of slope adjustment to geology by G.K. Gilbert. Equilibrium slope developed at *a* is maintained at times *b* and *c*.

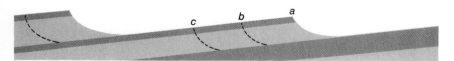

It is clear from this that the emphasis on process in modern geomorphology is not really new, nor is the idea that all parts of a landscape are mutually adjusted and interdependent so that they form a system in dynamic equilibrium. But geology at the time was not ready for such penetrating ideas.

Gilbert also believed that the character and effectiveness of processes may change with time. For example, his deduction that vertical erosion by rivers will eventually give way to lateral cutting led him to explicit descriptions of planation. These were later to be utilized in theories concerning the origin of pediments and rock-cut terraces as well as in the Davisian cycle.

In addition to all his other deeds (including a magnificent treatise on Lake Bonneville), Gilbert was a keen observer of processes themselves. He distinguished clearly between competence and capacity in rivers, made detailed observations of weathering phenomena, and correctly predicted the relative magnitude of sediment yields from regions of differing climates. In his later years, he recognized the pervasive nature of soil creep (Gilbert 1909) and climaxed a brilliant scientific career with a study ahead of its time of sediment transport by running water (Gilbert 1914).

Such a short review cannot do justice to Gilbert's influence on modern geomorphology. Hardly a concept in today's treatment of fluvial geomorphology did not have its beginnings in his incredible genius. However, many years passed before Gilbert received the wide acclaim he deserved. He was truly a man before his time.

A third member of the early western expeditions, C. E. Dutton (1841–1912), led some of the field parties into the region of the Colorado Plateau and contributed much to the geologic understanding of the Grand Canyon region. He was among the first to recognize the effect of isostasy on geologic processes, suggesting that a thick accumulation of alluvium was capable of depressing the Earth's crust. Dutton is best remembered, however, for his belief that subaerial erosion could reduce a landscape to its base level. Powell, who introduced the base level concept, considered it as a theoretical plane toward which the elements of erosion were working, but never suggested it was a physical entity that could actually be attained on a regional basis. Dutton, therefore, carried the base level concept to its extreme and in doing so paved the road to the concept of peneplanation. He put into the geomorphic grist mill the last ingredient needed by Davis to develop his grand evolutionary scheme.

The Davisian Era

If Gilbert was ahead of his time, William Morris Davis (1850–1934) was the right man at the right time. Davis's work did little to increase our knowledge of geomorphic processes, and some might argue that it could be ignored in a process-oriented book such as this. However, if it is important to understand where we are in this science, we must also understand where we have been. There can be no argument that Davis influenced geomorphic thinking more than any other person. His interpretations of landforms led American geomorphology away from its geologic roots into the domain of geography, and he dominated the field for nearly a half century.

Davis rose quickly from a rather insecure position as an instructor at Harvard University to become a world-renowned physiographer. The vehicle of his mercurial ascent seems to have been one great idea, the *cycle of erosion* (sometimes called the geographical cycle), which was woven into his thinking in one form or another throughout his life.

The concept of the cycle of erosion is a masterpiece of synthesis. As Davis readily admitted, the ideas needed to formulate a cyclic thesis were already available to him in the writings of Powell, Gilbert, and Dutton. His own contribution was to integrate beautifully the diffuse components into a model for the evolution of regional landscapes. According to Davis, the ideal cycle of erosion begins with rapid uplift of a land mass and is followed by a long period of tectonic quiescence. During the episode of stable conditions, the landscape progresses through a sequence of stages: youth, maturity, and old age. As a region evolves from one stage into another, as shown sequentially in figure 1.2, its characteristic features gradually change until nearly all the original mass has been removed by erosion. What remains has the form of a featureless plain, called a *peneplain*.

In youth, a region is characterized by a small number of streams, set in poorly integrated drainage systems, which are actively cutting downward in rather narrow valleys. High, broad divides occupy most of the regional topography. As the master stream nears its base level and drainage lines increase into well-defined networks, the region enters the mature stage. The main streams begin to erode laterally, and flood plains develop in their lower reaches. The increased number of rivers tends to dissect the region so that divides become narrow and most of the area is occupied by valley-side slopes. Rivers become graded (in Gilbert's original sense) and the entire system is coordinated to erode and transport debris efficiently from the area. Gradually, the valleys widen by lateral erosion, slope angles decrease, and relief is lowered until the region evolves into old age. In this stage the original land mass has been virtually consumed. A few very large rivers separated by low, broad divides flow sluggishly across the region. With this concept, any landscape could be described by its physiographic characteristics and placed in a relative time framework on the basis of its developmental stage.

As Chorley and his colleagues (1973) point out, the intellectual climate at the end of the nineteenth century was perfect for a concept like the cycle of erosion. The idea of biological evolution was permeating all disciplines, and "time" became almost synonymous with continuous and irreversible change. Davis's model was in vogue and, indeed, he made the analogy with organic evolution in his early presentations of the cycle. Gilbert's emphasis on equilibrium was obviously not in phase with this heady philosophy, and so his ideas found little popularity at the turn of the century.

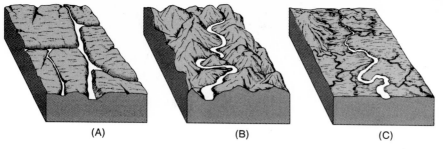

Figure 1.2. Schematic development of the Davisian cycle of erosion. (A) Youthful stage with high, broad uplands and narrow valleys. (B) Mature stage with narrow divides and predominance of slope. (C) Old age stage with peneplain truncating all bedrock except residual hills standing above the level of the peneplain.

Although Davis recognized that initial uplift was not necessarily rapid, he excluded such events from his model as being special cases that only confused the issue. He also regarded climatic changes and uplift during the cycle as aberrations that would interrupt the normal progress of erosion. Thus, much of his writing, and that of his followers, was devoted to explanations of landscapes that were complicated by not being allowed to complete the normal cycle.

In developing the model, Davis introduced his famous landform equation. Landforms are a function of three variables—structure, process, and time. It was evident, however, that he considered time (referring to stages, not absolute values) to be the most important. Landscapes progressing through a cycle showed fewer and fewer effects of the geologic framework and ultimately escaped its control completely. This view of landforms is diametrically opposed to Gilbert's belief that form reflects an equality of action between process and geology. In fact, Gilbert believed that the influence of geology could never be removed from a landscape:

> It is evident that if steep slopes are worn more rapidly than gentle, the tendency is to abolish all differences of slope and produce uniformity. The law of uniform slope thus opposes diversity of topography, and if not complemented by other laws, would reduce all drainage basins to plains. But in reality it is never free to work out its full results; for it demands a uniformity of conditions which nowhere exists. Only a water sheet of uniform depth, flowing over a surface of homogeneous material would suffice; and every inequality of water depth or of rock texture produces a corresponding inequality of slope and diversity of form. (Gilbert 1877, p. 115)

Gilbert's reference to reality strikes at the nub of the problem with the erosion cycle. There is nothing inherently wrong with the Davisian model if it can be substantiated by reality. The fact is that Davis openly admitted that his cycle or its ultimate stage could not be demonstrated. He states:

> . . .the scheme of the cycle is not meant to include any actual examples at all, because it is by intention a scheme of the imagination and not a matter for observation. . . .(Davis 1905, p. 152)

While it may be true that there are today no extensive peneplains still standing close to the sea level with respect to which they were denuded, the examples given in this and in the preceding section seem to me to prove that the earth contains many approximations to the peneplain condition, inasmuch as it preserves some excellent fossil peneplains; and that the stratigraphic as well as the physiographic method of investigation yields abundant and accordant evidence of their occurrence. (Davis 1899, pp. 232-33)

The Davisian approach, therefore, is purely deductive. Although Davis insisted he was examining the present landscape to explain the past (Judson 1960), the method employed by most denudational chronologists is to reconstruct a regional geologic and topographic setting at some point in the past to serve as a starting point for their analysis. The present landscape is then explained by deducing a sequence of evolutionary changes in topography until the landscape assumes its present form (Chorley et al. 1973). Basically, then, the past becomes the key to the present—a rather unusual use of the uniformitarian principle.

Until his twilight years Davis unrelentingly defended the correctness of his cyclic scheme and asserted that with proper interpretation any landform could be placed within its bounds. American dissenters were readily silenced by a stroke of the master's pen. Indeed, the greatest questions about the model came from abroad, especially the German school of geomorphology led by Albrecht and Walther Penck. They objected mainly on the grounds that Davis had not carefully considered the role of erosion during initial uplift, particularly when the rate of uplift was slow or moderate. In their analysis, "old" landscapes could be created during very slow uplift; therefore, flat upland surfaces, interpreted by Davis as rejuvenated peneplains, may have been like that before the cycle of erosion actually began.

The correctness or incorrectness of the Davisian model is not the important point, however, but Davis's influence on American geomorphology during the early part of this century. His dominance was so complete that the significant strides made in the explorations of the West and the growing number of detailed process studies made during the same period never culminated in a concerted drive to understand the physical side of geomorphology. Instead, most geomorphologists followed the more fashionable and philosophical Davisian path.

The Return to Gilbert

With the death of Davis in 1934, the cyclic theory lost its driving force, and geomorphologists began to question not only its basic tenets but also its relevance. In a symposium concerning the work of Walther Penck, Leighly states:

Davis's great mistake was the assumption that we knew the processes involved in the development of land forms. We don't; and until we do we shall be ignorant of the general course of their development.

In his eagerness to set up a general system, Davis jumped over the preliminary, necessarily painfully slow study of processes, and so left his system with an inadequate foundation. (Symposium 1940, p. 225)

Articles utilizing the cyclic concept were still common in geological journals through World War II (Judson 1960) but the tide was inexorably changing. Quantitative analyses of processes found a rebirth in studies by Rubey (1938), Bagnold (1941), and Horton (1945). New statistical descriptions of landscapes were introduced by Strahler (1950, 1952a) and his students and have continued to the present. At the same time, spurred by Leopold and Maddock's (1953) avant-garde approach to river processes, the U.S. Geological Survey produced a series of papers that laid the foundations for a revitalized geomorphology—one in which landforms are explained by examining the mechanisms of their origin rather than their supposed historical meaning. By the end of the 1960s almost every basic assumption in the cyclic theory had been seriously challenged. (For an excellent discussion, see Flemal 1971.)

In retrospect, disenchantment with the cyclic model left a vacuum in geomorphology. Most geomorphologists, having lost faith in the cyclic model, found that they had no conceptual structure within which they could work. To provide such a framework, Hack (1960b) suggested an alternative to the Davisian cycle by introducing the *dynamic equilibrium concept,* which, he admits, essentially returns to the Gilbert approach. Hack rejects the idea of stages and proposes that all landforms within a drainage basin are mutually adjusted to reflect an equilibrium between geology and the prevailing process. The equilibrium form will last as long as the controlling factors are not changed because all elements of the surface will downwaste at the same rate. Thus, in the ideal case, landforms become independent of time. In reality, uplift and/or climate change will alter processes, and downwasting will uncover different lithologic units. Changes do occur, but only in response to altered process or geology. Because a new equilibrium form will be established rapidly whenever changes occur, most topography should be adjusted to present conditions. Although few relict landforms should be found, they are not unusual. In fact, applying the equilibrium concept is often difficult because geomorphic responses to altered conditions do not always proceed at the same rates.

The dynamic equilibrium model diverges so radically from the Davisian cycle that the two may logically represent end members of a spectrum of geomorphic possibilities ranging from complete dependence to total independence of time. Several questions are critical. (1) Does time have an effect on landforms even where geology and climate remain constant? (2) Do slopes retreat by maintaining a constant angle or do they flatten progressively with time? (3) Can all parts of a basin be in mutual adjustment at the same time? Since each of these questions demands a time analysis, dynamic equilibrium may be as difficult to prove as the cycle of erosion.

Certainly not all geomorphologists have abandoned the Davisian approach, and many remain uncommitted. Different cyclic models have been proposed (King 1953), and others have been suggested to modify the steady-state requirement inherent in dynamic equilibrium. Bull (1975a, 1975b), for example, points out that a steady-state condition probably does not exist in most geomorphic situations because dependent variables cannot be independent of time if the independent variables, such as climate, tectonics, and human impact, are not constant. He suggests, therefore, that geomorphology commonly must be analyzed by comparing the rate of change of two variables within a system, a concept known as *allometry*.

Perhaps we will never know which model (if any) completely fits the complex nature of geomorphology. Candidly, it may not matter. What really counts is that geomorphologists have shifted their attention to process-oriented analyses. Only by explaining the mechanics of landform origin can geomorphologists provide useful information to other scientists. Thus, viewing landforms in terms of space rather than time is a healthy trend if geomorphology is to continue as a rigorous physical science.

The Basics of Modern Geomorphology

The Delicate Balance—A Systems Approach

The surface of the Earth marks the interface between the atmosphere, hydrosphere, lithosphere, and biosphere. Each of these taken separately would provide enough complexity to defy complete understanding. Add to this the complication of the different spheres' vigorous interaction at the surface, and you begin to realize the problem we face. Because each sphere contributes variables that play a role in developing the character of the Earth's surface, deciphering the precise origin or meaning of a landform is a challenge. To discuss geomorphology coherently, we must reduce the intertwined whole to its simplest components. This can best be achieved by perceiving landforms and processes as spatial entities that exist as steady-state phenomena when we view them. This does not mean that landforms and processes never change. They do. The physics of the Earth and atmosphere demands it. But if we stop time for an instant to understand landforms and process as static phenomena, we can then hope to determine how and why they change. In that sense, form and process reflect a momentary equilibrium, a delicate balance between driving and resisting forces.

The concept of equilibrium in geomorphology is not new. It was espoused by Gilbert, was given new life by Hack, and has quantitative roots in the work of Strahler. It was given structure, however, in Chorley's (1962) application of the *general systems theory* to geomorphology. Although the systems approach has been highly sophisticated, for our purposes it is best to consider landforms as part of an open system in which energy and mass are constantly supplied and removed. Losses and gains of energy or mass are kept in a steady state by continuous adjustment of forms within the system. Landforms serve as regulatory agents to balance gains and losses.

A system may be an actual physical entity composed of real elements within some area or volume boundaries (Howard 1968). A drainage basin, for example, is a system composed of many parts (slopes, valleys, flood plains, soils, rivers, etc.), each of which can logically be considered as a separate subsystem. The subsystems may have even smaller parts (soil profiles, stream channel cross sections) which themselves function as identifiable systems. The Earth's surface thus consists of a hierarchy of systems, each in instantaneous equilibrium.

Each system or subsystem can be defined by measurable variables or parameters which, taken together, indicate the character of the system at the time of measurement (velocity, slope angle, Fe content, etc.). Under equilibrium conditions these variables are totally adjusted to each other and to the external forces that provide or remove energy and mass. Realistically, exact equilibrium may never be attained in the steady state because each system responds to continuously changing external variables (variables outside the system boundaries) and most systems are interdependent. That is, changes in external variables cause reactions within systems, and a change of parameters in one system may require adjustments throughout the entire hierarchy. For example, assume that tectonic forces increase the total area of the ocean basins, resulting in a decline of sea level. This decrease in the ultimate base level causes entrenchment to begin in the lower reaches of the major rivers. The downcutting may be gradually propagated upstream, causing a similar erosional response in each tributary basin. Slopes are regraded to new elevations. Water table levels are lowered. In regions of carbonate rocks a lowered water table may cause a series of solution features, and eventually the surface may collapse. In short, one external change begins a chain reaction of adjustments in the intimately interrelated subsystems.

Furthermore, even if external factors could be held constant forever (which they cannot), some alteration of the system must occur with time. Because mass, in the form of sediment and dissolved solids, is continually being removed from regional systems, some changes in form are inevitable. However, changes in landforms (like all systems) reflect the net effects of external force on the defining variables. Each parameter within a system responds to the external controls at a different rate and in a different way. Even though mass is lost from the major system, each subsystem may adjust with different characteristics, and no regional evolutionary trends can be predicted.

The systems approach has these advantages:

1. It emphasizes the intimate relationship between process and form.
2. It stresses the multivariate nature of geomorphology.
3. It is a less rigid framework than the historical scheme because it recognizes that some forms may not have reached adjustment but owe their character to relict conditions. Late Wisconsinan terraces, for example, may owe their form to geomorphic controls different from those of the present.

Driving Forces and Resistance

With the systems approach, we can see that landforms represent some interaction between driving forces and resisting forces. Driving forces in geomorphology are climate, gravity, and other forces generated inside the Earth. Resistance is provided by the geologic framework. *Process* is something different, although its distinction from force may be rather ambiguous since we cannot have one without the other. Process may be considered as the method by which one thing is produced from something else, and as the vehicle by which a quantity of one system is transferred into, and participates in, the mechanics of another system.

In general, processes are either exogenic or endogenic. *Exogenic* processes operate at or near the Earth's surface and are normally driven by gravity and atmospheric forces. *Endogenic* processes function inside the Earth. Both types may sometimes be involved in the development of landforms. For example, the shape of a volcanic cone is the product of both endogenic volcanism and normal exogenic slope processes. Even though exogenic processes are systems, in our analogy they are the tools that the driving force applies against the resisting geology. How vigorously they are applied depends on the type of climate and the tectonic stability. How strongly they are resisted depends on the type of rocks involved. Exogenic processes, like any tools, are themselves altered by both the driving and the resisting forces. Which process does the most work and produces landforms most efficiently depends not only on the process itself but also on the driving force and the strength of the resisting framework. This is the general explanation for the incredible variation of landforms at the Earth's surface.

Thresholds   Any concept proposing equilibrium inherently implies a contrasting state of disequilibrium. If variations in external factors demand a response within the system, there must be a period of readjustment during which process and form are out of equilibrium. Landslides, subsidence, and gulley erosion are examples of disequilibrium generated when the variables of process and/or geology are altered so they can no longer maintain a balanced relationship. They represent events that occur as the interrelated systems attempt to reestablish a new equilibrium. Such events can happen suddenly or can proceed toward equilibrium over a long period of time, depending on how great the disequilibrium is and how much energy is involved. Obviously, the equilibrium state must have limits at which something tangible happens to the system.

The limits of equilibrium are critical conditions called *thresholds*. Even though they have not been identified in precise terms, thresholds must be expressible in values of the same parameters that define the systems. Theoretically, we should be able to determine the threshold conditions for any process working on any geologic setting. In practice, we are not even close to this stage of refinement.

Schumm (1973) calls a system's responses to variations of external factors *extrinsic thresholds*. Examples are numerous in nature; geologists will be most familiar with threshold velocity in streams, at which sediment movement begins.

The change in external variables (in this case the process) causes instability of the channel sediment. Other examples are found in responses to the fluctuating climate that characterized much of the Pleistocene epoch. A more subtle type of threshold, however, is the *intrinsic threshold*, where instability and failure of the system occur even though external variables remain relatively constant. The threshold conditions develop in response to gradual, often imperceptible, changes within the system. In many cases the threshold represents a deterioration of resistance rather than an increase in driving forces. For example, a region characterized by periodic heavy rains may have stable slopes for a long time, but continuous freeze and thaw or other soil-forming processes gradually reduce the cohesion of slope material. Eventually one storm, no more severe than thousands that have preceded it, triggers slope failure. This type of threshold, also called a geomorphic threshold, may be inherent in the development of landforms (Schumm 1973).

The ideas of disequilibrium and threshold conditions offer geomorphologists both a penetrating approach and a significant future. The disequilibrium state may be more important than the concept of equilibrium to scientists working at the Earth's surface. It is when thresholds are exceeded that things begin to happen, and many deleterious events at the surface may be nothing more than nature's way of reestablishing a geomorphic equilibrium. With that consideration, it becomes critical for geomorphologists to define threshold values for every environment and for all conceivable combinations of process and geology. Such information would be extremely important for future land management and could be the foundation for identifying natural hazards and predicting imminent disasters.

Obviously, we cannot derive the necessary data overnight. But the fact that it can be done was demonstrated by Patton and Schumm (1975) in a study of the Piceance Creek and Yellow Creek drainage basins in northwest Colorado. They were able to relate the incidence of discontinuous gulley erosion to the magnitude of the valley slope, with the results shown in figure 1.3. For any basin with an

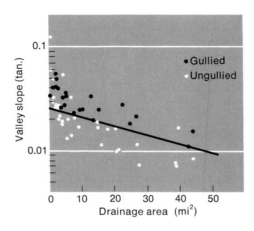

Figure 1.3. Threshold relationship between gullied and ungullied valley floors in several drainage basins of northwest Colorado. (From Patton and Schumm 1975, with permission of the Geologic Society of America)

area greater than 6.4 km² (4 mi²), a particular valley slope becomes a threshold condition. The regression line on the figure represents the slope at which gulley erosion will begin in any valley within the basins they studied. This known threshold condition can be used to predict which valleys are stable or unstable in that region, a valuable tool for future planning.

Cause and Effect   Why is the study of geomorphology significant to geologists? Ingrained in the systems approach and the threshold concept are the ingredients of cause and effect in nature, a dual relationship that is basic to geologic thinking. Cause and effect are essential components of geological history, where the effects are commonly preserved in the rocks and the cause becomes the target of our interpretation and investigation. What has been lacking in our reconstructions is an explicit understanding of how natural systems work. We know, for example, that Holocene climate changes were severe enough to upset the balance in fluvial processes. What is confusing is the diverse manner in which rivers responded to the same climatic trends (Ritter 1974). We must conclude that, for the very recent past, we commonly understand the cause better than the effect. How can we confidently infer how processes functioned 300 million years ago when, in fact, we cannot always predict their responses to modern stimuli? Certainly our insight does not increase as we contemplate the rocks rather than the process. We oversimplify the system in order to make any interpretation at all. There is nothing wrong with this mental process as long as we admit that our models are based on what we *think* about processes, not what we know. We cannot hope to comprehend processes by looking at rocks, for what we see are not processes but the results of processes, and this approach is no more scientific than the deductive scheme of Davis and his followers, in which processes were inferred from topography. What geology needs is a precisely defined understanding of modern processes; until geomorphologists provide it, our reconstructions of past events will be at best educated guesses.

The Driving Forces

Having suggested that processes are the methods by which energy is exerted on earth materials, we should briefly examine the driving forces in our systems. Although each of these forces has been detailed after long and careful study, we will treat them briefly and simply to fit our specific needs.

Climate

Radiation emitted from the sun is the major source of energy needed to drive exogenic processes. If an imaginary plane were placed at the outer limit of the atmosphere, perpendicular to the incoming rays of sunlight, it would receive 2.0 cal/cm²/min of radiant energy over its entire surface. This value, called the *solar constant*, represents the small fraction of the estimated 100,000 cal/cm²/min of energy produced by the sun that survives the long journey to the Earth. Since radiation affects only 25 percent of the earth-atmosphere system at any given time, the constant averaged over the entire surface is only 0.5 cal/cm²/min. As sunlight passes through the atmosphere, another segment of the radiation is reflected back into space by clouds, particulate matter in the atmosphere, and the Earth's surface. Therefore, the amount of energy absorbed in the system (*insolation*) and actually available for work averages about 0.30 to 0.35 cal/cm²/min over the entire globe.

The average insolation value varies greatly, however, with latitude (table 1.1) and with the seasons. Since the total heat budget does not change, the earth-atmosphere system must return to space as much heat as it receives, which it does in the form of long-wave, blackbody radiation. It is significant that although absorbed radiation decreases with increasing latitude, heat loss is fairly constant (table 1.1). This produces an obvious temperature differential between the equator and the poles, requiring a poleward transfer of heat in the oceans and, even more, in the atmosphere. Without this transfer a steady state could not be perpetuated, and the equator would get progressively hotter and the poles progressively cooler. The transfer of solar energy demands a series of complex processes that generate the various components of our weather. These processes tend to establish reasonably well-defined temperature and precipitation patterns for all portions of the Earth's surface. The average weather conditions at any place, considered over a long period of time, are called *climate*. Climate represents the net result of how solar energy is distributed in the earth-atmosphere system. In combination with the surface material of a region, it determines vegetation, weathering and soil-forming processes, and the hydrology. Combined with gravity, it controls glaciation, mass wasting, and fluvial processes.

Table 1.1 Annual heat balance and the transfer of heat in different latitude zones.

Zones of latitude (degrees)	Fraction of total area	Short-wave radiation absorbed (cal/cm²/min)	Long-wave radiation emitted (cal/cm²/min)	Poleward transport of heat across latitude parallels (cal/min)
0–20	0.34	0.39	0.30	
				57×10^{15} (20°)
20–40	0.30	0.34	0.30	
				77×10^{15} (40°)
40–60	0.22	0.23	0.30	
				50×10^{15} (60°)
60–90	0.14	0.13	0.30	
Weighted mean		0.30	0.30	

From Petterssen, S. 1964. Meterology. In *Handbook of Applied Hydrology*, edited by Ven T. Chow. Copyright © 1964 by McGraw-Hill, Inc. Used with permission of McGraw-Hill Book Company.

Heat Transfer   Some of the heat absorbed at the surface is transferred by *conduction* into exposed Earth materials (rock, soil, etc.) and into the air above the surface. How much heat is conducted into each medium and how deeply it penetrates depend on the physical properties of the air and surface matter. As Petterssen (1964) points out, these conductive properties can be linked to several important facts about thermal distribution. First, the high heat capacity of water makes large lakes and the ocean natural storage bins for heat. Second, the temperature ranges over continents are significantly greater than over the oceans. Third, the depth of heat penetration will be much larger in air or water than in solid, denser materials.

Considered separately in terms of radiation and of conduction, the Earth's surface and the atmosphere do not have balanced heat budgets. The surface gains more heat than it loses by these methods; the atmosphere loses more than it gains. What balances the thermal ledger on a local scale is a third process of heat transfer, *convection,* which causes hotter and lighter air (or water) to move toward zones of lower temperature. Air near the Earth's surface is warmer and so will rise into the cooler zones higher in the atmosphere. Some heat absorbed at the surface is used in the process of evaporation and transferred into the atmosphere in a latent form, along with the ascending water vapor. About 600 cal of heat is absorbed by air when one gram of water is evaporated. It is subsequently released in the atmosphere as sensible heat during condensation and precipitation. This heat usually is liberated at some distance from the point of evaporation, and therefore the redistribution of heat depends partly on air motions and the conditions needed to produce condensation.

The combined effects of these factors help explain the great differences between oceanic and continental climates. Continents heat and cool faster than oceans, with more extreme variations in temperature. Since the thermal character of the surface controls the heating of the adjacent air, we can expect temperatures over land and water to function the same way as the surface itself. The seasonality of mid-latitude continental climates can also be understood in terms of the relative rates of heating and cooling and the associated pressure changes of the land-water settings.

On a larger scale, the inequality of heat with latitude (table 1.1) requires a transfer of heat from the equatorial region to the poles. The precise mechanics of this transfer involves a series of complexly interrelated processes that are a basic concern of meteorology. Variables include world circulation patterns of air masses, vertical and horizontal pressure distribution, fronts, rotation of the Earth, and the distribution of landmass and oceans. In general, heat is transferred poleward by air motions, controlled by the average air pressure and wind patterns of the Earth.

As figure 1.4 shows, at latitudes near 30° (north and south), high-pressure centers dominate the circulation pattern and drive the persistent trade winds at low latitudes. Where the trade winds converge, an equatorial trough is created,

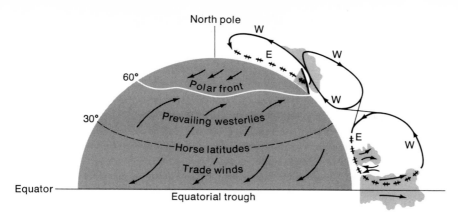

North pole

60°

30°

Equator

Polar front

Prevailing westerlies

Horse latitudes

Trade winds

Equatorial trough

Figure 1.4. Prevailing wind patterns and cellular air motions in the northern hemisphere. (Adapted from Rossby 1941)

and humid low-pressure air forms the rising limb of a giant convection cell. The air moves poleward and sinks, as the descending limb of the convection cell, along the zone of high pressure at 30° latitude. Sinking high-pressure air inhibits cloud formation and causes the predominance of desert conditions in this region. Thus, at low latitudes most heat is transferred poleward by convection, circulation, and the evaporation-precipitation process.

In the middle latitudes, heat migrates in association with cyclone and anticyclone wind circulation in a zone of unsteady westerly air flow ("prevailing westerlies"). Here cold polar air meets the air moving poleward from lower latitudes. Instead of mixing, the different air masses remain clearly defined and are turned into warm low-pressure masses rotating counterclockwise and cool high-pressure masses rotating the opposite way. In the middle latitudes, much of the poleward heat transfer is associated with atmospheric disturbances.

At higher latitudes, warmth is delivered poleward in the upper air by horizontal mixing, called *advection,* when warm air rides over southward-flowing polar air at the polar front.

Ocean currents also are important dispensers of heat. Warm ocean currents moving toward the poles flow under cold air and give off energy as latent heat that ultimately warms the overlying atmosphere. The distribution of heat is not complete in the sense that all regions attain equivalent temperatures. It is effective, however, in stabilizing temperature conditions on a regional basis, providing the Earth with a rather well-defined temperature pattern.

Precipitation is also controlled by the same processes that distribute temperature, and the pattern of precipitation generally follows the major zones of atmospheric circulation. Precipitation is greatest near the equatorial trough and least near the 30° sinking limb of the low-latitude convection cell (fig. 1.5). Secondary peaks of precipitation occur in the middle latitudes with the unsteady pressure cells of that region. Precipitation differs from temperature, however, in that it can only occur when large air masses are cooled. There must be some

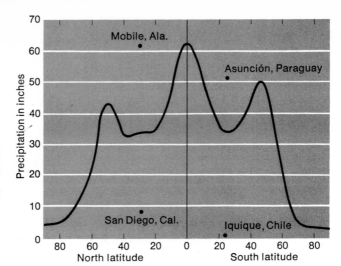

Figure 1.5. Generalized world distribution of average rainfall according to latitude. (From Petterssen, S.1964, Meterology. In *Handbook of Applied Hydrology,* edited by Ven T. Chow. Copyright © 1964 by McGraw-Hill, Inc. Used with permission of McGraw-Hill Book Company)

triggering mechanism that causes moisture-laden air to ascend and be cooled by the decreasing temperatures at higher elevations. The principal mechanisms that lift air masses and thereby initiate precipitation are convection, orographic effects, and frontal mechanics. In convection, warm air, being lighter than surrounding air, rises until condensation forms the familiar bulging shape of cumulus clouds. The orographic effect occurs when air masses are forced to rise over high mountain ranges. Frequent rain occurs on the windward side of the mountains, with a characteristic rainshadow on the leeward side. Frontal activity involves the interaction of low-pressure and high-pressure zones. A *front* is simply the line of contact between the moist, warm air of the low-pressure cells and the cool, dry air associated with high-pressure cells. At the frontal contact, warm air is forced to rise up and over the cooler air and so is cooled; the result is precipitation.

Climate and Landforms   Because of climate's vast effect on many surface phenomena, it was inevitable that scientists would attempt to group climates into some useful classification. Temperature and precipitation commonly are the prime variables used to define climatic boundaries, although vegetation and soil assemblages may serve as auxiliary factors. The most successful classification is an empirical system devised by Köppen in which five major climate groups are distinguished on the basis of observed temperature and precipitation values. Each major group can be divided into progressively smaller units with local variations such as seasonality. The Köppen approach has the great advantage of relying on measured data; climates can be defined by precise physical characteristics, and maps based on this system can provide rather detailed information about local or regional climates. The Köppen classification, however, tells us nothing about what causes the climate and therefore provides no genetic information. More recent attempts at classification employ variables that indicate the origin of climatic zones rather than simply placing them in numerical pigeonholes (Strahler 1965).

Geomorphologists should be more concerned with how well energy in the climatic regime is utilized in geomorphic work than with how the climate is classified. In other words, we need to know which geomorphic process will function most efficiently under any prevailing climate. The relationship between climate and process has been expressed by Peltier (1950), who designated certain areas as *morphogenetic regions*. In these regions, dominant processes can be ascertained and related to a specific climatic regime that is broadly defined by its prevailing temperature and precipitation. This concept has been followed by others, although the climatic variables used to distinguish the regime are sometimes different. L. Wilson (1968) objects to the term ''morphogenetic regions'' because it usually has been used to refer to a process and not to a tangible region of the earth. He suggests that the relationship between climate and process be called *climate-process systems,* and the relationship between climate, process, and landforms be called *morphogenetic systems.* The term ''morphogenetic region'' would be restricted to the case where an actual region is being considered.

Regardless of semantics, it is valuable in geomorphology to establish the relative importance of a particular process under a set of temperature-precipitation conditions. The concept is still useful even when it is recognized that landscapes are controlled by factors other than climate and that the details of morphogenetic systems are poorly understood. It serves as a viable first approximation of what landforms we can expect to find in any environment. In this sense it is helpful in detecting relict features, forms that developed under an earlier and different climate and have not yet adjusted to the present climatic regime. For example, moraines in central Illinois reflect a morphogenetic system that no longer exists; the landforms have not completely readjusted to coincide with modern conditions.

Figure 1.6 is a diagrammatic representation of six climate-process systems that ideally relate climate to a set of paramount processes. The diagram was derived by combining the relationships of several specific processes and climates; the arbitrary system names are meant to be descriptive of the climatic type. For instance (other factors being equal), we can expect that a region with a mean annual temperature of 0°C and mean annual precipitation of 500 mm will function as a periglacial climate-process system, and those processes that function most efficiently in such a climate will prevail. Each climate-process system in the figure can be converted into a morphogenetic system by defining the landforms that most commonly result from the processes involved, as is done in table 1.2.

In reality, minor heterogeneity of geology may produce subtle variations in process and in the topographic outcome. In addition, relict landforms and sea-

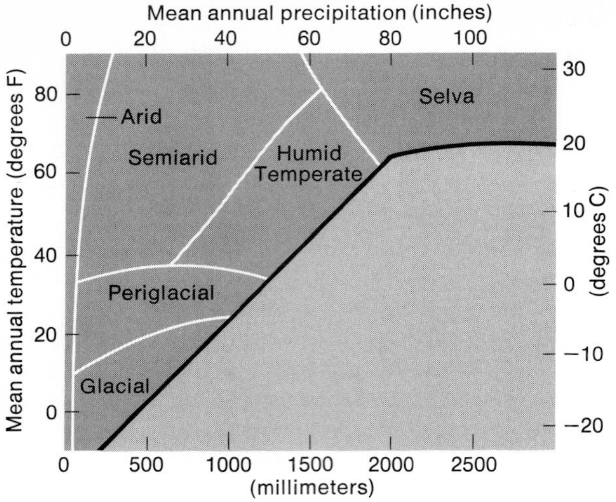

Figure 1.6. Six possible climate-process systems as suggested by Wilson (1968). Each set of temperature-precipitation values tends to drive processes that function most efficiently under those climatic conditions. (Copyright © 1968 by Dowden, Hutchinson, & Ross, Inc.)

sonally changing climates may complicate the expected form. Nonetheless, the morphogenetic approach does provide a reasonable framework for the analysis of landforms. Absolute servility to this approach, however, leads us away from the study of processes, since the method passes quickly from climate to form. This nonchalant leap from climate to landform does not mean that we understand the intermediate phase of process.

In addition to isolating the predominant processes, it is important to understand how climatic variables influence process mechanics. In most cases, solar energy must be converted into another form of energy to do geomorphic work. For example, although precipitation releases stored heat, it is not the temperature of raindrops that drives hydrologic processes. Evaporation and rising air provide water vapor with potential energy that is transformed to kinetic energy when rain begins to fall. A maze of reactions changes energy into various forms that ultimately drive the different geomorphic processes.

Figure 1.7 is a hypothetical flowchart demonstrating how climatic variables influence fluvial processes. Temperature and precipitation exert a direct control on the type and amount of vegetation, the geology (especially how rocks weather), and the hydrology (specifically, how water makes its way into the rivers). Each of these is, of course, also controlled by the type of rocks in the system. Once rainfall and heat strike the surface, they enter a new system in which geology is an extremely important independent variable. Any influence exerted on rivers can never be considered to be completely climatic, as its effect is greatly tempered by the geologic setting.

The result of these controls, including geology as an important partner, is to provide a river with a particular sediment concentration (the amount of sediment per unit volume of water). This value will be determined by how much sediment is yielded from basin rocks and the discharge of the river. The factors listed will produce a particular size of sediment and, in combination with the amount of

Table 1.2 Simple morphogenetic systems and their landscape characteristics.

System Name	Equivalent Köppen Climates[a]		Dominant Geomorphic Processes[b]	Landscape Characteristics[c]
Glacial	EF	Icecap	Glaciation Nivation Wind action (freeze-thaw)	Glacial scour Alpine topography Moraines, kames, eskers
Periglacial	ET EM D-c	Tundra Humid microthermal	Frost action Solifluction Running water	Patterned ground Solifluction slopes, lobes, terraces Outwash plains
Arid	BW	Desert	Desiccation Wind action Running water	Dunes, salt pans (playas) Deflation basins Cavernous weathering Angular slopes, arroyos
Semiarid (subhumid)	BS Cwa	Steppe Tropical savanna	Running water Weathering (especially mechanical) Rapid mass movements	Pediments, fans Angular slopes with coarse debris Badlands
Humid temperate	Cf D-a	Humid mesothermal	Running water Weathering (especially chemical) Creep (and other mass movements)	Smooth slopes, soil covered Ridges and valleys Stream deposits extensive
Selva	Af Am	Tropical Monsoonal	Chemical weathering Mass movements Running water	Steep slopes, knife-edge ridges Deep soils (laterites included) Reefs

[a]Equivalents are approximate, and not exhaustive.
[b]Processes are listed in order of relative importance to landscape (not in order of absolute magnitude). List is abbreviated.
[c]Both erosional and depositional forms are included. List is neither comprehensive nor definitive, merely suggestive.

From L. Wilson 1968. Copyright © 1968 by Dowden, Hutchinson & Ross, Inc.

discharge, will fix the proportion of the total load that the river must transport as bedload or suspended load.

As figure 1.7 shows, variations in sediment concentration (discharge and total load) and sediment size can trigger a multitude of fluvial responses. These responses are not exclusive of one another but may occur in a river simultaneously. In this way, the climatic variables may become manifest in a number of fluvial operations that are impelled through the intermediate controls of sediment and discharge. The entire procedure of applying the driving climatic energy involves an array of actions and reactions that are inherent in open physical systems.

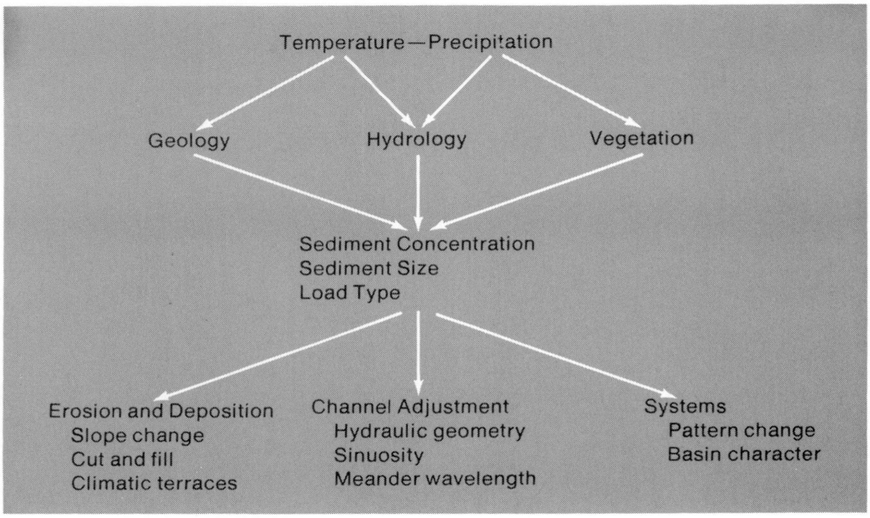

Figure 1.7. Hypothetical flow chart showing how climatic variables exert an influence on rivers. A change in climate alters sediment concentration, sediment size, or load type, requiring a response by the river system. Responses vary depending on local conditions.

The diagram shows: Temperature—Precipitation at the top, with arrows to Geology, Hydrology, and Vegetation. These flow to Sediment Concentration, Sediment Size, Load Type. These then flow to three groups:

Erosion and Deposition
   Slope change
   Cut and fill
   Climatic terraces

Channel Adjustment
   Hydraulic geometry
   Sinuosity
   Meander wavelength

Systems
   Pattern change
   Basin character

Gravity

The second major driving force, gravity, manifests itself in a myriad of both endogenic and exogenic geomorphic processes. Combined with the climatic engine, gravity determines the rigor of fluvial power, mass wasting, glaciation, tidal effects on coastal processes, and the movement of ground water. Internally, gravity bears directly on the process of isostasy, which tends to control the distribution of Earth materials of different densities, ultimately powering regional uplift. Gravity is ubiquitous, affecting all substances. The force of gravity is applied continuously in every system at, above, and beneath the surface and so can never be completely ignored in any consideration of process.

Sir Isaac Newton's classic work on the force of gravity was published in 1687, introducing his law of universal gravitation. Simply stated, the law says that there exists between any two objects a mutual attractive force which is a function of the two masses (m_1 and m_2), the distance separating them (r), and the universal gravity constant (G):

$$F = G \frac{m_1 m_2}{r^2}.$$

Thus, gravity attraction between two objects is an action-reaction phenomenon. Each body exerts a force on the other that is equal in magnitude but oppositely directed along a straight line joining the two bodies. Our main interest in gravity is how it affects geomorphology, especially surficial matter. The gravitational force exerted on surface materials is measured in terms of the amount of acceleration that the force imparts to any freely falling particle having mass. It is normally expressed by the equation

$$g = \frac{GM}{r^2}$$

where M is the mass of the Earth. In most scientific work g is assumed to be constant, having a value of 980 gals (a *gal* being a unit of gravity, which is 1 cm/sec/sec).

From this equation, it is obvious that g in fact cannot be a constant, as we normally assume, because it depends on several variable factors. The distance (r) changes because of topographic irregularities and because the Earth is not a perfect sphere. The density of the Earth materials is not evenly distributed and so may vary along the line connecting the masses. In addition, the rotation of the Earth introduces a counteracting force and causes a distinct latitudinal variation in gravity. Therefore, g is not distributed regularly at the Earth's surface. This fact is a justifiable concern of geophysicists because slight variations in gravity have real significance, especially as an exploration tool. However, the variation in gravity at the Earth's surface is so small compared with the total magnitude that for most exogenic analyses g can reasonably be considered to be constant, and this is normal practice in process analyses. On the other hand, the minor variations that reflect internal density or mass differences are extremely important in endogenic processes. We will discuss gravity again in chapter 2 when considering isostatic adjustments as a geomorphic process.

Internal Heat

Thermal energy, another force, is generated inside the Earth, primarily by radioactive decay and secondarily by friction caused by earth tides and rock deformation. The exact amount of heat available for geologic work is unknown because thermal characteristics deep within the Earth must be estimated from other physical properties (density, pressure, gravity, etc.) which have been determined from analyses of seismic waves. Internal heat can be measured directly only in deep wells or mine shafts; any postulated thermal distribution below the outer fringe of the crust is based on assumption, not observation. Not only are we uncertain about the physical and chemical properties of subcrustal rocks, but hypotheses about temperature distribution tend to involve us in consideration of how much heat the Earth obtained during its formation and early history. Because of the ambiguities involved, estimates of thermal gradients within the Earth vary considerably (see Wyllie 1971). Temperatures proposed for a depth of 1000 km, for example, differ by as much as 1500°C; even at a relatively shallow depth of 100 km, estimated temperatures vary by approximately 600°C.

Regardless of the many problems inherent in determining the vertical distribution of temperature, it is a fact that the Earth transmits to the surface about 2.4×10^{20} cal each year of its internal heat. The total amount of heat is minor compared with the heat received at the surface from solar radiation, but it does indicate that heat, no matter what its origin or gradient, is being transferred from place to place within the Earth. The mechanics of heat transfer is significant since the energy distributed drives internal geological processes. Like its atmospheric counterpart, internal heat is transferred by several methods. Conduction is very slow because of the low conductivity of silicate minerals, but nonetheless it is the dominant transfer mechanism in the crustal layers. Convection as a method is still hypothetical since it cannot actually be seen, but its presence is supported by observed tectonic features, such as the evidence for sea-floor spreading, that are virtually inexplicable without some type of convective overturn. Theoretically, convection is caused by temperature differences at depth (presumably in the mantle) that heat rocks locally and thereby create a less dense mass. The hot, light rocks rise toward the surface as cool, denser rocks are simultaneously sinking to replace the ascending mass. In this way, rock materials of different heat and density are continuously exchanged, following the path of a large convective "cell." The excessive heat at depth is carried along with the rising rock masses and released closer to the surface, efficiently transferring heat.

Measurements of heat reaching the Earth's surface are difficult and costly, and often they are affected by secondary factors such as ground water, variations in conductivity, and recent volcanism. In addition, measurements are not randomly spaced but tend to be concentrated in areas of some specific interest so that large regions exist for which little or no data are available. Nonetheless, the development of sophisticated instrumentation and the current interest in ocean tectonics have produced a storehouse of information that is beginning to yield a reasonable picture of surface heat flow. Except for local abnormalities, heat emerges from all parts of the Earth in amazingly equal amounts, with average continental and oceanic values differing by only 0.2 μ cal/cm^2 (Wyllie 1971). If radioactivity is the major source of internal heat, the equality of heat flow from continents and ocean floors would require an unusual distribution of radioactive minerals beneath the two environments unless the thermal condition were balanced by a convective process. Such a process may be demonstrated by examining heat flow for major physiographic regions of the Earth, as presented by Lee and Uyeda (1965) and shown in table 1.3. Note that heat flow from ocean ridges and trenches differs considerably from average values for entire ocean basins; ridge crests are abnormally high and trenches notably low. Heat may be actively rising under ocean ridges as part of a convective overturn, while the low heat values beneath the trenches presumably represent the descending limbs of the overturning cells. On continents, as one would expect, the lowest heat flow values occur in the very stable shield areas and the highest in the most recent orogenic belts and their associated regions of Cenozoic volcanism.

The transfer of internal heat plays a significant role in determining the major topographic framework of Earth. Heat transfer drives processes beneath the surface causing uplift and deformation, distributes rock masses of varying resistance, and controls the volume of ocean basins, thereby influencing the position of sea level. Precisely how or if heat flow relates to gravity distribution is debatable, but certainly the two forces combined represent a major geomorphic element.

Table 1.3 Heat flow values from major geologic features.

Geologic Feature	Average Heat Flow (μcal/cm^3/sec)
Land Features	
1. Precambrian Shields	0.92
Australian Shield	1.02
Ukrainian Shield	0.69
Canadian Shield	0.88
S. African Shield	1.03
Indian Shield	0.66
2. Post-Precambrian	
Non-orogenic areas	1.54
Europe	1.67
Interior Lowlands, Australia	2.04
Interior Lowlands, N. America	1.25
S. Africa	1.36
3. Post-Precambrian	
Orogenic areas[1]	1.48
Appalachian area	1.04
E. Australian highlands	2.03
Great Britain	1.31
Alpine system	2.09
Cordilleran system	1.73
Island arcs	1.36
4. Cenozoic volcanic areas[2]	2.46
Ocean Features	
1. Ocean Basins	1.28
Atlantic	1.13
Indian	1.34
Pacific	1.18
Mediterranean seas	1.20
Marginal seas	1.83
2. Ocean Ridges	1.82
Atlantic	1.48
Indian	1.57
Pacific	2.13
3. Ocean Trenches	0.99
4. Other ocean areas	1.71

Data from Lee and Uyeda 1965, with permission of the American Geophysical Union.
[1]Excluding Cenozoic volcanic areas
[2]Excluding geothermal areas

As pointed out earlier, landforms reflect a balance between the application of driving forces and the resistance of the material being worked on. Having reviewed the salient features of the driving forces in our systems, we should now examine the resisting elements, but exactly how to do so is rather perplexing. It is tempting simply to state that the resistance in geomorphic systems is geology— the geologic affect on geomorphology is so pervasive and so varied that any brief review of its role in determining process and form must be inadequate. A complete discussion of the geological control of geomorphology would require an analysis of every possible geologic framework in every possible climatic and tectonic regime. Although such an effort is impossible here, some general examples will show how geological resistance manifests itself in landforms and how it directly influences the processes that produce those landforms.

Lithology

The resisting force in geomorphology is implemented through the two major geologic variables, lithology and structure. The diverse origins of rocks create lithologies at the surface that differ vastly in their chemical and mineralogic compositions, textures, and internal strengths. In any given climate each rock type will respond to the processes of weathering and erosion in a different manner and at a different rate. With time and tectonic stability, high-standing landmasses commonly will be underlain by resistant rocks, and low-standing regions will be formed from rocks that are more susceptible to weathering and erosional attacks. These effects of differential weathering and erosion in landscape development are stressed in every introductory course in the basics of geology. In fact, we are conditioned early in our geological training to view regional topography as a mirror of gross lithology, tectonics, and geologic history. For example, the concept of physiographic provinces stresses this approach, causing us to think of geological controls in geomorphology as regional phenomena. It is worthwhile to emphasize, however, that geomorphic differentiations function at many levels, and even minor topographic effects can provide valuable evidence in a geomorphic interpretation. Mega-scaled differentiation produces regional features such as mountains and plains (fig. 1.8), and can be utilized in an erosional topography as a first approximation of the gross lithologic distribution. Within any large region of similar rock type, small lithologic discrepancies will also surrender to geomorphic processes and appear at the surface as minor landform deviations. These tiny blips in the general landscape provide critical information about geological history and exert important controls on subsequent geomorphic developments (figs. 1.9 and 1.10). Even on a microscopic scale, lithologic variations may have a distinct effect on the style of weathering (fig. 1.11), ultimately causing subtle topographic differences (Eggler et al. 1969). It seems certain that even long periods of erosion cannot completely erase the influence of minor lithologic abnormalities from the landscape (Flint 1963), although their appearance may be greatly subdued.

Figure 1.8. Mountains and surrounding plains, looking west-southwest from a point about one mile northeast of Boulder at an elevation of 2,152 meters. Boulder County, Colo., ca. 1934. (Photograph by T.S. Lovering, U.S. Geological Survey)

Figure 1.9. A topographic irregularity caused by differences in lithology—West Spanish Peak, from the northwest. Dikes cutting flatlying Eocene strata. Spanish Peaks quadrangle, Huerfano County, Colo. (Photo by R.C. Hills. From G.W. Stose 1901, U.S. Geological Survey, Folio 71, fig. 4)

Lithologic diversity must be considered on a variety of levels. Large areas underlain by crystalline rocks or sedimentary rocks may develop a distinct regional character, but smaller variations within the region are revealed in subtle topographic changes that often provide significant geologic and geomorphic information. The geomorphologist must be able to "read" these subtle topographic modifications in order to present a coherent interpretation of history and process.

Structure

Geologic structures that influence landforms also range in magnitude from large, area-wide tectonic styles to minor features that exert only local control (Lattman 1968). Structural influence is readily apparent only when the rocks and climate involved are conducive to differential weathering and erosion. In depositional environments, structures may be buried by thick accumulations of sediment that mask the surface expression of the underlying structure. Comparably, the internal structure may not be immediately evident in erosional topography formed in areas with distinctly similar lithology, such as shields or crystalline mountain cores, but minor structures still may produce a discernible topographic control (Flint 1963). Spacing of joints, for example, is recognized as a prime factor in the development of the longitudinal "staircase" profiles that characterize glaciated valleys in mountains held up by uniformly resistant rocks. The most likely lithologic environment to display structural control is a sedimentary sequence with alternating resistant and nonresistant units, such as the Valley and Ridge province of the Appalachian Mountains. There resistant sandstone and conglomerate layers form ridges that are separated by intervening valleys underlain by easily eroded shales and limestones. The regional topography reveals the pervading structure of plunging anticlines and synclines because the ridges cross the countryside in a sinuous pattern that shows the character of the underlying folds.

Figure 1.10. Variations in lithology as evidenced in cuestas formed by hard sandstones north of Galisteo Creek, N.M. The rocks in succession from left to right are Mansamo red beds, Morrison, Dakota, and Mancos (Galisteo Creek and the Santa Fe Railroad in foreground). (Photo by W.T. Lee, U.S. Geological Survey)

Effect on Process

Lithology and structure leave ubiquitous imprints on a landscape, and form is usually related in some intimate way to these variables. It is becoming increasingly clear, however, that these resisting elements also influence the geomorphic processes themselves. The effect may be disclosed in how the process is applied or in the amount of work it is able to accomplish. The resisting framework (geology), therefore, cannot be thought of as merely a passive member of a system, for its influence on process is an active factor in determining what the ultimate balanced landform will be. This exemplifies the fact that geomorphic systems are permeated with *feedback mechanisms;* the geology helps to predetermine how the process will work to mold the geologic framework into topography. Although the details of this interrelationship are poorly understood, a few examples will demonstrate how it works.

Figure 1.11. Disintegration of granitic boulders due to expansion of biotite grains. Boulders in terrace gravel near Red Lodge, Mont. (Photo by R.R. Dutcher)

1. Fluvial processes are directly dependent on the size of sediment produced by weathering and on the relative amounts of water and sediment delivered to the channel (fig. 1.7). Different lithologies produce widely divergent sediment yields; sedimentary and low-rank metamorphic rocks generally yield much larger volumes of clastic debris than igneous or high-rank metamorphics (Judson and Ritter 1964). Rivers adjust in different ways to variations in discharge and sedimentology, and thus the geology plays an integral role in determining how a river will do its work and what the character of the river will be. For example, streams in southern Illinois, which are tributaries to the Mississippi River, drain two distinctly different source areas, one glaciated and one unglaciated. In the region formerly covered by Illinoian ice, the bedrock is buried by a thick accumulation of till and loess. These materials provide streams north of the glacial boundary with a silt-clay load transported primarily in suspension. The streams in this zone are characteristically narrow, deep, and highly sinuous (fig. 1.12). South of the glacial border, in the western part of the Shawnee National Forest, chertiferous limestones are exposed, providing the streams with a coarse-grained gravel load moved primarily as bedload. These rivers are wide, shallow, and quite straight (fig. 1.13). The details of river adjustment to different load requirements have been documented by Schumm (1960, 1963b) and will be discussed in chapter 6. This example simply shows how the resisting material directly influences the type of load a river must transport and how the river adjusts its channel character to do the work required of it most efficiently.

Figure 1.12. Narrow, deep channel cut in loess and till, southern Illinois. (Photo by author)

2. Since lithology is an important determinant of relief, it directly controls slope angles and processes. In a flat-lying sedimentary sequence, a resistant unit commonly serves as a caprock, leading eventually to the typical butte or mesa form. This lithologic setting tends to create oversteepened slopes because the less resistant rocks beneath the capstone are backwasted more rapidly than the capping unit; with time, the lower slopes retreat farther than the upper slopes. This tends to increase the incidence of mass wasting, and the result is frequent slumping and other sudden mass movements (fig. 1.14).

3. Lithology and structure directly influence the degree of infiltration of rainfall, separating it into surface and subsurface components. Unless impermeable rocks are highly fractured or deeply weathered, most precipitation will run quickly off the slopes into stream channels. Streams tend to be flashy in

Figure 1.13. Shallow, wide channel adjusted to coarse gravel load. Shawnee National Forest, southern Illinois. (Photo by author)

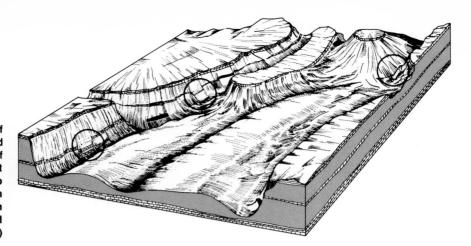

Figure 1.14. Slumping of bedrock due to oversteepened slopes produced by erosion of rocks with different resistance. (Names refer to local formations.) (From Bristol and Howard 1974 with permission of the Geological Society of America)

such areas, increasing to peak discharge and receding to normal flow very rapidly. In highly permeable surface material, much of the rainfall is absorbed into the underground system and discharged gradually into the river channels. Flooding is usually less severe, but high flow is maintained for a longer period of time. Thus, basin hydrology, drainage networks, and groundwater flow depend to a large degree on the surface lithology and structure. In addition, the mode of weathering processes may be governed by the geologic control on infiltration. A good example of this interaction was documented by Wahrhaftig (1965), who suggested that the stepped topography on the west slope of the southern Sierra Nevada resulted from granitic rocks weathering more rapidly where they are covered than where they are exposed.

4. Glacial erosion also depends heavily on the structural and lithologic setting. Resistant igneous and metamorphic terrains commonly show only minor effects of glacial action, while regions underlain by nonresistant rocks are profoundly marked by glacial processes. One needs only to compare the glaciated areas of New England with those in the Central Lowlands to visualize the lithologic control. In New England, drift is thin and morainal topography exists only in valleys that are virtually unmodified by glacial erosion. In contrast, the interior of the United States displays large morainal systems, thick accumulations of drift, and in general, a dominant, constructional topography. The difference between the two regions is due mainly to the surface lithology exposed prior to the onset of glaciation. The tough New England crystallines yielded very little material to the ice, resulting in a glacial system with plenty of energy but virtually no tools to erode with and no load to deposit.

Many other examples of the process-geology interrelationship are possible but not necessary in simply demonstrating that such a interdependence exists. Geology is a multipurpose factor in geomorphic systems because it not only resists the driving forces but also alters the processes involved. Precisely how processes apply energy on the geological framework and how efficiently they mold the surface form depends in part on (1) how readily rocks release their weathered and broken fragments to be used in their own destruction and (2) if and how the process can be altered internally to use its energy best in the performance of geomorphic work.

It is impossible to go further in this introduction without facing up to the question of time in geomorphic systems. The assumption has been implicit that time has no bearing on systemic operations, since landforms and processes have been categorized as time-independent phenomena. Yet it seems obvious that the continued removal of rock material by erosion demands a general decline in relief and an accompanying progressive decrease in potential energy. Some of the potential lost through erosion is restored by uplift and climate change, but unless systems are constantly revitalized in phase with erosion, there must be episodes when the landform-process balance is changing or nonexistent. Two very critical questions arise. First, is energy loss great enough to require a notable modification of landscape, or can minor adjustments of processes absorb the energy change without major alteration of the equilibrium form? Second, is the time interval between successive injections of increased energy short enough to offset energy loss before major adjustments in form occur, or are spasms of uplift and/or climate change separated by long periods during which widespread changes take place in process and landscape? These questions strike at the heart of geomorphology, and to answer them demands an interpretation of how time influences the development of landforms. The great, and sometimes bitter, debates in geomorphology revolve around the different answers. Like it or not, we cannot ignore the consideration of time.

In an attempt to reconcile divergent opinions about the importance of time in geomorphology, Schumm and Lichty (1965) suggested that time could be considered on three levels of magnitude (table 1.4), in which it may or may not be a significant factor in landscape development. They proposed basically that geomorphic variables, including time, assume different degrees of significance when they are examined over various periods of time, and that the distinction between cause and effect depends on what time span is being used in the analysis. During *cyclic time,* which involves long periods, time is significant because continuous denudation demands a concomitant change in landforms as regions progress toward a base level configuration. In this sense, landscapes are very dependent on time. With cyclic time we can consider very large systems as a whole, perhaps never reaching a completely balanced state, but forever approaching equilibrium.

A Consideration of Time

Table 1.4 Drainage basin variables and river variables as they behave during different time spans.

Drainage basin variables	Status of variables during designated time spans		
	Cyclic	Graded	Steady
Time	Independent	Not relevant	Not relevant
Initial relief	Independent	Not relevant	Not relevant
Geology (lithology, structure)	Independent	Independent	Independent
Climate	Independent	Independent	Independent
Vegetation (type and density)	Dependent	Independent	Independent
Relief or volume of system above base level	Dependent	Independent	Independent
Hydrology (runoff and sediment yield per unit area within system)	Dependent	Independent	Independent
Drainage network morphology	Dependent	Dependent	Independent
Hillslope morphology	Dependent	Dependent	Independent
Hydrology (discharge of water and sediment from system)	Dependent	Dependent	Dependent
River variables			
Time	Independent	Not relevant	Not relevant
Geology (lithology and structure)	Independent	Independent	Independent
Climate	Independent	Independent	Independent
Vegetation (type and density)	Dependent	Independent	Independent
Relief	Dependent	Independent	Independent
Paleohydrology (long-term discharge of water and sediment)	Dependent	Independent	Independent
Valley dimension (width, depth, and slope)	Dependent	Independent	Independent
Mean discharge of water and sediment	Indeterminate	Independent	Independent
Channel morphology (width, depth, slope, shape, and pattern)	Indeterminate	Dependent	Independent
Observed discharge of water and sediment	Indeterminate	Indeterminate	Dependent
Observed flow characteristics (depth, velocity, turbulence, et cetera)	Indeterminate	Indeterminate	Dependent

From Schumm and Lichty 1965. Used with permission of American Journal of Science.

Graded time is considerably shorter than cyclic time, although absolute values in years cannot be assigned to either. When geomorphic systems are analyzed on the graded time level, landforms exist in a state of dynamic equilibrium with the processes that mold them. At this level, time is not relevant, because form and process are balanced; however, the equilibrium may apply only in component parts of the system, such as a hillslope or stream segment. Schumm and Lichty argue that all parts of a regional system cannot be simultaneously in dynamic equilibrium, since the whole is still losing mass and therefore must be progressing toward the cyclic-time equilibrium just mentioned. Their contention represents a genuine departure from Hack (1960b) and Chorley (1962), who felt that all parts of a major system could be in a continuous state of dynamic equilibrium. So we arrive once more at the unresolved question of whether a loss in mass by erosion requires a dramatic change in form or whether landscapes can be denuded by losing relief but maintaining the same general exterior shape.

The third level of time, *steady time*, is reserved for even shorter time spans and, in Schumm and Lichty's scheme, applies to even smaller parts of the total system. In this time interval a true steady state is possible, since the time during which variables are being considered is too short for any sensible change in process or form.

To visualize the time distinction more clearly, let us examine a hypothetical drainage basin and the component subsystems (rivers) within it. If we observe a single cross section of any river, we can describe the channel morphology at that point by a group of parameters such as width, depth, slope, and shape. Any single flow event in the river may bring about temporary scouring or filling, which causes immediate changes of the channel parameters but does not affect them permanently; they will return to their previous state when the flow event passes. Measurements taken an hour or day apart will show different values for the variables, but they will always be internally consistent and apparently adjusted to their external controls. Significantly, observed sediment and water discharges through the cross section are dependent variables when viewed in this time sense because they are modified by changes in the channel configuration. For example, if the river scours its bed, the moving load will gain material and the channel area will increase, allowing a greater volume of water to be held within the banks. That is, both water and sediment discharge are temporarily affected by the channel changes. However, we are observing the river for only a very short period of time, and any equilibrium that we define is almost instantaneous (or "steady" in the terminology of Schumm and Lichty). Since no permanent changes can be expected on this temporal scale, we can justifiably say that processes and landforms are time-independent when considered in this sense.

Over longer periods, the steady-time channel measurements will vary with mean values of sediment and water discharge; the statistical relationship between the external and internal variables defines the equilibrium state for graded time. On this time scale (unlike steady time), sediment and water discharge function as independent variables to determine essentially the morphologic and flow characteristics of the stream reach. In graded time, changes in external variables due to modifications of climate or base level may require a new set of equilibrium conditions within the channel. If and when thresholds are exceeded, some significant change in the river is necessary and (again in contrast to steady time) the adjustment will be permanent unless still another change occurs in the external controls. Commonly the response takes the form of channel erosion or deposition, pattern adjustments, or modification of sinuosity.

The point is that the system is temporarily out of balance, and there exists a certain time interval during which the river is approaching a new equilibrium state. The time involved is intermediate between long-term cyclic time and instantaneous steady time. Absolute time in years is not the important element in recognizing graded time as a valid geomorphic concept. Its significance lies in the fact that, once dynamic equilibrium is achieved, disruption of the balance will be counteracted by each subsystem's ability to reestablish a new equilibrium quickly (in the geologic sense); graded time thus does not involve the continuous, progressive changes theorized in the Davisian scheme. In addition, all parts of the regional system may not be affected simultaneously or in precisely the same way.

If we consider our basin over cyclic time, the balance (or temporary imbalance) seen in steady or graded time becomes irrelevant. The inexorable loss of sediment and energy from the basin suggests that the system must be continuously approaching, but perhaps never attaining, an ultimate equilibrium condition. In this time framework, landforms within the basin should be progressively losing relief in phase with the deteriorating systemic energy. If the surface configuration also gradually changes, the Davisian cycle gains considerable credence.

Clearly a hierarchy of time exists; how one is able to interpret geomorphology depends, in a real sense, on which time scale is used. In most of this book we will examine geomorphology on a steady or graded time scale, assuming landforms and processes to be time-independent or only temporarily time-dependent phenomena.

Process—The Unifying Variable

It should be obvious by now that the main emphasis in this book is on the processes that fabricate the surface features of the Earth. What may not be clear is why this approach is more beneficial than using some other criterion, such as climate or time, as a central theme. As we have said, geomorphology stands at the interface between geology and many other disciplines that deal with surficial phenomena. Today geomorphologists must be aware of the problems facing hydrologists, civil engineers, pedologists, foresters, urban planners, and other specialists. And since those scientists are working in an environment underlain and partly controlled by the geologic fabric, they must be concerned with geologic concepts and problems. It follows that there must be a common interest uniting these apparently diverse fields, since they all function in the same place at the same time. It is the universal need to understand processes that is basic to all surficial disciplines.

Furthermore, it is important for all scientists to admit that the complex problems involving surficial systems are with us forever. These problems cannot be solved by a single person trained in the broad manner of nineteenth-century naturalists, nor can they be attacked by a person who understands everything about a small part of the problem but little about its entire scope. Instead, our surface problems will be resolved by teams of scientists and technicians, with each member contributing his or her own special knowledge and expertise to the solution.

What does a geologist trained in geomorphology have to offer in this concerted effort? The answer is that geologists understand better than any other group of scientists the complexity of the natural framework, the causes and effects in nature and, importantly, the significance of time. Then why are geologists commonly excluded from the planning of surficial projects even though the list of human-caused catastrophes—many of which could have been predicted with minimal geological knowledge—continues to grow? Clearly, if geologists intend to communicate with other scientists and engineers or hope to provide them with meaningful and sometimes critical information, other scientists must be convinced that we understand something about the systems in which they work. To do this we must analyze geomorphology on the common ground linking it with the other fields. Failure to change the image of geomorphology will surely leave geologists as merely interested and knowledgeable spectators of human surficial activities.

Summary

Prior to World War II, geomorphic studies were primarily descriptive and directed toward understanding the historical significance of landforms. The emphasis in this book is on the processes that operate in open systems to create landforms. This approach reflects a fundamental change in the philosophical basis of geomorphology that has occurred since World War II. Processes and the resulting landforms will be analyzed as balances between the driving forces (climate, gravity, etc.) and the resistance offered by the geologic framework that makes up the Earth's surface. Processes will be considered on short time scales rather than over geologically significant intervals.

Suggested Readings

This short list of references includes readings that will help you understand the emphasis used in this book.

Bull, W. B. 1975a. Allometric change of landforms. *Geol. Soc. America Bull.* 86: 1489-98.

Chorley, R. J. 1962. Geomorphology and the general systems theory. U.S. Geol. Survey Prof. Paper 500-B.

Flemal, R. C. 1971. The attack on the Davisian system of geomorphology: A synopsis. *Jour. Geol. Educ.* 19: 3-13.

Hack, J. T. 1960b. Interpretation of erosional topography in humid temperate regions. *Am. Jour. Sci.* Bradley Vol. 258-A: 80-97.

Schumm, S. A. 1973. Geomorphic thresholds and complex response in drainage systems. In *Fluvial geomorphology,* edited by M. Morisawa, pp. 299-310. S.U.N.Y., Binghamton, Pubs. in Geomorphology, 4th Ann. Mtg.

Schumm, S. A., and Lichty, R. W. 1965. Time, space and causality in geomorphology. *Am. Jour. Sci.* 263: 110-19.

Strahler, A. N. 1952a. Dynamic basis of geomorphology. *Geol. Soc. America Bull.* 63: 923-38.

In theory, exogenic processes unimpeded by opposing forces will gradually reduce the landscape to a rather dull, featureless surface with only minor topographic irregularities to interrupt its sameness. This procession to flatness must be a salient characteristic of regions that, considered on a cyclic time scale, are dominated by erosional processes. However, as the Earth's surface does possess relief, and has done so throughout geologic time, the Earth must be constructed in such a way that exogenic processes are not always preeminent. The surface is not a static environment but a locale where denudational processes have been repeatedly counteracted by the introduction of new mass into the system. Each influx of mass not only brings with it new potential energy and escalated relief, but also resets the cyclic erosional clock.

The mechanisms that offset progressive decrease in surface elevation lie in the endogenic processes. These processes are the concern of geomorphology because they provide several major ingredients needed for a complete geomorphic analysis. First, endogenic processes create the geological framework in which exogenic processes must work; that is, they distribute the Earth's resisting materials. Second, as suggested before, they determine original relief and with it the potential energy used by most exogenic processes to accomplish work. Third, they manifest the way in which internal energy is dissipated.

Overall, endogenic processes create the pristine topography in a geomorphic system. If we could somehow turn off the exogenic engines, the landscape would be clearly a surface expression of internal processes. Obviously, we cannot do this, but we also cannot ignore the endogenic imprint on landscapes, for often it was there before any erosional modification occurred. The imprint is made by the combined effects of diastrophism and lithology and is most easily recognized in major topographic units.

Endogenic Processes and the Resisting Framework 2

Topographic Orders

Geomorphologists have repeatedly attempted to classify surface physiography into a hierarchy of features, with each descending order of forms less significant in the overall appearance of the landscape. In the past, many approaches have been utilized as the basis for classification (Howard and Spock 1940), and new methods are constantly being devised (Strahler 1946; Cailleaux and Tricart 1956; Enzmann 1968; Fairbridge 1968). Table 2.1 represents a rather classical categorization of Earth physiographic orders, having as its primary divisions the continents and ocean basins. Using these as the major units is not simply a matter of convenience, for distribution frequency, plotted in figure 2.1, shows that most land surface exists at two persistent levels, which correspond to the continents and the ocean floors. The Earth's hypsometric curve (fig. 2.1) also indicates the physiographic importance of continents and the ocean floor; surfaces at elevations between sea level and 1 km account for 20.8 percent of the world's total area, and ocean bottoms between 3 and 6 km in depth comprise about 52.4 percent of the total surface area. The average elevation of continents is about 0.875 km, and the average depth of the ocean floor is approximately 3.729 km.

Continents   The rocks that make up continents are drastically different in mineralogic and chemical composition from those that form the ocean basins. They are richer in silica and less dense than the ocean rocks, so that continental crust is thicker and stands higher than the ocean floors. Exactly how this monumental difference between oceanic and continental crust developed is not certain because theories concerning the origin of continents lack hard data. The reconstruction of events that led to the formation of protocontinents depends entirely on indirect evidence of the Earth's early history. Most geologists accept the proposal that the Earth was formed by the accretion of solid particles that col-

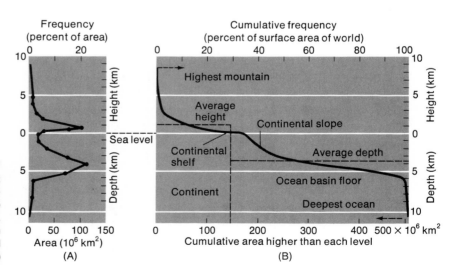

Figure 2.1. Distribution of areas of the Earth standing at different elevations. (A) Distribution frequency. (B) The Earth's hypsometric curve. (From Wyllie 1971. Used with permission of John Wiley and Sons, New York)

Table 2.1 Classification of the Earth's physiographic features according to their order of significance.

First order	Second order	Third order
Continents	Plateaus and plains	Basins
	Mountain systems	Domes
	Continental shelf and slope	Mountain ranges
		Rift valleys
Ocean basins	Abyssal plain	Rift zones
	Ocean ridges	Seamounts
	Trenches	Guyots
	Volcanic island systems	Volcanoes
		Plateaus

lided and coalesced under the influence of gravity. Presumably the process of accumulation and growth occurred well below the melting temperatures of the accreting materials; thus, the original Earth probably consisted of a mixture of cool silicate rocks and iron-nickel solids more or less evenly distributed throughout the mass.

Soon after the accumulation stage ended, the temperature must have risen drastically, apparently driven by radioactive energy, and differentiation of the lithic components began. During partial fusion, the less dense matter separated from the heavier, and the internal stratification of core, mantle, and crust was established. Most likely only one large continent was formed during this great fractionation event (Hess 1962; Dietz and Holden 1970); the lack of bilateral symmetry in the Earth's continents probably is a genetic characteristic. Hess (1962) suggests a single-cell convective overturn as the principal mechanism of the original differentiation. As figure 2.2 shows, during convection, the low-melting sialic constituents were extruded over rising limbs to form the primordial continent. Hess estimates that about one-half the Earth's present continental mass was formed during the original differentiation.

If this was the origin of continental crust, then the fragmentation and distribution of the modern continents and the accumulation of the rest of the continental crust must be the result of endogenic processes. Presumably these processes are still active today. Certainly the maintenance of high continents and low ocean basins represents a balance perpetuated by endogenic mechanics. As figure 2.3 shows, with the exception of Antarctica, large continents stand at higher elevations, a fact that must reflect some primary systemic operation.

Much of the total continental area today is covered by water where the ocean inundates the continental shelves and slopes. These regions are considered to be part of the continent because they are underlain by sialic crust. The abrupt change in slope at the edge of the continental shelf marks the transition from continental to oceanic crust and is normally considered as the continental margin. Margins can be classified on the basis of characteristics that reflect their degree

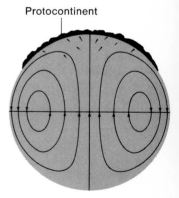

Protocontinent

Figure 2.2. Single-cell convective overturn of the Earth's interior. Continental material is extruded over the rising limb. (After Hess 1962. Courtesy, The Geological Society of America, Inc.)

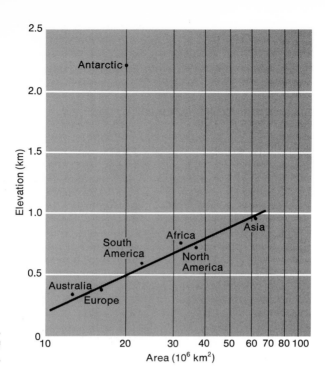

Figure 2.3. Relationship between area and elevation of the continents.

of tectonic stability. Stable margins, called *Atlantic-type* (Mitchell and Reading 1969; Heezen 1975), have broad shelf areas that terminate at the outer edge in a well-defined continental slope. Slopes, in turn, descend at 40–200 m/km until they merge at their base with the continental rise (fig. 2.4). Commonly, the continental slope is dissected by large submarine canyons which serve as passageways for turbidity-current transportation of sediment from the shelf to the continental rise and abyssal plain. In contrast, *Pacific-type* continental margins (Fisher 1975) do not show the ideal shelf, slope, and rise topography because the margin has been disturbed by recent or continuing tectonism. Atlantic-type margins are found in the Pacific Ocean basin and vice versa, so the terms carry no connotation of distribution. Regardless of nomenclature, it is obvious that the transition from oceanic to continental crust can take different topographic forms, which are an external expression of endogenic processes or the lack of them.

In the interior of continents, second-order topographic features consist mainly of shields, mountain systems, plateaus, and plains. A *plain* is a flat or gently undulating surface that is usually underlain by nearly horizontal strata and stands at a relatively low elevation. *Shields* are usually gently undulating surfaces of low relief and elevation marking the central core of most continents. Unlike plains, they are composed of crystalline rocks, most of which are very old. Generally speaking, a *plateau* is a relatively flat surface covering a vast area that

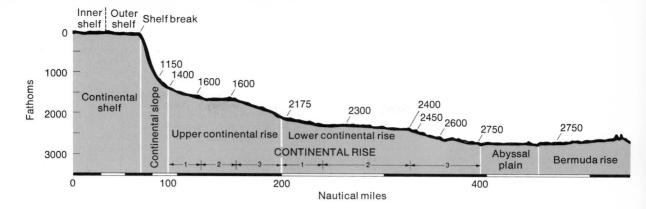

characteristically stands at a much higher elevation than the surrounding region. Plateaus also differ from plains: first, the surfaces of most plateaus are usually dissected by deep river canyons; and second, plains usually stand at lower elevations than plateaus. *Mountain systems* consist of individual mountains or mountain ranges. The system is often elongate and normally is significantly higher than the surrounding surface areas. Mountain systems can be distinguished from plateaus by a more rugged topography; however, highly dissected plateaus may possess considerable relief and are sometimes considered as mountain systems. In this sense, mountains have no tectonic connotation, and the mode of geological development has no bearing on their classification as physiographic features.

Figure 2.4. Profile of an Atlantic-type margin, the continental margin of northeastern United States between Georges Bank and Cape Hatteras, N.C. (From Heezen et al. 1959. Courtesy, The Geological Society of America, Inc.)

The Ocean Floor   Although our main concern in this book is with processes that affect continental masses, ocean floor topography has become extremely significant in the last two decades. Virtually uncharted prior to World War II, the topographic elements beneath the oceans were rather quickly documented when naval operations required such information. Postwar science became enamored with oceanography. The distribution of the major physiographic features listed in table 2.1 is a critical component in the analysis of ocean basin dynamics and, combined with associated geophysical and petrologic evidence, is a cornerstone of modern global tectonics. The features of greatest importance are the mid-ocean ridges and the island arc-trench systems near the continental margins.

Mid-ocean ridges form a belt of mountainous topography that extends continuously over a linear distance of about 56,000 km (fig. 2.5). The ridge zone varies in width but averages more than 1500 km, and peaks within its confines stand 1 to 3 km above the general level of the nearby *abyssal plain*. Commonly a rift valley is found near the central axis of the ridge zone; this feature probably indicates a grabenlike structure, which would require a tensional tectonic environment for its development. Significantly, where the mid-ocean ridge belt intersects a continent, the ocean valley seems to continue as a subaerial rift zone, such as those found in the Middle East (fig. 2.5). Presumably this indicates that the

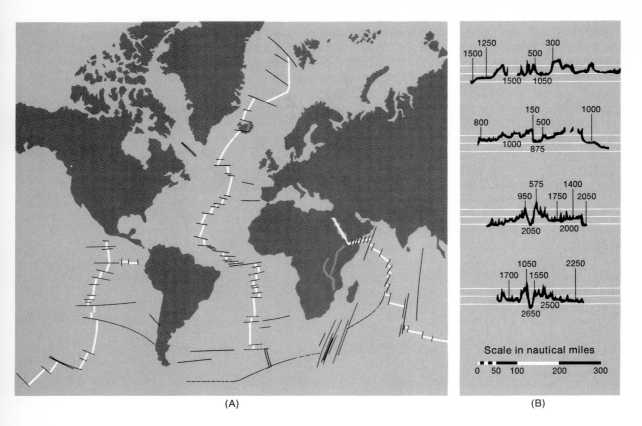

(A)   (B)

Scale in nautical miles

0   50   100   200   300

Figure 2.5. (A) Part of the
Earth's mid-oceanic ridge
system. (After Isacks et al.
1968, *Jour. Geophys. Res.*, 73,
fig. 3, © 1968 American
Geophysical Union) (B) Profiles across the mid-Atlantic
ridge system; numbers represent depth in fathoms. (From
Heezen et al. 1959. Courtesy,
The Geological Society of
America, Inc.)

extensional forces needed to form such features are part of a larger tectonic system that operates without regard to crustal lithology.

In detail the mid-oceanic ridge is not a single uninterrupted feature but a
series of individual ridge segments that have been displaced laterally from one
another (fig. 2.5). The planes along which the ridge components have been
shifted are called transform faults (J. T. Wilson 1965); they are unique in that the
direction of rock movement, as indicated by magnetic anomalies on the ocean
floor (Vine and Matthews 1963; Vine 1966), is opposite of that required to offset
the ridge. In an attempt to explain this phenomenon, Morgan (1968) projected
the transform faults on a spherical surface where relative motion between two
rigid crustal blocks can be described by rotational movements. According to
Morgan, if two adjacent blocks have large transform faults as common boundaries, the faults must lie on "circles of latitude" around a pole of rotation. The
velocity of angular motion must vary across the fault, since it would be
maximum at the "equator" and zero at the poles (fig. 2.6). Therefore, two
blocks moving in the same direction at different rates will show offsets with the
characteristics of transform faults.

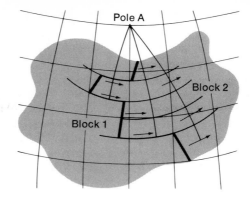

Figure 2.6. Relative motion between crustal blocks. All faults on the boundary between the blocks must be small circles concentric to a pole of rotation (Pole A). (From Morgan 1968. Used with permission of the *Journal of Geophysical Research*)

The mechanics of transform faulting, along with the anomalous magnetic distribution on the ocean floor, confirms the sea-floor spreading hypothesis first proposed by A. Holmes (1929) and later revitalized by Dietz (1961) and Hess (1962). In fact, Morgan's hypothesis of rigid-block rotation fits nicely the observed offsets of the mid-Atlantic ridge and agrees reasonably well with the spreading velocities determined from the magnetic anomalies.

The geophysical character of mid-oceanic ridges is extremely important in determining the relationship between the physiographic form of the ridge system and the endogenic processes that maintain it. In addition to being the regions of greatest heat flow (table 1.3), mid-oceanic ridges are seismically active and the site of many shallow-focus earthquakes. Reversal of the Earth's magnetic polarity has been used to show that new crustal material, mostly basalts, is being injected continuously along the ridge zone, subsequently spreading toward the ocean margins.

The second major topographic component of ocean basins is deep *trenches* that occur normally as curved gashes in the ocean floor. Trenches and other trough types are found most commonly in the Pacific basin, but they are also known to exist beneath the Indian and Atlantic oceans. They are the deepest portions of the oceans, some exceeding 10 km in depth. In cross-profile they are generally V-shaped, but have narrow, flat bottoms and, commonly, benches or terraces on the side slopes. They range from 20 to 120 km in width, while their lengths are measured in thousands of kilometers, producing the typical linear pattern (see Fairbridge 1966). Most deep trenches are located on the ocean side of arcuate chains of volcanic islands that mark zones of transition from oceanic to continental crust (fig. 2.7). The island arcs may consist of an inner and outer belt of active volcanoes where much of the extruded material is andesitic rather than basaltic.

As was shown in table 1.3, heat flow from the trench floors is the lowest of any part of the ocean. Trenches are also noted for extremely large negative gravity anomalies, indicating that they are zones of significant mass deficiencies. In

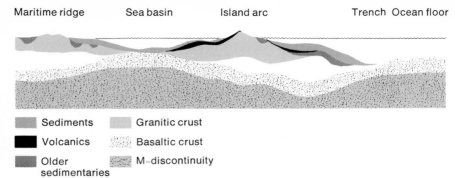

Maritime ridge   Sea basin   Island arc   Trench Ocean floor

Figure 2.7. Profile of tectonic framework in an island arc environment on west side of Pacific Ocean. (From Matsumota 1967. Used with permission of Elsevier Scientific Publishing Co.)

Sediments   Granitic crust

Volcanics   Basaltic crust

Older   M–discontinuity
sedimentaries

addition, the trench-island arc systems are seismically active but, unlike the mid-ocean ridges, earthquake foci occur along a zone dipping toward the continent and descending to considerable depth. This seismic belt, called the Benioff zone, marks the location of nearly all of the Earth's deepest quakes. Some foci are as much as 700 km beneath the surface.

The revelation of ocean floor topography allows us to consider a different classification of features that is more amenable to our discussion because it groups physiographic units that manifest more clearly the work of endogenic processes. Before introducing the classification, we must briefly examine the basis for its plausibility.

Diastrophism

The brief review of the Earth's major topographic components implied that they somehow reflect processes operating within the Earth. We must now attempt to understand how these processes function and how the geologic framework relates to their basic mechanics.

The processes by which the Earth's crust is deformed are collectively known as *diastrophism* and are commonly divided into two distinct tectonic styles. *Orogenic* processes culminate in the formation of structural mountains. Such mountain systems are typified by intense disruption of the included rocks due to folding and overthrusting; the effects are usually localized in narrow, elongated belts. In contrast, *epeirogenic* processes cause uplift or depression on a regional scale and proceed without internal disruption of original rock structures. Response to driving forces is rather passive in this type of deformation. Although gentle tilting of strata may accompany the vertical displacements, folding and thrusting are absent. Regional metamorphism, usually associated with orogenic deformation, is not present in areas affected by epeirogenic tectonics.

Although it is not specified in their original definitions, the different diastrophic styles connote adjustment to oppositely directed internal forces and, therefore, probably to different driving mechanisms. Orogenic deformations are usually associated with crustal shortening, suggesting a horizontal, compressional stress environment. Epeirogenic movements are vertical and probably entail little, if any, compressional stress.

Global Tectonics

Often in the history of science, efforts to synthesize existing knowledge have resulted in quantum leaps in the understanding of natural systems. One truly outstanding breakthrough began when earth scientists developed the techniques needed to examine the ocean floors and culminated in the late 1960s with the encompassing theory of plate tectonics. The allied concept of continental drift is not new—it was forcefully expounded by Wegener in the early part of this century and had its roots in many earlier writings (see Hallam 1973). General acceptance of the idea, however, came grudgingly after a long period of adamant pro-and-con rhetoric and only after the ocean floor surrendered key information that overwhelmed the negative arguments.

The growth of the plate tectonic model can easily be traced through a series of landmark papers (most of which have already been referred to) that took scientists from the concept of sea-floor spreading and its proof (Hess 1962; Dietz 1961; J. T. Wilson 1965; Vine and Matthews 1963; Heirtzler et al. 1968; McKenzie and Parker 1967) to the idea that the outer Earth consists of a group of rigid moving plates (Morgan 1968; Le Pichon 1968; Isacks et al. 1968). Although resistance to global tectonics persists (Meyerhoff and Meyerhoff 1972), it seems to be the only single theory that can reasonably integrate the geophysical, geological, and topographic elements of the Earth. Since many excellent accounts of the plate tectonic model are now available, there is no need for a lengthy discussion here. We will briefly review the basic tenets of the concept, since later discussions will utilize the model framework.

1. The *lithosphere* (approximately the outer 100 km of the Earth) is composed of a small number of rigid plates that vary in thickness and overlie a weaker, hotter zone in the upper mantle called the *asthenosphere*.
2. Plate margins are of three primary types. Along one type of margin new crust is created and added to the plate by injection of asthenospheric material. This margin is identified with the oceanic ridges. At the second margin type, lithospheric material is consumed by being forced downward into the asthenosphere to depths as great as 700 km. Normally the zones of descending rocks (called sinks or subduction zones) are associated with ocean trenches or mountain belts. The third type exists where two plates slide past one another. These mark zones where crust is neither created nor destroyed; they are usually related to major transform faults, such as the San Andreas Fault in California.
3. Entire plates may move with reference to fixed rotational axes, and rocks within plates spread from margins of crustal formation toward zones of subduction. Rates of lateral spreading, based on fossil, radiometric, and magnetic ages, are normally 2 to 5 cm a year (Maxwell et al. 1970; Heirtzler et al. 1968). The total volume of new crust created and destroyed is about equal over any long time period.
4. A single plate may contain both ocean basins and continents; continental drift is thus a manifestation of plate motion. Continents are carried passively

in the upper parts of the lithosphere, moving along with other components of the plates.

5. The endogenic processes that cause plate motion are not known, but several possibilities have been suggested. Convective overturn involving slow creep in crystalline rock (solid flow) is most frequently cited as the probable cause. Hot rocks in the asthenosphere rise along ocean ridges as ascending limbs of the thermal conveyor, and cold material sinks in the subduction zones. Morgan (1972) suggests that plates may be affected by plumes of heated material, roughly 150 km in diameter, rising from the lower mantle to the asthenosphere at a rate of about 2 m per year. As the material spreads out in the asthenosphere, it exerts stresses on the bottom of the plates that are great enough to influence plate motion. These "hot spots" are in fixed positions and may be partly responsible for volcanic island chains, such as Hawaii, that form as plates migrate across the hot spot areas. Morgan calculates that only 20 such plumes may account for as much as one-half of the Earth's heat flow, and implies that such a plume distribution requires the entire mantle to overturn once every 2 b.y. Although convection in the form of cells or plumes shaped like cumulus clouds is theoretical, it seems to be the most logical process to move the plates.

The global tectonic model provides a reasonable framework in which to discuss the relationship between landforms and endogenic processes. Even though precise relations are not clear because the processes functioning in the mantle are distant from their external results, we can establish a morphotectonic hierarchy of landforms based on the plate model (see table 2.2). Morphotectonic analysis is the basis of structural geomorphology, an important consideration in many European studies, especially those in the Soviet Union. Here we will use the morphotectonic units to facilitate our discussion of how surface features reflect the diastrophic processes. Our chief concern is to focus attention on the continental landforms that are subjected incessantly to the exogenic processes.

Table 2.2 Classification of the Earth's landforms based on a plate tectonic model, according to their order of significance.

First Order	Second Order	Third Order	Fourth Order
Plates	Plate margins	Trenches	Intermontane basins
		Ridges	Mountain ranges
		Orogenic zones	Volcanoes
		Island arcs	Rift zones
	Plate interiors	Platforms (plains, plateaus)	Domes
			Basins
		Shields	Guyots, volcanoes, seamounts
		Abyssal plains	
		Shelves	

Assuming that plate margins are the sites of orogenic deformation, we can scrutinize mountain belts as the results of horizontal compressional stresses generated during physical contact between plates moving in opposite directions. The geologic framework of the interacting margins determines the lithologic and structural character of the mountains ultimately produced, while the endogenic processes involved are clearly related to the plate mechanics. In contrast, the plate interiors, called *cratons,* are not affected by the intense tectonism experienced at the margins; deformation in those regions is epeirogenic. The resulting morphotectonic units on continents are *shields* and *platforms.* Platforms are areas of undeformed or gently tilted sedimentary strata that overlie the crystalline basement; physiographically, they are usually described as plains or plateaus. Within platform regions, domes and basins may reflect broad undulations of the crystalline surface underlying the sedimentary rocks. As described earlier, shields are regions where the crystalline rocks, mostly Precambrian, are exposed at the surface. Shields are commonly thought of as the central core of a continental block. The complex history recorded in the shield rocks includes evidence for recurring major orogenies; and so shields may defy morphotectonic classification, as they contain elements of both orogenic and epeirogenic diastrophism. In general, however, the vertical displacements in the cratons are probably not related to the stresses generated at the plate margins and so must reflect endogenic processes other than those associated with plate movement.

Plate Margins, Orogenic Forces, and Mountains

Scientists have pondered the origin of mountains ever since they first recognized that rocks in mountain belts were structurally different from those in other areas. The intense folding and overthrusting displayed within mountainous regions led geologists to realize that significant crustal shortening was involved in their formation, but the transition from this insight to the cause of the deformation was hindered by our ignorance of the Earth's interior. Initially, a progressively cooling and shrinking Earth was suggested to explain the needed compressional stress, but this idea was rejected after the perception of continuous addition of internal heat by radioactivity and the acceptance of a cold origin for the Earth. Other proposals met an equally unsatisfactory fate.

Geologists also noted that abnormally thick accumulations of sediment deposited in elongated troughs were somehow precursors of mountains. Until recently, most analyses of mountain building revolved around the question of how these geosynclines evolved into orogenic belts. Through time, the precise meaning of "geosyncline" was confused by its extension to mean any thick sediment accumulation regardless of geographical position, thus including deposits in the continental interior that did not ultimately become part of orogenic mountain systems. Although grand schemes such as the geosynclinal cycle integrated the sedimentary and metamorphic styles in orogenic belts and accounted sensibly for the sequence of geological events in the transition from a geosyncline to a mountain, the fundamental driving mechanics remained a mystery. The advent of the plate tectonic theory, however, forced geologists to reevaluate geosynclines and mountain building in light of the new global model. Mitchell and Reading (1969) suggest that modern geosynclines are located near continental margins where thick sediment accumulation is continuing today. They also recognize several geosynclinal varieties based on different marginal conditions, each geosynclinal type being marked by a particular assemblage of sedimentary and volcanic rocks (table 2.3).

The significance of the Mitchell-Reading format is not their terminology, but their attempt to relate stratigraphic assemblages to a specific plate boundary and their assertion that any geosynclinal type can permute into any other if the differential movement between ocean and continent is somehow altered. In this way, the development of marginal types and their associated geosynclines can be related to the plate theory. Temporal changes in the position of subduction zones and in the location of continents on the lithospheric plates provide the explanation for the enormous complexity of rock sequences seen in mountain systems.

It would be naive to suppose that the plate theory would spur a new conception of geosynclines and continental margins without geologists taking the next logical step of integrating mountain belts into the model. In fact, young mountains and island arcs are intimately associated with the seismicity and volcanism found at plate margins where the lithosphere is being actively consumed. Since the salient characteristics of old mountains are similar in most respects to those in recent orogenic zones, it is almost anticlimactic to suggest that they have a common origin and that the driving forces in their genesis lie in the mechanics of plate motion. However, shifting subduction zones and the variable distances of plates from poles of rotations introduce a degree of uncertainty into the reconstruction of ancient orogenic events. These factors combined with the ability of plates to break apart and recouple in a different form further increase the complexity of the analyses. Obviously there is room for a variety of interpretations of the mountain-building/plate-tectonic relationship. Until more is known about the Earth's interior it is doubtful that a consensus will emerge. On the other hand, it seems unlikely that someone will argue away the apparent link between orogeny and continental margins.

Dewey and Bird (1970) suggest that mountain belts develop from Atlantic-type continental margins (see table 2.3). Deformation occurs in one of two ways, resulting in two primary types of mountains: *cordilleran-type* and *collision-type*. Although collision and cordilleran mountains form by radically different mechanics, there is no compelling reason to believe that a given mountain system evolved exclusively through one method. In the Appalachians, for example, cordilleran mechanics functioned in the early (Ordovician) orogenic phases, while later (Devonian) tectonism involved collision (Bird and Dewey 1970). Any mountain belt that evolves over long periods of time may have orogenic episodes of both types and lithologic and structural regions representative of each.

Table 2.3 Classification of geosynclinal types based on marginal conditions and rock assemblages.

Characteristic Rock Types	Atlantic-Type Geosynclines		Andean-Type Geosynclines		Island Arc-Type Geosynclines		Japan Sea-Type Geosynclines
	Miogeosyncline Continental Shelf and Coastal Plain	Eugeosyncline Continental Rise, Abyssal Plains, Oceanic Rise	Arcuate Mountain Chain on Continental Margin	Trench	Volcanic Island Chain	Trench	Continental Margin in Restricted Basin
Shallow marine and coastal plain clastics	Abundant		Rare				Abundant (3)*
Carbonates	Abundant		Rare		Locally abundant		Locally common
Interbedded pelagics, tholeiitic lavas, and ultrabasics		Common preflysch (1)*		Abundant preflysch (1)*	Rare preflysch (1)*	Abundant preflysch (1)*	Present if basin floor is oceanic (1)*
Tholeiitic volcanic turbidites		Rare (1)*					
Compositionally mature turbidites		Abundant flysch (2)*		Rare to common flysch (2)*			Abundant flysch (2)*
Calc-alkaline volcanics and minor intrusions			Rare to abundant (2)*		Abundant (2)*		Tuffs
Calc-alkaline volcanic turbidites				Common to rare (2)*	Abundant flysch (2)*	Common flysch (2)*	Rare
Continent-derived coarse clastics			Abundant molasse (3)*				
Intermediate or acidic plutons			Common		Common		
Nature of underlying crust	Continental	Oceanic	Continental	Oceanic	Intermediate	Oceanic	Intermediate, modified oceanic and/or subsided continent

*(1), (2), (3) = upward stratigraphic sequence at any one locality in each geosyncline.

From Mitchell and Reading, 1969. *Journal of Geology 77,* p. 632 and The University of Chicago Press.

Shields and Continental Accretion   The origins of mountain systems are associated with two other phenomena that recur at plate margins. First, the formation of some mountains appears to expand the volume and area of continental crust. The increase, if real, comes from plutonic intrusions and partial melting of geosynclinal sediments and metamorphics to create a granitic mass, a process known as *anatexis*. The increment of growth involved in any orogenic event probably depends on how much sialic matter is released by partial melting during the heating coincident with subduction. Second, although structural deformation and concurrent igneous activity take place at or near sea level, mountain belts, even young ones, stand at high elevations relative to sea level and especially to the level of the ocean floor. Processes other than those causing the deformation are therefore needed to lift the granitized mass after the orogenic event is over. The net result of these two factors is the theory of *continental accretion:* the hypothesis that, through time, continents have increased in size when new sialic crust, born in the fury of orogeny, is united with the pre-orogenic continental block by passively rising to its level. If accretion exists, it is the method by which the final 50 percent of continental mass has been added to the crust since the initial segmentation of the Earth's interior.

The evidence for continental accretion lies mainly in the fact that mountain belts are usually oldest in the center of continents and become progressively younger toward the present continental margins (fig. 2.8). The successively younger mountain chains are delineated by radiometric dates of granites intruded during each orogenic phase. Presumably, then, the sialic crust that composes platform basements and much of the shield area has experienced a series of overlapping orogenies occurring along earlier continental margins. Detailed studies in shield regions show that volcanic activity was prevalent during the orogenic phases and that volcanism probably functioned in much the same way as the present island-arc-continental margin processes (Goodwin 1973, 1975). The volcanic belts in the Canadian Shield, for example, resemble the modern island arc environment in many ways (Goodwin 1975). In addition, if ultramafics are related to a deep sea-mantle origin, belts of these rocks in the western United States demonstrate that there was a westward-expanding, North American continent during much of Phanerozoic time (Maxwell 1975).

It seems clear that orogenic deformation on some continents has shifted its position through geologic time. What is not clear, however, is whether the succeeding mountain-building episodes necessarily indicate a continuous growth of continental crust on a worldwide basis. The pattern demonstrated by figure 2.8 may not represent additions of totally new crust, because some of the sialic matter intruded during each orogeny must have been reworked from older continental crust located at the plate margin (Muehlberger et al. 1967). At any given time, the continents may have been larger than the position of the dated rocks indicates, and part of the "new" crust is merely a reincarnation of older continental material destroyed by subduction. Considering this possibility, estimates

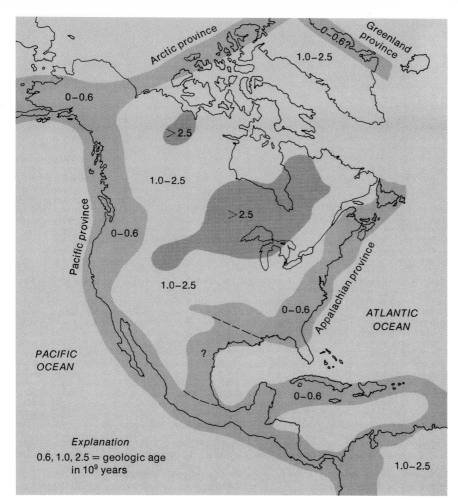

Figure 2.8. Ages and distribution of rocks in North American geologic provinces. Provinces are defined by major granite-forming episodes and mountain-building events. (From Engel, A.E.J., 1963, *Science*, Vol. 140, pp. 143-52, fig. 2. Copyright © 1963 by the American Association for the Advancement of Science)

of the total volume of continental accretion derived from differentiating the mantle may be too high.

In a more drastic analysis, Wise (1972, 1975) proposes that neither the total volume nor the area of the continents and oceans has changed appreciably over the last 2.5 b.y. Instead, continental margins serve as regulatory agents of a constant-volume system in which no net accretion or depletion of continents has occurred since the Earth's early history. Wise bases this proposal on a rather convincing argument that the freeboard of the North American continent has been within ± 60 m of a normal value of 20 m above present sea level for more than 80 percent of post-Precambrian time. Assuming a constant area for the

Earth's surface and a constant thickness of continents through time (Condie 1973), progressive accretion of continental mass could occur only at the expense of the ocean basin area. It is reasonable to expect that continually shrinking ocean basins could hold less and less of the total volume of the ocean water and that, therefore, the freeboard of continents would have increased with time. But the remarkable consistency of freeboard values argues against this, and as Wise points out, the maintenance of a volume-area-isostasy balance is a "master equilibrium of tectonics." If Wise's analysis is correct, it vividly demonstrates the theme of this book—that the Earth's systems, even internally, exist in a beautifully integrated dynamic equilibrium.

In brief, orogeny can change the external shape of continents through time, but the total amount of the world's continental crust may not have increased for 2500 m.y. Extensive transgressions of the ocean onto the continents represent temporary perturbations of a large-scale equilibrium between the volume and the area of first-order Earth features. Equilibrium is probably reestablished by vertical isostatic movements, especially near continental margins. The timing of orogenic events and the exact role of plate mechanics in the overall scheme are far from being resolved (Engel et al. 1974). However, we can say with assurance that shield regions owe their character to both orogenic and epeirogenic processes.

Plate Interiors, Isostasy, and Epeirogeny

In chapter 1 we briefly examined the force of gravity and the various factors that determine its effect on a body at any location. The net force on any mass is the vector sum of all gravitational attractions acting on it. Each body, therefore, possesses a discrete amount of potential energy because mutual attractions can be transferred into a kinetic form that is capable of doing work. Unfortunately, all the attracting elements influencing any particular mass are not applied in the same direction; thus, precise calculations of gravity should be resolved into separate components operating in the orthogonal (x, y, z) directions. This complicated procedure can be simplified by viewing gravity as an energy field consisting of horizons of equal potential (U) in which the attractive force is defined by

$$F = - \,\text{grad}\, U.$$

In this field, F is everywhere normal to a series of surfaces, each of which includes only points with equal potential and therefore a constant value of U. In such a model the value of F can be expressed in terms of energy, and its magnitude, perpendicular to the equipotential surfaces, is

$$F = -\frac{dU}{dr} = \frac{Gm}{r^2}$$

where dU is the change in potential over the distance dr. By integration,

$$U = \frac{Gm}{r}$$

where U goes to zero when r is infinity.

In using this model, sea level becomes an extremely important equipotential surface in the Earth's gravitational field. Even though the surface may be slightly distorted because of local factors, the inflections are small in amplitude compared with the radius of the Earth and limited in areal extent.

The sea level equipotential surface is called the *geoid,* which on land is usually defined by the water level in a series of imaginary canals cut through the solid mass. The Earth's surface topography is referred to as the geoid because any elevation is determined by extending upward a succession of planes that are parallel to sea level. At any point to be measured, a surveying instrument is set tangent to the geoid, with the tangential line being the perpendicular to a vertical plumb line over the site. Like any body, however, the direction of the plumb bob itself is the vector sum of all the gravitational forces acting on it and so may not be perfectly normal to the Earth's center of gravity. To resolve this complication, geodesists utilize a second surface called the *spheroid,* which is a mathematical representation of sea level with all irregular influences removed. Essentially, the spheroid is the *hypothetical* sea level surface of an Earth with no lateral variations in density or topography, and with a vertical change in density that is uniform from the center of gravity to the surface. With such a mass distribution, gravity would vary consistently from pole to equator, and its theoretical sea level values can be easily calculated. The differences between the predicted values of gravity, calculated under the above assumptions, and the actual measured values are called *gravity anomalies;* they indicate the departure of the geoid from the spheroid (fig. 2.9).

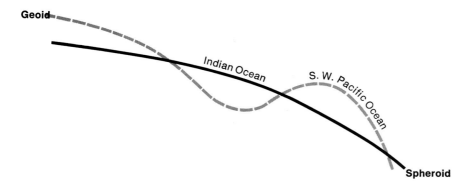

Figure 2.9. Generalized relationship between the spheroid and the geoid along a line between east Africa and the Pacific Ocean near Australia.

Because few gravity measurements are made at sea level, most observations must be reduced into various components and corrected before they can be compared with the spheroidal value. The influence of elevation is such that gravity will decrease with increasing height because the distance factor *(r)* becomes larger. However, this upward decrease in gravity does not consider the effect of the mass between sea level and the observation point, and therefore, it is called the *free-air anomaly*. The attraction caused by the material above sea level can be approximated if the intervening rocks are thought of as a horizontal plate of constant density with a thickness equal to the height of the observation point. The discrepancy in gravity caused by this mass is called the *Bouguer anomaly*. Data concerning the two anomalies reveal a decided tendency for Bouguer values to be strongly negative in mountainous regions and strongly positive in the oceans. This suggests that rocks under mountains have somewhat lower densities than expected, while rocks beneath the ocean basin have higher than normal densities. More significantly, most Bouguer values become more negative with higher elevations, demonstrating a most important principle in geomorphology: that surface topography is somehow related to the internal distribution of mass.

Isostasy   The idea that topography is influenced by the distribution of mass within the Earth is not new. It was expressed by Leonardo da Vinci, and its concrete formulation as a hypothesis arose from analyses of data obtained in the mid-nineteenth-century land surveys in India. Dutton (see chapter 1) introduced the term *isostasy* to define the internal process involved. In essence, the concept of isostasy requires that at some depth beneath sea level the pressure exerted by overlying columns of rock will be the same, regardless of how high the various columns stand above sea level. Mountains, ocean basins, shields, etc., are balanced with regard to the total mass overlying each area at some internal level called the *depth of compensation*. This isostatic equilibrium is maintained by internal adjustments that are not clearly understood, and the mechanics may differ depending on the scale of topography being supported at depth. In reality, most topographic blocks, local or regional, are not perfectly equalized with respect to one another; vertical movements of crustal segments are inherent in the attempt to establish equilibrium. When isostatic balance is disrupted by erosion or tectonics, a counteraction by isostasy is required to restore equilibrium.

Two ideas about the conditions represented by isostatic equilibrium were proposed over a century ago (fig. 2.10), and most analyses of gravity anomalies since then have been designed to test or refine these basic conceptual models. The prime difference between the two models lies in the assumed density of the rocks that support the surface topography from below. In the Pratt-Hayford model, equilibrium occurs at the same depth, and surface irregularities are explained by variations in rock densities between the adjacent vertical columns (fig. 2.10 A). Regions underlain by light rocks stand higher than those over heavier material because additional thickness is needed to provide equal lithostatic pressure at the depth of compensation. In contrast, the Airy-Heiskanen model of isostasy assumes a constant density of rock columns (fig. 2.10 B). Each vertical block is compensated locally; high elevations must be held up, like icebergs in the ocean, by deeper roots. Under mountains this flotation is maintained by a thicker crust which is needed to support the excess mass above sea level.

By employing either of the above theories, a correction of measured gravity due to isostasy can be calculated, and when this is subtracted from the Bouguer anomaly, a residual value called the *isostatic anomaly* remains. Regardless of which compensatory method is utilized, a negative isostatic anomaly indicates an undercompensation, so at that locality the surface should have a tendency to rise. Similarly, positive isostatic anomalies should portend sinking since they indicate overcompensation.

This rather circuitous route leads us to the realization that isostasy is the endogenic process that causes epeirogenic diastrophism. Isostasy also is responsible for maintenance of the topographic relationship between large blocks of the Earth's crust. For example, the relief between ocean basins and continents probably reflects the isostatic balance established because the crustal thickness and density of rocks underlying the two areas are different. The Airy-Heiskanen model seems to fit this relationship rather well (fig. 2.11).

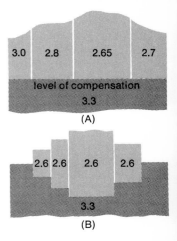

Figure 2.10. Two basic concepts of isostasy. (A) Pratt-Hayford model suggests that equilibrium occurs at the same depth, and surface irregularity is due to different densities of crustal blocks. (B) Airy-Heiskanen model assumes constant density, so that higher blocks must be supported by deeper roots.

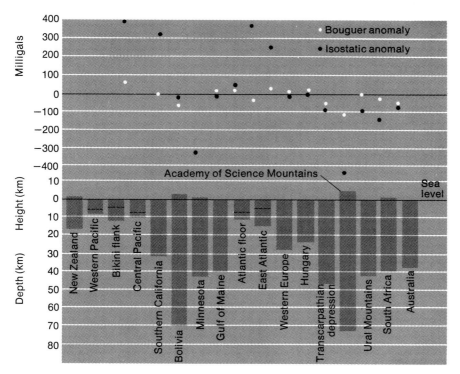

Figure 2.11. Crustal sections over various oceans and continental regions. Bottom of columns represents the Moho. Various gravity anomalies are shown. (Reprinted with permission from G.D. Garland, *The Earth's Shape and Gravity,* 1965, Pergamon Press, Ltd.)

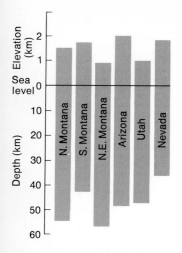

Figure 2.12. Crustal segments in parts of western United States. (Reprinted with permission from G.D. Garland, *The Earth's Shape and Gravity,* 1965, Pergamon Press, Ltd.)

But in looking at crustal segments within an individual continent, the Airy relationship between topographic height and crustal thickness is not nearly so convincing (fig. 2.12). In the United States, some regions of high elevation have crustal thicknesses less than those in adjacent regions of lower elevation (compare Nevada and southern Montana with Utah and northeastern Montana). In the same areas, however, the isostatic anomalies are very small, indicating that compensation has been achieved but apparently not by the Airy mechanism. The failure of the Airy model to explain the isostatic equilibrium in western North America prompted Hsu (1965) to suggest that lateral variations of density in the upper mantle may be sufficient to explain topographic irregularities, thereby eliminating the need for an internal process that controls isostasy by thickening or thinning the crust. For example, the Basin and Range area has an average elevation of 2 km and is underlain by a mantle with a density of 3.3 g/cm^3. Assuming an 80-km depth of compensation, Hsu calculates that a temperature flux that alters the mantle density to 3.45 g/cm^3 would demand a subsidence of 2 km and would result in the region's being submerged. An opposite effect may have caused the original uplift of the area.

In the central interior of the United States, tectonic segments are inversely related to Bouguer anomalies; that is, structural highs (arches and domes) reflect strongly negative Bouguer values while basins show neutral or slightly negative values (fig. 2.13). Because the entire region is isostatically balanced and crustal thickness is fairly constant, it is increasingly evident that even the crust has significant lateral density variations. The area necessary for isostasy to work in the platform domain is somewhat conjectural. Woollard (1962) proposed that compensation is probable for geologic features wider than 250 km but that smaller features would generally be supported by the strength of the crust and so

Figure 2.13. Relationship between Bouguer gravity values and the elevation of the Galena Dolomite in various regions of the Midwest. (From McGinnis 1966, in Illinois Geological Survey, Rpt. of Inv. 219)

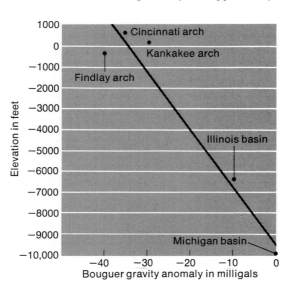

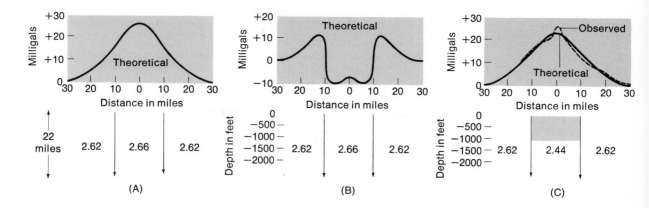

(A)

(B)

(C)

would be lowered or raised along with the larger regional adjustments. It may be, however, that even local structures such as small fault systems are displaced vertically as individual isostatic components as shown in figure 2.14. The salient point is that initial relief in plate interiors, on any scale of magnitude, may be formed by isostatic adjustments to minor variations in crustal densities. Undulations of the surface by uplift or subsidence do not always involve horizontal forces, and common features such as domes and intracratonic basins can form in response to only small divergences from average density values in the Precambrian basements.

Figure 2.14. Interpretation of the gravity field over a graben in northeastern Illinois.
(A) Gravity field resulting from density differences. (B) Field over graben without sediment. (C) Theoretical and observed gravity over the graben filled with sandstone. (From McGinnis 1966, in Illinois Geological Survey, Rpt. of Inv. 219)

Rates of Vertical Displacement   Our main interest in diastrophism is how its mechanics is transferred to the surface domain where uplifted rocks can interact with and directly affect the externally driven exogenic processes. The major influence of tectonics is to create relief through vertical displacements of the surface, for such movement produces energy that is available for use in exogenic processes. Geologists intuitively recognize this effect because enormous rates of deposition have been observed in the geologic record during and following a major mountain-building episode. What is difficult to ascertain from these events, however, is how much uplift or depression of the surface is associated with the orogeny. The enigma is due to the fact that the geologic time scale lacks the precision needed to calculate a meaningful uplift rate, and the geologic evidence is often ambiguous. For example, assume that a shallow-water marine deposit containing fossils of Late Pliocene age is now exposed at an elevation of 3000 m. We may state confidently that 3000 m of uplift has occurred since the Late Pliocene, even though we cannot be certain how far below sea level the animals lived or whether the Pliocene sea level was exactly the same as sea level today. More troublesome is the question of how to translate the assumed 3000 m uplift in terms of rate with units of velocity such as m/yr. Two problems are inherent in such a calculation. First, how many absolute years are represented by the Late Pliocene age of the marine fossils? Is the deposit 2 m.y. old or perhaps 5 m.y.? Obviously the calculated rate of uplift depends on one's interpretation of an inadequately known time interval. For the second problem, assume that an absolute age of 2.2 m.y. is given for the deposit by radiometric dating of inter-bedded volcanics. Here again large plus-or-minus errors must be considered. But assuming the date is accurate, the calculated rate of uplift then becomes 3000 m in 2.2 m.y. or .00136 m/yr. Although the numbers may seem precise, they say very little about the actual rate of uplift because there is no way of knowing whether the vertical motion was continuous over the entire time interval or whether is occurred in one catastrophic spurt sometime between the Late Pliocene and the present. The calculated rate assumes a constant and continuous uplift for 2.2 m.y. and so is a minimum value. If the entire uplift was accomplished during a limited segment of the total time, the rate may have been much higher.

Meaningful uplift rates are important in geomorphology because they determine whether a surficial system can remain in equilibrium during the uplift event. If rates of uplift exceed by far the prevailing rates of denudation, the system will be pushed into disequilibrium. How long it will take to reestablish a new balance depends on how radical the difference is between the rates of uplift and of denudation. Schumm (1963c) discusses the complications involved in making such an analysis because although uplift rates are significantly higher than denudation rates, the uplifts occur in short, spasmodic bursts rather than as long, continuous events.

Because of the problems inherent in determining a precise uplift rate from long-term data, table 2.4 is presented to demonstrate rates of vertical displacement measured in the modern setting or based on data from late glacial and postglacial times. The data presented were chosen at random and should not be considered as being complete. Some of the represented movement is caused by isostatic adjustment to the unloading of ice and/or water or the loading of sediment and water. Other motion is generated in orogenic regions that presumably are experiencing the effects of active tectonism. Based on this limited sample, then it is interesting to note that the rates of vertical displacements are high regardless of the tectonic environment. For example, the uplifts in Fennoscandia and central North America, undoubtedly isostatic rebounds initiated by deglaciation, are within an order of magnitude of those in the California and Alaskan regions, which are probably undergoing orogenic deformation. It also appears that all vertical movement is considerably greater than maximum rates of denudation, which will be discussed later. Schumm (1963c) suggests that a reasonable average rate of uplift is about 750 cm/1000 yr. The values presented in table 2.4 are also complicated by the fact that rates for isostatic rebound decline progressively as the region gets closer to equilibrium. They will be very high immediately following the removal of the excess weight of ice or water but may be very low when equilibrium is nearly reestablished (see Gutenberg 1941; Crittenden 1963).

Table 2.4 Rates of vertical displacement.

Area	Time	Rate (cm/1000 yrs.)
Fennoscandia[1]	Recent	(+) 1100
Hudson Bay[2]	Recent	(+) 1700
Lake Superior[1]	Recent	(+) 500
Lake Mead[3]	Modern	(−) 1200
Lake Bonneville[4]	Late glacial—Modern	(+) 1200-60*
East Coast U.S.[5]	Modern	(+) 200 − (−) 500
California Coast Ranges[6]	Late glacial—Recent	(+) 500-800
Alaska[7]	Modern	(+) 2400

(1) Gutenberg 1941 (2) Walcott 1972 (3) Longwell 1960 (4) Crittenden 1963 (5) Fairbridge and Newman 1968 (6) Bandy and Marincovich 1973 (7) St. Amand 1957
*Represents declining rates during last 20,000 years

The significant geomorphic aspects of isostatic uplift can be summarized briefly:

1. Almost all regions within the continents are in some form of isostatic equilibrium.
2. Isostatic anomalies calculated by the Airy-Heiskanen method vary in the United States with the geologic province (Woollard 1966). In the western United States, negative isostatic anomalies are associated with basins and positive values with uplifts. These appear to be related to tectonics that tend to displace large regions from their natural equilibrium position. Regions such as the Basin and Range and the Colorado Plateau are characterized by negative anomalies. In contrast, major basins in the central United States are characterized by positive *isostatic* anomalies, and uplifts by negative values that are readily explained by variations in mean crustal density. There seems to be no universal mechanism governing isostasy on the North American continents, and mass distribution at depth may be defined at places by either the Airy or the Pratt model (Woollard 1966). It is clear, however, that minor lateral variations of density exist in both the crust and the upper mantle of continental masses, suggesting a liaison with the Pratt model. The balance between larger Earth segments—such as entire continents and ocean basins—is readily explained by the Airy hypothesis.
3. Structural features and initial relief in platform regions are formed by vertical epeirogenic movements associated with density variations in the basement rocks. Presumably post-orogenic uplift in plate-margin mountains is also related to isostatic compensation.
4. Redistribution of mass at the surface by erosion and deposition, glacier development, thrusting, etc., requires vertical movement of the underlying rocks to reestablish isostatic equilibrium.

5. The driving force behind isostasy is gravity, which is responsive to a heterogenous distribution of rock density. The cause of density variations and the precise mechanics of isostatic compensation are very poorly understood, but the process and its effect on geomorphic systems are real.

6. Rates of uplift are normally high compared with rates of denudation.

Before concluding our consideration of diastrophism in geomorphology, it seems appropriate to present a different perspective of tectonics. Although the plate theory is supported by voluminous suggestive data, when it is applied to the origin of mountain belts it is just a theory. It may be an excellent hypothesis, but a number of geologists remain unconvinced of its correctness. Some scientists believe strongly that vertical tectonics are much more important in the entirety of mountain building than we care to admit (Beloussov 1970, 1972), and some reasonable evidence exists to support that stance (Qureshy et al. 1974). The main contention in this concept is that deformation of the sedimentary pile occurs *after* rather than before vertical movement—folds and thrusting are due to gravity sliding of the sedimentary pile off the uplifted mass. In light of these objections, we should proceed in our investigation of plate mechanics in a carefully conceived, scientific manner. Tossing caution to the wind in the exhilaration of big, new ideas may be scientific fun, but it commonly leads to shaky foundations for future work.

The Lithologic Factor

The importance of diastrophism as just discussed is to create the structural character of rocks and to provide geomorphic systems with initial relief and the potential energy associated with it. The second major input of endogenic processes is to generate and distribute the variety of rock types in the surficial domain. More than 90 percent of the Earth's crust is composed of igneous and metamorphic rocks. Combined with diastrophic processes, the internal mechanics that produces these crystalline rocks determines the resisting framework in geomorphology, at least in the primitive state. The distribution of sedimentary rocks, which are also resisting material, is the result of exogenic processes; it functions only after the internal machinery builds an original setting that can yield sediments.

In a general way topography depends on the mineralogy and texture of rocks because (1) they control the magnitude of resistance and (2) variations in mineral density require the isostatic adjustments that vertically differentiate crustal blocks. In addition, volcanic activity constructs the unique landform of volcanoes so familiar to even beginning geologists. In that sense, endogenic processes do much more geomorphic work than simply creating blocks of resisting matter waiting to be molded into topographic form.

Our main concern with igneous and metamorphic rocks is not the details of their crystallization; that topic belongs to the petrologist. We should understand, however, where geomorphically significant rock types are found and why they are so dispersed. To do this we must seek the source of magma and, in addition, explore the relationship between tectonism, metamorphism, and plutonic intrusion.

Petrologists are now certain that magma originates by local melting of rocks in the upper mantle or deep within the crust. Before you call this an obvious statement, recognize that until sophisticated geophysics became available, many geologists believed that magma was residual liquid from a cooling and crystallizing molten Earth. As late as 1962, some scientists believed that magma stemmed from a worldwide, semiliquid basalt layer in the lower crust.

A great unanswered question facing petrologists is how many primary magma types exist, and various models have been suggested ranging from one to an almost infinite number. Attach to this confusion the raging granite controversy, and you have the ingredients for individual interpretation and the basis for great petrological debates. The advent of high temperature-pressure experimentation began to narrow the possibilities, however, and certain limiting constraints were imposed on the conditions of magma generation. We can now say that two primary magmas have significance in the geomorphic sense. A basalt-type magma generated in the upper mantle is essentially mafic in composition because it issues from the melting of ultrabasic rocks. Another magma, formed in the crust by anatectic processes, is rich in silica and deficient in ferromagnesian components because it is derived from low-melting minerals. Although petrologists commonly recognize two or more primary basaltic magmas, the chemical and mineralogic differences between them probably have little geomorphic consequence. The magma fused in the mantle can be hybridized into a wide range of liquid derivatives by the processes of fractional crystallization and assimilation. The result is volcanic rocks of infinite chemical variety but rather limited areal extent.

Volcanism and Volcanic Landforms

Volcanism is nothing more than a surface manifestation of the internal processes that create and mobilize magma. Although volcanoes are spectacular in eruption and unique in topographic form, describing examples of active volcanoes is not necessary here. Most physical geology texts do this, and excellent treatises on volcanoes and volcanic landforms are available (Bullard 1962; Macdonald 1972; Green and Short 1971). It may be pertinent, however, to examine briefly how internal variables control the type of volcanic eruption and the ensuing topographic form.

The violence of a volcanic eruption is determined mainly by the composition of the magma and the amount of gas in it. By affecting the viscosity of the magma, these factors influence the observed differences between continental and oceanic volcanism. Even though they are highly fluid, basaltic magmas are produced in both continental and oceanic environments. The more viscous, high-silica lavas that crystallize as andesites, dacites, and rhyolites are generally restricted to continents or marginal island arcs. In orogenic zones, where most viscous lava occurs, the erupting magma also contains more gas than do the true oceanic types. The combined effect of higher gas content and greater viscosity creates a tendency toward more explosive eruptions. This trend is magnified in continental volcanoes where violent explosions are commonplace; tephra, rather than flowing lava, is the dominant volcanic material.

The major surface landforms constructed by repeated eruptions from the same vent or different vents within the same region are related to the fluidity of the lava. *Lava plains* and *plateaus* are extremely flat surfaces, both continental and oceanic, that have been aggraded by overlapping flows of fluid lava with a mafic composition. Although the nature of the vents is not clear, they probably are fissure types distributed over wide areas or a series of unconnected pipe vents. The dominant characteristic of lava plains and plateaus is the enormous volume of lava extruded and spread over a vast surface area. For example, in the Columbia River Plain of Washington and Oregon, areas greater than 200,000 km^2 are known, and in the Deccan Plain of India, more than 500,000 km^2. Oceanic plains can be even more extensive (Kuno 1969). Commonly the total thickness of the extrusive rocks exceeds 2000 m—great enough to bury a mountainous terrain. In fact, in the Columbia River Plain, peaks of buried mountains (called steptoes) may project through the flat surface as isolated ''islands'' of older rock.

The surfaces of lava plains and plateaus show minor perturbations where broad, shield-type cones rise 30–60 m above the general level. However, individual flows, 2 to 50 m thick, can extend for hundreds of kilometers from the cones, blending imperceptibly into lava issued from fissured vents to create a surface that normally slopes less than 1°. Lava plains are almost always composed of basaltic rocks because silicic magma tends to be too viscous for the long distance of flow necessary to create a planar topography.

Some silica-rich volcanics do underlie flat plains, but these rocks did not crystallize from lava and the surface extends over much smaller areas. Silicic plains develop when incandescent volcanic glass is erupted within dense clouds of gas capable of flowing for considerable distances. The welded tuffs (ignimbrites) resulting from the ash flows are common in the cordilleran of the United States, especially in the southwest. Field evidence there suggests that the ash was vented from both linear and arcuate fissures and central pipes. Wherever cones are noted, they resemble the basaltic types except that the summit regions are shattered by collapse caldera, and the side slopes are higher (2°–3°).

A second major volcanic landform is the broadly rounded profile of *shield volcanoes*, typified by the topography of Hawaii and Iceland. The sides of these volcanoes slope from 2° to 10° and merge gradually, at their base, with the adjacent ocean floor (fig. 2.15). The massive Hawaiian shields rise at least 4800 m above their bases, and some (Mauna Loa, Mauna Kea) have an absolute relief greater than 9 km, making them among the largest topographic mountains of the world.

Figure 2.15. Profile and crater area of shield volcanoes, Hawaii. Summit caldera of Kilauea volcano in foreground. Mauna Kea in background shows gentle slopes typical of a shield volcano. (Photo by U.S. Geological Survey)

Shield volcanoes are always built up from fluid, basaltic magma; tephra is only a minor part of the erupted material, and explosive eruptions are rare. The magma in the Hawaiian chain moves surfaceward in rift zones that generally parallel the Hawaiian Ridge (topographic high in the central Pacific). The rift zones usually contain hundreds of fissures that the ascending magma uses as vents for eruption. The Hawaiian rifts extend discontinuously for 3500 km across the center of the Pacific plate to a point where they abruptly bend to join the Emperor Chain (see map, fig. 2.16). This linear group of seamounts, similar in all aspects to the Hawaiian volcanoes, continues northward for another 2500 km.

The Hawaiian eruption rates, estimated between 0.05 and 0.1 km^3 a year, (Moore 1970; Swanson 1972) are the greatest known on Earth. There seems to be little question that the primary Hawaiian magma originates in the upper mantle (Eaton and Murata 1960) and, furthermore, that no genetic relationship exists between the volcanoes and the age or structure of the adjacent sea floor (Dalrymple et al. 1973). The oldest dated volcanics in the Hawaiian-Emperor system are 46 m.y. in the Koko Seamount, located 300 km north of the pronounced bend in

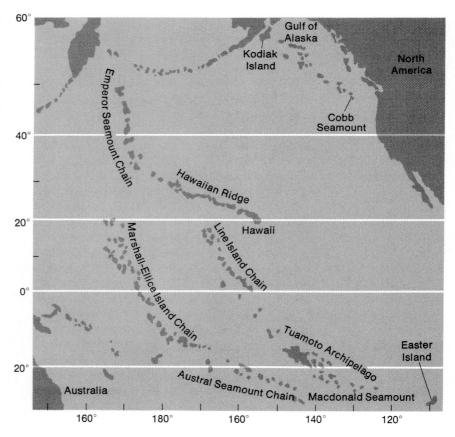

Figure 2.16. Seamount and island chains in the Pacific Ocean. (From Dalrymple et al. 1973. Reprinted by permission, *American Scientist,* journal of Sigma Xi, The Scientific Research Society of North America)

the linear system. Because the volcanic rocks get progressively younger to the southeast, the youngest occurring where volcanism is still going on, the ocean floor here, generally recognized as Cretaceous in age, is considerably older than any of the Hawaiian-Emperor volcanics. The plate-interior location of the island chains precludes an orogenic origin for the volcanic activity. This apparent lack of coincidence between plate tectonics and volcanism has prompted the hypothesis that volcanic action here occurs over a roughly circular hot spot in the upper mantle, about 300 km in diameter. Volcanism occurs as the Pacific plate rides over the heated region, and successive generations of volcanoes are carried away from the heat source. This comfortably explains the increasing age northwest along the linear Hawaiian chain if one assumes a fixed position for the thermal high. Actually, the precise origin of the melting spot and whether it is fixed or mobile are still unanswered considerations (Dalrymple et al. 1973).

The Icelandic shields commonly are smaller in all dimensions than the Hawaiian volcanoes. Although they are not related to orogenic events, they do originate in a plate-margin environment. The magma probably rises along the mid-Atlantic ridge in tensional fractures associated with the forces that move the plates in opposite directions.

Most other volcanoes take the form of a *composite cone,* so named because the cone is constructed of interbedded lava flows and layers of tephra. The flows are usually blocky and the tephra mostly cinder or ash, but the characteristics of all components may vary with the viscosity of the ascending magma. Composite cones, like the one in figure 2.17, are more peaked than shield volcanoes, com-

Figure 2.17. Steep slopes of composite volcanoes. Mount Rainier, Washington, in foreground. (From Fiske et al. 1963)

monly rising several thousand meters above a narrow circular base. Most develop from a single pipe vent. Side slopes are typically high, ranging from 10° to 35°, because the lack of fluidity in the magma leads to a more localized deposition. Most of the world's continental volcanoes are of this type; in North America famous examples are Mt. Rainier (Washington) and Mt. Shasta (California).

Rock Associations and Tectonic Relationships

The distribution of igneous rock types is closely allied to the composition of the primary magmas, and therein lies the importance of seeking their ultimate source. Many igneous rock types with similar resisting capabilities have been distinguished on the basis of minor mineralogic and textural variations. Such detailed classifications have important ramifications for the petrologist, but they have little significance in geomorphology and could easily lead us into a semantic quagmire. However, certain rock suites are related to particular magmatic roots and, in general, show an affinity to a distinct tectonic framework. In addition, field studies show that most discrete rock associations form during a definite segment of an orogenic cycle.

Table 2.5 divides igneous rocks into types that could be expected to form in a particular phase of an orogenic cycle. In the *pre-orogenic* stage, volcanic activity is associated with rocks of the ophiolite suite, which presumably reflect the trench environment. The volcanics are usually abnormally high in Na_2O; combined with a pillow structure, this points to contamination of a normal basalt as it

Table 2.5 Most common rock assemblages formed during, before, and after an orogeny.

	Volcanics and Hypabyssal Intrusives	Tectonic Setting and Character	Plutonics	Tectonic Setting and Character
Pre-orogenic or Early orogenic	Na-rich basalts (spilites) and trachytes (keratophyres)	Ocean trenches	Peridotite and serpentine	Island arcs, trenches, mid-ocean ridges
Synorogenic	Basalt, andesite, rhyolite	Continental margins, subduction zones	Granite, granodio-rite, syenite, anorthosite	Continental margins; during folding and metamorphism
Post-orogenic	Alkaline and thoeliitic basalt, diabase, lamprophyres	Mountain belts, normal fault zones, rift zones	Granites, syenite, gabbroic layered intrusions	Regions of crustal stability near orogenic zones but after folding and metamorphism

crystallizes in a submarine environment. It should be emphasized, however, that most oceanic basalts are not rich in sodium, which only reinforces the ambiguity surrounding the origin of these rocks.

Associated with these early orogenic volcanics is the emplacement of peridotite or serpentine plutons. These invade the island arc-trench system and commonly mark the core of the islands. Serpentinized peridotite, however, has also been dredged from the mid-oceanic ridges. Peridotites contain about 80 percent olivine and 20 percent pyroxenes derived from a melt that probably crystallizes at temperatures between 1400°C and 1200°C depending on pressure conditions and precise chemistry. Paradoxically, field evidence suggests that many peridotites were emplaced at much lower temperatures. The final movement of these bodies, therefore, was probably the result of tectonics rather than of intrusion in a fluid state.

In the *synorogenic* phase (table 2.5), tectonic forces at the plate margins give rise to a variable suite of volcanic rocks ranging in composition from basalts to rhyolites. The suite is dominated by somewhat basic andesites and aluminum-rich basalts, although more felsic varieties are present. The modern analog of this assemblage is definitely along tectonically active continental margins, but the diverse compositions make the source and history of the parent magma somewhat uncertain. Suggestions for the magma source vary from the upper mantle to an anatectic fluid, and both fractional crystallization and assimilation have been employed as auxiliary factors to explain the variations in composition. The aluminous basalts and pyroxene-rich andesites are probably oceanic, but more siliceous types clearly have a continental heritage. For example, the acidic volcanics in the Cretaceous-Tertiary sequence of the eastern cordillera almost certainly developed from a continental crust, and they are intimately related to miogeosynclinal-type sediments. They were also associated with normal faulting which presumably vented explosive eruptions of the tuffs, ashes, and other pyroclastics so common in that area.

During orogeny, large granitic batholiths are emplaced that evidently relate to the high-grade metamorphic rocks developed in response to regional tectonic forces. Some synorogenic batholiths are definitely emplaced during or before the main episode of folding because they display a strong gneissic texture (Simonen 1969). In the western United States, large batholiths such as the Sierra Nevada, the Idaho, and the Boulder make up major portions of the exposed crystalline rocks. These complex bodies probably grew by the joining together of smaller plutonic bodies with variable mineralogy, texture, and age. The general similarity of age and tectonic environment between the massive anatectic batholiths and the felsic volcanics is more than coincidence, and a common origin has been suggested for the two.

Mineralogically, the batholithic complexes are characterized by a few abundant types; plagioclase feldspar (An_{20-45}) is generally most prominent, constituting 15–85 percent of the rocks, and potassium feldspars (5–70 percent) and quartz (<10->50 percent) are the chief ancillary types. Mafic minerals are normally biotite and hornblende, which range from 5 to 20 percent in alkalic granites and from 25 to 50 percent in diorites.

A special type of synorogenic plutonic rock is anorthosite, which occurs in masses ranging from tiny intrusive bodies to large batholiths such as the Adirondack Mountains of New York. These rocks consist of more than 95 percent calcic plagioclase and, like most unique igneous rocks, their origin is the subject of much debate. It is established that they develop in regions of high-grade metamorphism but, oddly, all large anorthosite plutons are Precambrian in age. Hyndman (1972) suggests that the most likely origin for these bodies is the partial fusion of ultrabasic rocks or fractional crystallization of a highly mafic magma.

In *post-orogenic* rocks, the most spectacular assemblages are flood basalts of thoeliitic or alkaline composition. These include such widely separated masses as the Hawaiian shields and the great plateau basalts (described earlier). The unifying feature of the flood basalts, regardless of their location, is the extraordinary fluidity of the extruded lava. The flood basalts are probably related to hypabyssal intrusives that are also abundant in the post-orogenic phase. The intrusives are typified best by the diabase dikes and sills of Triassic age that intrude the rocks of the Appalachian Mountains. They do, however, occur at many places throughout the world and are not restricted to any particular time interval. Intrusive bodies vary in thickness from centimeters to massive units like the Palisades sill (New Jersey) which exceeds 300 meters. The magma producing diabase sills and dikes is the same as that which supplies thoeliitic and alkaline flood basalts, and indeed, the intrusives of the Appalachians can often be related, temporally and spatially, to nearby volcanic flows of similar composition.

Other post-orogenic intrusives with chemistry similar to that of alkaline olivine basalts have been recognized, but they may have distinctive textural and mineral characteristics. In southeast Colorado, for example, two granite stocks of the Spanish Peaks area have been invaded by a swarm of radiating lamprophyric dikes (fig. 2.18). The lamprophyres differ from normal basic intrusives only because they are proliferated with euhedral phenocrysts of mafic minerals.

Plutonic rock assemblages that form in the post-orogenic phase consist of nepheline syenites, small granite plutons, and large gabbroic layered intrusions. The syenites are rather uncommon and usually occur as small intrusive masses on the continents adjacent to the orogenic zones. They appear in areas of tectonic stability or regions of only moderate deformation such as zones of block faulting or gentle folding. Their geometry is variable; they occur as small dikes and sills, laccoliths, stocks, and small batholiths. Syenites are composed mainly of alkali feldspars and Na-rich pyroxenes or amphiboles; chemically they are undersaturated in SiO_2 and high in Na-K and FeO/MgO. Although many origins have been proposed, post-orogenic syenites probably result from differentiation of an

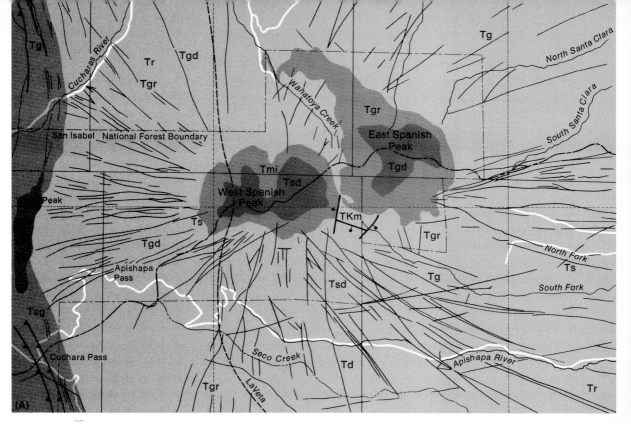

(A)

(B)

Figure 2.18. (A) Map of radial dike swarm in the Spanish Peaks region, Colorado. (From R.B. Johnson 1968, U.S. Geol. Survey Prof. Paper 594G, Plate 1) (B) Parallel walls formed by indurated sandstone on margins of a lamprophyre dike that forms trough. View is southward. Huerfano County, Colorado. (From R.B. Johnson 1968, fig. 9)

alkaline basalt magma. The fractionation, however, is influenced by the greater time available for the process to work in the relatively stable environment.

Gabbroic layered intrusions include a wide range of rock types in the sequence, but they normally have ultrabasics (peridotite) at the base, changing systematically to a gabbro at the top. They range in size from minor stocks to immense batholiths; the Bushveld intrusion in South Africa, for example, is 340 km by 680 km by 8 km.

Crustal and Surface Rock Distribution

In geomorphology we are concerned with the modern resisting framework, regardless of its history. It is important to gain an overall picture of the crustal and surface rock distributions as they presently exist. After all, the Earth's surface represents the battlefield between conflicting endogenic and exogenic forces. The distribution of rock types in North America, specifically in the United States, provides the resisting system that will be used in the geomorphic analyses that follow.

Table 2.6 synthesizes several estimates of the bulk chemical composition of the Earth's lithosphere. As expected, the chemistry of continental crust is higher in silica and K_2O than that of oceanic crust, and lower in CaO, MgO, and total iron. Such a chemical distribution can be converted into reasonable estimates of the volume-percentage of common rock types and their modal mineral composition (table 2.7). The significance of these analyses is to emphasize that the resisting framework in geomorphology basically entails only two igneous and metamorphic rock suites, and approximately ten mineral varieties. The crust consists primarily of a silicic assemblage (granites, gneisses, schists, granodiorites, and diorites) that makes up 48 percent of the crustal volume and a mafic association that constitutes about 43 percent. Obviously the silicic group is

Table 2.6 Weight percent of common elements in the Earth's lithosphere.

	Continental Crust			Oceanic Crust		Total Lithosphere	
	Polder-vaart 1955	Pakiser and Robinson 1966	Ronov and Yaroshevsky 1969	Polder-vaart 1955	Ronov and Yaroshevsky 1969	Polder-vaart 1955	Ronov and Yaroshevsky 1969
SiO_2	59.4	57.8	61.9	46.6	48.7	55.2	59.3
TiO_2	1.2	1.2	0.8	2.9	1.4	1.6	0.9
Al_2O_3	15.5	15.2	15.6	15.0	16.5	15.3	15.9
Fe_2O_3	2.3	2.3	2.6	3.8	2.3	2.8	2.5
FeO	5.0	5.5	3.9	8.0	6.2	5.8	4.5
MgO	4.2	5.6	3.1	7.8	6.8	5.2	4.0
CaO	6.7	7.5	5.7	11.9	12.3	8.8	7.2
Na_2O	3.1	3.0	3.1	2.9	2.6	2.9	3.0
K_2O	2.3	2.0	2.9	1.0	0.4	1.9	2.4

Table 2.7 Abundance of rock and mineral types in the Earth's crust.

Rocks	% of crustal volume	Minerals	Modal %
		Quartz	12
Sands	1.7	K-feldspar	12
Clays and shales	4.2	Plagioclase	39
Carbonates	2.0	Micas	5
Granites, gneiss and crystalline schist	36.9	Amphibole	5
		Pyroxene	11
Granodiorite and diorite	11.2	Olivine	3
		Clay	4.6
Syenite	0.4	Calcite and dolomite	2.0
Basalt, gabbro, amphibolite, eclogite	42.5		
		Magnetite	1.5
Peridotite, dunite	0.2	Others	4.9
Total	99.1	*Total*	100.0

Modified from Ronov and Yaroshevsky, *Geophysical Monograph* 13, pp. 37–57, 1969, copyrighted by American Geophysical Union.

plutonic or metamorphic in origin and is dominantly continental; the mafic types are overwhelmingly volcanic and rooted beneath the oceans.

The crust beneath the conterminous United States, however, is more mafic than one might guess (table 2.8). Pakiser and Robinson (1966) point out that, based on seismic velocities, the total U.S. crust is 54 percent mafic by volume (55 percent by weight). In addition, they show that the mafic content is considerably greater in the provinces of the eastern United States. In general, the eastern regions have a crust that is predominantly mafic, and the western provinces a crust that is mostly silicic. In each region, the outer, silica-rich part of the crust has a fairly constant thickness of 19 km. The lower mafic zone of the crust, however, thickens drastically from 15 km in the west to 25 km in the east. Thus, the compositional difference in the United States is caused by either an abnormally thick mafic zone in the eastern crust or an abnormally thin mafic zone in the western crust. The sedimentary rocks, included as silicic crust, are of no volumetric significance when compared with the total crustal mass.

If Pakiser and Robinson are correct, it is even more interesting to examine the igneous rocks exposed at the surface in the Appalachian and Cordilleran regions (table 2.9). In the Appalachians, where the crust is predominantly mafic (as noted in table 2.8), the surface igneous rocks are overwhelmingly calc-alkalic, plutonic rocks. Of the rocks of this type indicated in table 2.9 (84.5 percent of the total), 96 percent of the plutons are granites. In contrast, the igneous rocks exposed in the Cordilleran system are mainly extrusive (63.6 percent), and of these 77 percent are basaltic or andesitic in composition. The thick mafic crust in the eastern United States supports a surface rock assemblage that is

Table 2.8 Volume percentage and chemical composition of silicic and mafic crust in the United States.

Western Provinces	Silicic	Mafic
California coastal region	75	25
Sierra Nevada	50	50
Pacific NW (coastal)	28.6	71.4
Columbia Plateau	22.3	77.7
Basin and Range	66.7	33.3
Colorado Plateau	62.5	37.5
Rocky Mountains	62.5	37.5
Average	56.4	43.6
Eastern Provinces		
Interior Plains and Highlands	40.0	60.0
Coastal Plain	57.2	42.8
Appalachian Highlands and Superior Upland	37.5	62.5
Average	42.6	57.4
Total United States	46.3	53.6

	Western	Eastern
SiO_2	60.0	57.1
TiO_2	1.1	1.3
Al_2O_3	15.1	15.2
Fe_2O_3	2.3	2.3
FeO	4.9	5.7
MgO	4.5	5.6
CaO	6.3	7.5
Na_2O	3.0	3.0
K_2O	2.0	2.1
Total	99.2	99.8

From Pakiser and Robinson, *Geophysical Monograph* 10, pp. 620-26, 1966, copyrighted by American Geophysical Union.

Table 2.9 Area of different igneous rocks exposed in the Appalachian and Cordilleran regions of the United States (percent).

	Cordilleran	Appalachians
Plutonic Rocks		
Calc-alkalic Rocks (granite, granodiorite, quartz monzonite, quartz diorite, diorite, gabbro, anorthosite)	33.6	84.5
Alkalic Rocks (syenite, monzonite, others)	0.4	neg.
Ultramafic Rocks (peridotite, pyroxenite)	0.5	neg.
Hypabyssal Intrusives		
Calc-Alkalic Rocks (porphyries, quartz diabase, diabase)	1.5	7.4
Alkalic Rocks (porphyries)	0.3	0.2
Extrusive Rocks		
Calc-alkalic Rocks (basalt, dacite, andesite, rhyolite)	63.4	7.9
Alkalic Rocks (trachyte, latite, phonolite, others)	0.2	neg.
Total	99.9	100.0

Adapted from Daly 1933. Used with permission of McGraw-Hill Book Co.

dominantly granitic, in contrast to a silicic crust supporting mafic surface rocks in the west. Whether this relationship reveals some fundamental endogenic mechanism is conjecture; but it is tempting to speculate that the different isostatic characteristics of the two regions noted earlier are somehow intimately involved with the distribution of crustal and surface lithologies.

The spatial distribution of igneous and metamorphic rocks is complicated by the fact that orogenic and other tectonic settings change their position through time. Former orogenic belts, originally on the continental margin, may now be part of the stable plate interior. Old mountain belts and massive continental shields may therefore contain rock assemblages that are incongruous with their modern quiescent tectonics. As a result, any or all of the rock associations discussed above may now appear in any continental setting.

In North America, sedimentary rocks make up most of the exposed materials (table 2.10) even though they are only a minor constituent of the total crustal volume. Their ultimate source, however, is older igneous, metamorphic, and sedimentary rocks, and so their chemistry and mineralogy reflect changes induced by exogenic geomorphic processes. Geomorphology, therefore, becomes an important link in the rock cycle, explaining why the chemistry of metamorphic rocks and of geosynclinal sedimentary rocks is so similar (Ronov and Yaroshevsky 1969). Exogenic processes determine the chemistry of geosynclinal sediments, and metamorphism takes place in a closed chemical system. Based on average chemistry, however, Ronov and Yaroshevsky point out that the creation of granites by anatexis of geosynclinal sediments requires addition of silica and alkalis to the system (table 2.11). The precise source of these substances is not clear.

The wide areal distribution of sedimentary rocks undoubtedly causes a surface mineral composition different from that shown in table 2.7. At the surface, quartz and feldspars are dominant and probably exist in equal amounts (feldspar 30 percent, quartz 28 percent); calcite and dolomite increase to about 9 percent; and clay minerals and micas become much more significant, rising to approximately 18 percent of the surface material (Leopold et al. 1964).

Table 2.10 Rocks exposed at the surface of the North American continent (expressed as % of area).

	Gilluly 1969	Blatt and Jones 1975
Sedimentary	61.5	52
Volcanic	8.2	11
Plutonic	3.8	6
Metamorphic and total P€	26.5	31

Table 2.11 Chemical composition of sedimentary rocks and associated volcanics in platform and geosynclinal environments compared with the total granitic crust.

Platform	SiO_2	TiO_2	Al_2O_3	Fe_2O_3	FeO	MgO	CaO	Na_2O	K_2O	
Sands	75.75	0.49	6.90	2.55	1.31	1.43	3.40	0.58	1.83	
Clays	55.09	0.86	16.30	4.17	1.87	2.46	4.75	0.75	3.01	
Carbonates	9.80	0.18	2.54	0.85	0.54	6.83	38.93	0.24	0.76	*
Evaporites	3.02	0.05	0.76	0.27	0.13	6.16	28.25	13.06	0.28	†
Volcanics	49.22	1.53	15.74	3.33	8.02	6.11	10.00	2.51	0.73	
Average	49.21	0.65	10.88	2.97	1.65	3.37	12.28	0.81	2.11	
Geosynclines										
Sands	62.93	0.52	12.12	2.30	3.20	2.28	5.69	1.92	1.69	
Clays and shales	55.76	0.71	17.56	3.61	3.35	2.52	4.08	1.27	2.76	
Carbonates	13.30	0.14	2.70	0.43	0.94	2.92	42.49	0.56	0.49	*
Evaporites	3.02	0.05	0.76	0.27	0.13	6.16	28.25	13.06	0.28	†
Volcanics	55.62	1.00	16.12	4.17	4.52	4.22	6.91	3.33	2.02	
Average	50.00	0.65	13.69	2.98	3.21	2.99	11.42	1.83	2.00	
Granitic Crust	63.94	0.57	15.18	2.00	2.86	2.21	3.98	3.06	3.29	

*missing % CO_2
†missing % CO_2, SO_2, Cl

Adapted from Ronov and Yaroshevsky, *Geophysical Monograph* 13, pp. 37-57, 1969, copyrighted by American Geophysical Union.

Summary

The major elements that make up the resisting framework in geomorphology are created by diastrophism and igneous activity. These endogenic processes relate nicely to the plate tectonic theory, and major topographic components can be classified according to their environment of origin as occurring along plate margins or in plate interiors. Along plate margins, tectonic forces produce orogenic mountain belts containing highly deformed rocks, volcanism of various types, and large plutonic intrusions. In the plate interiors, topography seems to be controlled by vertical epeirogenic displacements driven by isostatic adjustments. Isostasy also establishes and maintains relief between the ocean basins and the continents and probably uplifts mountains after orogeny has occurred. The type of volcanic rocks and landforms found in any region depends on the viscosity of the magma that is produced; viscosity is controlled by the chemistry and gas content of the fluid.

Suggested Readings

Suggested readings to provide greater detail about the topics discussed are included in the following short bibliography.

Carmichael, I.; Turner, F.; and Verhoogen, J. 1974. *Igneous petrology*. New York: McGraw-Hill.

Dalrymple, G.; Silver, E.; and Jackson, E. 1973. Origin of the Hawaiian Islands. *Am. Scientist* 61: 294-308.

Dewey, J. F., and Bird, J. M. 1970. Mountain belts and the new global tectonics. *Jour. Geophys. Research* 75: 2625-47.

Garland, G. D. 1965. *The Earth's shape and gravity*. London: Pergamon Press.

Green, J., and Short, N. 1971. *Volcanic landforms and surface features*. New York: Springer-Verlag.

Hallam, A. 1973. *A revolution in the earth sciences*. London: Oxford Univ. Press.

Mitchell, A., and Reading, H. 1969. Continental margins, geosynclines, and ocean floor spreading. *Jour. Geology* 77: 629-46.

Wise, D. U. 1975. Continental margins, freeboard and volumes of continents and oceans through time. In *The geology of continental margins,* edited by C. Burk and C. Drake. New York: Springer-Verlag.

Woollard, G. P. 1966. Regional isostatic relations in the United States: In *The Earth beneath the continents,* edited by J. Steinhart and T. Smith. Am. Geophys. Union, Geophys. Monograph 10: 557-94.

In chapter 2 we examined the processes that create and distribute the rocks
and minerals included in the surficial geomorphic framework. We are now ready
to focus on the exogenic processes that mold the framework into recognizable
topographic forms. The first step is to recall that most of the Earth's surface is
not composed of solid rock but is underlain by the unconsolidated remains of
thoroughly altered rock. The fresh rocks and minerals that once occupied the
outermost position reached their present condition of decay through a complex of
interacting physical, chemical, and biological processes, collectively called
weathering. Weathering progressively alters the original lithologic character
until what finally remains in the space of the former rock is an unconsolidated
mass consisting of (1) new minerals created by the weathering processes,
(2) minerals that resisted destruction, and (3) organic debris added to the weathered zone.

Since every mineral species has, by definition, a unique chemical composition or atomic arrangement, it is not surprising that each type resists or responds
to weathering in a special way. Considering the wide variety of climates driving
the processes and the almost endless array of rock structures and mineral types,
the precise mechanics of weathering could easily be beyond our comprehension.
Nature, however, has simplified our task because, as indicated earlier
(table 2.7), the bulk of the Earth's crust is composed of a surprisingly small
number of mineral varieties with an equally limited chemistry. With this advantage, we can explore weathering efficiently even though the systems involved are
exceedingly complex.

Weathering is usually divided into separate domains of chemical processes
(*decomposition*) and physical processes (*disintegration*). The distinction between

Chemical Weathering and Soils 3

the two is real because the processes of disintegration involve no chemical reactions but simply produce smaller particles from larger ones. Nonetheless, the two realms of weathering operate simultaneously and, in fact, each may directly affect the character and rate of the other. For example, breaking a large rock into smaller particles increases the total surface area and thereby accelerates the chemical attack on the material. Conversely, expansion of minerals by chemical processes may exert enough internal stress to hasten the disintegration of the rock. Realistically, then, the physical and chemical functions of weathering may be so intimately intertwined that to consider them as unique processes is mostly a matter of convenience. In this chapter we will deal with only the chemical and biological aspects of weathering, which are dominant in the development of soils. Disintegration will be examined in the next chapter when we consider the stability of slopes, because many of the processes that break rocks apart are also important in the erosion of unconsolidated slope material by mass wasting.

A soil is the residuum that results from the application of weathering over an extended period of time. Given the proper conditions, a distinct layering, called the *soil profile,* will develop in the residual material, and its characteristics will directly reflect the weathering and soil-forming processes. Although both geologists and pedologists use soil profiles as a basis for classification, other scientists use the term "soil" in different ways. To an engineer, for example, a soil consists of any accumulation of unconsolidated debris; it may include material such as alluvial deposits that have not experienced the effects of weathering. Agricultural experts may be interested in only the upper part of a soil profile, which is the segment that supports vegetation. Clearly, the definition and classification of soil depend on what information an investigator wishes to gain from it and on what ways such data will be used. The farmer, of course, wants increased crop production, and the engineer requires knowledge of physical properties, such as bearing strength, to design buildings and other constructions judiciously. The geologist, however, utilizes soils and other weathering phenomena as clues to the intricacies of geologic history and the relative age of unconsolidated deposits (Porter 1975). Although most soils are young in the geological sense, some profiles have been preserved in the older record, and these provide critical evidence about environmental conditions at the time of formation. Mineral compositions of many sedimentary rocks reflect the combined tectonic and climatic conditions during their origin, and a correct interpretation of such rocks requires a knowledge of weathering and soil-forming processes. Landforms are often recognized as relict features because their soil character is inconsistent with the prevailing modern climate. Many other examples could show that we utilize an understanding of soils in many aspects of geology. The study of chemical weathering and soils is not simply an adventure into esoteric geomorphology, for a working knowledge of weathering processes is essential to any scientist interested in the surface environment.

Since rainwater is usually mildly acidic, chemical weathering can be visualized as a neutralizing process whereby minerals assimilate hydrogen ions and/or water, and release cations to the soil liquid. Complicating this simple model are two facts: (1) rainwater is not chemically pure but contains a variety of different cations and anions captured from the atmosphere; and (2) organic processes, involving metabolism of microorganisms and decay of vegetal matter, add gases and organic acids to the system. These organic functions are of such importance that some authors (Carroll 1970) suggest that chemical weathering proceeds in two stages. The first stage, driven primarily by inorganic processes, is called *geochemical weathering* and produces rotten rocks or *saprolites*. The second stage, called *pedochemical weathering,* leads to the formation of soils from the saprolitic material; it is chiefly a biologically controlled phenomenon.

As rainwater percolates into exposed rock material, the first important response commonly is the breaking apart of the structures of the parent minerals. As water surrounds a mineral or penetrates its structure through micro-openings and cleavages, chemical reactions occur that tend to disrupt the mineral's orderly atomic arrangement. Jenny (1950) pointed out that atoms along exposed mineral surfaces are not satisfied electrically and so may attract the dipolar water molecules. If the attraction causes water to dissociate into H^+ and OH^-, these will bond to the exposed ions of the mineral, as in figure 3.1. Hydrogen is then in the proper position to replace mineral cations, thereby releasing them to the surrounding fluid. This process may have a profound effect on the pH of the liquid, as hydrogen is progressively depleted and hydroxyls are concentrated. Experiments observing the change in pH when different minerals are pulverized in distilled water and carbonic acid (Stevens and Carron 1948) show that an equilibrium pH is eventually attained. Presumably this condition is reached when the number of cations removed is balanced by an equal number reentering the mineral structure. These *abrasion pH* values are significant in that they help determine the fluid's effectiveness in attacking other minerals. The values of abrasion pH for common minerals (table 3.1) demonstrate that minerals with

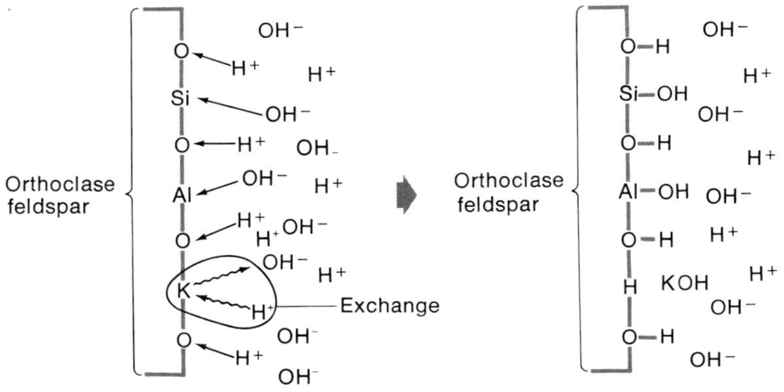

Decomposition

Figure 3.1. Ion exchange and chemical bonding take place as surface of orthoclase feldspar comes in contact with a solution containing dissociated H^+ and OH^- ions.

Chemical Weathering and Soils 81

Table 3.1 Abrasion pH values for common minerals.

Mineral	Composition	Abrasion pH
Kaolinite	$Al_2Si_2O_5(OH)_4$	5, 6, 7
Boehmite	$AlO(OH)$	6, 7
Gibbsite	$Al(OH)_3$	6, 7
Gypsum	$CaSO_4 \cdot 2H_2O$	6
Hematite	Fe_2O_3	6
Montmorillonite	$(Al_2Mg_3)Si_4O_{10}(OH)_2 \cdot nH_2O$	6, 7
Quartz	SiO_2	6, 7
Zircon	$ZrSiO_4$	6, 7
Chlorite	$(Mg,Fe)_2Al_4Si_2O_{10}(OH)_4$	7, 8
Muscovite	$KAl_3Si_3O_{10}(OH)_2$	7, 8
Biotite	$K(Mg,Fe)_3AlSi_3O_{10}(OH)_2$	8, 9
Calcite	$CaCO_3$	8
Anorthite	$CaAl_2Si_2O_8$	8
Orthoclase	$KAlSi_3O_8$	8
Albite	$NaAlSi_3O_8$	9, 10
Dolomite	$CaMg(CO_3)_2$	9, 10
Augite	$Ca(Mg,Fe,Al)(Al,Si)_2O_6$	10
Hornblende	$Ca_2Na(Mg,Fe^{+2})_4 (Al,Fe^{+3},Ti)_3$ $Si_6O_{22}(O,OH)_2$	10
Olivine	$(Mg,Fe)_2 SiO_4$	10, 11

Data from Stevens and Carron 1948.

alkali or alkaline earth elements in their composition tend to produce high pH liquids, while other minerals yield acidic fluids.

The original mineral is not necessarily completely destroyed. The process may proceed layer by layer from the external surface, or it may break the mineral into many small pieces, each retaining the structure of the original material. This is especially true when hydrogen can penetrate the interior of the lattice, where its high charge-to-radius ratio creates a completely unstable condition. The end product of weathering may be directly influenced by the manner in which the minerals are destroyed. Some silicate minerals, for example, break into molecular chains that are easily recombined with available cations to form layered clay minerals.

The second important response in chemical weathering relates to how easily ions are released from the parent structure, and how mobile or immobile they are after their liberation. Considerations here are whether released ions or groups of ions are readily fabricated into new, stable minerals, or whether they can be removed in solution by percolating soil fluids. The ultimate fate of released particles depends on what they are and on the characteristics of the fluid into which they are released. The equilibrium state, for example, can only be attained in a closed system, for continuous addition and removal of water surrounding the decomposing mineral will alter the pH, carry some of the released ions away, or provide new elements that can combine chemically with those escaping the mineral. The final result is determined by a myriad of interreactions, depending on how the original structure breaks apart and how mobile the ionic or molecular particles are under the physical and chemical constraints within the weathering zone.

Processes of Decomposition

As just suggested, the magnitude and direction of chemical weathering depend on the composition and structure of the minerals being attacked, how they break apart, and how mobile the constituents are in the weathering environment. Various combinations of mineral types and chemical processes can result in a variety of final products. The common chemical reactions involved in decomposition are *oxidation and reduction, solution, hydrolysis,* and *ion exchange.* Each process plays a particular role in the overall scheme of chemical alteration, although all function simultaneously in the weathering zone.

Oxidation and Reduction   Oxidation occurs when an element loses electrons to an oxygen ion. The process tends to occur spontaneously above the water table where atmospheric oxygen is readily available; therefore, most elements at the Earth's surface exist in an oxidized state. Below the water table the environment is generally reducing (Loughnan 1969); however, high concentrations of organic matter may cause local reducing conditions to occur above the water table.

The ease of oxidation depends on the redox potential (Eh), the magnitude of which is controlled by the abundance of organic matter and the accessibility of free oxygen. In most soils, Eh values range from -350 to $+700$ (fig. 3.2),

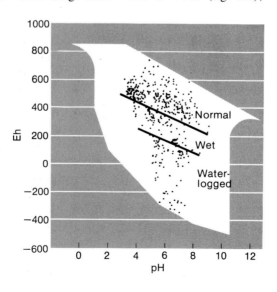

Figure 3.2. Eh-pH characteristics of soils. Eh values in millivolts. (From Baas-Becking et al. 1960. Used with permission of The University of Chicago Press, © 1960 by the University of Chicago)

sufficient to keep the majority of common elements in their oxidized state. However, some elements, such as aluminum, change from a reduced to an oxidized form with considerable difficulty. The most common elements affected by fluctuations of Eh are iron, manganese, titanium, and sulfur. Iron is easily oxidized to the ferric (Fe^{+3}) state:

$$2Fe^{+2} + 4HCO_3^- + \tfrac{1}{2}O_2 + 2H_2O \rightarrow Fe_2O_3 + 4H_2CO_3.$$

Its appearance as crustations on grains or as reddish-brown stains along fractures usually signifies the first tangible evidence of decomposition.

Solution   The process of solution is critical in chemical weathering because when atoms are dissolved from a mineral, the structure becomes unstable. However, the precise way in which a mineral collapses varies with its crystalline structure and the mobility of its constituent ions. When an atom is removed from its parent mineral, it may remain in solution and be taken completely out of the system by the downward moving fluids. This depends on the concentrations of reactable ions in the fluid and on other chemical characteristics of the medium.
    Solution is usually exemplified by the reaction of calcite and acid:

$$\text{(calcite) (carbonic acid)}$$
$$CaCO_3 + H_2CO_3 \rightarrow Ca^{+2} + 2(HCO_3)^-.$$

The role of CO_2 in the solution of calcite is critical and is especially significant in the development of karst topography. The CO_2 factor will be discussed in chapter 12 as we examine the processes that form karst regions. Here we simply emphasize that most common elements and minerals are soluble to some degree in normal groundwaters, where pH values usually range from 4 to 9 (fig. 3.3). In the weathering of silicate minerals, note that silica is soluble under all normal groundwater conditions. The degree of solubility is rather low (≈ 6 ppm) when the silica is contained in quartz (Morey et al. 1962) but increases considerably (≈ 115 ppm) in amorphous silica (Morey et al. 1964). Aluminum oxides are virtually insoluble under normal groundwater conditions, and iron in the ferric state can be dissolved only by rather acidic fluids. It is not surprising, therefore, that mature weathering profiles in humid climates should be characterized by the presence of abundant ferric iron and aluminum. Other cations are readily soluble, and their removal to the water table concentrates the least soluble constituents in the weathered zone.

Hydrolysis   The reaction between mineral elements and the hydrogen ion of dissociated water is called *hydrolysis*. Chemically, hydrolysis involves a reaction between a salt and water to produce an acid and a base; it is probably the most important mechanism in breaking apart structures of the silicate minerals. During the process, metallic cations are separated from the mineral structure and replaced by H^+ which is held in the original aluminosilicate complex (fig. 3.1). Most of the replaced cations are soluble in natural waters.

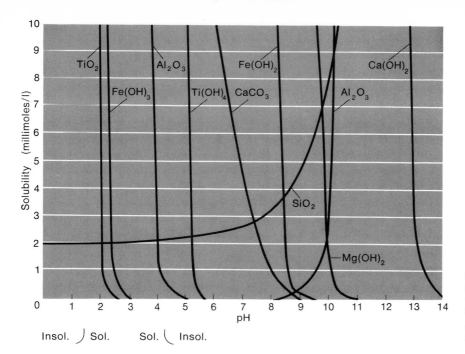

Figure 3.3. The relationship of solubility of common substances to various pH conditions. (From Loughnan 1969. Used with permission of American Elsevier Publishing Co., N.Y.)

In addition to freeing cations, hydrolysis usually produces H_4SiO_4, HCO_3^- and OH^-, all of which can be considered to be in solution (Birkeland 1974). In the common case where water and acid attack an aluminosilicate mineral, the reaction leaves the H^+ embraced in segments of the original structure, and these are subsequently recombined into clay minerals. The hydrolysis of orthoclase feldspar shown here and in figure 3.1 demonstrates this response:

$$\text{(orthoclase)} \qquad\qquad \text{(kaolinite)}$$
$$2KAlSi_3O_8 + 2H^+ + 9H_2O \rightarrow H_4Al_2Si_2O_9 + 4H_4SiO_4 + 2K^+.$$

The removal of the metallic cation will proceed as long as free hydrogen ions are available, easily replaced cations are present, and the solvent has not reached saturation with respect to the ion being liberated. Continuous introduction of fresh water and the development of organic acids will assure a ready source of H^+. Since OH^- is carried downward to the groundwater table by percolating water, it seems unlikely that highly alkaline water can ever be produced in open systems with constantly moving water. Therefore, the normal product of a continuously leached system is a residue in which all mobile cations have been freed from the original mineral and carried in solution to the water table. From there they are delivered to the regional streams. Clearly, the effect of hydrolysis decreases as clays depleted of cations become the dominant aluminosilicate in the weathering zone. In fact, as more clays form, they commonly become colloidally suspended in the fluid and may adsorb H^+ to their surfaces. This may cause the liquid to become more acidic rather than more alkaline as hydrolysis proceeds.

Ion Exchange   Ion exchange is the substitution of ions in solution for those held by mineral grains. Although all minerals possess some capability for ion exchange, the process is most effective in clay minerals. The ions to be exchanged are held on the surfaces of clays because unsatisfied charges, exposed hydroxyl groups, and isomorphic substitutions such as Al^{+3} for Si^{+4} have given the clays an overall negative charge. Cations adhere to the mineral surfaces in an attempt to neutralize the charge.

Each clay species has a different propensity for adsorbing cations, called its *cation exchange capacity* (c.e.c.), which is expressed as the number of milliequivalents per 100 grams of clay. When the adsorbed ion is hydrogen, the c.e.c. directly influences the pH value that the clay assumes. Kaolinite, for example, takes on a pH of 4–5 under complete adsorption, while H^+-montmorillonite attains a pH value as low as 3. Colloidal suspensions of these clays create acids that are capable of attacking other minerals. In soils, colloids of organic compounds produce similar acids because the c.e.c. of organic matter is usually rather high, ranging from 150 to 500 (Birkeland 1974).

Ion exchange is governed by the composition and pH of the interstitial water as well as the type of ion in the exchangeable position. In general, strongly acidic water allows H^+ to replace metal cations of the parent minerals, but this tendency changes as the water becomes neutral. At higher pH values, the mineral cations may remain in the exchangeable position or, in fact, H^+ may be replaced by metallic cations. In soils, the pH of the soil-water mixture is an indicator of the number of cations held in the exchange position by clays. This characteristic is commonly referred to as a *percentage of base saturation,* meaning the percentage of exchange sites occupied by cations other than hydrogen. The higher the percentage of base saturation, the higher the pH of the soil-water complex because less H^+ is held by the clays.

Mobility

The extent to which chemical weathering will alter the parent mineralogy depends largely on the relative mobilities of the constituent ions. Some ions are easily removed (high mobility) from the weathering system under normal groundwater conditions, while others are relatively difficult to remove (high immobility). The presence of highly mobile ions in a mature weathering zone indicates that some factor is impeding the transfer of the ion from the system. Orthoclase feldspar, for instance, will hydrolyze to kaolinite (as shown earlier) if all the potassium is lost in the process. If some potassium is retained, the clay product will be illite rather than kaolinite, as shown in the following reaction:

(orthoclase)   (illite)
$$3KAlSi_3O_8 + 2H^+ + 12H_2O \rightarrow KAl_3Si_3O_{10}(OH)_2 + 6H_4SiO_4 + 2K^+.$$

Accordingly then, since K^+ is a mobile ion, its immobile behavior in the formation and preservation of illite must be linked to the prevailing character of the fluid or to an incomplete breakdown of the orthoclase which traps the K^+ in molecules of the original structure. In either case, the easily removed potassium is rendered immobile and remains in the weathered zone.

The relative mobilities of common cations are as follows, in order of decreasing mobility:

$$(Ca^{+2}, Mg^{+2}, Na^+) > K^+ > Fe^{+2} > Si^{+4} > Ti^{+4} > Fe^{+3} > Al^{+3}.$$

Such a mobility distribution is probably related to a parameter known as the *ionic potential,* which is expressed as the ratio of the valence (Z) to the ionic radius (r). In general, Z/r is a useful first approximation of mobility because very mobile ions have Z/r values less than 3; those forming immobile precipitates have values between 3 and 9.5; and those forming soluble, complex anions are usually greater than 9.5 (fig. 3.4). However, as each cation can be immobilized by external factors, those factors in a sense greatly influence how effectively the major processes of chemical weathering will work. If all processes do function without hindrance, the mobile ions will be depleted from the system and the immobile ions will be progressively concentrated. The major external factors that control mobility include leaching, pH, Eh, fixation and retardation, and chelation.

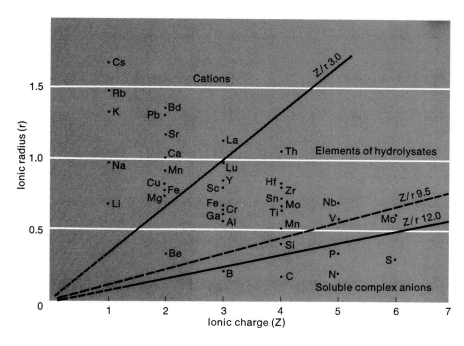

Figure 3.4. Grouping of some common elements according to their ionic radius (*r*), ionic charge (*Z*), and ionic potential (*Zr*). (From Gordon et al. 1958)

Leaching   The most important factor influencing ion mobility is the amount of water leaching the weathering zone. How frequently rainwater flushes through the system, and in what quantity, depends mostly on the climate but also on the permeability of the material. The significance of leaching is several-fold: (1) it removes in solution the constituents that have been separated from minerals by hydrolysis and ion exchange and, by providing new hydrogen, allows these processes continuously to alter the original material toward an ultimate degraded condition; (2) it directly affects the pH of fluids surrounding minerals and thereby helps determine which elements will remain in solution; and (3) it provides the mechanism by which dissolved ions and clays are transferred from higher levels to lower levels in the weathering zone. This introduces new components to each level and so may engender the chemical environment needed to precipitate new minerals.

In general, the absence of continual leaching makes the weathering zone function like a closed system. Since ions removed from the parent minerals are held in the surrounding fluid, the exchange processes will continue only until an equilibrium condition is established between the ions in the fluid and those in the mineral. For all practical purposes, further chemical weathering is impossible beyond this stage, and mobile ions and easily decomposed minerals will remain in the system. In this way, the retardation of leaching, usually due to insufficient rainfall, is a prime factor causing immobility of ions in a weathering system. It is not surprising, then, that mobile constituents are abundant in arid-climate soils. On the other hand, in humid regions where leaching is continuous, the same mobile substances are removed entirely. Even very resistant parent materials such as quartz can be chemically altered to a drastic extent if high-volume leaching continues over a long period of time. Loughnan and Bayliss (1961) reported a truly amazing example of how a kaolinitic sandstone with 90 percent quartz was altered under a tropical climate to a residuum that contained only 5 percent quartz (fig. 3.5).

The Influence of pH   Figure 3.3 shows that solubility of most common elements is directly related to the pH of the percolating fluid. Stated in another way, the mobility of ions is partly controlled by pH. Although rainwater reaching the surface is slightly acidic, processes operating beneath the surface can considerably alter the original pH value. Thus, pH is not an independent variable in weathering but may, in fact, be dependent on the type and degree of inorganic and organic processes within the weathering zone. (Such an interdependence represents another of the feedback mechanisms so prevalent in geomorphic systems.)

In geochemical weathering the pH is influenced by leaching, by the c.e.c. of residual minerals, and by the composition and structure of the parent minerals (Loughnan 1969). The effect of leaching and cation exchange capacity can be shown diagrammatically by considering the progressive weathering of a basalt,

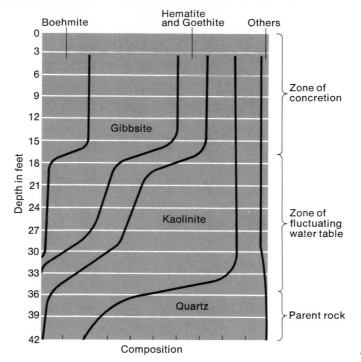

Figure 3.5. Characteristic mineralogy of weathering zone formed on kaolinitic sandstone under intense chemical weathering. Note decrease in quartz and increase in stable clays at top of profile. (From Loughnan and Bayliss 1961, *American Mineralogist* 46:211)

shown in figure 3.6. With continued leaching, the pH is related almost inversely to the type of clay produced during various stages of the weathering. Assuming complete mobility of Ca, Mg, and Na, the first clay developed is an H^+-montmorillonite with a high c.e.c. value (70–100 meq/100 g). The high c.e.c. allows the clay to adsorb many of the hydrogen ions made available by the continuous leaching, and the colloidal acid formed causes a pronounced drop in pH. As leaching continues, however, montmorillonite is degraded to kaolinite, which becomes the dominant clay mineral. The much lower c.e.c. of kaolinite permits less hydrogen adsorption, and the pH rises. The final product, bauxite, can adsorb virtually no H^+ and the pH climbs to an almost neutral condition. Obviously, pH is dependent on the other factors.

It is important to remember that organic substances also form colloidal suspensions with very high cation exchange capacities. Although the values vary with climate and the type of organic matter, it is not uncommon for these humic acids to lower the pH below 4.

The Eh Factor   The effect of Eh on ion mobility is due to the fact that oxidation-reduction reactions are reversible. When soils become waterlogged, free oxygen is excluded and strongly reducing anaerobic bacterial action begins,

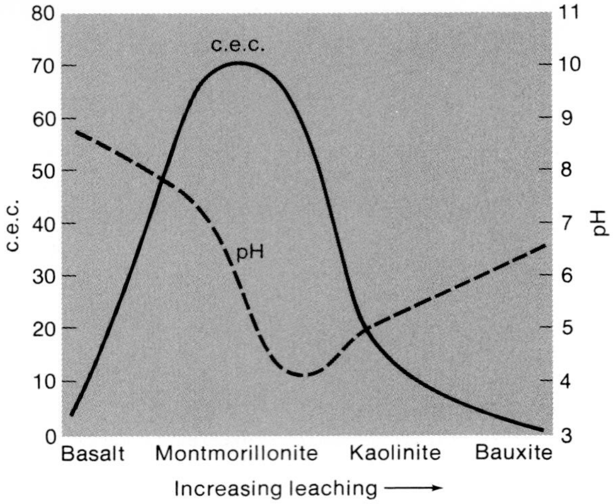

Figure 3.6. Relationship of cation exchange capacity (c.e.c.) and pH with continuous weathering of a basalt. Transition from basalt to bauxite represents gradual change in mineralogy with increased leaching. (From Loughnan 1969. Used with permission of American Elsevier Co., N.Y.)

generating lowered Eh values that, in turn, cause some elements to revert to their reduced forms. In some cases, ease of reduction is also dependent on pH (Connell and Patrick 1968). Such transitions have a direct influence on the mobility of certain elements that may be relatively insoluble in one form and easily dissolved in the other. Removal of iron oxides attached to clays, for example, can result from only a minor lowering of Eh, which transforms ferric iron to the more soluble ferrous type (Carroll 1958). Apparently, a fluctuating water table can have important ramifications on the redox potential (Eh) and with it the relative mobility of important ions.

The significance of Eh in decomposition and ion mobility can be briefly stated as follows:

1. Ferrous iron (Fe^{+2}) commonly binds silica tetrahedra in the structures of silicate minerals. The oxidation of the iron to the ferric state requires impossible internal adjustments and the lattice structure is destroyed (Carroll 1970).
2. The by-products of the oxidation process may facilitate the decomposition of other, more stable, minerals. For example, the oxidation of pyrite produces sulfuric acid, as shown:

$$4FeS_2 + 14H_2O + 15O_2 \rightleftharpoons 4Fe(OH)_3 + 8H_2SO_4.$$

    This acid lowers pH and will react with any nearby mineral susceptible to attack by acid. Groundwater in regions near zones of sulfide mineralization is usually very acidic.
3. As indicated before, the solubility or insolubility of some elements in groundwater having a normal pH (fig. 3.3) is directly controlled by whether the elements exist in oxidized or reduced form. This is especially true for iron and titanium.

Fixation and Retardation   Some cations, especially potassium, have a tendency to be retained or fixed in the weathered zone, so that their mobility is considerably lower than might be expected. The fixation of potassium may account for its low concentration in seawater and the widespread distribution of illite in sedimentary rocks. The precise mechanism involved is poorly understood, but it is probably related to the fact that potassium silicates (muscovite, K-feldspars) are more resistant to chemical weathering than are other minerals with similar lattice structures. It is known that potassium is most readily fixed in clay minerals with expandable lattices such as chlorite, illite, micas, vermiculite, and montmorillonite. Why this should be or why potassium is so much more susceptible to this phenomenon than other ions is not clear, but Wear and White (1951) suggest that the unique size of the K^+ ion provides the most stable structure when combined with oxygen in layered silicates.

Ion mobility may also be slowed or retarded without complete fixation. Wollast (1967) demonstrated that the initial stage of orthoclase hydrolysis is accompanied by the creation of a thin surface layer of amorphous $Al(OH)_3$ and silica. Although these substances are transferred from the coating layer to the surrounding liquid, the fluid is quickly saturated with respect to aluminum. Silica, and presumably other ions, continue to diffuse through the layer, but the rate diminishes because the distance from fresh feldspar to the fluid increases continuously as the aluminum-enriched sheath thickens. The retarding coat is not formed at pH values less than 5 because $Al(OH)_3$ is soluble under those conditions.

Chelation   The process of chelation represents one of the most dramatic effects on the mobility of ions. Lehman (1963) defines chelation as "the equilibrium reaction between a metal ion and a complexing agent, characterized by the formation of more than one bond between the metal and a molecule of the complexing agent and resulting in the formation of a ring structure incorporating the metal ion." To those of us unversed in organic chemistry, this means that metallic ions that are extremely immobile under normal conditions can be mobilized by reacting with complexing agents and be vertically transported as part of the compound.

Most complexing agents involved in chelation are organic compounds nurtured in soils by alteration of humus into a plant acid called fulvic acid (Wright and Schnitzer 1963), although other organic processes also produce chelating complexes. Lichens, for example, secrete such materials (Schatz 1963). The complexing agent ethylenediaminetetraacetate (EDTA) is by far the most completely understood chelator, and its structure (fig. 3.7) shows how the metallic ion is bonded and held.

Although the chemistry of the chelating process is far from clear, its importance in weathering has been recognized for several decades (Schatz et al. 1954). During this time several excellent studies have demonstrated that EDTA and other chelators may play a dominant role in mobilizing iron and aluminum under

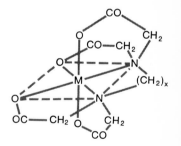

Figure 3.7. Structure of complexing agent EDTA. Metallic ion (M) is bonded within structure. (Reproduced from Soil Science Society of America *Proceedings* 27:169 (1963), by permission of the Soil Science Society of America)

pH conditions that are not conducive to Fe and Al solution (Atkinson and Wright 1957; Wright and Schnitzer 1963; Schalscha et al. 1967). Using several different chelating agents Schalscha and his co-workers (1967) were able to extract iron from a variety of minerals. Significantly, their study revealed no correlation between pH and the amount of liberated iron. EDTA released more iron from magnetite and hematite, for example, than did hydrochloric acid, even though the HCl solution was more acidic. On the other hand, HCl removed more iron from goethite, augite, epidote, and biotite.

In soils, chelating agents become soluble in water as they are oxidized (Wright and Schnitzer 1963). Iron and aluminum are locked in the ring structure and carried downward with the percolating water. The downward movement continues until the entire mass is flocculated because of small changes in ionic content of the soil water, or until the complex is broken by microbial action. In either case further downward movement is curtailed, and the iron and aluminum are redeposited.

Even before soils begin to develop, chelation can be important in the alteration of the bare rock. Jackson and Keller (1970) showed that the presence of the lichen *Stereocaulon vulcani* accelerated the chemical weathering of volcanic rocks in Hawaii. Rocks covered by lichen growth had a thicker weathering crust than lichen-free rocks, and the crust was enriched in iron and depleted in titanium, silicon, and calcium. Rocks devoid of lichens were measurably less affected by chemical weathering.

The Degree and Rate of Decomposition

To utilize soils in geological studies, it is important to know what the end products will be in a completely altered system. Geologists are, therefore, concerned about what material will remain when no more chemical reactions are possible, given the constraints of climate, vegetation, and rock type. If such a condition can ever be reached, a steady state will have been established between the driving forces (decomposition) and the resisting materials, and although the weathering zone may get progressively thicker, the upper parts of the profile are degraded to their ultimate form. In thoroughly leached soils, all minerals remaining presumably will be stable and all mobile ions will be gone. Thus, if we have some idea as to what final products are expectable in any climate, we can estimate the degree of weathering (how far the material has progressed toward the steady state) by observing the mineral and chemical composition of the soil.

Mineral Stability   In considering the degree of weathering, a logical first question to ask is what minerals will be most rapidly destroyed in a thoroughly leached open system and, conversely, which types will remain if a steady state is attained. Perhaps the first analytical attempt to answer that question was provided by Goldich in 1938. In a detailed study of the weathering of several varieties of igneous and metamorphic rocks, he suggested that mineral stability of the rock-forming silicates is directly related to their order of crystallization as determined earlier by Bowen (1928). Quartz, being the last to crystallize, forms under the lowest temperature conditions and therefore should be most stable in the surface environment. High-temperature minerals such as olivine and pyroxene are least stable and weather most rapidly. With the exception of muscovite and the plagioclase feldspars, the Goldich stability series (table 3.2) reflects the number of oxygen atoms shared in the silicate structure. Quartz and orthoclase, for example, have four shared oxygens in their structure, making their internal bonding strength much higher than that of a mineral like olivine which shares no oxygens. Keller (1954) in fact suggested that bonding energy may provide a useful basis on which to estimate mineral stability.

Historically, the study of heavy minerals has intrigued sedimentologists and geomorphologists, mainly because a unique heavy mineral suite can be utilized as evidence in a variety of geological problems. Determining provenance, making correlations, and establishing relative age of deposits are some kinds of investigations in which heavy mineral analyses can provide key information. Such minerals are easily separated from other less dense varieties, and most are distinct enough to be identified without difficulty. The geologic use of heavy mineral analyses, however, demands some idea of whether the assemblage of minerals is complete or whether some unstable varieties have been removed by weathering or intrastratal solutions. Pettijohn (1941) addressed himself to the problem of heavy mineral stability by comparing the frequency of occurrence of various mineral types in recent sediments with their frequency in older sedimentary rocks. This led him to suggest an order of persistence for the common heavy minerals, which is shown in table 3.3. Since Pettijohn, other analyses of heavy mineral stability have been made, but except for minor variations, they agree rather closely with his ordering.

Still another guide to mineral stability is the *weathering potential index* or WPI (Reiche 1943), which is calculated from the chemical analysis of a mineral or rock by the following formula:

$$\text{WPI} = \frac{100 \times \text{mols}(CaO + Na_2O + MgO + K_2O - H_2O)}{\text{mols}(SiO_2 + Al_2O_3 + Fe_2O_3 + CaO + Na_2O + MgO + K_2O)}.$$

Table 3.2 Weathering stability of the common silicate minerals.

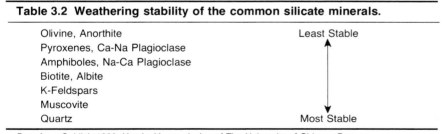

Olivine, Anorthite	Least Stable
Pyroxenes, Ca-Na Plagioclase	
Amphiboles, Na-Ca Plagioclase	
Biotite, Albite	
K-Feldspars	
Muscovite	
Quartz	Most Stable

Data from Goldich 1938. Used with permission of The University of Chicago Press.

Table 3.3 Order of persistence of some common heavy minerals.

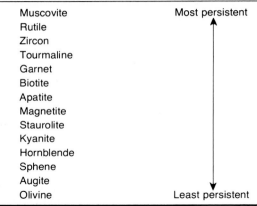

Muscovite	Most persistent
Rutile	↑
Zircon	
Tourmaline	
Garnet	
Biotite	
Apatite	
Magnetite	
Staurolite	
Kyanite	
Hornblende	
Sphene	
Augite	↓
Olivine	Least persistent

After Pettijohn 1941. Used with permission of The University of Chicago Press.

Table 3.4 Weathering potential indices for common minerals. [a, b]

Forsterite[d]	66	Biotite[c]	22
Olivine[c]	54	Leucite[d]	17
Wollastonite[d]	50	Albite[c]	13
Enstatite[d]	50	Orthoclase[c]	12
Diopside[d]	50	Quartz[c]	0
Tremolite[d]	40	Sillimanite[d]	0
Augite[c]	39	Muscovite[c]	−10.7
Hornblende[c]	36	Analcite[d]	−17
Talc[d]	29	Pyrophyllite[d]	−20
Nepheline[d]	25	Kaolinite[d]	−67
Anorthite[d]	25	Boehmite[d]	−100
Epidote[d]	23	Gibbsite[d]	−300

[a] After Reiche (1943).
[b] Italics signify common minerals listed in the Goldich sequence.
[c] Average value obtained by Reiche.
[d] Calculated from theoretical formula.

From Loughnan 1969. Used with permission of American Elsevier Publishing Co., N.Y.

Minerals and rocks with low stabilities have high WPI values (table 3.4), and their presence in a soil indicates either immaturity or leaching insufficient to remove the unstable minerals.

The results of these various methods suggest that the degree of chemical weathering can be estimated from the mineral assemblage contained in a soil. In detail, however, the correlation between the different approaches is far from perfect, and the factors that control mineral stability are not clearly defined. The relationship between stability and silicate structure recognized by Goldich fits generally, but not precisely, the stability hierarchy determined by other procedures. The plagioclase feldspars, for example, show considerable variation in stability even though they all have the same framework structure. Micas also show the same type of discrepancy. In addition, zircon appears to be an extremely persistent heavy mineral, yet it is structurally the same as olivine, a notably unstable species. The WPI of analcite ($NaAl\,Si_2O_6\cdot H_2O$) perhaps best illustrates the confusion surrounding mineral stability. A member of the notoriously unstable zeolite family, analcite has a framework silicate structure and a WPI value lower than quartz or muscovite. Why it is so easily destroyed by weathering when all the indices predict its high stability is simply not understood.

Thus far we have stressed the stability of minerals that are components of the original rock, commonly referred to as *primary minerals*. It is well to remember, however, that the processes of chemical weathering play a dual role. In addition to destroying primary minerals, they also create new minerals by recombining or reprecipitating materials liberated from the parent rocks. These *secondary minerals,* born within the weathered zone, are distinctly more stable than their primary ancestors because they reach equilibrium in the temperature-pressure environment of the soil rather than in the magmatic or metamorphic conditions that created the original crystals. The most common secondary products of weathering are clay minerals and amorphous hydrous oxides of iron, aluminum, silica, and titanium.

Clay minerals are hydrous aluminum silicates in which silica tetrahedra and aluminum octahedra are bonded together into a layered atomic structure. A silica tetrahedron has one silicon atom surrounded by four oxygen atoms; an aluminum octahedron consists of a single aluminum atom bonded to six oxygen atoms. In clays the individual tetrahedrons and octahedrons are linked together in planes, forming distinct sheets or layers typified by either a tetrahedral or an octahedral structure. Most clays are either a 1:1 layer silicate or a 2:1 layer silicate (table 3.5). A 1:1 structure, exemplified by the mineral kaolinite, has as its fundamental building block one layer of silica tetrahedra and one layer of aluminum octahedra. In 2:1 mineral structures, such as those of the micas, illite, and montmorillonite, one aluminum octrahedral layer is positioned between two layers of silica tetrahedra. Other clay varieties do form, but they either represent combinations of the basic types (mixed-layer) or originate under very special weathering conditions (chain silicates).

Differences in stability are present even within the realm of clay minerals. Jackson and his colleagues (1952) proposed a distinct sequence of clay mineral development in which muscovite or illite is the initial product of weathering, and, assuming effective leaching, progressively degrades through montmorillonite to a final kaolinitic clay (fig. 3.8). The suggestion that kaolinite should be the most common end product under normal conditions has received support from evidence other than mineral indices. Feth and his co-authors (1964) found that the composition of groundwater draining the granitic rocks of the Sierra

Table 3.5 Classification of the clay minerals.

Structure	Minerals	Remarks
1:1	Kaolinite	
	Dickite	
	Nacrite	
	"Fire-clay mineral"	Disordered kaolinite
	Metahalloysite ($2H_2O$)	Dehydrated; $d(001) = 7\text{Å}$
	Halloysite ($4H_2O$)	Contains interlayer water; $d(001) = 10\text{Å}$
	Allophane	Unordered or slightly ordered grouping of SiO_2, Al_2O_3 with H_2O
	Mica group	
2:1	Muscovite, 2M	Dioctahedral; layers bonded together with K^+ ions; generally detrital
	Illite	Dioctahedral; layers partially bonded together with K^+ ions; common mica of soils and sediments
	Glauconite	Marine sediments; somewhat like illite; mostly diagenetic
	Celadonite	Vugs in basalt
	Biotite	Trioctathedral; alters readily
	Vermiculite	Trioctahedral
	Montmorillonite group	
	Many species	Units separated by a water layer containing exchangeable cations that neutralize the structure; montmorillonites expand in organic liquids
	Chlorite group	Trioctahedral
	IIb polytype	Common detrital variety
	Ib, polytype	Diagenetic variety, low energy form
	Chlorite	Dioctahedral
	Chlorite, swelling	Corrensite, sudoite
	Chamosite	Generally a 7Å chlorite
Chain silicates	Sepiolite	
	Palygorskite (attapulgite)	
Mixed-layer clay minerals	Rectorite	Regular; give an integral series of basal spacings from a long $d(001)$ spacing
	Montmorillonite-chlorite	
	Corrensite (see chlorite)	
	Mica-chlorite	Random; do not give an integral series from $d(001)$
	Vermiculite-chlorite	
	Mica-vermiculite	

From Carroll 1970, table 6. Used with permission Plenum Publishing Corporation.

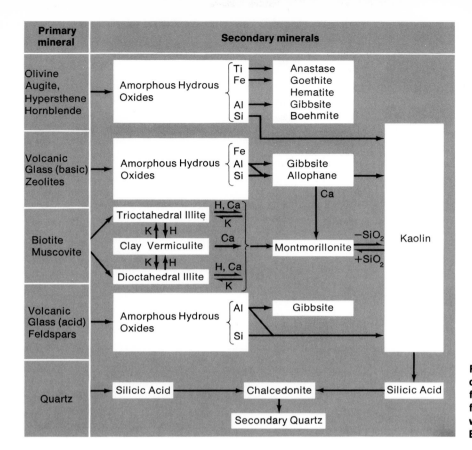

Figure 3.8 Weathering sequence of primary rock-forming minerals. (Adapted from Loughnan 1969. Used with permission of American Elsevier Publishing Co., N.Y.)

	Most likely to dominate in
Illite	Podzols; Red Desert
Montmorillonite	Desert; Chestnut; Chernozem; Brunizem; Prairie
Kaolinite	Laterites; Red–yellow podzols

Figure 3.9. Common clay minerals and the Great Soil Groups in which they are most likely to be dominant.

Nevada was chemically in equilibrium with kaolinite. This suggests that other clay minerals in the weathered residuum are unstable and will ultimately change to kaolinite. Assuming that a sequential development of clay types is correct, the type and amount of clay should provide a reasonable basis for estimating the degree of weathering in a soil; such estimates have in fact been made (Barshad 1964). It should also be apparent that specific clay minerals will tend to be dominant in mature soils developed under a particular set of climatic conditions (fig. 3.9).

Estimates Based on Chemical Analyses   The degree of chemical weathering has also been gauged on the basis of total chemical analyses. The most common approach utilizing such data is a direct comparison of the fresh parent rock with the saprolite or soil derived in situ from it. To make an analytical comparison of the two, Al_2O_3 is usually considered to be constant since its mobility is extremely low under normal conditions. The weight percent values of oxides in the soil are recalculated by using the conversion factor $\% Al_2O_3$ (fresh rock) / $\% Al_2O_3$ (weathered material) as a multiplier of each constituent (Birkeland 1974). Table 3.6 demonstrates this method by showing a hypothetical comparison of a saprolite developed from a parent metagabbro. The recalculated values indicate the degree of weathering, and relative gains and losses from the original composition can be used to compare nearby exposures or different horizons within the same soil profile. Maximum errors in this method occur when the assumption of a completely constant aluminum value is incorrect. As mentioned earlier, aluminum may be mobilized by chelators or humic acids, and so it is likely that some aluminum will be lost from the upper weathering zones and some gained in the lower horizons. Nonetheless, the method provides a good first approximation of the relative degree of weathering, even though absolute gains and losses are suspect because of the aluminum problem.

In regional studies, complete chemical analyses may be abandoned in favor of comparing molar ratios (percent oxides/molecular weight) such as SiO_2/Al_2O_3 (Ruxton 1968). The *leaching factor* (Jenny 1941), shown below, is an example of the molar ratio approach:

$$L.F. = \frac{(K_2O + Na_2O)/Al_2O_3 \text{ of weathered horizon}}{(K_2O + Na_2O)/Al_2O_3 \text{ of parent material}}$$

Table 3.6 Gains or losses of chemical constituents in a hypothetical example of weathering.

	Original rock	Saprolite	Adjusted %*	Loss or Gain*
SiO_2	50.3	41.3	24.78	−25.52
Al_2O_3	18.3	30.6	18.3	0
Fe_2O_3	2.5	11.3	6.78	+ 4.28
FeO	9.4	1.2	0.72	− 8.68
MgO	5.5	0	0	− 5.5
CaO	12.6	0	0	−12.6
TiO_2	1.1	0.1	0.06	− 1.04
H_2O	0.2	14.6	8.76	+ 8.56
Total	99.9	99.1	59.40	−40.5

*To obtain loss or gain multiply each value in the saprolite column by 0.6. Subtract the adjusted percentage from the original % in the fresh rock.

Al_2O_3 (fresh)/Al_2O_3 (saprolite) = 18.3/30.6 = 0.6

Such ratios have the advantage of allowing comparisons to be made between soils and saprolites that develop from different parent materials.

The rate of chemical weathering is also commonly estimated by chemical analyses of water that has filtered through the rock and soil system. In general, these studies rely on the assumption that the parent material is the only source of dissolved constituents in the water. However, because rainwater is not chemically pure, some knowledge of its composition is required before valid conclusions about solution within the weathering profile can be made. Cleaves and his colleagues (1974), for example, estimate that about 37 percent of dissolved solids in a small Maryland stream were introduced into the system as part of the precipitation. Land treatment programs such as irrigation and fertilization also can add ions to the analyses.

If corrections can be made for ions gained from precipitation, the dissolved load in rivers should represent the weight loss of rock material from a drainage basin. Weights can be converted into volumetric terms and, when considered over a reasonable period of time, rates of chemical weathering can be expressed in terms of surface lowering, or denudation. Hembree and Rainwater (1961) showed significant variations in chemical degradation in different parts of the Wind River Range (Wyoming) that depended on whether they were underlain by sedimentary rocks or granitic crystallines. In the Piedmont region of Maryland, differences in the type of crystalline rocks have a pronounced control on the rate of chemical denudation (Cleaves et al. 1974). In larger areas the rate of chemical denudation also appears to be a function of rock type and climate (Livingstone 1963; Judson and Ritter 1964; Strakhov 1967), increasing in magnitude in warm-humid regions. Rapp (1960) demonstrated, however, that solution may be the dominant erosional process even in arctic regions, and therefore, chemical weathering is a factor that simply cannot be ignored in the overall scheme of landscape development.

Soils

Weathering processes that continue over an extended period of time result in an unconsolidated mass of soil that is measurably different from the original rock in its physical and chemical properties. Soils also contain a significant amount of organic material that is added progressively to the system as decaying vegetation and microorganisms. In addition, pronounced layering develops in the weathered mass during the transition from being simply decomposing rock or alluvium to being a true soil. The vertical arrangement of the layers constitutes a diagnostic property of soils known as the *soil profile*. The profile extends from the surface downward to the fresh parent material. The time in absolute years needed to form a soil profile, as well as the perfection of the profile's development, varies widely with the intensity of the weathering processes and the character of the original material. Nevertheless, where the soil profile is well developed, its character reflects the environment under which it formed and serves as the basis for classification of soils and interpretation of paleo-conditions.

The Soil Profile

In its simplest form, the soil profile can be visualized as consisting of three main layers, usually designated as the A, B, and C horizons. The A horizon is normally considered to be the thin, dark-colored surface layer where organic matter is concentrated and where clays and mobile components are continuously leached downward or *eluviated*. The C horizon is usually thought of as the underlying parent material which is essentially unmodified by the soil-forming processes. The B horizon, therefore, becomes the transitional zone between the A and C horizons and historically has been considered as an *illuviated* zone, i.e., a zone of accumulation and concentration of the material brought down from the A horizon. Each horizon, defined on the basis of its physical and chemical properties, shows enough internal variation to require subdivisions indicating some special trait is present. These are shown diagrammatically in figure 3.10.

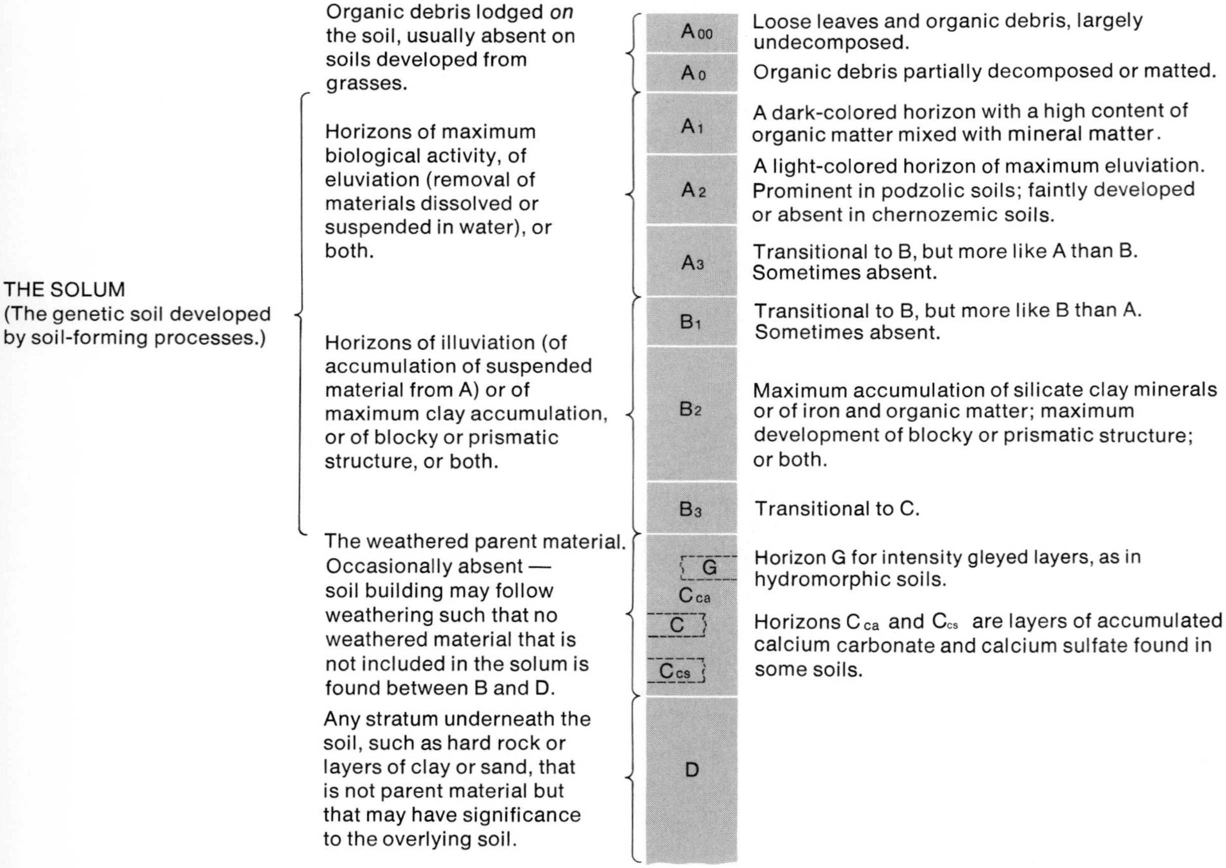

Figure 3.10. A hypothetical soil profile showing all the principal horizons. It will be noted that horizon B may or may not have an accumulation of clay. Horizons designated as C_{ca} usually appear between B3 and C. The G horizon may appear directly beneath the A. (From Soil Survey Staff 1951)

THE SOLUM
(The genetic soil developed by soil-forming processes.)

Organic debris lodged *on* the soil, usually absent on soils developed from grasses.

Horizons of maximum biological activity, of eluviation (removal of materials dissolved or suspended in water), or both.

Horizons of illuviation (of accumulation of suspended material from A) or of maximum clay accumulation, or of blocky or prismatic structure, or both.

The weathered parent material. Occasionally absent — soil building may follow weathering such that no weathered material that is not included in the solum is found between B and D.

Any stratum underneath the soil, such as hard rock or layers of clay or sand, that is not parent material but that may have significance to the overlying soil.

A_{00} — Loose leaves and organic debris, largely undecomposed.

A_0 — Organic debris partially decomposed or matted.

A_1 — A dark-colored horizon with a high content of organic matter mixed with mineral matter.

A_2 — A light-colored horizon of maximum eluviation. Prominent in podzolic soils; faintly developed or absent in chernozemic soils.

A_3 — Transitional to B, but more like A than B. Sometimes absent.

B_1 — Transitional to B, but more like B than A. Sometimes absent.

B_2 — Maximum accumulation of silicate clay minerals or of iron and organic matter; maximum development of blocky or prismatic structure; or both.

B_3 — Transitional to C.

G, C_{ca} — Horizon G for intensity gleyed layers, as in hydromorphic soils.

C — Horizons C_{ca} and C_{cs} are layers of accumulated calcium carbonate and calcium sulfate found in some soils.

C_{cs}

D

A variety of properties can be used to distinguish the horizons and zones of a soil profile. In addition to characteristics already discussed, such as pH, Eh, and c.e.c., the most important criteria are color, texture and structure, organic content, and moisture characteristics (Birkeland 1974). *Color* in soils is an indicator of high organic content (black, dark brown), ferric iron (yellow-brown to red), or a concentration of SiO_2 or $CaCO_3$ (light grey to white). Small amounts of a pigmentor can cause rather intense discoloration, however, and so color alone may be a poor index of the total quantity of the pigmenting substance. *Texture* is simply the relative proportions of different particle sizes in a soil horizon, analogous to the property of sorting as used by geologists (fig. 3.11). *Structure* in soils, however, is a unique characteristic in that it designates the shape developed when individual particles cluster together into aggregates called *peds* (fig. 3.12). In clay-rich soils, structures may play an extremely important geomorphic role by providing the only open avenues for downward percolation through an otherwise impermeable soil. *Organic matter* in soils consists mainly of dead leaves, branches, and the like, called *litter,* and the amorphous residue,

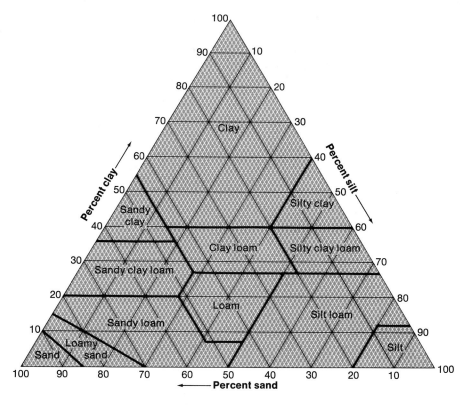

Figure 3.11. Percentages of clay (<0.002 mm), silt (0.002-0.05 mm), and sand (0.05-2.0 mm) in basic soil textural classes as defined by the U.S. Department of Agriculture. (From Soil Survey Staff 1951)

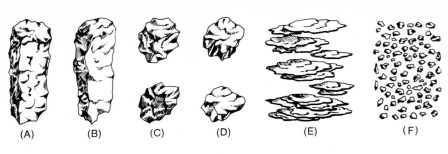

(A) (B) (C) (D) (E) (F)

Quantity ⟶

90
80
70
60
50
40
30
20
10

A

B

| 0° | 5° | 10° | 15° | 20° | 25° | 30° | 35° | 40° | 45° C |
| 30° | 40° | 50° | 60° | 70° | 80° | 90° | 100° | 110° | 120° F |

- - - A = organic raw material production in humid climate

⸺ B = destruction of organic matter in an aerated environment (aerobic)

called *humus,* that develops when litter is decomposed. Litter may form at mean annual temperatures as low as freezing, but its optimum production occurs at about 25°–30°C and decreases rapidly above those levels. Microorganisms that convert litter to humus begin to function at temperatures slightly above freezing (5°C), but the optimum temperature for their life activities may be as high as 40°C (fig. 3.13). It is significant that at temperatures between 0° and 25°C, humus is produced in abundance, but above 25°C little if any humus is accumulated. Humus has an important effect on soil formation because it includes chelators, increases water absorption, and has a high c.e.c. In addition, the development of humus releases CO_2 in high concentrations, leading to unusual amounts of carbonic acid within the humic zone and an associated lowering of the pH. The total quantity of water that can be held in a soil is called the *available water capacity* (AWC). This characteristic is very difficult to estimate in the field, and the lack of a consistent laboratory procedure for its computation has led to some confusion as to its relevance (Salter and Williams 1965). Nonetheless, by combining this parameter with the bulk density (dry weight of soil/unit volume), an estimate of the depth of wetting can be made (see Birkeland 1974). Such information is significant in that it relates to many soil properties, especially those affected by redistribution of material during downward percolation of water.

Available water capacity can be calculated if the moisture content at an upper limit (*field capacity*) and a lower limit (*permanent wilting point*) are known. The field capacity is determined by allowing a saturated sample to drain by gravity for at least 48 hours, by which time the remaining water content is held by adhesion to mineral and organic particles. After field capacity is reached, water can still be taken from a soil by the normal functions of plants. If plants continue to extract water to the extent that no more can be removed, the vegetation wilts. The water remaining in the soil is held with tensional stresses too high for the plants to break, and this amount of water represents the permanent wilting point. Both field capacity and permanent wilting point are expressed as a weight percentage according to the following equation:

$$P_w = \frac{W_s - W_d}{W_d} \times 100$$

where P_w is moisture percentage, W_s is total soil weight, and W_d is weight of soil after drying at 105°C. The available water capacity is simply the difference between the moisture content at field capacity and that at the permanent wilting point.

Horizons   Assuming that some criteria are distinct enough to identify a soil horizon or zone, the pertinent consideration then becomes what nomenclature should be used to convey that information. In recent years the simple designation of A and B horizons as zones of eluviation and illuviation has been seriously questioned. Soil scientists have found that some soils contain more than one eluvial or illuvial horizon, obviously necessitating that either an eluvial zone be placed in the B horizon or an illuvial zone be placed in the A horizon. In addition, it is now clear that concentrations of clays or sesquioxides are not always caused by illuviation but may result as a lag when other materials are preferentially removed. Because of these and other complications, soil processes and characteristics are often incompatible with the general sense of the A, B, C nomenclature; in response, the Soil Conservation Service of the U.S. Department of Agriculture proposed new terms to indicate the diagnostic soil horizons (Soil Survey Staff 1960). The horizons established by the S.C.S. may not correlate with the older system and, in fact, their definitions are so precise that laboratory analyses may be required before certain zones or horizons can be identified. The result has been somewhat chaotic, for some scientists still use the older notations while others have accepted the newer terms.

In an attempt to rectify the confusion, Birkeland (1974) suggests a horizon nomenclature that adopts parts of both systems; it is shown in table 3.7. The

Table 3.7 Nomenclature of soil horizons.

Horizon*	Characteristics
O	Organic matter above mineral soil. Must have > 30% organic content if mineral fraction contains > 50% clay minerals, or > 20% organics if no clay minerals.
A	Mixture of humus and mineral matter. Organic content lower than requirements of O horizon. May be characterized as *mollic* (dark, > 1% organic, base saturation > 50%; usually grassland vegetation); *umbric* (same as mollic but base saturation < 50%; forest vegetation); *ochric* (light color, < 1% organic; semiarid vegetation).
E	Beneath A or O horizons. Low organic, often less sesquioxides or less clay than lower horizon. Usually light colored. Equivalent of A2.
B	Underlies O, A, or E. May be characterized as *argillic* (high clay); *natric* (high concentration of exchangeable sodium); *spodic* (translocated organics and sesquioxides); *oxic* (hydrated oxides of Fe and Al, secondary silicates including 1:1 clays); *cambic* (well-developed soil structures and/or intense oxidation, often more red than underlying horizons).
K	Subsurface horizon. High concentration of carbonate forming continuous horizon. Often laminated and impervious.
C	Unconsolidated material beneath soil that may or may not be parent material of the soil. May show some weathering.
R	Consolidated bedrock underlying a soil.

*Horizons can be subdivided into subhorizons, indicated by arabic numbers such as B1, A2, B12, etc.

Adapted from *Pedology, Weathering, and Geomorphological Research* by Peter W. Birkeland. Copyright © 1974 by Oxford University Press, Inc. Reprinted by permission.

near-surface master horizons are called O or A depending on the amount of organic matter they contain. The A horizon can be further defined as *mollic, umbric,* or *ochric* on the basis of criteria established in the S.C.S. system. The E horizon is comparable to the A2 unit of the old system (see fig. 3.10) and, when present, underlies the O or A horizon. It is characterized by intense leaching that removes Fe^{+3} or organic coatings from the mineral particles and, therefore, it usually is bleached gray in color. The B horizon has characteristics that may reflect any or all of the weathering processes discussed earlier. It commonly has a high clay content due to illuviation or to mineral growth in situ, reddish hues, iron and aluminum concentrations as sesquioxides, and stable primary minerals. The properties of the B horizon, however, can change markedly with variations in the fundamental soil-forming controls such as climate and parent material. The fact that the B horizon has such variable traits prompted the descriptive subterms *argillic, natric, spodic, oxic,* and *cambic*, each of which indicates a special property of the preserved horizon (see table 3.7). The lowermost master horizons, C and R, exist below the B horizon. The C horizon has no properties typical of the overlying horizons, but it has been affected by weathering, as features such as oxidation show. It is composed of unconsolidated material that may or may not be like the material from which the soil presumably formed (Birkeland 1974). The R horizon is simply consolidated bedrock beneath the soil.

The K horizon is similar to the petrocalcic horizon of the S.C.S. As originally proposed (Gile et al. 1965), the term is reserved for those horizons in which the spaces between the skeletal soil grains are so impregnated with carbonate that the morphology of the zone is determined by this secondary material. Although the K horizon is not usually formed near the surface, its depth and its position with respect to other horizons vary. It differs from other carbonate-rich zones in that the accumulation of carbonate is great enough to form a laterally continuous layer. In some cases the layer becomes completely plugged, forming an impermeable barrier to downward percolation and causing water to pond on its upper surface. This results in the precipitation of a laminated carbonate zone at the top of the K horizon. This zone is unique in that it contains virtually no skeletal grains as a framework (Gile et al. 1966). Lattman (1973), however, has shown rather convincingly that some "petrocalcic layers" may form in ways unrelated to the soil-forming processes. Therefore, some care should be exercised when interpreting these horizons.

The master horizons can be further subdivided by the use of letters to designate special features (table 3.8) or numbers to denote specific zones. For example, a zone transitional between the A and B horizons may have characteristics typical of both. Its designation as A3 or B1, both of which are transitional zones, depends on which master horizon it most closely resembles. The B2 zone is reserved for that portion of the horizon where the diagnostic properties identifying the B horizon are most prominently displayed. A second number is sometimes utilized to stress minor variations in the soil. Figure 3.14 demonstrates the use of the standard nomenclature to describe a soil profile.

Table 3.8 Some common descriptive symbols to be used in conjunction with major soil horizons.

Symbol*	Meaning
b	buried soil horizon
ca	accumulation of alkaline earths, commonly calcium
cs	accumulation of calcium sulfate
g	strong gleying
h	illuvial humus
ir	illuvial iron
m	strong cementation
p	plowing
t	illuvial clay
x	fragipan character

*Symbols used with other profile designations. For example B2t, B1h, Cca.

From Soil Survey Staff 1960.

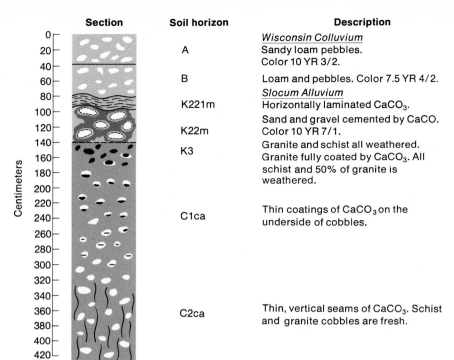

Section	Soil horizon	Description
		Wisconsin Colluvium
	A	Sandy loam pebbles. Color 10 YR 3/2.
	B	Loam and pebbles. Color 7.5 YR 4/2.
		Slocum Alluvium
	K221m	Horizontally laminated $CaCO_3$.
	K22m	Sand and gravel cemented by CaCO. Color 10 YR 7/1.
	K3	Granite and schist all weathered. Granite fully coated by $CaCO_3$. All schist and 50% of granite is weathered.
	C1ca	Thin coatings of $CaCO_3$ on the underside of cobbles.
	C2ca	Thin, vertical seams of $CaCO_3$. Schist and granite cobbles are fresh.

Figure 3.14 Calcareous soil profile developed on unconsolidated deposits near Boulder, Colorado. Note designations of the soil horizons and the descriptions. (From Baker 1975. Used with permission of Springer Verlag, Inc.)

Pedogenic Controls and Classification

Soils can be grouped together systematically on the basis of profile similarities. In general, these result from a unique combination of certain factors that control the magnitude and types of soil-forming processes. The external factors of climate, biota, topography, parent material, and time have long been recognized as the prime controls, and are commonly expressed as the soil equation (Jenny 1941):

$$S \text{ or } s = f\,(cl,\ o,\ r,\ p,\ t\dots)$$

where S = soil, s = any soil property, cl = climate, o = biota, r = topography, p = parent material, and t = time. Although the soil equation cannot be solved in quantitative terms (for discussion see Yaalon 1975), the relative effect of each factor has been determined, with some difficulty, by analyzing field sites where four factors are held essentially constant and one is allowed to vary (Jenny 1941). In addition to the obvious problem of locating soils in which all the factors behave independently, other difficulties arise because many soils are polygenetic; that is, they developed under more than one set of controlling factors. Nonetheless, a qualitative understanding of which factor is dominant in any physical setting can indicate the kind of soil development we should expect to find in a particular region.

On a large scale, climate appears to be the most important factor controlling the soil-forming processes, but within any regional climate other factors may be responsible for subtle variations within the generalized profile. Locally, therefore, each factor may play a significant role in determining boundaries of soil units to be used for mapping purposes. The basic unit in soils mapping (analogous to the *formation* in geologic work) is the *soils series*. This fundamental unit is defined by a number of profiles that exhibit similar properties and is named for some geographic entity such as a nearby county or town. Lines representing series boundaries are placed on a map where the characteristics that define the different series are no longer predominant. Because profiles are gradational as one factor replaces another as the primary influence on the soil-forming processes, the precise location of a series boundary requires some judgment on the part of the investigator. Such decisions are usually fortified by a working knowledge of variations in the pedogenic factors. Indeed, where changing factors produce soils that vary systematically according to topography or parent material, the entire association is called a *soil catena*. The catena recognizes the effect of differences in soil factors and the influence of past history on the soil development. In this way, a soil catena allows pedologists to find, in the complex of local series, both reasons for distinguishing an individual series and common bonds that relate one series to another.

Like a geologic formation, a soils series can be divided into progressively smaller units, called *types* and *phases,* that are identified by minor variations in features such as texture, angle of slope, and salinity. When correctly compiled, a soils map provides extremely useful geomorphic information, and its study is perhaps a logical first step in the analysis of regional or local geomorphology.

Classification   The classification of soils above the series level is, like all classifications, simply an attempt by someone systematically to group together and name soils that exhibit pronounced similarities. The trait or traits being classified serve as the philosophic basis for the groupings. The choice of diagnostic traits depends on what the classifier deems important, and since individuals disagree on that point, no classification will satisfy all members of the pedologic community.

The first real attempt at soil classification came from Russian soil scientists in the late nineteenth century, especially from the work of Dokuchaiev and his students. Although fraught with inconsistencies, Dokuchaiev's scheme had great influence on American pedologists. His groups were defined in part by climate and vegetation, and until recently these genetic criteria were accepted as the basis of all soils classifications used in the United States. In fact, many Russian terms are still integrated in American soils nomenclature.

In the United States, C. F. Marbut was the leading force in efforts to systematize soils. Marbut's classification, developed over a span of years in the 1920s and 1930s, was based on characteristics that are present only in "mature" soils. The system, therefore, had no place for soils that were not fully developed.

Actually Marbut himself was inconsistent on this point when, for example, he included immature alluvial soils with the pedocals because he assumed that alluvium in a dry climate would develop a $CaCO_3$ horizon with time. Clearly he was classifying the climate, not the soil. The term *pedocal* generally refers to soils having a calcium carbonate horizon; *pedalfers* are those soils having an iron- or aluminum-rich B horizon.

A revised classification was developed in 1938 (Baldwin et al. 1938) because of the problems inherent in Marbut's system. Its final format (Thorp and Smith 1949) consists of three main *orders* of soils which in turn are subdivided into *suborders* and finally into *Great Soil Groups* (table 3.9). *Zonal* soils include Great Soil Groups that have properties reflecting the influence of climate and vegetation on the soil-forming processes. Their character is best revealed in mature, well-drained soils located in relatively flat upland areas. In contrast to the zonal soils, *intrazonal* soils are those whose properties are dominated by some subregional factor such as relief or parent material. *Azonal* soils have no well-defined profile because they are very young or are formed on steep slopes or resistant parent matter, both of which retard the normal profile development.

As mentioned before, the Soil Conservation Service completely revised the descriptive nomenclature and the classification of soils in 1960 (Soil Survey Staff 1960). It is important to note that this revision is not simply a formulation of new class names (although that occurred) but represents a fundamental disagreement with the philosophical basis of earlier classifications. Until the new system was devised, all classifications were essentially genetic in scope; that is, the major soil classes were based on climatic and vegetal factors. Although the subdivisions were linked to observable aspects of the profile, these were not explicitly defined, and soil scientists inescapably allowed their knowledge of climatic and vegetation distributions to influence decisions about placing a soil in a particular group. In many cases, therefore, pedologists were classifying the genetic factors and not the tangible resulting soil. The S.C.S. system is nongenetic. It requires no climatic interpretation but is based on very specific, often quantitative, criteria. We will describe it briefly, without becoming mired in its details.

Soils in the S.C.S. system are grouped into 10 *orders* distinguished by the major horizons in their profiles (tables 3.10, 3.11). The orders are subdivided into *suborders* (table 3.10), defined by a diagnostic physical or chemical soil property which in some cases may require quantitative laboratory data to be recognized. The terms used to designate a particular suborder represent the combination of two syllables; the prefix indicates the diagnostic property of the suborder and the suffix reveals the order, since it is composed of several key letters of the order name (tables 3.11, 3.12). An Argid, for instance, is an Aridisol with an argillic horizon, and a Udent is a moist Entisol with a low to moderate organic content. Further subdivision into *great groups* and *subgroups* is made on the basis of even more detailed properties than those differentiating the suborders. For example, within the Udent suborder, a great group characterized by a low

Table 3.9 Soil classification in the higher categories.

Order	Suborder	Great Soil Groups
Zonal soils	Soils of the cold zone.	Tundra soils.
	Light-colored soils of arid regions.	Desert soils. Red Desert soils. Sierozem. Brown soils. Reddish-Brown soils.
	Dark-colored soils of semiarid, subhumid, and humid grasslands.	Chestnut soils. Reddish Chestnut soils. Chernozem soils. Prairie soils. Reddish Prairie soils.
	Soils of the forest-grassland transition.	Degraded Chernozem. Noncalcic Brown soils.
	Light-colored podzolized soils of the timbered regions.	Podzol soils. Gray-Brown Podzolic soils. Red-Yellow Podzolic soils.
	Lateritic soils of forested warm-temperate and tropical regions.	Reddish-Brown Lateritic soils. Yellowish-Brown Lateritic soils. Laterite soils.
Intrazonal soils	Halomorphic (saline and alkali) soils of imperfectly drained arid regions and littoral deposits.	Solonchak, or Saline soils. Solonetz soils. Soloth soils.
	Hydromorphic soils of marshes, swamps, seep areas, and flats.	Humic Gley soils. Alpine Meadow soils. Bog soils. Low-Humic Gley soils. Planosols. Ground-Water Podzol soils. Ground-Water Laterite soils.
	Calcimorphic soils.	Brown Forest soils. Rendzina soils.
Azonal soils		Lithosols. Regosols (includes Dry Sands). Alluvial soils.

Modified from Soil Survey Staff 1960 and Thorp and Smith 1949

Table 3.10 Orders and suborders of the Soil Conservation Service classification.

Order	Suborder	Order	Suborder
ENTISOL	Aquent	SPODOSOL	Aquod
	Psamment		Humod
	Ustent		Orthod
	Udent		Ferrod
VERTISOL	Aquert	ALFISOL	Aqualf
	Ustert		Altalf
INCEPTISOL	Aquept		Udalf
	Andept		Ustalf
	Umbrept	ULTISOL	Aquult
	Ochrept		Ochrult
ARIDISOL	Orthid		Umbrult
	Argid	OXISOL	
MOLLISOL	Rendoll		
	Alboll	HISTOSOL	
	Aquoll		
	Altoll		
	Udoll		
	Ustoll		

From Soil Survey Staff 1960.

Table 3.11 Formative elements in names of soil orders in the Soil Conservation Service classification.

No. of order[1]	Name of order	Formative element in name of order	Derivation of formative element	Mnemonicon and pronunciation of formative elements
1	Entisol	ent	Nonsense syllable.	recent.
2	Vertisol . . .	ert	L. *verto*, turn.	invert.
3	Inceptisol .	ept	L. *inceptum*, beginning.	inception.
4	Aridisol . . .	id	L. *aridus*, dry.	arid.
5	Mollisol . . .	oll	L. *mollis*, soft.	mollify.
6	Spodosol . .	od	Gk. *spodos*, wood ash	Podzol; odd.
7	Alfisol	alf	Nonsense syllable.	Pedalfer.
8	Ultisol	ult	L. *ultimus*, last.	Ultimate.
9	Oxisol	ox	F. *oxide*, oxide.	oxide.
10	Histosol . . .	ist	G. *histos*, tissue.	histology.

[1]Numbers of the orders are listed here for the convenience of those who became familiar with them during development of the system of classification.

From Soil Survey Staff 1960.

Table 3.12 Formative elements in names of suborders in the Soil Conservation Service classification.

Formative element	Derivation of formative element	Mnemonicon	Connotation of formative element
acr	Gk. *akros*, highest.	acrobat	Most strongly weathered.
alb	L. *albus*, white.	albino	Presence of albic horizon (a bleached eluvial horizon).
alt	L. *altus*, high.	altitude	Cool, high altitudes or latitudes.
and	Modified from *Ando*.	Ando	Ando-like.
aqu	L. *aqua*, water.	aquarium	Characteristics associated with wetness.
arg	Modified from argillic horizon; L. *argilla*, white clay.	argillite	Presence of argillic horizon (a horizon with illuvial clay).
ferr	L. *ferrum*, iron.	ferruginous	Presence of iron.
hum	L. *humus*, earth.	humus	Presence of organic matter.
ochr	Gk. base of *ochros*, pale.	ocher	Presence of ochric epipedon (a light-colored surface).
orth	Gk. *orthos*, true.	orthophonic	The common ones.
psamm	Gk. *psammos*, sand.	psammite	Sand textures.
rend	Modified from Rendzina.	Rendzina	Rendzina-like.
ud	L. *udus*, humid.	udometer	Of humid climates.
umbr	L. *umbra*, shade.	umbrella	Presence of umbric epipedon (a dark-colored surface).
ust	L. *ustus*, burnt.	combustion	Of dry climates, usually hot in summer.

From Soil Survey Staff 1960.

temperature is a Cryudent, the prefix "cry-" inserted to indicate coldness. Subgroup nomenclature requires an additional descriptive word placed before the great group name. A Cryudent with a weakly developed spodic horizon becomes a Humodic Cryudent.

It is obvious even from a brief description that the S.C.S. system allows for extremely detailed classification of soils. However, with 10 orders, 29 common suborders, and countless great groups and subgroups, the number of possible combinations of terms is enormous; the system requires an intimate understanding of soil properties for its successful use. Much of the resistance to the new classification has arisen because it is far too cumbersome and detailed to satisfy the needs of scientists other than professional pedologists. In addition, a complete soil description requires many laboratory analyses that are not readily available, especially to those scientists who are using soils as ancillary evidence in studies primarily concerned with other problems.

The choice between available soil classifications depends on how specific the needs of the investigation are. The essentials of the new and old classifications have been presented so that those interested in the details can refer to the original sources and choose which system best satisfies their needs and temperament. As the theme of this book is processes in geomorphology, it seems more appropriate here to understand the general trends of soil development as they reflect soil-forming processes than to become lost in a consideration of how they are classified. Furthermore, table 3.13 shows that there is a reasonable correlation between the new and old systems, especially at the order level. In fact, the S.C.S. recognizes the significance of climate and vegetation in its classification:

> The differentiae selected for the orders tend to give a broad climate grouping of soils. The zonal soils of Baldwin, et al., tend to be grouped with their associated intrazonal soils in many of the orders. The bias in favor of properties associated with or produced by widely differing climates is deliberate, and possibly is in part the result of our experience with earlier classifications. The product of Dokuchaiev's thinking has been proven useful and has been widely accepted by several generations of pedologists. The only alternatives which could be found were tested in earlier approximations and found wanting. (Soil Survey Staff 1960, p. 12)

The general trends of soil development under specific climatic and vegetal controls can be recognized in either classification scheme. These trends, called *pedogenic regimes,* probably represent the aspect of soil development that has the greatest significance for geologists. The use of Great Soil Groups in the following discussions is a matter of convenience because these terms are more familiar to geologists. It is not a suggestion to ignore the S.C.S. terminology;

Table 3.13 Soil orders of the Soil Conservation Service classification and approximate equivalents of the Great Soil Groups.

Orders in S.C.S. classification	Approximate equivalents
1. Entisols	Azonal soils, and some Low Humic Gley soils.
2. Vertisols	Grumusols.
3. Inceptisols	Ando, Sol Brun Acide, some Brown Forest, Low-Humic Gley, and Humic Gley soils.
4. Aridisols	Desert, Reddish Desert, Sierozem, Solonchak, some Brown and Reddish Brown soils, and associated Solonetz.
5. Mollisols	Chestnut, Chernozem, Brunizem (Prairie), Rendzinas, some Brown, Brown Forest, and associated Solonetz and Humic Gley soils.
6. Spodosol	Podzols, Brown Podzolic soils, and Ground-Water Podzols.
7. Alfisols	Gray-Brown Podzolic, Gray Wooded soils, Noncalcic Brown soils, Degraded Chernozem, and associated Planosols and some Half-Bog soils.
8. Ultisols	Red-Yellow Podzolic soils, Reddish-Brown Lateritic soils of the U.S., and associated Planosols and Half-Bog soils.
9. Oxisols	Laterite soils, Latosols.
10. Histosols	Bog soils.

From Soil Survey Staff 1960.

geologists should recognize that future soil maps produced in the United States will utilize the new classification, and for purposes of communication alone a rudimentary understanding of that scheme will become increasingly necessary. The new terminology is included in parentheses to indicate the S.C.S. groups that correlate with the more familiar soil names.

Pedogenic Regimes

Table 3.9 shows that most Great Soil Groups are assigned to suborders that indicate certain combinations of climate and vegetation. Three prominent regimes emerge in the classification: podzolization, laterization, and calcification. Since the soil-forming processes in each culminate in the general characteristics we can expect to find in most common climatic zones, understanding their mechanics should provide us with a solid framework to interpret soils. A widespread podzolic profile in an arid zone, for example, is a pronounced deviation from the predictable soil and therefore must be explained by factors other than the modern regional climate.

Podzolization   The processes that result in the removal of iron and/or aluminum from the A horizon and their accumulation in the B horizon are referred to as *podzolization*. Podzolization is common in humid-temperate climates, especially in soils that develop under forest vegetation. For example, most soils in the eastern half of the United States where precipitation exceeds 70 cm annually are podzolized to some extent. To the west where precipitation is less than 56 cm, another pedogenic regime (calcification) becomes dominant. Podzolization produces a continuous translocation of mobile substances as they are released from parent minerals. Because forest vegetation uses less alkalis and alkaline earths than do grasses or shrubs, most soluble ions (Ca, Na, Mg) are rapidly leached, and colloids of all types adsorb H^+. Therefore, podzolic soils (Spodsols, Alfisols, Ultisols), like all pedalfers, are notably acidic. In fact, Hunt's (1972) grouping of modern soils according to their acidity or alkalinity fits rather nicely the climatic distribution of podzols and other great soil groups.

Where podzolization is very active, an eluvial E horizon may be found overlying an illuvial, spodic B horizon. This bleached zone represents the ultimate form of podzolization because almost all the Al^{+3} and Fe^{+3} is removed and the residuum is highly concentrated in silica. Although the soil in a true podzol is acidic, the E horizon can be developed where the pH of the zone is still within the insolubility range of Al^{+3} and Fe^{+3}. Thus, most of the translocation of these insoluble ions involves chelation rather than dissolution by colloidal acids.

Beneath the zone of maximum eluviation, podzolic soils are usually brown or grey-brown, but as the temperature increases they grade progressively into yellow and red colors (Ultisols; see table 3.13). The greater humus content in cooler regions, combined with the translocation and redeposition of organic matter, accounts for the drabber color in the higher latitudes. However, as the pro-

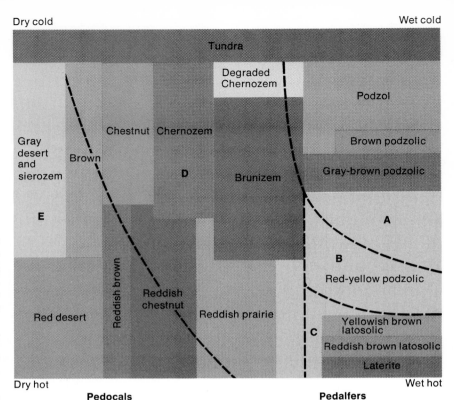

Dry cold   Wet cold

Tundra

Degraded
Chernozem

Podzol

Chestnut Chernozem

Brown podzolic

Gray
desert
and   Brown
sierozem

Gray-brown podzolic

D

Brunizem

E

A

B

Red-yellow podzolic

Reddish brown

Reddish
chestnut

Red desert

Reddish prairie

Yellowish brown
latosolic

C

Reddish brown latosolic

Laterite

Dry hot   Wet hot

Pedocals   **Pedalfers**

Figure 3.15. Distribution of Great Soil Groups on basis of climate. Superimposed are generalized boundaries between major orders of the new Soil Conservation Service classification system: *A*, Spodsols and Alfisols; *B*, Ultisols; *C*, Oxisols; *D*, Mollisols; *E*, Aridisols. (After Millar et al. 1958. Used with permission of John Wiley & Sons, Inc.)

duction of humus declines with increased temperatures, the ferric content rather than the organic matter becomes the predominant factor in determining the color of the B horizon. The characteristic color of podzolic profiles, therefore, changes from north to south in the eastern United States. Figure 3.15 shows these changes diagrammatically.

The end products of podzolic weathering include resistant primary minerals, iron and aluminum sesquioxides or hydrates, and kaolinitic clays. Most of the mobile ions have been removed from the system in solution, but some cations may be retained by fixation in the clay mineral structures. Clays accumulate in the B horizon in any or all of three ways: (1) the component polymers of clays can be carried in solution from the A horizon and recombined in the B horizon; (2) clays can be leached as already formed minerals and physically accumulated; or (3) clays can form in situ by chemical alteration of parent minerals that were originally in the position of the B horizon.

Laterization   The group of processes resulting in laterites (Oxisols) and other lateritic soils constitute a pedogenic regime called laterization. The conditions needed for these processes to function are high rainfall and temperature, intense leaching, and oxidation. These conditions commonly exist in tropical regions, especially where wet and dry seasons alternate and drainage is good. It is now clear, however, that laterization does not necessarily require a tropical climate since rapid leaching of the proper lithologies can form lateritic soils in a wide variety of climates, including subarctic. Thus, to consider laterites (Oxisols) as *the* zonal soil of the tropics may not be absolutely correct.

Laterization is marked by two essential differences from podzolization: (1) organic accumulation is inhibited; and (2) silica is leached from the system in addition to the common mobile ions, leaving abnormally high concentrations of hydrated oxides of iron and aluminum in the soil. In some cases silica combines with available alumina to form kaolinitic clays, but where silica is more extensively leached the excess alumina is present as gibbsite $Al(OH)_3$ or, more rarely, boehmite ($AlO(OH)$). The type of parent material and the organic content exert a direct influence on whether the clays will be kaolinitic or gibbsitic. Where granites are the parent rocks, less silica will be removed during laterization because it exists as quartz rather than as amorphous SiO_2. Lateritic soils developed on granites, therefore, tend to have more kaolinite in their profiles; quartz may be a notable end product. Where basalts are the parent rocks, gibbsite tends to be more dominant because the more soluble amorphous silica is thoroughly leached, leaving little available to form the kaolinite. By observation, when limestone is the parent material, the main clay mineral seems to be boehmite.

In a pure sense, the term *laterite* is reserved for a nodular or continuous layer of oxides or hydroxides of iron and/or aluminum. It was originally used to describe earthy saprolites in India which could be cut into blocks and, because they dehydrated upon drying, could be used as building materials. Over the years, however, the term has been used to describe an array of soils and even some materials of nonpedogenic origin. Soils depleted of silica and enriched in iron and alumina are commonly called laterites even if the diagnostic laterite layer is missing. Many workers now prefer the term *latosol* (Oxisols) to define soils that are not true laterites but have obviously been affected by laterization (see Paton and Williams 1972).

Profiles of true laterites do not unequivocally fit the classification schemes discussed above. Because of sparse humic development, A horizons may be extremely thin or, in some cases, completely missing. In fact, the original definition of laterite included only the B horizon. A section through a typical laterite revealing the common horizons is shown in figure 3.16. In most laterites the uppermost zone is the laterite layer, consisting of hydrated iron or aluminum with an unusual texture that can be oolitic, pisolitic, concretionary, or vesicular. Between the lateritic layer and the parent material a thick, clay-rich segment is normally present. This segment can be divided into two horizons called the *mottled zone* and the *pallid zone*. Red mottles or vertical stringers of iron oxide, set in a white clay matrix, characterize the mottled zone. In contrast, the pallid zone is very low in iron content and is typically blanched. In reports describing laterites, the material beneath the pallid zone varies from fresh bedrock to a deeply

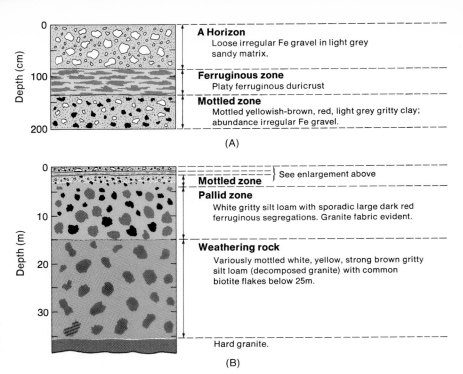

Figure 3.16 A typical lateritic soil profile. Entire profile (B) extends to a depth of 35 m; (A) represents only the upper 2 meters of the profile. (After Gilkes et al. 1973. Used with permission of Oxford University Press)

weathered saprolite, each of which has been proposed as the parent source of the overlying laterite horizon. It is important to realize that the mottled or pallid zones are not always present and that in some cases the laterite horizon lies directly on top of fresh bedrock. The thickness of the entire profile varies depending on what horizons are present but as indicated in figure 3.16, lateritic soils commonly extend to a depth of 15 meters or more.

Gilkes and his co-workers (1973) detailed the mineralogy of the laterites depicted in figure 3.16 and concluded that there is a distinct mineral zonation in the profiles. Oddly, the mineral zones are not correlative in a simple way to the morphologic horizons but tend to transcend horizonal boundaries. Immediately above the parent material a thin zone exists in which only the least stable minerals have been destroyed. Proceeding to the surface, however, each successive mineral zone indicates more pronounced alteration until, in the upper 50 cm, hematite and boehmite are the dominant residual products.

Laterization is perhaps the most poorly understood of the pedogenic regimes. We know that iron and aluminum can be dissolved in acidic fluids, yet the environments of their accumulation are very acidic; precisely how these substances are concentrated is not clear. Much controversy has arisen over interpretations concerning the environment of laterite development. Most workers now believe that part of the laterite profile, especially the mottled zone, forms between the upper and lower limits of a fluctuating water table. The pallid zone is probably saturated at all times and has lost its iron through solution. The iron, however, may have migrated upward through the oxidized portions of the profile instead of taking the usual downward movement associated with soil development.

The inferred mobilization of iron and aluminum raises the major controversy among students of laterites, namely, the source of the sesquioxides that make up the laterite layer. Some investigators insist that the concentration of sesquioxides occurs in situ as a residuum when all nonlateritic components are mobilized and removed. Others feel that iron and aluminum are taken into solution at one place and subsequently transported to the site of the lateritic layer where they are concentrated by deposition (for discussion see deSwardt 1964).

In light of these controversies, it may be worthwhile to stress again that although laterization should function efficiently in warm, humid climates, laterites (Oxisols) are not necessarily restricted to the tropics. Such a binding inference can lead to rather hapless paleoclimatic interpretations, since intrasols in cool climates can be lateritic and podzols (Spodsols, Alfisols, Ultisols) are not uncommon in the tropics. In addition, historical attempts to relate laterite formation to a late stage in the classical cycle of erosion are also fraught with difficulties. Indeed, deSwardt (1964) suggests that the laterites in parts of equatorial Africa were developed on a surface with as much relief as the present landscape, perhaps hundreds of meters. How all these constraints and interpretations relate to the fact that most true laterites are ancient rather than modern soils (Hunt 1972) is simply not known. The only truth that seems to emerge about laterites is that the confusion concerning their origin is great and will not be erased until we understand in detail the laterizing processes.

Calcification   The third pedogenic regime, calcification, functions in subhumid to arid climates where precipitation is insufficient to drive the soil water downward to the water table. Ions mobilized in the A horizon are reprecipitated in the B horizon where zones of $CaCO_3$ are commonly developed. The depth of the carbonate zone is a function of the annual precipitation (fig. 3.17) and in general represents a first approximation of the vertical extent of leaching. Arkley (1963), however, showed that this relationship holds only in regions without pronounced seasonality of rainfall or orographic effects. Nonetheless, for our purposes it seems useful to visualize the depth of the carbonate zone as a gross index of the magnitude of calcification. The pedocal soils derived from those processes all possess a K horizon or a calcium carbonate zone, but its position within the soil profile rises proportionately with decreasing rainfall and/or the efficiency of the calcifying processes.

Calcification takes place most efficiently under a grass or brush vegetation because such plants utilize large quantities of alkali and alkaline earths in their life processes. When the plants die, the mobile ions are returned to the soil as the

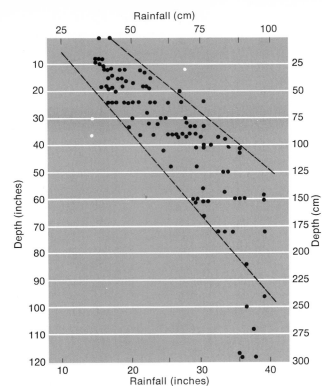

Figure 3.17. The depth of carbonate zones in soils as related to mean annual rainfall. (From Jenny and Leonard 1939, *Soil Science* 38:367, fig. 2. Used with permission of the Williams and Wilkins Co., Baltimore)

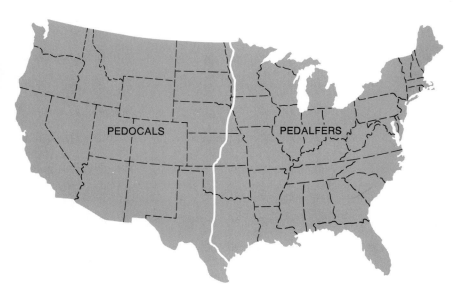

Figure 3.18. Generalized distribution of pedalfers and pedocals in the United States. (From Jenny 1941. Used with permission of the McGraw-Hill Book Co.)

litter is decomposed. The amount escaping this recycling process depends on the precipitation, the temperature, and the related bacterial activity. The resulting profile differences are best exemplified by considering the distribution of pedocals within the United States, shown in figure 3.18, and the associated Great Soil Groups, shown in figure 3.15.

The Chernozem group (Mollisols) occurs mainly in the temperate zones of the United States, especially in the northern Great Plains states (Dakotas and Nebraska). In these soils the profile (fig. 3.19) usually has an undifferentiated, black A horizon ranging from 30 to 120 cm in thickness, which reflects the high accumulation of humus in cooler temperature zones. The B horizon is usually light-yellow to brown, and carbonate may exist there as disseminated material or in a distinct layer at the base of the horizon. In more arid areas the Chestnut and Brown soils (Mollisols, Aridisols) become dominant. These groups are similar to the Chernozems except that the A horizon is thinner and less rich in organic material, and the carbonate zone is higher in the profile. Desert soils (Aridisols) are normally very immaturely developed; $CaCO_3$, if present as a secondary accumulation, will be near or at the surface. Generally, then, in cooler areas with less than 64 cm of annual precipitation, increasing aridity is shown in the profiles by a progressively less developed A horizon and a carbonate zone rising closer to the surface.

Where the temperature is higher, humus production is less and a red hue becomes more prominent in the calcified soils. This gives rise to the Reddish

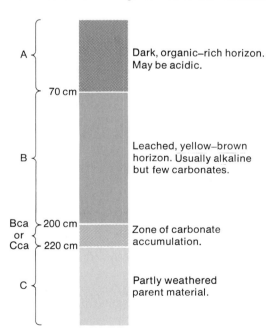

A — 70 cm

Dark, organic–rich horizon. May be acidic.

B

Leached, yellow–brown horizon. Usually alkaline but few carbonates.

Bca — 200 cm
or
Cca — 220 cm

Zone of carbonate accumulation.

C

Partly weathered parent material.

Figure 3.19. Generalized profile of a Chernozem (Mollisol) in eastern Great Plains region of the United States. Greater aridity would decrease the size of the A horizon and raise the carbonate zone to a higher level. Depth values will vary with local conditions.

Chestnut, Reddish Brown, and Red Desert soil groups (Mollisols, Aridisols) found in the lower latitudes of the pedocal regions (fig. 3.15). Moving eastward from the Chernozem regions, a transitional soil between a Chernozem and a Podzol is common. These soils, called Prairie or Brunizem (Mollisols), lack the precipitation and vegetal cover needed to generate pronounced podzolization, but their profiles do not show a carbonate zone.

Because of the immobility of the soluble ions, most clay minerals formed in the calcification regime are montmorillonite or illite types. In some cases kaolinite is abundant near the top of the profile where slightly acidic soil water is present, especially where humus is abundant. The retention of the alkaline earths, however, is not favorable for the development of kaolinite in the deeper zones.

Significance of Weathering and Soils

Our discussion thus far has been a very brief treatment of the basic processes involved in weathering and soil development. How is such information important and relevant to the study of geomorphology? The significance of weathering and soils in the surficial realm manifests itself in two major categories of geomorphic work which, for lack of ingenuity, we can call historical geomorphology and physical geomorphology.

Historical Geomorphology

As the term is used here, *historical geomorphology* is essentially the procedure of reconstructing past geomorphic events. It is a bona fide part of the geomorphic science, for any complete study of a specific area demands such an analysis. Past geomorphic events may be directly responsible for the type and magnitude of modern processes because the framework within which those processes function is partly inherited from the results of earlier geomorphology. Although geomorphology leaves its mark in the entire geologic record, the Pleistocene events, being most recent and dramatic, have left the greatest imprint on the present geomorphic setting. Determining the sequence of Pleistocene events and the environments in which those events occurred is often facilitated by detailed studies of weathering and soil.

If a sequential history is desired, it becomes clear that putting the pieces together requires some estimate of the relative ages of deposits and features within the region being investigated. Weathering and soils phenomena are almost always used as indices of relative age when radioactive dates are not available. A variety of characteristics have been used in this manner, including (1) the thickness of weathering rinds on boulders (Porter 1975); (2) the percentage of decomposed or disintegrated boulders contained in the soil profiles developed on alluvial deposits (Nelson 1954; Flint and Fidalgo 1964; Eggler et al. 1969); (3) mineral analyses (Dryden and Dryden 1946; Bhattacharya 1963; Willman et al. 1963, 1966); and (4) the relative thickness and maturity of soil profiles developed on similar parent material in the same climatic zones (Richmond 1962).

The use of soils in historical geomorphology is complicated by the fact that soil-forming factors have not been constant with time. This is especially true in areas that have been affected by the pronounced climatic fluctuations of alternating glacial and interglacial episodes. As the climate changes, the dominant factors of soil formation in any given area must also change accordingly. Thus, many soils preserve in their profiles characteristics that reflect more than one set of soil-forming factors. These *polygenetic soils* (sometimes called *complex soils*) complicate historical geomorphology because thickness and maturity of soils are reliable indices of age only when the conditions developing the soils have been maintained continuously. When conditions change, the properties of the initial soil are supplanted by new characteristics; because such alterations vary in the degree of completeness, complex soils are very difficult to correlate with soils in other areas. A pre-Wisconsinan soil, for example, will accrue properties related to the controlling factors at the time of its formation. A change in those factors at some later time, perhaps post-Wisconsinan, will obviously cause the original soil properties to be out of phase with the new soil-forming environment. For all practical purposes, the pre-Wisconsinan soil becomes the parent material on which the post-Wisconsinan climate and biota are working, and a younger soil profile is superimposed on the older one. Separating the older soil from the younger one is a demanding field problem which, in this example, would require lateral tracing of the complex soil to a locale where Wisconsinan deposits intervene. On those deposits the post-Wisconsinan soil will be present, but the pre-Wisconsinan profile will be absent or buried beneath the Wisconsinan material; figure 3.20 illustrates an actual example.

Soils that form on a landscape of the past are called *paleosols*. They can be of three types: (1) *buried* soils are developed on a former landscape and subsequently covered by younger alluvium or rock; (2) *relict* soils were not subsequently buried but still exist at the surface; and (3) *exhumed* soils were at one time buried but have been reexposed when their cover was stripped by erosion (Ruhe 1965). Buried soils, like that in figure 3.20, are immediately recognized as paleosols. Relict and exhumed soils are much more difficult to identify, because it must be proved that their properties are inconsistent with the modern

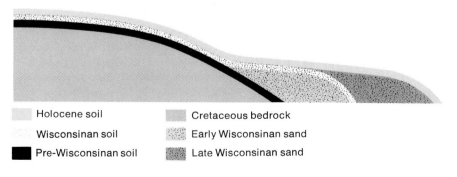

Figure 3.20. Superimposed soil profiles near Denver, Colorado. Lateral tracing reveals unweathered parent materials of different ages and separates the soils that formed at various times. (After Hunt 1954)

Holocene soil

Wisconsinan soil

Pre-Wisconsinan soil

Cretaceous bedrock

Early Wisconsinan sand

Late Wisconsinan sand

environment or that their age is the same as the paleosurface on which they rest. Some additional geomorphic or stratigraphic data are usually needed to substantiate those requirements.

The development of complex soils is especially striking near the boundary between two zonal soil orders. There fluctuations of climate and vegetation cause both zones to occupy the same geographic position alternately, and the type profile of each may be superimposed on the other. In the geographical center of a zone, however, a climatic change may not be severe enough to alter perceptibly the soil-forming factors, and the combined profile may not reveal the zonal shift. In western piedmont regions where climatic zones are vertically telescoped, the complexing process is very likely to recur. In these areas, the rainshadow effect places humid mountain zones only a short horizontal distance from adjacent arid basins; even minor climatic change can produce dramatic maladjustments within the soil profile (Bryan and Albritton 1943). For example, figure 3.21 illustrates a profile revealing a red-brown, slightly argillic horizon overlain by a calcium carbonate zone, which is essentially impossible to explain under one soil-forming regime since the leaching needed to concentrate the clay could not have precipitated $CaCO_3$ at a higher level. In addition to the horizontal relationships such as argillic and calcic zones, the types of clay minerals present in a soil generally reflect the climatic conditions under which they formed (Birkeland 1969), but they may not be a particularly useful indicator of relative age between various soils (Birkeland and Janda 1971).

Finally, the detailed study of modern and ancient soils has encouraged geologists to utilize the stratigraphic relationships of soils to decipher Pleistocene and Recent history. The avant-garde study of soil stratigraphy in the La Sal Mountains of Utah (Richmond 1962) exemplifies this technique and illustrates how soils can be a powerful weapon in the arsenal of the Pleistocene geologist.

Figure 3.21. Complex soil formed on Early Wisconsinan outwash gravel near Roberts, Montana. B horizon was formed by weathering during Early Wisconsinan/Late Wisconsinan interstadial. Bca developed during Holocene when climate became arid.

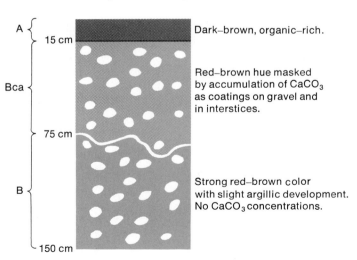

A
15 cm

Dark–brown, organic–rich.

Bca

Red–brown hue masked by accumulation of $CaCO_3$ as coatings on gravel and in interstices.

75 cm

B

Strong red–brown color with slight argillic development. No $CaCO_3$ concentrations.

150 cm

However, much of the correctness of soil stratigraphic work depends on the common belief that soils form in distinct, relatively short episodes separated by long periods when little if any weathering or pedogenesis takes place (Morrison 1967). This concept of soil-forming intervals has been seriously challenged (Birkeland and Shroba 1974). If future work determines soil formation to be a continuous rather than a spasmodic process, much of the stratigraphic work that utilized soils as the basic criteria will have to be reevaluated.

Physical Geomorphology

The term *physical geomorphology* is being used here in an informal sense simply to indicate the physical processes operating in the modern environment. Soil characteristics affect these geomorphic processes in a myriad of ways, and thus a unique controlling property may assume extreme importance in the mechanics of a surficial system. In fact, the influence that soils exert on the entire physical system may be their greatest contribution to geomorphology. Once a particular soil property is identified as a controlling influence on a physical process, soils mapping and historical geomorphology take on added significance since they explain and document the distribution of the factor involved. Critical information needed for environmental control or for regional and local planning can be provided by a multipronged geomorphic analysis. Indeed, integrated soil and geological studies are now being used to help planners understand the limitations, hazards, and resources of the areas they are considering (McComas et al. 1969).

In urban areas where pressures of growth and development seem to be inevitable, a recognition of soil properties can have most important and beneficial consequences. Baker (1975) reports a typical case in which urban planning is highly dependent on the integrated soil-geology concept. In the Boulder, Colorado area, three types of parent material form the geologic framework: a terrace or pediment gravel, loess, and the Pierre Shale. Soils developed on each material have distinctive engineering properties determined primarily by whether clay minerals included in the soil have the potential for swelling; figure 3.22 shows these effects. The pediment gravels usually display the profile shown in figure 3.14, although minor variations occur with the age of the alluvium. In general, clays are present in the gravelly profile, but they do not form a decided argillic horizon nor do they tend to expand. Stresses generated by the weight of buildings are absorbed at the contacts of coarse particles in the soil, so that few foundation problems arise in construction on the pedimented surfaces. Soils on the Pierre Shale, however, are an engineering nightmare. Plastic clays are present in abundance and are concentrated in the B horizon. As the clays become wetted, their swelling and expansion exerts differential pressure on the overlying material, and building foundations tend to fail. Soils developed on the loessal parent material present a different problem. The clays in these soils are not prone to swelling, and it might be expected that construction in the loess areas would be devoid of problems. The soils, however, have a well-developed prismatic struc-

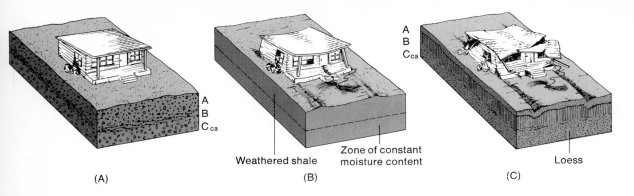

| (A) | (B) | (C) |

Weathered shale   Zone of constant
                   moisture content

Loess

A
B
C ca

A
B
C ca

A
B
C ca

Figure 3.22. Conditions for building foundations near Boulder, Colorado, relate to differing types of parent material and soil profiles. (A) Pediment gravel provides excellent building support; (B) swelling of clays in weathered shale produces differential movement; (C) differential settling in loess is caused by solution of CaCO₃. (From Baker 1975. Used with permission of Springer Verlag, Inc.)

ture in the B horizon which funnels soil water to the lower Cca horizon (fig. 3.23). As the water destroys the intergranular strength of the calcic unit, differential settling causes footings of buildings to rotate, and the structures may collapse. The chain of events culminating in failure may be initiated by something as innocuous as watering the lawn.

The importance of soil development can also be seen in its effect on the hydrologic system. Accelerated illuviation sometimes gives rise to clay pans or other zones of low permeability in a soil. This not only lowers infiltration rates, but may also trigger flash flooding. The process is most obvious when the stricture is close to the surface, because the total available pore space above the impermeable layer quickly fills with rainwater, and any additional precipitation will immediately take the form of direct runoff. An excellent example of this relationship has been noted on the alluvial fans adjacent to Las Vegas, Nev. (Cooley et al. 1973). In that environment, highly cemented calcic zones, usually less than 60 cm below the surface, create barriers to infiltrating water that falls as precipitation on the fans or traverses them in shallow stream channels. Water crossing the fans in the stream washes would normally be influent from the channel bottom, thereby reducing the downfan discharge, but the impervious soils prevent this. Even where channels are incised below the calcic horizons, their sides are often made impermeable by case-hardening (Lattman and Simonberg 1971), and the bed material is turned into a solid mass of calcified breccia by a process called gulley-bed cementation (Lattman 1973). The result is to make Las Vegas subject to flash flooding even when no rain has actually fallen on the city itself. In addition to the flooding problem, the calcified layers also prevent water from recharging the underground aquifers that serve as the major source of water for the area.

In many arid regions a similar but more extensive form of cementation binds the near-surface fringe of alluvial deposits into resistant zones called *duricrusts* (Woolnough 1927). Normally the cementing material is CaCO₃, Fe₂O₃, or SiO₂, giving rise to the terms *calcrete, ferricrete,* and *silcrete* to describe the hard

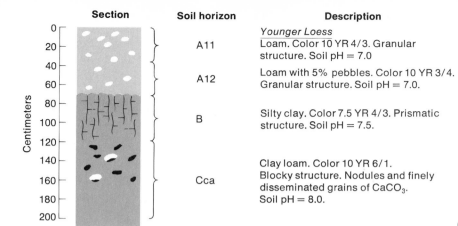

Section	Soil horizon	Description
		Younger Loess
	A11	Loam. Color 10 YR 4/3. Granular structure. Soil pH = 7.0
	A12	Loam with 5% pebbles. Color 10 YR 3/4. Granular structure. Soil pH = 7.0.
	B	Silty clay. Color 7.5 YR 4/3. Prismatic structure. Soil pH = 7.5.
	Cca	Clay loam. Color 10 YR 6/1. Blocky structure. Nodules and finely disseminated grains of $CaCO_3$. Soil pH = 8.0.

Figure 3.23 Soil profile developed on loess near Boulder, Colorado. Prismatic structure in loess provides infiltration avenue for surface water, and the resulting solution may cause foundation problems. (Compare with fig. 3.22c.) (From Baker 1975. Used with permission of Springer Verlag, Inc.)

carapace developed by each of the indurating types. Duricrusts are significant in physical geomorphology because they are usually more resistant than the deposits or rocks that are not encrusted. The difference in resistance creates relief because the duricrust armors the more erodible materials, allowing the crusted areas to stand as residual hills or mesas when the adjacent, noncrusted regions are lowered more rapidly (see Stephens 1964; Mabbutt 1965; Goudie 1973 for greater detail). Duricrusting also affects the intensity of surface runoff because infiltration of rainwater is drastically reduced as the amount of cementing material increases.

The importance of soils in geomorphology could be documented in an almost endless array of case histories. Quite obviously, planners who fail to recognize soils as part of the physical system are flirting with disaster. It should be equally clear to all surficial scientists, including geologists, that soils simply cannot be ignored in their work.

Summary

The processes of chemical weathering (hydrolysis, oxidation-reduction, solution, ion exchange) alter the exposed portion of the geologic framework and, combined with organic processes, produce soils. The degree of chemical change depends on how mobile the ions of the parent minerals are under the external and internal controls on the weathering mechanics. In regions with abundant precipitation, highly mobile ions are usually removed from the weathered zone unless the original mineral structure is incompletely broken down and elements such as potassium are fixed in the system. In contrast, where leaching is minimal, mobile ions (Ca, Na, Mg) are concentrated in the weathered zone. Immobile ions (Fe^{+3}, Al^{+3}) may be transposed in very acidic ground waters or by special organic processes such as chelation. The mobility of most substances is also dependent on the pH and Eh of the weathering environment. The type of clay mineral formed in the weathering zone is usually a good indicator of the intensity of decomposition.

Soils are described and classified according to the soil profile. The character of the soil profile varies with parent material, climate, biota, topography, and the length of time involved in its formation. Three major pedogenic regimes—podzolization, laterization, and calcification—produce the dominant soil groups; however, changes in the controlling pedogenic factors may result in complex soils that show evidence of forming under more than one pedogenic regime. Soils are important elements in reconstructing geomorphic history, and they directly influence other surficial processes.

Suggested Readings

The following references provide greater detail about the topics discussed in this chapter.

Baker, V. R. 1975. Urban geology of Boulder, Colorado: A progress report. *Environmental Geol.* 1: 75-88.

Birkeland, P. 1974. *Pedology, weathering, and geomorphological research.* London: Oxford Univ. Press.

Carroll, D. 1970. *Rock weathering.* New York: Plenum Press.

Cooley, R.; Fiero, G.; Lattman, L.; and Mindling, A. 1973. Influence of surface and near-surface caliche distribution on infiltration characteristics and flooding, Las Vegas area, Nevada. Project Report 21. Reno: Univ. of Nevada, Desert Research Inst.

de Swardt, A. M. J. 1964. Lateritisation and landscape development in parts of equatorial Africa. *Zeit. f. Geomorph.* 8: 313-33.

Hunt, C. B. 1972. *Geology of soils.* San Francisco: W. H. Freeman.

Loughnan, F. 1969. *Chemical weathering of the silicate minerals.* New York: American Elsevier.

Schalscha, E.; Appelt, H., and Schatz, A. 1967. Chelation as a weathering mechanism—I. Effect of complexing agents on the solubilization of iron from minerals and granodiorite. *Geochim. et Cosmochim. Acta* 31: 587-96.

Soil Survey Staff 1960. *Soil classification, a comprehensive system—7th approximation.* Washington, D.C.: U.S. Dept. of Agriculture, Soil Conserv. Service.

The transformation of rocks into unconsolidated debris is the prime geomorphic contribution of weathering and soil-forming processes. Whether the debris produced by weathering will resist erosion depends on the balance between the internal resistance of the materials and the magnitude of the external forces acting on them. The relative resistance of any natural substance is partly reflected in the character of the slope that develops on it. Extremely steep slopes, for example, can be maintained for long periods only if the underlying rock or soil is so tightly bound together that the forces and agents of erosion cannot lower the slope angle. On the other hand, gentle slopes in regions of low relief and elevation may be stable for relatively long time spans even if the underlying material is very friable. Clearly then, slope characteristics provide us with useful information only when we understand the erosive processses attacking them.

In a large sense, the evolution of landscapes is the history of regional slope development. The formation of these slopes encompasses a multitude of geomorphic processes, and the properties of slopes reflect in subtle ways the temporal effect of these processes on the resisting framework. Thus, interest in slopes and slope-forming processes crosses the entire range of geomorphic thinking, from the analysis of a modern stability problem to the abstraction of geologic history.

The mechanics of slope erosion are in many ways closely related to the processes of physical weathering because the forces that disintegrate rocks and minerals simultaneously lower the internal strength of the unconsolidated cover. We will begin, then, with a brief review of physical weathering.

Physical Weathering, Mass Movement, and Slopes 4

Physical Weathering

Physical weathering culminates in the collapse of parent material and its diminution in size. The continuing breakdown of rock takes place when stress is exerted along zones of weakness within the original material. These zones may be planar structures such as bedding or fractures which, upon rupture, produce fragments whose size and shape are controlled by the spacing of the planes. In other cases, failure may occur along mineral boundaries, resulting in an accumulation of particles similar in size and shape to the original rock texture. Although stresses are generated in different ways, the common bond that unites all processes of disintegration is that in every case a force within the material itself is responsible for its destruction.

The stress field involved in disintegration results from either expansion of rocks or minerals themselves or pressure generated by growth of a foreign substance in voids within the lithologic fabric. In each method the direction of the principal stress may change according to the process involved, but the most pronounced disintegration invariably occurs where the adjacent rocks exert the least confining pressure. Intuitively, then, we should expect disintegration to be most pernicious near the surface, where static load from overburden is minimal and fractures are abundant and closely spaced. With increasing depth, confining pressure increases, fractures are less common, and the disintegrating processes become less effective.

Expansion of Rocks and Minerals

Thermal Expansion   Rocks and minerals expand in response to several phenomena that can rightfully be considered as agents of physical weathering. There seems to be little doubt that the application of intense heat can cause physical disruption of rocks. The low thermal conductivity of rocks prevents the inward transfer of heat, allowing the external fringe of a rock mass to expand significantly while little, if any, change occurs below the outer few centimeters. Differential stresses are produced by this thermal constraint, and the rock exterior spalls off in plates or wedges 1–5 cm thick. As figure 4.1 illustrates, this process functions during forest fires; in semiarid forested mountains of the western United States, it may be the dominant process of physical weathering (Blackwelder 1927). Whether or not insolation can drive the process has been debated for many years. Many geologists have gradually, if not grudgingly, accepted the premise that diurnal temperature fluctuations are not severe enough to produce thermal spalling (Twidale 1968b). Ollier (1963, 1969), however, has reaffirmed the suggestion that insolation is a viable cause of thermal expansion, even though experimental studies (Griggs 1936a,b) support the opposite conclusion. The major unanswered question is whether repeated applications of small thermal stresses over a long period of time can induce progressive rock fatigue and thereby lower the rock's internal strength. Until the time factor is adequately understood, we cannot be absolutely certain whether insolation has any geomorphic significance.

Unloading   Expansion of large segments of rock masses occurs when confining pressure is released by erosion. As denudation removes overburden, the stress squeezing the underlying framework is lowered, and rocks tend to split into widely spaced sheets, 1–10 m thick, that are oriented perpendicular to the direction of pressure release. The sheeting tends to mirror the surface topography and, since outer sheets are relatively easy to erode, the process helps perpetuate the surficial configuration because subsequent sheets develop with a similar orientation. Although other processes aid in the removal of the sheets, it can be readily documented that the original formation of the fractures is a pressure-release phenomenon. Rock bursts in deep mines, for example, are explicit proof that something as simple as excavation of tunnels can trigger a rapid expansion of surrounding rocks. In the natural setting, postglacial entrenchment of the Vaiont River in Italy permitted valleyward expansion of rocks and produced a joint system parallel to the valley sides (fig. 4.2). Hack (1966) demonstrated that arcuate patterns of streams, ridges, and vegetal types in the eastern United States are probably controlled by the position of curved sheets that dilated during erosion of crystalline rocks.

Figure 4.1. Spalling of granitic boulder caused by heat expansion during forest fire. Beartooth Mountains, Mont. (Photo by R.R. Dutcher)

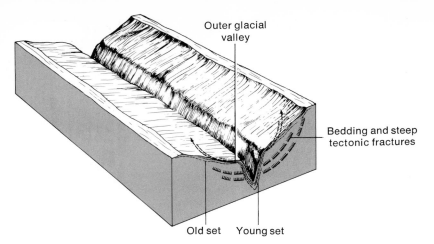

Outer glacial valley

Bedding and steep tectonic fractures

Old set   Young set

Figure 4.2. Expansion joints produced by pressure release during valley entrenchment. Vaiont River valley, Italy. (From Kiersch 1964, *Civil Engr.* **34, n. 3, p. 35)**

Hydration and Swelling   Expansion also occurs when minerals are formed or when they are altered by the addition of water to their structure. Although the process begins as a chemical process called *hydration,* its physical side is particularly obvious when clay minerals containing layers of OH or H_2O are formed. The creation of the layered structure expands the minerals and propagates stress outward from the clay particle. Clays such as bentonite (Na-montmorillonite), which do not have a fixed OH or water layer in their structure, have the capacity to absorb water into the mineral during periods of wetting. The swelling produced by wetting exerts the same outward stress as during clay formation. Most clays show the trait to some extent, but the percentage of expansion depends on the mineral type plus a myriad of other factors (table 4.1). Montmorillonite clays, for example, drastically lose their swelling capacity when sodium is replaced by some other cation (Mielenz and King 1955). Upon drying, the expanded clays lose part or all of the absorbed water, initiating an alternating swelling and shrinking sequence associated with episodes of wetting and drying. In contrast, the well-ordered hydrated clays have a stable structure, and destruction of the OH or water layer occurs only when the mineral is heated to at least 300°C. The disintegrating effect of these clays, therefore, occurs during their formation but, in contrast to the swelling clays, is exerted continuously until relieved. The use of hydrated clays as agents of disintegration is a one-shot affair, however, because once their expansive stress is released in a disintegrating event there is no way to reinstate the internal stress.

The effect of mineral expansion has been clearly demonstrated in the physical breakdown of granites in arid or semiarid regions (Wahrhaftig 1965; Eggler et al. 1969). In these settings, the major weathering product of granite is a coarse, angular mass of rock and mineral fragments called grus, shown in figure 4.3, in which feldspars are often unaffected by decomposition. In the Laramie Range of Colorado and Wyoming, the sequence of grus development started during the

Table 4.1 Expansion of common clay minerals by hydration.

Clay Mineral	% Expansion
Ca-Montmorillonite	
Forest, Miss.	145
Wilson Cr. Dam, Colo.	95
Davis Dam, Ariz.	45-85
Na-Montmorillonite	
Osage, Wyo.	1400-1600
Illite	
Fithian, Ill.	115-120
Morris, Ill.	60
Tazewell, Va.	15
Kaolinite	
Macon, Ga.	60
Langley, N.C.	20
Mesa Alta, N.M.	5

Adapted from Mielenz and King 1955, with permission of the California Division of Mines and Geology.

Figure 4.3. Formation of grus by disintegration of granitic boulders. Upper part of photo shows grus matrix developed when boulders break apart. (Coin is silver dollar.) (Photo by R.R. Dutcher)

Precambrian with formation of hematite by high-temperature oxidation along cleavage planes in the biotite (Eggler et al. 1969). Although this expanded the biotite in the direction of the c-axis, the stress was not sufficient to cause disintegration. It did, however, weaken the biotite's ability to resist further geomorphic attack. Subsequent near-surface weathering produced clays from the biotite with as much as 40 percent increase in volume, and the stress generated by this expansion shattered the granite into grus.

It also seems certain that hydration of salts within pores of building stones and concrete develops sufficient stress to cause extensive spalling (Winkler and Wilhelm 1970). According to E. Winkler (1965), a similar process almost destroyed Cleopatra's Needle, an obelisk that was brought from Egypt to New York City in 1880. Salts trapped in spaces within the red-granite monument did not hydrate until they were placed in the humid climate of the eastern United States, but since then significant spalling has taken place.

In most humid regions, the process of mineral expansion manifests itself in different end products. Rocks are peeled off to produce curved surfaces; the process on a large scale is called *exfoliation* and on a smaller scale *spheroidal weathering*. Even though the resulting large domelike masses or rounded boulders (shown in figs. 4.4 and 4.5) are probably in part a function of pressure

Figure 4.4. Northeast side of Half Dome taken from the subsidiary dome at the northeast end of the rock mass, revealing exfoliation on a gigantic scale. In the foreground is an old shell disintegrating into undecomposed granite sand. Yosemite National Park, Mariposa County, Cal. (Photo by F.C. Calkins. From Matthes 1930, plate 49)

release, it seems certain that water and mineral alteration are intimately involved (Gentilli 1968). Spheroidal boulders are formed because edges and corners of lithologic blocks are weathered more rapidly than flat surfaces, a phenomenon especially apparent where the parent rock has been fractured into a blocky framework by perpendicular joint sets. The relatively fresh spheroidal cores are usually surrounded by a zone of disintegrated flakes and spalls that is enriched in secondary clay minerals. Simpson (1964), for example, found that the clay matrix in weathered graywacke increased by 5–10 percent in the spalled zone and also contained abundant vermiculite, an expandable clay not present in the fresh rock. Evidence such as this seems to indicate that outward expansion caused by the development of clay minerals peels off the fresh rock layer by layer, working progressively inward from the surrounding joint openings.

Growth in Voids

A second group of processes generate stress when some substance grows in spaces within the rock. The pressure gradient differs from that in the processes explained above because it is the openings that are expanded, not the parent minerals or rocks. Because all spaces are not expanded simultaneously or with equal magnitude or direction, the resultant pressures differ locally, and the entire

Figure 4.5. Gabbro boulder showing spheroidal weathering. Himalaya Mine, San Diego County, Cal. (Photo by W.T. Schaller, U.S. Geological Survey)

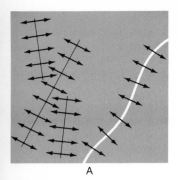

A

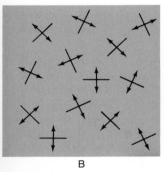

B

Figure 4.6. Stresses generated in rock material by growth of foreign substances in fractures. (A) Direction of expansion along fractures. (B) Stress field resulting in highly fractured rock.

Figure 4.7. Pebble fractured by growth of calcite along planes of weakness. Near Roberts, Mont.

system is burdened with a differential stress field (fig. 4.6). Such a pressure distribution is conducive to fracturing or granulation; the processes responsible for its development are probably the dominant agents of disintegration.

Plants and organisms aid in the disintegrating processes, but their greatest effect usually occurs after the parent rock has already been converted into soil. Plant roots commonly grow in fractures of the parent rock and physically pry the solid material apart. Nonetheless, compared with other processes, rootlet growth is of minor consequence.

Frost Action and Crystallization   The most pervasive processes of physical weathering involve forces generated by crystallization of ice (frost action) or other minerals in rock spaces. In a perfectly closed system, water increases 9 percent in volume upon freezing and almost certainly produces hydrostatic pressures that exceed the tensile strength of all common rocks. Frost action is most effective when the rock is saturated prior to the freezing event. If more than 20 percent of the available pore space is empty, the expansion pressure upon freezing may be less than rock strength, and shattering will not occur (Cooke and Doornkamp 1974). Some evidence exists to suggest that the intensity of frost action is related to the structure of the pores rather than simply the percentage of pore space. In a system containing a variety of pore sizes, ice crystals will preferentially grow in large pores rather than smaller ones (Everett 1961).

Minerals can also grow in rock spaces, with results similar to those of frost action, as figure 4.7 shows. Most commonly the process functions when percolating fluids evaporate within the pores, giving rise to supersaturated conditions and eventual precipitation of minerals. The pressures exerted in crystallization are probably greater than those produced by ice, but their absolute values depend on the concentration of the ionic constituents in the solution. The most common precipitates are sulfates, carbonates, and chlorides of very mobile cations (Ca, Na, Mg, K), and the process is therefore more prone to operate in arid and semiarid regions where the ions are rendered immobile by insufficient leaching.

The Significance of Water

Even a short review of physical weathering makes clear the importance of water in the disintegrating processes. Hydration, frost action, crystal growth, and swelling all require water as a basic component of the system. The amount of water need not be great. Many believe, for example, that even thin films of condensing dew in desert regions may be infinitely more destructive than insolation (Twidale 1968b). Therefore, it seems fair to expect a direct relationship between climate and the prevalence of disintegration.

Peltier (1950) utilized mean annual temperature and precipitation to predict relative intensities of physical and chemical weathering; figure 4.8 shows these data. Physical weathering should be dominant where precipitation is readily available and the mean annual temperatures are near or below freezing. Presumably this analysis equates with the importance of frost action as a mechanical tool and with the fact that frost action is preeminent in those areas having the most freeze-thaw cycles during the year. The frequency of freeze-thaw events has been detailed for the United States by R. Russell (1943) and L. Williams (1964).

Where unusual local problems exist, the regional climatic characteristics may have little significance. In those cases it may be extremely important to understand in detail the climate-lithologic-weathering system, and a more sophisticated approach than those reviewed above will be necessary. An excellent example of this point was provided by Weinert (1961, 1965). In the eastern part of South Africa, the parent Karoo dolerite has been altered into a mature soil that is unsatisfactory for maintaining road foundations. In the western part of the region, this mature soil is not present. Instead, hydration of the micas has apparently disintegrated the dolerite into a grus that has considerable internal strength and is quite sound from an engineering viewpoint. Weinert found that the boundary between the sound and unsound surface materials (as defined by engineering properties discussed in the next section) could be located by the distribution of evaporation and precipitation in the area. He was able to express numerically the climate conditions that control the type of weathering (and the engineering properties) within that region by the following equation:

$$N = 12 \frac{E_J}{Pa}$$

where E is potential evaporation from the Meyer formula (see Kazmann 1972); Pa is annual precipitation; and J is warmest month (in South Africa, the month of January). The areas dominated by disintegration have N-values greater than 5, while those areas where N is less than 5 are characterized by decomposition and poor soils for road construction. Although the numbers and relationships are unique for that area and cannot be applied elsewhere, Weinert's study demonstrates how detailed analyses of this type can be useful to society.

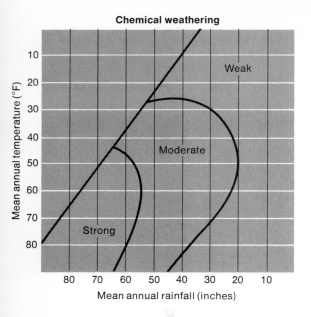

Chemical weathering

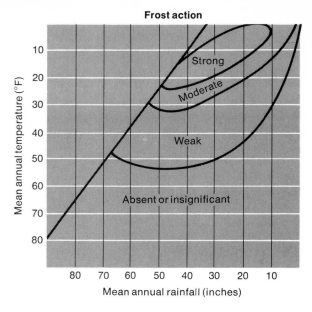

Frost action

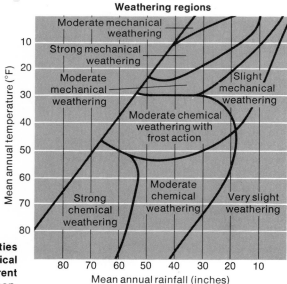

Weathering regions

Figure 4.8. Relative intensities of physical and chemical weathering under different temperature-precipitation conditions. (Reproduced by permission from the *Annals of the Association of American Geographers,* vol. 40, 1950, L. Peltier)

The resistance of unconsolidated debris to the forces of erosion is dependent on the physical properties of the material. In a real sense these properties determine whether slopes developed on any substance will be stable or, if they fail, the manner and rate of the resulting sediment movement. In addition, the physical properties help determine the shape of the slope profile when and if it attains an equilibrium condition. Clearly, then, understanding the forces of erosion is not the only important goal in slope analyses because the material itself directly influences the resulting process and landform. It is rather disconcerting to find that most geologists have only a vague knowledge of the basic physical properties of soil, which have been identified through years of study by engineers, even though these characteristics directly control the mechanics of many geologic hazards. We will briefly discuss these fundamental properties before we scrutinize the slope-forming processes.

Shear Strength

The properties of matter that resist the stresses generated by gravitational force are collectively known as the *shear strength*. Some resisting factors, such as permeability and vegetal cover, are indirect in that they affect the force of water moving across and through the material. However, the shear stress imposed by gravity cannot be substantially altered by the surface material; it is the resistance to the gravitational force that varies drastically with the characteristics of the mass and usually determines whether or not the material will fail (Carson and Kirkby 1972). The detailed analysis of internal strength is an extremely complex procedure, well beyond the scope of this book. For our purposes, we can safely say that shear strength of any material derives from three components: (1) its overall frictional characteristic, usually expressed as the angle of internal friction; (2) the effective normal stress; and (3) cohesion. These factors determine shear strength by the well-known Coulomb equation,

$$S = c + \sigma \tan \phi$$

where S is shear strength (in units of stress), c is cohesion, σ is normal stress, and ϕ is the angle of internal friction.

Internal Friction   Internal friction is composed of two separate types: *plane friction*, produced when one grain slides past another along a well-defined planar surface, and *interlocking friction*, which originates when particles are required to move upward and over one another. These are shown graphically in figure 4.9. The angle of plane friction is approximated by the angle at which a block will begin to slide over another along the plane separating them. Once sliding begins, the frictional angle actually decreases slightly, requiring that a distinction be made between a static and a dynamic angle. In addition, the plane friction angle varies considerably with smoothness of the plane surface, moisture at the contact, mineralogy, and other factors.

Physical Properties
of Unconsolidated
Debris

Physical Weathering, Mass Movement, Slopes 137

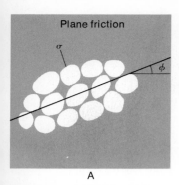

Plane friction

σ

ϕ

A

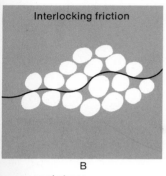

Interlocking friction

B

σ = normal stress
ϕ = angle of internal friction

Figure 4.9. Two types of internal friction. (A) Plane friction, in which resistance occurs along a well-defined plane that may cut through individual grains. Angle of plane friction is approximated by angle at which sliding begins. (B) Interlocking friction, in which resistance occurs around particle boundaries.

Interlocking friction is usually greater than plane friction because extra energy must be used to move interlocked grains in an upward direction. The angle of interlocking friction also varies with mineralogy and moisture and, it would seem, must be affected by the density of packing within the mass. In loose particulate matter of any size, the angle of repose should approximate the angle of internal friction, but Carson and Kirkby (1972) point out that a platform holding a cone of debris at its angle of repose can be tilted up to 10° before the slope fails. This suggests that the angle of repose may be somewhat less than the maximum angle at which a slope can stand.

Effective Normal Stress   The importance of normal stress is its capacity to hold material together, thereby increasing the internal resistance to shear. In theory, normal stress acting perpendicular to a shear surface (fig. 4.9) is absorbed by the underlying slab at the point of contact between grains. In reality, some of the shear surface is occupied by openings filled with air or water. Since pore pressure exists in these interstitial spaces, they tend to support part of the normal stress. The total normal stress (σ) therefore includes two elements, effective normal stress (σ') and pore pressure (μ), such that:

$$\sigma = \sigma' + \mu.$$

When considering internal friction, effective normal stress (stress exerted at the solid-to-solid contacts across the shear surface) is the critical parameter rather than total normal stress.

The value of pore pressure (μ) has a direct bearing on the effective normal stress because it can add to or detract from the total stress value. For example, figure 4.10 shows a distinct water table in unconsolidated material. At some level X below the water table, a hydrostatic pressure is exerted that is equal to the specific weight of the water (γw) times the vertical distance (h) between the water table and the level of X. As Terzaghi (1936) pointed out, this pore pressure will be directed upward and will provide relief from the overburden stress equal to the magnitude of water pressure. But in the unsaturated zone above the water table, some of the water will be prevented from moving downward because it is attached to particles by surface tension. Simply stated, this attached moisture increases the weight of the soil. Relating water content to the effective normal stress, three possible situations can be envisioned.

1. In a completely dry soil, the pore pressure is atmospheric and μ is zero. Therefore, the effective normal stress and the total normal stress are the same since

$$\sigma' = \sigma - 0.$$

2. Below the water table, pore pressure is positive ($>$ atmospheric pressure), causing the effective normal stress to be lower because

$$\sigma' = \sigma - \mu.$$

3. Above the water table, μ is negative and the effective normal stress is higher:

$$\sigma' = \sigma - (-\mu).$$

Because the effective normal stress directly influences internal friction, it is clear that dry or partially saturated soils, especially those with a high clay content, should have greater shear strength and stand at higher slopes than equivalent materials that are thoroughly saturated.

Cohesion   Figure 4.11 demonstrates the relationship between shear strength and effective normal stress. The graph indicates that as the effective normal stress increases, the values of shear strength also rise. The relationship between the two variables defines a straight line that passes through the origin of both axes. The angle between the line and the abscissa represents the angle of internal friction. The material represented in figure 4.11A has no discernible strength when the effective normal stress decreases to zero, a condition that is common in coarse, unconsolidated detritus. Solid rocks, however, possess shear strength even when σ' is removed (fig. 4.11B) because the constituent particles are packed or cemented together. The strength revealed here is called *cohesion*, a factor that presumably is unaffected by normal stress.

Clay-rich soils also have some cohesion, presumably because adsorption of ions and water by clay minerals creates a binding structure among the particles. The cohesive strength depends on the attractive force between the particles and the lubricating action of the interstitial liquid. As Grim (1962) points out, the molecules of the inner layers of water adsorbed to the surfaces of clays are oriented by the electrical charge of the minerals; because of this, fluidity and lubrication are not possible when moisture content is low, even though water is present. For clays to become plastic and exert a lubricating action, the adsorption of water layers must continue until the outermost ones can be held but are no longer fixed in a rigid, oriented position. Fortunately, simple tests can indicate how much water a soil can absorb before it begins to behave like a plastic substance; or, if more water is added, when the substance will lose all its cohesion and become a muddy fluid.

Atterberg Limits   If water is gradually added to a dry, pulverized soil, the voids fill and the mixture becomes increasingly more plastic in its behavior. As more water is added, however, the cohesion decreases, and when all the pores are filled, any further input of water results in complete destruction of the internal fabric. Atterberg (1911) suggested two simple tests to indicate the transition from the solid to the plastic state and from the plastic to the liquid state; they are called respectively the *plastic limit* and the *liquid limit*. The two limits are expressed as moisture contents determined by the weight of contained water divided by the weight of the dry soil. The range of water content between the two limits is called the *plasticity index*. The tests used to determine the limits were refined by Casa-

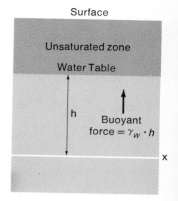

Figure 4.10. Buoyant force produced by hydrostatic pressure beneath a water table.

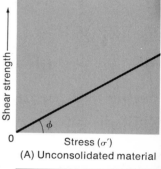

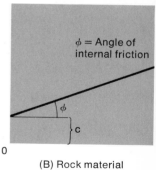

Figure 4.11. Relationship between shear strength and effective normal stress. (A) Unconsolidated material has no shear strength when normal stress is zero. (B) Rock material has shear strength (c) from cohesion even where no normal stress is present. (From Carson and Kirkby 1972. Used with the permission of Cambridge University Press)

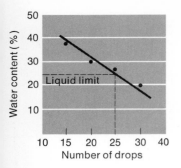

Figure 4.12. Determination of liquid limit by standard testing. Liquid limit is water content of sample that flows when dropped in a testing cup 25 times.

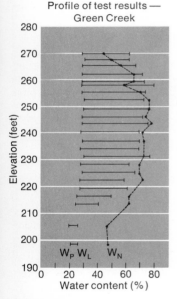

Figure 4.13. Variation with depth of soil test results in the Leda clay, St. Lawrence River valley. Natural water content commonly is greater than the liquid limit in these sensitive soils. W_P is the plastic limit; W_L is the liquid limit; W_N is measured water content. (From Legget 1967. Used with permission of the Geological Society of America)

grande (1932) and are now standard procedure in engineering studies of soils (A.S.T.M. 1971).

The plastic limit is arbitrarily defined as the lowest water content at which soil can be rolled into threads 1/8 inch (.32 cm) in diameter without the threads breaking into pieces. To find this value, a soil sample is mixed with water until it can be rolled into a ball. The sample is then rolled into a 1/8-inch (.32 cm) thread, split into 6 or 8 pieces, remolded into a ball, and rolled again. This procedure is repeated until the sample crumbles, at which time the pieces are weighed, dried at 110°C, and reweighed.

The liquid limit is arbitrarily defined as the water content at which two halves of a soil cake will flow together for a distance of 1/2 inch (1.27 cm) along the bottom of a groove separating the two halves when a testing cup is dropped 25 times from a distance of 1 cm at the rate of two drops a second. (A.S.T.M. 1971). In this test a thoroughly mixed sample is flattened in a special cup and grooved with an instrument designed for that purpose. The cup is attached to an engine that drops the cup onto a base at a controlled rate until the two halves of the cake flow into contact. The number of drops and the water content are recorded; as the procedure is repeated, the data can be plotted as a straight line on a semilogarithmic graph, as shown in figure 4.12. The intersection of the flow curve at the 25-drop ordinate represents the liquid limit.

The values of Atterberg limits are affected by a myriad of auxiliary factors. First, they are related to the types of clay minerals included in the soil, although the precise limits for any particular clay species vary appreciably (see table 4.2). In general, plastic and liquid limits are higher for montmorillonite clays than for illites or kaolinites, mainly because the montmorillonites are able to disperse into very small particles with an enormous total water-adsorbing area. The type of exchangeable cation in the montmorillonite, however, may engender wide varia-

Table 4.2 Average Atterberg limits in clays saturated with common cations.

	Montmorillonite[1]		Illite[2]		Kaolinite[3]		Halloysite[4]	
	Plastic Limit	Liquid Limit	Plastic Limit	Liquid Limit	Plastic Limit	Liquid Limit	Plastic Limit	Liquid Limit
Na[+]	91	442	36	65	27	41	42	46
K[+]	63	173	41	75	33	52	45	48
Mg[++]	59	164	39	84	29	50	54	60
Ca[++]	68	155	39	86	31	54	48	60

[1] 4 samples
[2] 3 samples
[3] 2 samples
[4] 1 sample 2H₂O, 1 sample 4H₂O

Adapted from Grim 1962. Used with permission of McGraw-Hill Book Company. Data from White 1955.

tions in the limit values, especially in the case of Na^+ and Li^+. Plasticity indices vary in a similar manner. Second, limit values tend to increase when the particle size is smaller. The *activity*, defined as the ratio of the plasticity index to the abundance of clays (Skempton 1953), shows that plasticity increases proportionately with the percentage increase of clay-sized particles. As the activity increases, soils tend to have a lower resistance to shear, and so the parameter has some engineering significance. Finally, drying of a soil reduces its plasticity because shrinkage during the dehydration process brings particles closer together. The attraction between particles is so strengthened that penetration by water is difficult. Conversely, repeated wetting combined with only partial drying may increase plasticity (Grim 1962).

Although the tests for Atterberg limits are rather unsophisticated, the results are consistent when determined by more than one analyst, and they are geologically significant. They have been related empirically to the clay mineralogy of the soil (Casagrande 1948; Seed et al. 1964). More important, the moisture content in fine-grained material is a guide to its internal strength. Some soils, called *sensitive soils*, exist with natural water contents above their liquid limits (fig. 4.13), an apparent inconsistency with the limit concept. The fact is, however, that some soils develop an open ''honeycomb'' structure (fig. 4.14) that is capable of holding water in excess of its liquid limit. Although the structure is unstable, in an undisturbed soil the material will be solid and possess some strength. The disruption of the internal structure by erosion, earthquake shocks, or other phenomena will cause the excess water to be released, and the solid material will become a fluid. A simple test for the liquid limit can reveal the presence of sensitivity and the potential for dangerous mass wasting.

Compression and Response to Stress

Compression refers to the change in character of a soil when it is confined laterally and subjected to increased normal stress. Usually the response manifests itself by a decrease in the *void ratio* (e) of the soil and a settling of the soil surface by what engineers call ''primary consolidation.'' The void ratio is defined as the ratio of the total volume of voids to the total volume of solids in a soil; it is somewhat similar to porosity. When load is applied to a soil sample, the soil consolidates rapidly at first as pore water is squeezed out. The rate decreases with time until it reaches an equilibrium between the water content (void ratio) and the pressure. The load may be increased and the procedure repeated until the relationship between e and pressure is established over a wide range of load values. Figure 4.15 shows the results of such a test.

The compression characteristics for different clay minerals are rather distinct. Montmorillonites, especially bentonite (Na-montmorillonite), show a large reduction in void ratios with only a small applied stress, but increased pressure thereafter causes little volume reduction. In contrast, kaolinites are considerably less compressible. More interesting, perhaps, is that at high, constant pressure

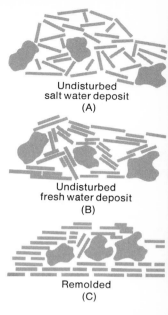

Undisturbed
salt water deposit
(A)

Undisturbed
fresh water deposit
(B)

Remolded
(C)

Figure 4.14. Open soil structures in sensitive clays. Structures in (A) and (B) allow water content to exceed liquid limit. Disturbance of this structure causes temporary fluidity until substance is remolded (C). (From Lambe 1953, p. 315-38)

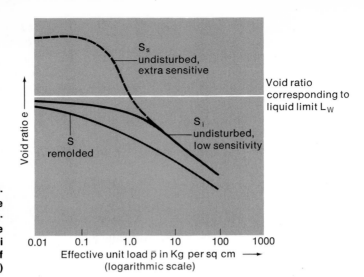

Figure 4.15. Relationship between effective load and the void ratio (e) in unconsolidated but cohesive materials. (From Terzaghi 1955. Used with permission of Economic Geology)

Figure 4.15. Relationship between effective load and the void ratio (*e*) in unconsolidated but cohesive materials. (From Terzaghi 1955. Used with permission of *Economic Geology*)

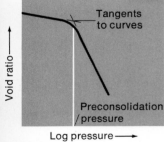

Consolidation test result

Figure 4.16. Determination of preconsolidation load from the results of a consolidation test relating pressure and void ratio. (From Legget 1967. Used with permission of the Geological Society of America)

for seven days, montmorillonites consolidated slowly during the initial stages, but the rate increased with time. Kaolinites, however, behaved in the opposite way, with 50–60 percent of the total consolidation occurring in the first minute (Samuels 1950). Grim (1962) explains these differences by the way each mineral holds water. Montmorillonites hold large amounts of water both in the oriented form attached to the mineral and in interstitial pores. Even a minor pressure increase releases the pore water, causing the void ratio to drop quickly. Since relatively large amounts of water are held close to the small particles, consolidation starts slowly and increases only after the oriented water breaks up. Kaolinite, in contrast, holds only a small amount of water at its liquid limit, most of it in pores.

A complicating factor in the analysis of strength is the fact that many (if not most) natural unconsolidated materials have been compressed at some time before they are tested in the laboratory. It is known that placing a stress on sediment reduces its later compressibility under smaller loads. *Preconsolidated sediments,* therefore, are those that have been previously squeezed by a load greater than the one currently being applied. The preconsolidation load can be determined from the curve of the consolidation test (fig. 4.16) by projecting tangents of the two segments of the curve until they intersect. The stress value at the point of intersection can be considered as the earlier load. The value has significant application both in engineering and in geological studies. For instance, it can be used to approximate the thickness of ice that earlier rested on soils (W. Harrison 1958), to calculate overburden removed by erosion, or to provide information on sea-level fluctuations (Kenney 1964).

It should be pointed out that instruments such as the penetrometer are available to measure the resistance to penetration at an actual field site; however, these measure bearing capacity, not shear strength. In addition, they do not indicate pore pressure and cannot separate strength into c and ϕ components even though the measurements can be translated into strength terms by empirical equations. In the laboratory, detailed analyses of shear strength are done by the direct-shear test, the triaxial compression test, or the uniaxial (unconfined) compression test. (The details of these tests are provided in most textbooks of soils mechanics.)

Measurement of shear strength is a complicated task, made additionally difficult because some techniques do not yield certain critical data. In the direct shear test, for example, pore pressure cannot be determined. It should also be kept in mind that laboratory tests are designed to measure stress at the moment of failure, which is analogous to shear strength. Natural materials, however, especially soils, begin to deform well below the stress level that causes rupture (fig. 4.17) and may continue to deform at a rate that increases steadily with additional stress. Ultimately the material ruptures at a critical stress called the breaking strength.

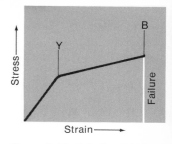

Figure 4.17. Relationship between stress and strain. Permanent deformation begins at Y (yield stress) but rupture of the substance does not occur until the stress value reaches B (breaking strength). Material behaves like a plastic substance between Y and B.

Mass Movements of Slope Material

The melange of slope debris movements can be differentiated into major groups on the basis of whether the material is transported as a coherent mass or whether the "soil" components move as individual particles. The mode of transport and, ultimately, the slope profile depend on the relationship between the physical properties of the material and the stresses produced within the system.

Mass movements are of three primary types: slides, flows, and heaves, each having distinctive characteristics (Carson and Kirkby 1972). In slides, cohesive blocks of material move on a well-defined surface of sliding, and no internal shearing takes place concurrently within the sliding block. In contrast, *flows* move entirely by differential shearing within the transported mass, and no clear plane can be defined at the base of the moving debris. The velocity in flows tends to decrease from the surface downward. In *heave,* the disrupting forces act perpendicular to the ground surface by expansion of the material. This movement does not in itself provide a lateral component of transport, but it facilitates slow, downslope movement by gravity and serves as an important forerunner of more rapid mass movements.

Although each of these movements could theoretically function alone, it seems certain that all are involved to some extent in most natural slope failures. Many processes can in fact only be explained by some combination of the primary types of movement. As figure 4.18 shows, mass movements can be thought of as multifarious events. The location of a particular process near a corner of the triangle in the figure indicates the dominance of the primary movement occupying that corner; the closer a term is to the corner, the more dominant that type of movement. Superimposed on the diagram are lines that represent the relative transport velocities of each process and the usual water content within the debris being moved.

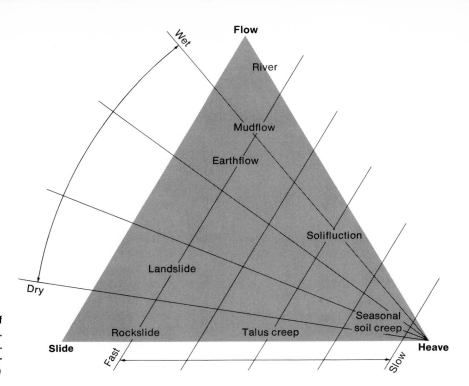

Figure 4.18. Classification of
mass movement processes.
(From Carson and Kirkby 1972.
Used with permission of Cambridge University Press)

The various mass movements are alike in that all begin when the shear stress
tending to displace material exceeds the resisting strength. As long as slope material maintains its internal resistance at a level greater than the driving impetus,
no slope failure will occur. Stability, therefore, represents some balance between
driving forces (shear stress) and resisting forces (shear strength), and can be
expressed as a safety ratio:

$$G_s = \frac{\text{resisting force}}{\text{driving force}} \, .$$

G_s values greater than 1 connote slope stability, but as the ratio approaches unity
a critical condition evolves and failure is imminent. Clearly, any factor that lowers the safety ratio (see table 4.3) can trigger mass movement, and this tendency
can be produced by increasing the driving force, lowering the resistance, or both.
Theoretically, failure occurs when $G_s = 1$; this value is an excellent example of a
geomorphic threshold. In reality, however, because of imprecise measuring
techniques, some failures have occurred when the G_s value was slightly positive,
and some slopes have been maintained temporarily with small negative values.
Once failure does occur, the type of movement depends on precisely how the
forces interact with one another.

Table 4.3 Factors that influence stress and resistance in slope materials.

Factors that increase shear stress
- Removal of lateral support
    - Erosion (rivers, ice, wave)
    - Human activity (quarries, road cuts, etc.)
- Addition of mass
    - Natural (rain, talus, etc.)
    - Human (fills, ore stockpiles, buildings, etc.)
- Earthquakes
- Regional tilting
- Removal of underlying support
    - Natural (undercutting, solution, weathering, etc.)
    - Human activity (mining)
- Lateral pressure
    - Natural (swelling, expansion by freezing, water addition)

Factors that decrease shear strength
- Weathering and other physicochemical reactions
    - Disintegration (lowers cohesion)
    - Hydration (lowers cohesion)
    - Base exchange
    - Solution
    - Drying
- Pore water
    - Buoyancy
    - Capillary tension
- Structural changes
    - Remolding
    - Fracturing

After Varnes 1958, with permission of the Transportation Research Board

Heave

Heave is instrumental in the process of creep, the almost imperceptibly slow movement of material in response to gravity. *Seasonal creep* or *soil creep* is the downslope movement of regolith that is aided periodically by the heave mechanism. No continuous external stress is placed on the mass; it moves under gravity when its cohesion and frictional resistance are spasmodically lowered. The process functions in the upper several feet of the soil, and its effect decreases rapidly with depth. The phenomenon of soil creep was first recognized in the latter part of the nineteenth century, and its ubiquity was gradually accepted as its effects were observed in the field. Historically, evidence that suggests a soil creep influence on slope materials has included downslope curvature of bedding (fig. 4.19), stone lines, downslope growth of trees or tilting of structures, and accumulations of soil upslope from a fixed obstruction (see Young 1972). Recently, however, observations and measurements of seasonal creep have become more sophisticated. Precise surveying methods and trenches such as Young pits (Young 1960) are fairly standard techniques in current studies of the process. A complete review of the common methods employed to measure soil creep can be found in Selby (1966) and *Revue de Géomorphologie Dynamique* (1967).

Figure 4.19. Creep in vertical Romney shale. Western Maryland Railroad cut one mile west of Great Cacapon. Washington County, Md. (Photo by G.W. Stose. From U.S. Geol. Survey Folio 179, plate 15, 1912)

Although burrowing animals and vegetation may cause random disturbances in soils, their effects are minor compared with the heave produced by swelling or freezing and thawing. In the heave mechanism, expansion disturbs soil particles perpendicular to the ground surfaces; when the soil contracts, the vertical attraction of gravity acts on the particles. The expansion-contraction cycle, therefore, adds a lateral component to particle motion in any soil having an inclined surface. Because gravity is reasonably uniform over the Earth's surface, the distance of transport in each heave event, and presumably the rate of creep in any climate, should vary with the slope angle and should decrease with depth beneath the surface. Schumm (1967a) has demonstrated a significant correlation between the rate of surficial rock creep and the sine of the slope angle, but documentation of this relationship for fine soils or below the surface is lacking. Actually, as figure 4.20 shows, the contraction event is never perfectly downward but usually moves in a direction about midway between the normal and the vertical. As the lateral distance traveled in each heave is less than would be theoretically predicted, a clear relationship between slope angle and creep rate may be difficult to demonstrate. Detailed measurements (Kirkby 1967), however, have shown the creep rate to decrease with depth (fig. 4.21). Presumably this relates to the lower frequency of heaving at depth and to the greater difficulty of expansion with increasing overburden. In any case, below a depth of 20 cm movement ceases or becomes drastically smaller (Young 1960; Kirkby 1967).

Velocity measurements in soils under humid temperate climates indicate that the upper 5 cm of soil moves downslope at an average rate of 1 mm a year, but variations occur with differences in slope angle, moisture content, position on the slope surface, and measuring technique. The particle size also introduces variability in the rate because fine-grained clays tend to swell more upon wetting, and frost heaving is greatest in silty material. Nevertheless, the few available measurements seem to show that soil creep in humid temperature areas proceeds at a volumetric rate between 0.5 and 6.0 cm^3/cm/yr. The volumetric rate is calculated as the volume of soil moved annually across a plane set perpendicular to the surface and parallel to the contour of the slope, for a unit horizontal distance along the plane. In semiarid regions the rate seems to be somewhat higher. In arctic climates a special kind of creep process called *solifluction* is extremely important in the geomorphic scheme; it is discussed in chapter 11.

A second type of creep called *continuous creep* (Terzaghi 1950) is fundamentally different from seasonal creep in that (1) it is driven by gravity alone, (2) it may affect consolidated rock, and (3) it can function at levels well below the surface. Continuous creep is the strain response to stress that is generated by the weight of overburden. It begins at the yield stress and continues even though no

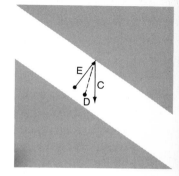

Figure 4.20. Movement of near-surface material by heaving. During expansion (E), particle is displaced perpendicular to the surface. During contraction (C), particle settles in a vertically downward direction under influence of gravity. Actual movement is shown by line D.

Figure 4.21. Rate of creep as it relates to type and depth of material. Rate in all soils decreases with depth. (From Kirkby 1967, *Journal of Geology*, © 1967 University of Chicago Press. Used with permission of the University of Chicago Press)

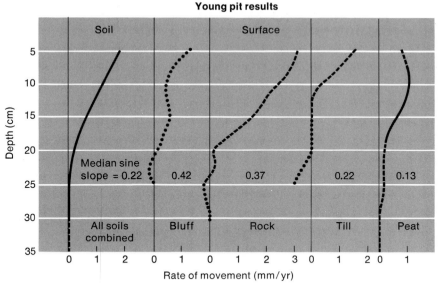

• • • • • Based on 6-10 measurements
- - - - - Based on 11-20 measurements
———— Based on more than 20 measurements

additional stress is placed on the material. Continuous creep is especially pronounced where rocks or semiconsolidated materials with low yield stress (fig. 4.17) are overlain by stronger substances. For example, a weak clay unit sandwiched between resistant strata is prone to deform by continuous creep. Excavations of any kind through that rock sequence will reduce the lateral confining pressure on the clay unit, and it will begin to flow. Creep of this type is not important in terms of the volume of material it moves or the distance of transport, but it is very significant as a precursor of rapid, sometimes catastrophic, mass movements. In many cases landslides are immediately preceded by accelerated creep; persons who intervene in the natural setting can trigger these events by not recognizing the potential for continuous creep (Kiersch 1964).

The heave mechanism is also an essential element in some rapid mass movements, especially falls. Falls in both rock and soils involve a single mass that travels as a freely falling body with little or no interaction with other solids. Movement usually is through the air, although occasional bouncing or rolling may be considered as part of the motion. Rock falls are most common where the parent material is well jointed and a steep slope is developed on the rock face. The fractures are enlarged progressively by heaving, mainly in the form of freezing and thawing, until the gravitational force exceeds the internal resistance. Undercutting of the rock or soil face by erosive agents acting at the base of the material accelerates the process. The removal of the subjacent support tends to increase tension in the overhang and so helps to create and expand incipient cracks.

Slides

As in all classification, rapid mass movements are grouped according to the classifier's opinion as to what aspects of the phenomena are most important. Because most rapid movements are not observed as they occur, their fundamental properties of motion must be interpreted after the event. This, combined with the fact that no clear-cut distinction can be made between the primary modes of transport, has resulted in a number of viable classifications of rapid mass movements (Sharpe 1938; Ward 1945; Varnes 1958; Hutchinson 1968). The classification prepared by Varnes (1958) is adopted here as figure 4.22 because it relates well with our process orientation. The classification is based primarily on the type of material being moved and the primary type of movement.

As defined earlier, slides are slope failures that are initiated by slippage along a well-defined planar surface. The sliding mass is essentially undeformed; however, it may partially disintegrate during the sliding motion, giving rise to flow movement in the latter phase of the event. The plane of sliding may be shallow and approximately parallel to the ground surface as in the case of *rockslides* and *debris slides,* or it may penetrate to some depth as a concave surface along which *rotational slip* (fig. 4.23) or *slumping* may occur.

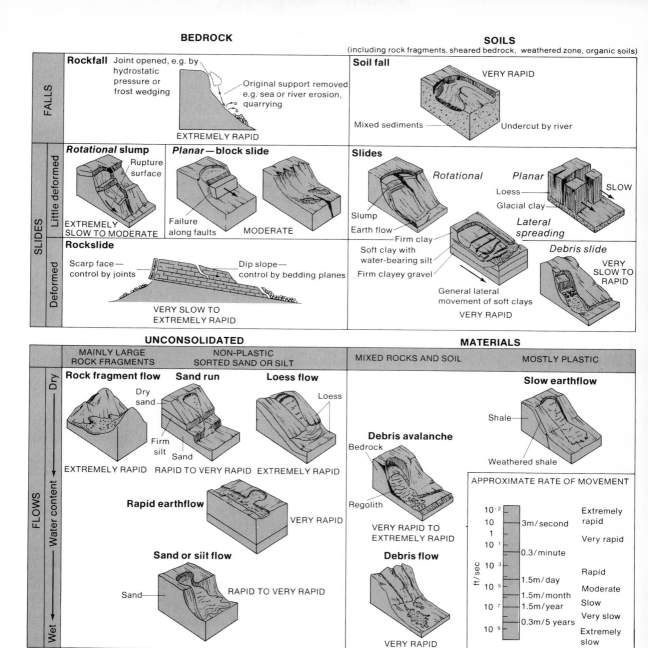

Figure 4.22. Classification of landslides. (After Varnes 1958. Used with permission of the Transportation Research Board)

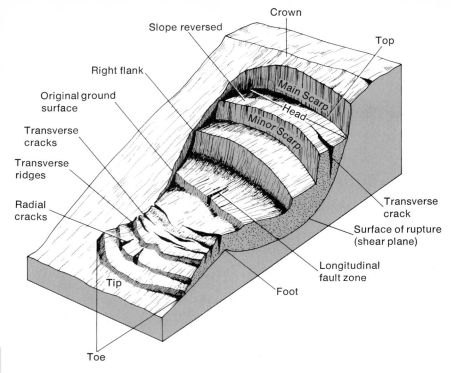

Figure 4.23. Features of a rotational slide. (After Varnes 1958. Used with permission of the Transportation Research Board)

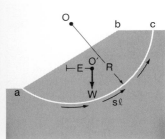

Figure 4.24. Driving and resisting forces in sliding phenomenon. Potential surface of sliding (ac) is arc of circle with radius OR. Driving force is gravity (W) acting downward from center of gravity (O′) operating around lever arm with length E. Resistance is shear strength (s) operating over length (ℓ = ac) around lever arm R. (From Terzaghi 1950. Used with permission of the Geological Society of America)

The theory of sliding, shown graphically in figure 4.24, a basic topic in textbooks of soils mechanics, was presented to geologists years ago by Terzaghi (1950). Consider a slope (ab) underlain by homogeneous unlithified regolith. At some depth a surface can be defined where the ratio between resistance and shear stress will be minimum. In homogeneous material this surface (ac in fig. 4.24), called the *potential surface of sliding,* is assumed to be an arc of a circle with its center at O and a radius of length R. The weight (W) is suspended from the center of gravity (O′) of the mass above ac and acts at the lever arm with length E. This tends to produce a rotation of the block abc around the axis O and represents the driving force in the stability regime. The driving force is resisted by the shear strength (s) imposed along the length (l) of the arc ac and acting at a lever arm R. Thus, the safety ratio can be considered as

$$G_s = \frac{\text{resisting force}}{\text{driving force}} = \frac{slR}{EW}.$$

As long as G_s is greater than 1 at the potential surface of sliding, the slope will be stable. If an increase in the driving force or a decrease in resistance brings the ratio to unity, sliding will ensue with an outward and upward rotating motion. Commonly the driving force is increased by an addition of mass to the sliding block, but other factors can produce the same effect. Earthquakes, for example, generate a horizontal mass force n_gW (fig. 4.25) that passes through the center of gravity (O') and acts at a lever arm with length F. The safety ratio is accordingly reduced during an earthquake to

$$G_s = \frac{slR}{EW + n_gFW}.$$

Steepening of the slope by erosion or human activity also increases the driving force because (as shown in figure 4.26) the length of the lever arm E is increased. Actually this is a rather simplistic view since steepening also reduces the shear strength by complicated changes in cohesion, pore pressure, and effective normal stress (Terzaghi 1950; Carson and Kirkby 1972).

The sliding phenomenon can also be produced by a variety of events that reduce the internal resistance of the debris. From observation, sliding usually occurs after prolonged or exceptionally heavy rainfall, indicating that the lowering of resistance is predominantly a function of water. In the past, the water effect was interpreted to be lubrication along the sliding surface. Terzaghi (1950), however, refutes this notion by pointing out that water applied to many common minerals such as quartz is actually an antilubricant. Furthermore, most soils in humid regions contain more than enough water to cause lubrication at all times, yet they also fail after rainstorms. Clearly water affects strength in other ways. You will recall that shear strength is a function of cohesion (c), effective normal stress (σ') and friction (ϕ) such that

$$S = c + (\sigma') \tan \phi.$$

The response of these factors (c, σ', ϕ) to wetting is significantly more important in the initiation of slippage than is lubrication. For example, the rise of the water table or the piezometric surface, which accompanies all prolonged rainfalls, may be the most common culprit in sliding. As the water table rises, the pore pressure (μ) at any point within the saturated mass increases, ultimately resulting in a decrease of effective normal stress (σ') and the concomitant reduction of shear strength.

Rockslides are usually associated with major structural features within the rock such as the stratigraphy of the rock sequence or joint patterns. Massive rock units normally have several prominent joint sets, which are related to regional tectonics, as well as a superimposed network of randomly spaced and oriented fractures. Prior to the creation of the joints these rocks have great shear strength associated with the high cohesion in lithified materials. Once joints begin to form, however, the process becomes self-generating because shear stresses are concentrated preferentially on the unfractured zones of solid rock. With time, more and more of the original rock is consumed by jointing, and eventually the near surface becomes a cohesionless mass of densely packed angular blocks (Terzaghi 1962). At this stage, except for particle size, the rock mass

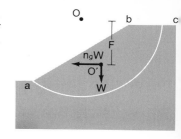

Figure 4.25. Landslide produced by earthquake. Additional driving force (n_gw) is applied horizontally through the center of gravity (O′) and operates around lever arm with length F.

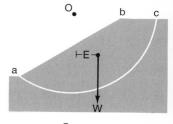

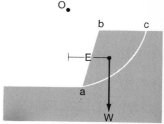

Figure 4.26. Landslide triggered by steepening of slope ab. This increases the length of lever arm E and adds to driving force.

resembles an aggregate of dry sand in which the cohesion within the individual particles has little bearing on slope stability. The shear strength of the total mass stems entirely from internal friction because the cohesion across the joint openings is zero. Under these conditions, the steepest slope that can be maintained depends on the pattern of jointing and the orientation of the pattern relative to the slope. The slope angles may vary from 30° to vertical but according to Terzaghi (1962) should average about 70°. The angle developed presumably represents a critical value and should be maintained if the rocks are unweathered and the slopes retreat by sliding.

Rockslides can be divided into two types: *rock avalanches* and *slab failure* (Carson and Kirkby 1972). Both obey the same mechanics, and they differ only in the amount of fracturing within the rock and the angle on the potential surface of sliding (fig. 4.27). In slab failure, cracks develop where a rock mass expands because the horizontal confining pressure is removed, allowing strain to proceed outward in the direction of the pressure release. In that sense, the process in rocks is similar to pressure-release sheeting. As lateral stress is removed, a tensional zone develops in the upper part of the mass, and cracks form within the zone. The tensional fractures penetrate to a depth that is controlled by the strength of the material. Because rocks usually have high tensile strength, fracturing does not penetrate the total depth of the tensional zone. This is not always the case in unconsolidated substances.

The stability of the outer slab depends on the depth of the fracture relative to the height of the unconfined surface that is undergoing expansion. Equations have been derived to predict the maximum height attainable before slab failure occurs (Terzaghi 1943), but predictive models are not always completely successful. For example, in one study (Lohnes and Handy 1968), tension cracks probably did not appear until slab failure was imminent. As a result, the unsupported face was considerably higher than would have been possible had the cracks developed at an earlier stage. In addition, Terzaghi (1962) considered the worst possible case—the weakest rocks—and found that most rock types could probably stand vertically up to heights of 1300 m. The observation that few vertical cliffs stand at this height even in stronger rocks indicates that fracturing drastically reduces the height that can be maintained on an unsupported rock face.

Rock avalanches occur when the joint network becomes essentially continuous down to the potential surface of sliding. An avalanche differs from slab failure in that it involves the entire mass above the sliding surface whereas slab failure includes only the material outside the outermost continuous joint; figure 4.27 shows both events. Both types of rockslide differ from rock falls in that the rocks fail only when fractures intersect the potential surface of sliding. Falls, on the other hand, can occur above the potential slide plane. Both heaving and sliding processes may be involved in some natural mass movements (Schumm and Chorley 1964). In fact, the incidence of sliding is so complicated that Terzaghi (1950) was able to list 19 possible causes.

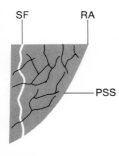

Figure 4.27. Differences in rockslides caused by slab failure and by rock avalanche. Slab failure (SF) involves only the outer sheet of rock, which splits along one prominent fracture. A rock avalanche (RA) occurs when fracture system penetrates to the potential surface of sliding (PSS).

Flows

In true flows, the movement within the displaced mass resembles closely that of a viscous fluid, in which the velocity is greatest at the surface and decreases downward in the flowing mass. In many cases, flows are the final event in a movement begun as a slide, and the distinction between the two is usually unclear. For most types of flow, abundant water is a necessary component, but dry flows called *rock fragment* flows by Varnes (1958) do occur when rockslides or falls increase drastically in velocity and lose their identity as a unitized mass. When rocks slide or fall down a steep slope, the material disintegrates as it crashes into the relatively flat surface at the base of the slope. From there the rocks travel as masses of broken debris (called *sturzstroms* in Germany), moving with enormous velocities over the gentler slopes in the piedmont or valley bottom. For example, the wet, mud-soaked sturzstrom at Mount Huascarán, Peru, sped at approximately 400 km a hour over a distance of 14.5 km, wreaking destruction along its path and killing some 80,000 persons (Ericksen et al. 1970; Browning 1973).

The exact mechanism required to move these large bouldery masses (usually greater than 1 million m^3) for such long distances is the subject of some controversy. Shreve (1966, 1968) proposes that the material slides continuously on a layer of compressed air that was trapped beneath the debris as it came to the bottom of the steep slope. The air-lubrication hypothesis has been challenged by Hsu (1975) who suggests that sturzstrom movement is primarily a flow phenomenon, reviving a conclusion made earlier by Heim (1932). In most cases the highest strata involved in the fall are contained in the rear portion of the deposited debris. Shreve correctly indicates that this distribution negates viscous flow as the transporting mechanism, because in that process the uppermost layers are transported at higher velocities and so would be farther downstream in the blocky debris. The deposits, however, also have geometric features similar to those formed by lava flows and glaciers. Hsu therefore feels that the debris moves by flow, but that the mechanism differs from viscous transport in that individual particles are dispersed in a dust-laden cloud, and the kinetic energy driving the flow is transferred from grain to grain as they collide and push one another forward. The entire mass moves simultaneously until all the original energy is dissipated by friction from the particle collisions.

Flow by this process would explain the distribution of the source rocks within the deposits. It would also permit great distances of transport because the frictional resistance decreases when grains are immersed in a buoyant interstitial fluid that reduces the effective normal stress. Variations in flow distance and velocity in different events probably depend on the properties of the interstitial substance.

In soils, the transition from debris slides to debris flows also requires an increase in water content or another buoyant substance. In the initial phase, the mass breaks into progressively smaller parts as it moves downslope (fig. 4.22) even though the velocity of the advance may be slow. If the mass is wet, a *debris avalanche* may be generated as a long, narrow flow that extends well beyond the foot of the slope. The erosive nature of the movement commonly begins gulley formation on relatively undissected slopes, examples and descriptions of which can be found in Hack and Goodlett (1960) and Rapp (1960). *Debris flows* usually result from heavy rains or the sudden melting of frozen soils. Torrential rain is especially effective in producing debris flows where vegetation has been stripped from a deep soil on moderate or steep slopes. For example, the annual southern California pattern of summer brush fires followed by torrential winter rains produces excessive runoff that converts the unbound soil into a high-density flow (Sharp and Noble 1953). These flows, following preexisting drainage, have tremendous erosive and transporting power, and may cause damage many kilometers downvalley while simultaneously eroding the sediment-yielding slopes in the headward reaches.

Earthflows involve movement of fine-grained slope material and range from slow to rapid. In slow earthflows, the original failure of the slope is usually in the form of slump, often when the mass becomes saturated with groundwater. As explained earlier, a rising water table and pore pressure tend to lower shear resistance, and slippage results. If the slumped mass is relatively wet, it may slowly bulge forward at its front by viscous flow and take the form of tongues, superimposed piles of rolled mud, or bulbous toes (fig. 4.28). This movement may continue at a slow pace for many years until some stability is finally reached. Rapid earthflows normally involve highly sensitive or "quick-clay" soils. Although very heavy rain can start the movement, the properties of sensitive soils allow any type of slope failure to produce a subsequent rapid earthflow. Most earthflows, in fact, can be traced back to bowl-shaped scars indicative of sliding. The primary slip, even if the motion is small, may remold the soil and cause it to lose much of its original undisturbed strength as it releases the water in excess of the liquid limit. The material changes instantaneously from a plastic solid to a viscous liquid and flows downslope away from the slipped zone. Some authors (Varnes 1958; Young 1972) consider this type of movement to be a true mudflow.

To summarize, the distinction between slides and flows is often rather nebulous. Several generalizations can be proposed, however, to help put mass movements in some reasonable perspective:

1. Most flows rise naturally as the final stage of a movement that begins as a slide. For slope stability analyses, therefore, it is probably more important to understand the mechanics of sliding and the factors that might produce it.
2. The mobility or rate of mass movements depends to a large degree on the amount of water or other buoyant substances in the displaced material. This is particularly true in flow movements.

(A)

(B)

Figure 4.28. Types of flows.
(A) Lobate earthflow (center of photo). Near Roberts, Mont. (Photo by R.R. Dutcher)
(B) Hebgen Lake earthquake. Upstream view of Madison slide, showing the west edge of the slide debris, blocked highway, and dry river bed below the slide. Madison County, Mont., August 1959. (Photo by J.B. Hadley. U.S. Forest Service photo, fig. 54 in U.S. Geol. Survey Prof. Paper 435-K 1964) (C) A slump-earthflow caused by reactivation of an old stabilized slump block. The slump-earthflow follows the outline of an old slump block apparently subsidiary to the stabilized landslide on the left side of the active area. Charles Mix County, S.D. (From C.F. Erskine, U.S. Geol. Survey Prof. Paper 675, fig. 23, 1973)

(C)

155

Morphology of Mass Movements

It is appropriate to ask how we can reconstruct the mode of mass transfer, especially since subtle transitions from one mechanism to another are common, and most interpretations of movement characteristics are made after the event is over. Unless some concrete relationship exists between the surface configuration of the displaced material and the genetic process, we are facing an insoluble dilemma. Fortunately some evidence has been presented to suggest that the desired morphologic relationship is real. In a study of 66 landslips in New Zealand, Crozier (1973) arranged the common types of mass movements into five primary process groups: *fluid flow* (mudflows, debris flows, debris avalanches), *viscous flow* (earthflow, bouldery earthflow), *slide-flow* (slump/flow), *planar slides* (turf glide, debris slides, rockslides), and *rotational slides* (earth and rock slumps). Each of the 66 landslips studied was described quantitatively by seven morphometric indices, listed in table 4.4 and illustrated in figure 4.29. The relationship between each process group and the index values was tested statistically to ascertain whether the correlation was significant enough to warrant use of morphometry as a genetic determinant.

Crozier found that the classification index (D/L) was the best indicator of the process group, reaffirming Skempton's (1953) assertion about the importance of this parameter. As one would expect, the D/L value decreases markedly with greater flow (table 4.5) because the displaced material will flow farther downvalley than it would if moving as a sliding block. Some uncertainty will remain, however, unless the classification index is used in conjunction with other indices.

Table 4.4 Morphometric indices used to determine process of mass movement.*

Index	Description
Classification	D/L—maximum depth of displaced mass prior to its displacement over maximum length
Dilation	W_x/W_c—width of convex part of displaced mass to concave; indicates lateral spreading
Flowage	$(W_x/W_c - 1) \times L_M/L_c \times 100$—$L_m$ is length of displaced mass; L_c is length of concave segment.
Displacement	L_r/L_c—L_r is length of the surface of rupture exposed in concave segment. Low value indicates instability.
Viscous Flow	L_f/D_c—L_f is length of bare surface on displaced material, D_c is the depth of the concave segment.
Tenuity	L_m/L_c—Indicates how dispersed or cohesive the material is during displacement.
Fluidity	Amount of flowage expected from particular type of material on distinct slope. Varies with water content.

After Crozier 1973 with permission of *Zeitschrift für Geomorphologie*.
*Compare figure 4.29.

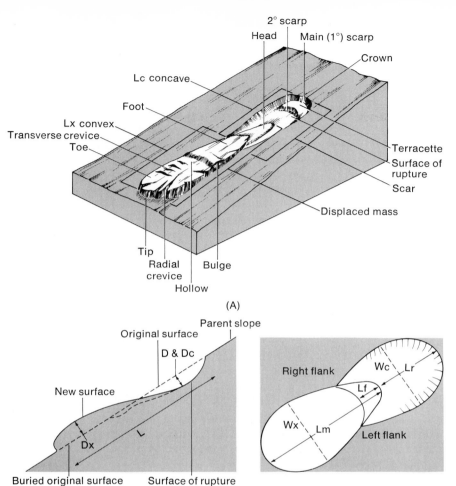

Figure 4.29. Different morphometric indices shown in diagram can be used to describe and distinguish processes of movement: (A) Landslip terminology; (B) longitudinal section; (C) plan view. (From Crozier 1973. Used with permission of *Zeitschrift für Geomorphologie*)

Table 4.5 Average values of the depth/length ratio in different types of landslips as calculated for different areas.

Type of landslip	Average D/L ratio
Flows	1.58
Planar Slides	6.33
Rotational Slides	20.84

Adapted from Crozier 1973. Used with permission of *Zeitschrift für Geomorphologie*.

Importantly, a definite inverse relationship was found between D/L and four other morphometric indices (flowage, tenuity, dilation, fluidity), each of which is presumably controlled by the water content of the material during its movement.

The relative mobility of sturzstroms may also be estimated from the geometry of the deposits. Heim (1932) observed that the distance traveled by a sturzstrom is a function of the height of the rockfall, the size of the mass, the characteristics of the mass, and the characteristics of the route followed. Later Shreve (1968) considered the H/L ratio to be an *equivalent coefficient of friction*, since in sliding H and L are related by

$$H = \tan \phi \, L$$

where $\tan \phi$ is the coefficient of friction and H and L are respectively the height of fall and the horizontal distance of the movement. The H/L ratio should, therefore, indicate resistance and be inversely proportional to the mobility of the movement. Hsu (1975), however, concluded that the equivalent coefficient of friction is somewhat misleading, and he introduced another parameter called the *excessive travel distance* (Le) to reveal mobility characteristics. This factor is defined as the ''horizontal displacement of the tip of a sturzstrom beyond the distance one expects from a frictional slide down an incline with a normal coefficient of friction of $\tan 32°$ (0.62).'' It is expressed as

$$Le = L - H/\tan 32°.$$

In general Le seems to correlate positively to the volume of the fallen mass.

Although more work is needed before a clear relationship between morphology and process can be defined, the morphometric approach exemplified by these studies holds real promise. Predictions of slope stability, possible modes of failure, and areas that might be affected are potential benefits if we can understand how previous movements occurred in any given region.

Water Erosion on Slopes

A second major kind of slope debris movement involves the transport of individual particles rather than large blocks of debris. Sediment is moved on the surface of slopes by raindrop impact (splash) and by overland flow (wash). In the wash process, the flow presumably is shallow and spread evenly across the slope as a uniform sheet; however, all hillsides have enough irregularity to ensure that flow crossing the surface will take many forms, ranging from true sheet flow to flow concentrated in permanent channels. At what precise stage the slope process of overland flow becomes a river process is a rather arbitrary decision. Most workers would probably place the transition at the point when flow is concentrated in a distinct channel that occupies the same position for a long time, and when the major hydraulic action is confined within the channel instead of being exerted over the slope surface. Small ephemeral channels that frequently shift their position would be considered part of the slope wash regime.

Rainsplash Erosion

A raindrop possesses a considerable amount of kinetic energy, derived from its mass and the velocity it attains during its fall. Under the influence of gravity, a raindrop accelerates until its force is equal to the frictional resistance of the air, the speed at that point being called the *terminal velocity*. As the distance needed to attain this condition is very short, most rain strikes the surface at its terminal velocity, although the absolute speed varies with wind, turbulence, drop size, etc. In high-intensity rains, drops usually reach a maximum size of approximately 6 mm and a terminal velocity of about 9 m/sec. The impact of such rain can directly displace into the air particles as large as 10 mm in diameter and, by undermining downslope support, can indirectly set even larger pebbles in motion. The amount of soil moved by splash depends on (1) the kinetic energy of the raindrops, (2) the type of soil exposed, and (3) the steepness of the slope. Free (1960), for example, found that splash loss varied as $E^{0.9}$ for a silt loam soil and $E^{1.46}$ for a sand, where E is the kinetic energy. Over a five-year period the total splash loss from the sandy surface was calculated at 1600 tons/acre, an amount three times greater than the loss from the loam, probably because the fine-grained soil had greater cohesion. Clearly, splash is not a negligible phenomenon; in fact, Ellison (1947) suggests that on very steep, highly permeable soils splash may cause as much downslope transportation as surface flow.

In addition to direct transportation, splash has several other erosion-inducing effects on the soil. By detaching particles, it destroys the structure of the soil and breaks apart resistant aggregates of clays. These physical processes make the soil much more susceptible to erosion by surface flow. Furthermore, as splash disperses the clays, they tend to form a fine-grained crust as they settle back on the surface. This crust forms a semipermeable barrier that reduces infiltration and promotes runoff, thereby increasing soil loss by overland flow.

Wash

Most natural slopes are too irregular to permit a uniform flow of water over the entire surface; flow is deeper over depressions and shallower over flat reaches or high spots. The variable depth of flow produces differences in the eroding and transporting capabilities of the water so that, in detail, *wash* does not imply that a regular sheet of debris is being carried continuously down the slope surface. In areas where sheet flow might be possible, only fine-grained particles can actually be moved efficiently, and those only if the surface has been prepared for erosion by rainsplash or weathering processes that reduce cohesion. In areas of concentrated flow, larger sediment can be moved, but the ability to erode depends more on the hydraulic force of the water and less on the condition of the surface.

When rainfall and flow become intense, small shallow channels may be formed which periodically shift their position so that in the long run, erosion is more or less even across the slope. In fine-grained soils, a set of well-defined subparallel channels, called *rills,* is usually formed. Rills vary in size with the erodibility of the soil, but normally they are only several centimeters wide and deep. Heaving and other processes can obliterate these tiny channels in periods between high rains, especially in highly seasonal climates where rain may be lacking for months at a time. The periodic destruction of rills allows new channels to form in an entirely different location and ensures less than equal lowering of the entire slope surface. Some rills escape this spasmodic destruction by entrenching to greater depths, a difficult task that only a few of the largest rills accomplish. These "master" rills become relatively permanent and eventually evolve into true rivers.

In soils that are sandy or coarser, the channels are usually braided because the material, although easily eroded, is transported with difficulty. Sediment commonly accumulates as temporary bars within the channel, and these subdivide the channel and the flow into a multitude of small passageways. Most braided channels are wider (up to 5 m) and deeper (1–10 cm) than rills, but they too change their position regularly because the bar deposits require a continuously shifting channel environment.

Total Soil Loss

Knowing total amount of soil that can be eroded by splash and wash is important in many scientific disciplines. Various attempts have been made to relate soil loss to some combination of climatic variables; the most successful of these is probably the rainfall-erosion index (EI_{30}) employed by Smith and Wischmeier (1962) to predict the potential for soil removal. This index, representing the product of kinetic energy (E) and the greatest average intensity of rain in any 30-minute period during a storm (I_{30}), serves as a useful guide for recognizing areas likely to have accelerated erosion. Realistically, however, the climatic factor can provide only half the information needed to understand soil loss, since resistance to erosion is equally significant in determining how much soil can be removed from a given slope. The resisting factors, mainly vegetation, topography, and the physical strength of the exposed soil, can combine in a multitude of ways to produce considerable variation in resistance within a climatic regime.

Vegetation not only binds soil particles together but also protects the surface from the high-velocity impact of raindrops. In addition, it lowers the discharge and velocity of overland flow because plants intercept and use the greater portion of any rainfall, decreasing the amount left over for runoff. The effect of vegetation as a deterrent to erosion varies with both density and type of natural cover (Langbein and Schumm 1958), and invariably the destruction of a pristine vegetal mat by cultivation drastically increases the amount of sediment lost from the slopes (Wolman 1967).

The strength properties of the exposed soil are usually referred to as the *erodibility*—an estimate of the ease with which the soil can be eroded. Analysis of erodibility is extremely difficult, and its usefulness as a factor in soil loss studies is complicated by the fact that many investigators use different criteria to express its magnitude. For example, in a study of English soils, Chorley (1959a) considers erodibility to be related to the product of permeability and shear resistance (measured by a penetrometer) such that

$$I_e = \text{shear resistance} \times \text{permeability}^{-1}$$

where I_e is the *index of erodibility*. Although I_e values correlate well with the relief maintained by the geological formations in that area, Bryan (1968) questioned the significance of the index because shear resistance probably relates more to long-term stresses associated with mass movements than it does to surface processes like raindrop impact. Bryan evaluated several of the more common indices used to estimate erodibility and concluded that none is particularly good. In his opinion, the best guide may simply be the weight percentage of water stable aggregates (W.S.A.) greater than 3 mm in diameter. Even this index, however, is suspect because little information is available to demonstrate that the aggregates remain stable under the impact of high-velocity rain.

Assuming that all factors can be assessed within reasonable limits of uncertainty, the total soil loss by water erosion should follow the general equation

$$A = RK(LS)CP$$

where A is average annual soil loss; R is rainfall; K is the erodibility factor; LS the slope length-steepness factor; C the cropping and management factor; and P the conservation factor (for details see Smith and Wischmeier 1962). The precision of such an equation is questionable, but it probably can provide useful ballpark data for predicting soil loss.

Slope Profiles and Evolution

Measurements of natural slopes in a variety of climates have revealed the interesting fact that slope angles appear to be concentrated in groups with rather small ranges of values (see Young 1972; Carson and Kirkby 1972). Most pronounced are those that cluster at 43°–45°, 30°–38°, 25°–29°, 19°–21°, 5°–11° and 1°–4°. Although any slope angle is possible, the frequency of these recurring groups is tantalizing to geomorphologists because it probably reflects the underlying control of the great geomorphic variables—time, lithology, climate, and process.

Each of these groups has well-defined maximum and minimum values, which have been termed *limiting angles* by Young (1972) and *threshold angles* by Carson and Kirkby (1972). The general interpretation of the angular distribution is that angles within any group represent a stability regime for slopes formed in a particular climatic and lithologic setting. Under those conditions, threshold values can be exceeded if the intrinsic properties of the parent material are altered or if the climate changes. When threshold values are reached, any further change requires a fundamental response in the system that adjusts the slope angle and places it within a different stability group. Exactly why and how slopes adjust, however, is a debatable question. According to one hypothesis, groups and their

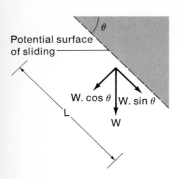

Potential surface of sliding

$W \cdot \cos \theta$ $W \cdot \sin \theta$

L

W

Figure 4.30. Driving and resisting forces on a slope. θ **is the angle of the potential surface of sliding, L is the length of that surface, and W is the weight of the material above the surface. (Adapted from Carson and Kirkby 1972. Used with the permission of Cambridge University Press)**

limiting angles represent characteristic angles for the processes that are eroding the slopes. In the other major theory, the change from one group to a lower one represents a normal event in the local geomorphic evolution.

Process Model

If we consider the solid rock mass shown in figure 4.30, the shear stress (τ) acting at a potential surface of sliding is given as

$$\tau = W \cdot \sin \theta.$$

The shear strength at the failure plane is

$$S = c \cdot L + W \cdot \cos \theta \cdot \tan \phi$$

where S is the shear strength and L is the length of the potential failure plane. This stress relationship is the general form used in the analysis of shallow slide phenomena. Depending on the cohesive properties of the rock and the rate of undercutting, a slope angle somewhere between the vertical and the substance's angle of repose will develop. As the material fractures, it gradually becomes a cohesionless mass (as explained earlier), and since the safety factor at the potential surface of sliding is 1,

$$1 = \frac{W \cdot \cos \theta \cdot \tan \phi}{W \cdot \sin \theta}$$

and therefore the angle on the surface of sliding is equal to the angle of internal friction because

$$\frac{W \cdot \sin \theta}{W \cdot \cos \theta} = \tan \phi$$
$$\tan \theta = \tan \phi$$
$$\theta = \phi.$$

The fractured rock material is eroded by slides and falls and accumulates near the rock mass as *scree* or *talus*. The talus slopes are extended upslope (fig. 4.31) and eventually mask the original rock face. In these accumulates ϕ angles are uncommonly high (43°–45°) because the mass is densely packed and the rock fragments are interlocked (Carson and Kirkby 1972). If the void percent is large, little pore pressure will develop and the stable angle of slope will correspond to the ϕ angle. Continuous breakdown of the scree deposits without much clay formation would produce a sandy matrix that should stand near the angle of repose for cohesionless sands, approximately 35°.

There is ample evidence that in a wide variety of climates a weathered mixture of rock rubble and soil called taluvium (talus + colluvium) underlies slopes between 25° and 28° (Melton 1965b; Robinson 1966; Young 1961). As the original talus deposits are progressively broken down by weathering, the mass gradually loses its open-pore framework. During times of abundant water

Figure 4.31. Upslope extension of talus slopes.

and high water tables, the material attains positive pore pressures that reduce the effective normal stress by buoyancy. This obviously lowers shear strength and changes the relationship between internal friction and the potential failure surface. Thus, the recurrence of slope angles at 25°–29° may lie in the mechanics associated with saturated soils. As summarized by Skempton (1964), cohesionless materials subjected to pore pressures are likely to experience shallow landsliding along failure planes that *approximate*

$$\tan \theta = \tfrac{1}{2} \tan \phi.$$

Assuming an original ϕ angle of 45° for coarse scree deposits and 35° degrees for a sandy mantle, the stable slope developed on these materials when they become saturated would be about 26° and 19° respectively. In clay-rich soils the ϕ angle is much lower, and stable slope angles are considerably less.

If one accepts these mechanical principles, it is reasonable to agree with the general model of slope development by mass movement envisioned by Carson (1969). He proposes that instability in slopes requires the progressive replacement of steep slopes by more gentle ones. In most regions the landscape goes through more than one phase of instability, but the exact number depends on the characteristics of the rocks and how they ultimately break down. In the initial stage, a steep rock cliff is replaced by scree slopes or slopes developed on thoroughly fractured rocks. This phase might be followed by a change to taluvial slopes, and eventually to the gentle slopes formed on clay-rich soils. Each slope is only temporarily stable, for as weathering changes the mantle's properties, and pore pressures vary, the mass reaches its slope threshold value. Further change causes the slope to adjust rapidly into a new stability range consistent with the revised properties of the mantle. Because of the variability of soil properties and pore pressures, any limiting angle values are possible, even though they apparently cluster in recurring groups. The types of material, the number of instability phases, and the threshold values combine in any area to control the progression of slope development. The net effect of the variables is eventually to form slopes that have long-term stability with respect to rapid mass movements; at that point, creep and surface water erosion become much more significant as slope processes. The scheme also suggests that form is related to the lithology and the type of processes operating in a particular climate.

Geomorphic Evolution Model

The other model is based on the assumption that any slope becomes more gentle with time and that groups of recurring angles represent stages of a cyclic landform evolution. Steep slopes presumably correspond to early stages in the cycle, and less steep slopes to later stages. In any area, vastly different slope angles formed on the same rock type (fig. 4.32) might be interpreted as evidence that the region has gone through more than one erosional epicycle (Young 1961).

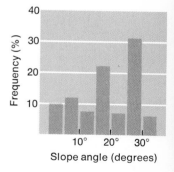

Figure 4.32. **Slope distribution on hypothetical sandstone. Several dominant angles might indicate cyclic erosion.**

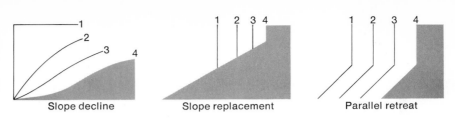

Slope decline   Slope replacement   Parallel retreat

The two models offered to explain the interesting frequency distribution of slope angles unfortunately lead us back to the fundamental controversy about the meaning of regional landscapes that we discussed in chapter 1. In one explanation, slope angles are time-dependent phenomena that represent stages in geomorphic cycles. In the other, slopes and their angles are viewed as time-independent features related to the properties of the slope materials and the mechanics of the dominant slope processes.

The cyclic hypothesis raises the question of how slope angles respond to continued erosion. Do slopes progress through thresholds by flattening with time? Or can one set of limiting angles be replaced by a lower set when the steeper slope retreats parallel to itself and the gentler slope extends as a basal component? Three main theories of slope evolution have been suggested: slope decline, slope replacement, and parallel retreat (fig. 4.33). In *slope decline,* the steep upper slope erodes more rapidly than the basal zone, causing a flattening of the overall angle. It is usually accompanied by a developing convexity on the upper slope and concavity near the base. Slope decline alone cannot in fact explain a concave profile on the lower slope unless some deposition occurs at the base. In *slope replacement,* the steepest angle is progressively replaced by the upward expansion of a gentler slope developed near the base. This process tends to enlarge the overall concavity of the profile, which may be in either a segmented or a smoothly curved form. Slopes evolving by *parallel retreat* are characterized by the maintenance of constant angles on the steepest part of the slope. Absolute lengths of slope parts do not change except in the concave zone, which gets longer with time. Although other models of slope evolution have been devised, most syntheses of regional landform development invoke these models or some basic combination of them.

The Rock-Climate Influence

Humid-Temperate Regions   The most common slope profile in humid-temperate regions is a distinct, convex upper slope and a concave lower slope. Contrary to some beliefs, straight slope segments do occur in regions with a humid-temperate climate, and some profiles do contain steep cliff faces. Most cliff faces, however, are ephemeral in the sense that as soon as undercutting ceases, a talus slope forms which will extend upslope until it covers the original cliff wall (fig. 4.31). If the lithology of the rock sequence underlying the slope is not uniform, cliff faces may persist because resistant units are maintained as caprocks when the weaker underlying strata retreat faster, essentially undercutting the stronger rocks.

Convex upper slopes are usually interpreted as a function of soil creep; the lower concavity probably results from soil wash, although not all slopes have this segment, particularly when there is active erosion at the slope base (Strahler 1950). The convexo-concave profile is most likely to be attained after mass movements have produced a long-term angular stability. At this stage, creep and wash become the dominant slope processes; the straight segment, representing stability of taluvial material, is gradually diminished in size.

Semiarid and Arid Regions   Semiarid and arid climates tend to engender slope profiles that are more angular than those found in humid-temperate regions, even though the same convex, straight, and concave segments may be present (fig. 4.34). Steep cliffs usually are present above a straight, debris-covered segment that normally stands at angles between 25° and 35°. At the base of the straight segment a pronounced change in slope occurs, and angles decrease over a short distance to less than 5°, a normal slope for most desert plains. The limited vegetal cover and low precipitation associated with arid zones assure that mass movements occur at higher angles and that creep is subordinated to wash. As a result the upper slope convexity, so prominent in humid regions, is much less pronounced.

Straight segments are maintained by the wash process, which is accelerated on the sparsely vegetated surfaces. Unlike similar segments in humid climates, these usually have only a thin veneer of rock debris. They are not, then, slopes of accumulation such as talus slopes but instead probably represent true slopes of transportation, on which the amount of debris supplied to the straight segment from the cliff face or by weathering of the underlying rocks is removed in equal quantities to the desert plain. The angle of slope represents some balance between the processes that break debris down and the actual transporting mechanism (Schumm and Chorley 1966). Most students feel that a general relationship between particle size and slope angle can be demonstrated.

Although other climatic regimes have characteristic slope forms, in most cases they are produced by the same mechanics that operates in the humid-temperate or arid zones. In the periglacial environment a special influence is

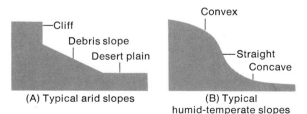

Figure 4.34. Typical slope profiles in (A) arid regions and (B) humid-temperate regions.

—Cliff
Debris slope
Desert plain

(A) Typical arid slopes

Convex
—Straight
Concave

(B) Typical humid-temperate slopes

exerted by magnified frost activity; a more extensive treatment of that environment is presented in chapter 11.

Lithology   The lithologic influence on slopes is shown in both declivity and profile shape. Coherent rocks tend to support steeper slope angles and, with equal cohesion, the more massive the bedding, the steeper the slopes. Where strata contain alternately weak and resistant rocks, an irregular profile may develop, and resistant units will assume higher than normal angles where they overlie weaker rocks. Resistance, however, is not defined by intrinsic properties of a particular rock type but is a relative feature determined by how rapidly slopes developed on the rock retreat and whether the rock stands relatively high in the local topography (Young 1972). Therefore, it is not so much the rock itself that determines resistance, but whether the slopes formed over the rock are controlled by processes of weathering or processes of removal. If a slope is weathering-controlled, resistance is related to how rapidly the rock is weathered; it is a direct function of the rock properties. In removal-controlled slopes, the resistance is attributable to the rate at which regolith can be eroded; the properties of the weathered mass, such as its infiltration capacity (Melton 1957) and the type and magnitude of the erosional processes, become important in slope development. For these reasons, the resistance of a particular rock type and its influence on slopes can be reversed if the rock is located in different climates. For example, the characteristics of slopes formed on limestones in humid climates contrast markedly with those developed in arid climates.

Summary   The processes of physical weathering tend to break rocks and unconsolidated debris into smaller particle sizes. The force needed to accomplish this disintegration is provided by expansion resulting from unloading, hydration of minerals, or growth of foreign substances in spaces within the parent material. Many important processes of disintegration require the presence of water. Physical weathering, combined with gravity, is instrumental in determining the type and rate of mass movements; ultimately it has a direct bearing on the slopes developed in any region.

Mass movements occur as slides, flows, and heaves, or by water-induced transport of surface debris. The magnitude and type of mass movement are partly dependent on the physical properties of the parent material. Shear strength (a function of internal friction, effective normal stress, and cohesion) determines how vigorously any substance will resist the force attempting to produce mass movement. Thus, slope failure or other mass movements can result from an increase in shear stress (driving force), a lowering of shear strength (resistance), or both. Physical weathering tends to decrease the shear strength of materials and thereby helps to initiate mass movements and control the form of the resulting slopes. Two models of slope evolution have been proposed to explain the recurrence of groups of slope angles. Climate and lithology interact to influence slope profiles.

The following references are suggested to provide greater clarity and detail concerning the topics discussed in this chapter.

Suggested Readings

Carson, M., and Kirkby, M. 1972. *Hillslope form and process.* London: Cambridge Univ. Press.

Cooke, R. U., and Doornkamp, J. 1974. *Geomorphology in environmental management.* London: Clarendon Press.

Crozier, M. J. 1973. Techniques for the morphometric analysis of landslips. *Zeit. f. Geomorph.* 17: 78-101.

Grim, R. 1962. *Applied clay mineralogy.* New York: McGraw-Hill.

Legget, R. 1967. Soil: its geology and use. *Geol. Soc. America Bull.* 78: 1433-60.

Ollier, C. D. 1969. *Weathering.* Edinburgh: Oliver and Boyd.

Terzaghi, K. 1950. Mechanism of landslides. In *Application of geology to engineering practice,* edited by S. Paige, pp. 83-123. Geol. Soc. America Berkey Vol.

Varnes, D. J. 1958. Landslide types and processes. In *Landslides and engineering practice,* edited by E. Eckel, pp. 20-47. Washington: Highway Research Board Spec. Rept. 29.

Young, A. 1972. *Slopes.* Edinburgh: Oliver and Boyd.

Having examined the processes that break down rocks and transport weathered debris down slopes, we are now ready to consider the activity of rivers, which not only erode and convey sediment but also are primarily responsible for developing the valley levels to which slopes are graded. Two basic truths about rivers were realized long before geomorphology emerged as an organized science: (1) streams form the valleys in which they flow, and (2) every river consists of a major trunk segment fed by a number of mutually adjusted branches that diminish in size away from the main stem. The many tributaries define a network of channels that drain a discernible, finite area recognized as the *drainage basin* or *watershed* of the trunk river.

Each basin is separated from its neighbor by a divide, and so basins serve as excellent fundamental units of geomorphic systems. Any feature, fluvial or otherwise, within a basin can be reasonably considered as an individual subsystem of the basin, having its own set of processes, geology, and energy gains and losses. Furthermore, because it is possible to measure the amount of water entering a basin as precipitation and the volume leaving the basin as stream discharge, hydrologic events can be readily analyzed on a basinal scale. Similarly, most of the sediment produced within the basin limits is ultimately transported from the basin via the trunk river. Thus, considered on a long temporal scale, the rate of lowering of the basin framework can be estimated. The mechanics of fluvial processes usually reflect some balance between the sediment to be transported and the water available to accomplish the task.

In this chapter we will consider the questions of how streams develop their channels and valleys and how they evolve into adjusted networks with characteristics that seemingly obey fundamental hydrophysical laws.

The Drainage Basin— Development, Morphometry, and Hydrology

5

The Stream Network

Most geologists are introduced to watersheds by learning that drainage patterns or individual stream patterns often mirror certain traits of the underlying geology, as figure 5.1 and table 5.1 show. The gross character of patterns is a useful tool in structural interpretation (Howard 1967) and as a first approximation of lithology; a search of topographic maps and aerial photographs for a distinctive network arrangement is a logical first step in the study of regional geology. Until recently this application of network properties was most geologists' prime, if not only, interest in drainage systems. Prior to World War II, most basins were described hydrologically in qualitative terms such as "well-drained" or "poorly-drained," or they were connoted descriptively in the Davisian scheme as being youthful, mature, or old. The mechanics of how river channels or networks actually form was understood (or misunderstood) in equally vague terms by both geologists and hydrologists. Realizing this early twentieth-century view of streams and drainages, it is startling to examine the avant-garde approach presented by R. E. Horton in 1945. His attempt to explain stream origins in mathematical terms and to describe basin hydrology as a function of statistical laws can be cited as the birth of quantitative geomorphology. Much of the standard geomorphic analysis of drainage basins has its roots in Horton's original work; his classic paper was instrumental in the rise of a new breed of geomorphologist.

Initiation of Channels

Imagine a hypothetical slope that is part of an undulating surface topography developed by weathering and mass movements. As precipitation falls on the slope surface, the water is absorbed into the ground at a rate called the *infiltration capacity*, which is a function of a number of factors such as soil texture and structure, vegetation, and the condition of the surface. As long as the infiltration capacity exceeds the rate at which rainfall strikes the surface (rainfall intensity), all the incoming water will be infiltrated and none will flow down the surface as

Figure 5.1. Basic drainage patterns. Descriptions are given in table 5.1. (From A.D. Howard 1967. Used with permission of American Association of Petroleum Geologists)

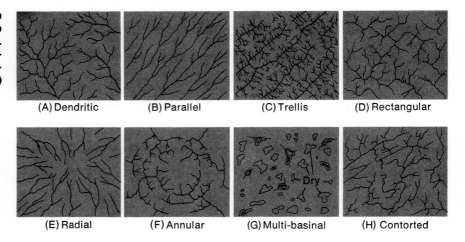

(A) Dendritic   (B) Parallel   (C) Trellis   (D) Rectangular

(E) Radial   (F) Annular   (G) Multi-basinal   (H) Contorted

Table 5.1 Descriptions and characteristics of basic drainage patterns illustrated in figure 5.1.

Basic	Significance
Dendritic	Horizontal sediments or beveled, uniformly resistant, crystalline rocks. Gentle regional slope at present or at time of drainage inception. Type pattern resembles spreading oak or chestnut tree.
Parallel	Generally indicates moderate to steep slopes but also found in areas of parallel, elongate landforms. All transitions possible between this pattern and type dendritic and trellis.
Trellis	Dipping or folded sedimentary, volcanic, or low-grade metasedimentary rocks; areas of parallel fractures; exposed lake or sea floors ribbed by beach ridges. All transitions to parallel pattern. Type pattern is regarded here as one in which small tributaries are essentially same size on opposite sides of long parallel subsequent streams.
Rectangular	Joints and/or faults at right angles. Lacks orderly repetitive quality of trellis pattern; streams and divides lack regional continuity.
Radial	Volcanoes, domes, and erosion residuals. A complex of radial patterns in a volcanic field might be called multiradial.
Annular	Structural domes and basins, diatremes, and possibly stocks.
Multibasinal	Hummocky surficial deposits; differentially scoured or deflated bedrock; areas of recent volcanism, limestone solution, and permafrost. This descriptive term is suggested for all multiple-depression patterns whose exact origins are unknown.
Contorted	Contorted, coarsely layered metamorphic rocks. Dikes, veins, and migmatized bands provide the resistant layers in some areas. Pattern differs from recurved trellis in lack of regional orderliness, discontinuity of ridges and valleys, and generally smaller scale.

After Howard 1967. Used with permission of American Association of Petroleum Geologists.

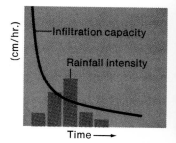

Figure 5.2. **Infiltration capacity and rainfall intensity plotted against time. Infiltration capacity decreases with duration of storm. Runoff occurs only when rainfall intensity is greater than infiltration capacity.**

runoff. This is why porous and permeable surfaces, such as those in sandy coastal plains, may experience heavy rains with no apparent surface flow. The infiltration capacity, however, is not constant; as figure 5.2 shows, in any precipitation event it usually starts with a high value that decreases rapidly in the first several hours of the storm and then more slowly as rainfall continues, until it finally attains a reasonably constant minimum value. The infiltration capacity changes because surface conditions are being changed, especially as aggregated soil clumps are broken apart and surface entry by the water becomes more difficult as pores become clogged with the clay particles. In addition, the infiltration capacity can be only as great as the lowest transmission rate in subsurface horizons. In prolonged storms, for example, a clay pan or a caliche zone in the B horizon may determine the ultimate infiltration rate. In the interval between rains, the infiltration capacity rises again as the surface dries and regains its aggregated structure, and the storage space within the soil increases as water gradually drains downward. The frequency of rainfall therefore becomes a significant factor, because a rapid succession of rains without enough time intervening will prevent the infiltration capacity from returning to its original high value. Rains of relatively small intensity can then trigger disastrous floods.

When rainfall intensity exceeds the infiltration capacity, runoff occurs, and only then does erosion become possible (fig. 5.3). For a stream channel to develop, the erosive force (F) of the overland flow must surpass the resistance (R) of the surface to being eroded. According to Horton (1945), as overland flow begins to traverse the slope, the force it exerts on a soil particle depends on the slope angle, the depth of the water, and the specific weight of the water such that

$$F = \gamma \frac{d}{12} \sin \theta$$

where γ is the specific weight, d the depth, and $\sin \theta$ the slope angle. Actually the force (F) represents a shear stress exerted parallel to the surface by the water. The stress progressively increases downslope because the depth rises as more and more water is added to the volume of the overland flow. Depending on the size of the slope material, a threshold force value is eventually reached at some point on the slope where $F > R$, and particles are dislodged or entrained.

The resisting factor is largely a function of the vegetal cover and, less importantly, the nature of the surface. Vegetation intercepts raindrops before they can strike the surface, thereby preserving cohesion in the aggregated-clay soil structures. In addition, rootlets tend to bind soil particles, and litter often serves as a protective mat above the surface material. Vegetation also inhibits the free flow of water and retards its velocity. In areas devoid of vegetal cover, the surface commonly develops a hard crust as it dries in the direct sunlight, which provides a high initial resistance to erosion. This may be destroyed during a storm, indicating that resistance, like the force of overland flow, may vary as a rainfall event progresses. It should also be noted that the factors that determine resistance are essentially the same ones that control infiltration capacity.

Erosion by overland flow begins when F exceeds R, taking the form of a series of shallow, parallel rills that are oriented perpendicular to the slope contours. Slight variations of the surface topography produce a greater depth of flow and more erosive force in the low spots. Erosion is accelerated at those points, and a rilled surface results rather than a slope that discharges water as an unconfined sheet. The actual point where rill formation begins depends on how efficiently the force of overland flow increases as the water moves down the surface. This ultimately relates to the infiltration capacity, the rainfall intensity, and the resulting rate of runoff (the *runoff intensity*).

Assuming a constant slope and runoff intensity, the distance between the watershed divide and the upper position of rills is a measurable segment called

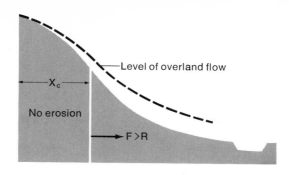

Figure 5.3. Hypothetical slope showing overland flow. No erosion occurs until the force of overland flow (*F*) exceeds the resistance of the surface material (*R*). Upslope from that point no erosion occurs. *Xc* is the distance from the divide to point where erosion begins. (After Horton 1945. Courtesy of the Geological Society of America)

the *critical length* (X_c); the surface between X_c and the divide is recognized as a "belt of no erosion" (figs. 5.3 and 5.4). Horton considered the critical length to be the most important single factor in the development of stream networks, but its significance can perhaps be appreciated more fully on a local scale because it is highly sensitive to changes in the factors that control erosion (fig. 5.5). Assume, for example, that a farmer wishes to plant additional row crops in an area within the belt of no erosion. As the land is cleared of its original vegetal cover and replaced by the crops, the resistance and infiltration capacity are drastically lowered. Under prevailing precipitation more runoff occurs, the force of overland flow increases, and X_c is shortened. The narrowing of the belt of no erosion allows rills and gulleys to form in the newly cropped region. Depending on their depth, these may make the area impassable for the heavy equipment needed for plowing and harvesting.

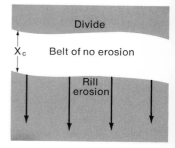

Figure 5.4. Map view of critical length (*Xc*) and belt of no erosion. Rill erosion starts downslope from belt of no erosion when *F>R*. (Compare fig. 5.3.)

On most slopes with convexo-concave profiles, the upper reaches are flatter than the middle portions, and so most of the upper areas fall within the belt of no erosion; the critical length usually occurs in the steeper zones further downslope. In another model, however, called *saturated throughflow* (M. J. Kirkby and Chorley 1967; M. J. Kirkby 1969), water moving downslope in the upper zones of the soil is considered to be an important factor in the development of stream channels. Although flow velocities are less here than in overland flow, nearly saturated soils experience throughflow during precipitation long before overland flow even begins. This is significant because throughflow can form rills, especially in the lower concave portions of a slope where saturated conditions are common. In this way, rills may develop in a position that is completely unrelated to the drainage divide or the critical length of overland flow. In humid regions where soils are frequently saturated, the throughflow process may extend rills and gullies upslope with little, if any, assistance from overland flow. Distinguishing between these two processes in a headwardly expanding rill is a difficult task because the headcut in a rill exposes a substratum with different, and commonly lower, resistance. Overland flow may also expedite upslope rill growth by undercutting, even though the surface resistance is greater than the eroding force of the flow.

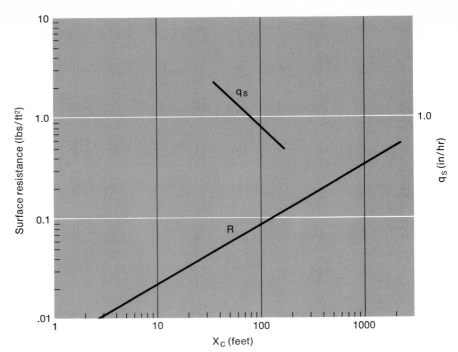

Figure 5.5. Relationship of the critical length (X_c) with resistance (R) and runoff intensity (q_s). (q_s is the amount of water flowing over the surface during a given interval of time.) X_c versus q_s plotted at constant R; X_c versus R plotted at constant q_s. (Data from Horton 1945)

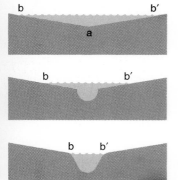

Figure 5.6. Stages in development of a rill. More rapid erosion occurs at the position of *a* than at *b* or *b* ' because water is deeper. (From Horton 1945. Courtesy the Geological Society of America)

Micropiracy and Cross-grading

The formation of rills does not explain stream channels because, as discussed earlier, a rilled surface is still part of a slope-wash system. Their development, however (shown in fig. 5.6), is the first necessary step toward a true river. It is now generally recognized that rills initiated on a slope cannot long remain as parallel unconnected channels. Deeper and wider rills develop where the length over which erosion can occur is the greatest. These master rills carry more water and, because of the greater depth, they undergo downcutting until all the flow is contained within the channel and the rill becomes a tiny stream. Because they become slightly entrenched, master rills capture adjacent rills when bank caving or overtopping during high flow destroys the narrow divides between them. The repeated diversion of rills, a process called *micropiracy,* tends to obliterate the original rill distribution, and gradually the initial slope parallel to the master channel is replaced by slopes on each side that slant toward the main drainage line.

The development of new slope direction in accordance with the master channel, called *cross-grading* by Horton (1945), is depicted in figure 5.7. In the final stage, only one stream, confined in the master rill channel, crosses the slope. The side slopes presumably develop a new rill system graded to the position of the initial stream, and the process repeats itself, culminating in a secon-

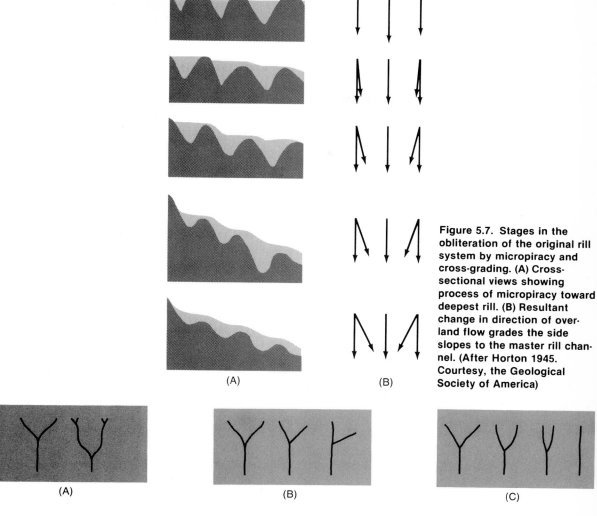

(A)   (B)

Figure 5.7. Stages in the obliteration of the original rill system by micropiracy and cross-grading. (A) Cross-sectional views showing process of micropiracy toward deepest rill. (B) Resultant change in direction of overland flow grades the side slopes to the master rill channel. (After Horton 1945. Courtesy, the Geological Society of America)

(A)   (B)   (C)

dary master rill serving as an incipient tributary. Each smaller tributary evolves in a similar way until the network of streams takes form.

The network pattern develops by repeated division of single channel segments into two branches, a process known as bifurcation. Schumm (1956) suggests that the angle between the limbs of a bifurcated channel probably evolves in one of three possible ways (fig. 5.8): (1) both limbs grow headward while preserving the original angle at their juncture; (2) one branch straightens its course and becomes dominant; or (3) the angle on steep slopes progressively decreases until the branches reunite into a single channel. Any or all of these procedures might

Figure 5.8. Development of bifurcation angles. (A) The original angle is preserved. (B) One branch becomes dominant. (C) Angle decreases and branches merge into one channel; occurs on steep slopes. (From Schumm 1956. Courtesy of the Geological Society of America.)

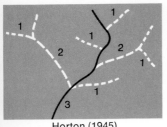

Horton (1945)

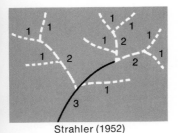

Strahler (1952)

Shreve (1967)

Figure 5.9. Methods of ordering streams within a drainage basin.

be found in the evolution of a network, constrained only by the fundamental erosive controls and the geologic framework. Divides between adjacent basins are predetermined by the extent to which streams can expand headward. Because critical length varies with resistance, the areal extent of the uneroded uplands partially reflects the geology and its history. Within the basin itself, smaller interfluves may be present where cross-grading has not operated. These small areas parallel the main stream and preserve the slope of the original surface.

Basin Morphometry

One of Horton's most dramatic contributions was to demonstrate that stream networks have a distinct fabric, called the *drainage composition,* in which the relationship between streams of different magnitude can be expressed in mathematical terms. Each stream within a basin is assigned to a particular order indicating its relative importance in the network, the lowest order streams being the most minor tributaries and the highest order, the main trunk river.

Figure 5.9 shows several methods of ordering streams. Horton's cumbersome method was refined by Strahler (1952a) so that stream segments rather than entire streams become the ordered units. As the figure shows, a segment with no tributaries is designated as a first-order stream. Where two first-order segments join they form a second-order segment; two second-order segments a third-order segment, and so forth. Any segment may be joined by a channel of a lower order without necessitating an increase in its order; i.e., third-order segments may have an infinite number of second- or first-order tributaries. Only where two segments of equal magnitude join is an increase in order required. The apparent inconsistency in Strahler's method of not accounting for all tributaries is removed in the network analysis conceived by Shreve (1967). He considers streams as links within the network, with the magnitude of each link representing the sum of the link numbers of all tributaries that feed it. That is, networks in which the downstream segments are of the same magnitude have equal numbers of links within their basins. Shreve's link system gives a number that at any point within the basin is equal to the number of first-order streams upstream from that point. Regardless of any deficiencies it may have, and even though methods other than those shown in figure 5.9 have been suggested (Scheidegger 1965; Woldenberg 1966), the Strahler system of ordering has been used in most network analyses.

Horton's original work and the studies by Strahler and his students have spurred a plethora of basin analyses in recent years, and some are demonstrably practical. It is now generally recognized that in every basin a group of measurable properties exist that define the linear, areal, and relief characteristics of the watershed (table 5.2). These variables seem to be statistically related to stream orders, and various combinations of the parameters obey mathematical laws that hold for a large number of basins. Two general types of numbers have been used to describe basin morphometry or network characteristics (Strahler 1957, 1964, 1968). *Linear scale* measurements allow size comparison of topographic units.

Table 5.2 Common morphometric relationships.

Linear morphometry

Stream number in each order (N_o)	$N_o = R_b^{s-o}$
Total stream numbers in basin (N)	$N = \dfrac{R_b^s - 1}{R_b - 1}$
Average stream length	$\bar{L}_o = \bar{L}_1 R_L^{o-1}$
Total stream length	$L_o = \bar{L}_1 R_b^{s-1}\left(\dfrac{u^s - 1}{u - 1}\right)$ where $u = R_L/R_B$
Bifurcation ratio	$R_b = N_o/N_{o+1}$
Length ratio	$R_L = \bar{L}_o/\bar{L}_{o+1}$
Length of overland flow	$\ell_o = \dfrac{1}{2D}$

Areal morphometry

Stream areas in each order	$\bar{A}_o = \bar{A}_1 R_a^{o-1}$
Length-area	$L = 1.4A^{0.6}$
Basin shape	$R_F = \dfrac{A_o}{L_b^2}$
Drainage density	$D = \dfrac{\Sigma L}{A}$
Stream frequency	$F_s = \dfrac{N}{A}$
Constant of channel maintenance	$C = \dfrac{1}{D}$

Relief morphometry

Relief ratio	$R_h = H/L_o$
Relative relief	$R_{hp} = H/p$
Relative basin height	$y = h/H$
Relative basin area	$x = a/A$
Ruggedness number (Melton 1957)	$R = DH$

s = order of master stream
o = any given stream order
H = basin relief
P = basin perimeter

Adapted from Strahler 1958. Courtesy of the Geological Society of America.

The parameters may include the length of streams of any order, the relief, the length of basin perimeter, and many other measures. The second type of measurement consists of *dimensionless numbers,* often derived as ratios of length parameters, that permit shape comparisons of basins or networks. Length ratios, bifurcation ratios, and relief-length ratios are common examples. Table 5.2 gives the various linear, areal, and relief relationships, which are discussed in the sections that follow.

Linear Morphometric Relationships   The establishment of stream ordering led Horton to realize that certain linear parameters of the basin are proportionately related to the stream order and that these relationships could be expressed as basic laws of the drainage composition. Many of the linear morphometric laws are a function of the *bifurcation ratio* (R_b), which is defined as the ratio of the number of streams of a given order to the number of the next higher order. The primary use of the bifurcation ratio is to allow rapid estimates of the number of streams of any given order and the total number of streams within the basin. Although the ratio value will not be constant between each set of adjacent orders, its variation from order to order will be small, and a mean value can be used. A mean value is emphasized in the linear relationship revealed by a semilogarithmic plot of order versus number (fig. 5.10). The R_b value can be determined from such a plot because the regression coefficient is the logarithm of

Figure 5.10. Relation of stream order to the number and mean lengths of streams in the Susquehanna River basin. (After Brush 1961)

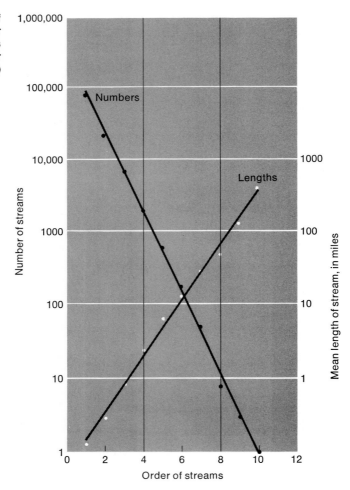

R_b. Also, as Horton pointed out, the number of streams in the second highest order is a good approximation of R_b. When geology is reasonably homogeneous throughout a basin, R_b values usually range from 3.0 to 5.0.

The length ratio (R_L) is similar in context to the bifurcation ratio; it is the ratio of the average length of streams of a given order to those of the next higher order. The length ratio can be used to determine the average length of streams in an unmeasured given order ($\bar{L}_o$) and their total length (L_o). The combined length of all the streams in a given basin is simply the sum of the lengths in each order. For most basin networks, stream lengths of different orders plot as a straight line on semilogarithmic paper (fig. 5.10), as stream numbers also do.

The laws of stream numbers and lengths have been repeatedly verified in studies subsequent to Horton's work (Schumm 1956; Chorley 1957a; Morisawa 1962; Selby 1967; Chose et al. 1967; and many others) and are now firmly established.

Areal Morphometric Relationships   The equity among linear elements within a drainage system suggests that areal components should also possess a consistent morphometry, since dimensionally area is simply the product of linear factors. The fundamental unit of areal elements is the area contained within the basin of any given order (A_o). It encompasses all the area that provides runoff to streams of the given order, including all the areas of tributary basins of a lower order as well as interfluve regions. Schumm (1956) demonstrated (fig. 5.11) that basin areas, like stream numbers and lengths, are related to stream order in a geometric series.

Area has also been employed to manifest a variety of other parameters (shown in table 5.2), each of which has a particular significance in basin geomorphology. One of the more important factors involving area is the drainage density (D), which is essentially the average length of streams per unit area and as such reflects the spacing of the drainageways. The drainage density is of interest here because it is directly controlled by the interaction between geology and climate. As these two factors differ from region to region, wide variations in D can be expected (table 5.3). In general, resistant surface materials or those with high infiltration capacities have widely spaced streams and, consequently, low drainage densities. As resistance or surface permeability decreases, runoff is usually removed in a greater number of closely spaced channels, and D tends to be much higher. As a rule of thumb, where geology and slope angles are the same, humid regions develop a thick vegetal mat that increases resistance and infiltration and thereby perpetuates a lower drainage density than would be expected in more arid basins. Geology and the slope angles produced on it are also major contributors to the prevailing density, since rocks have different resistances and steep slopes tend to increase the density ratio. Methods for rapid estimation of drainage density have been devised (McCoy 1971; Mark 1974).

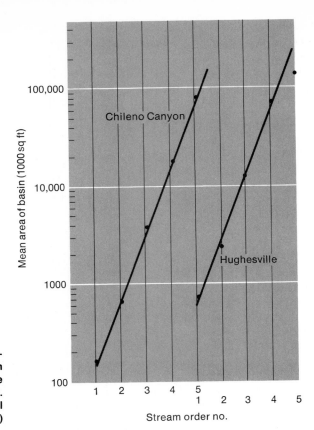

Figure 5.11. Relationship between stream order and mean basin area in two drainage basins. (After Schumm 1956. Courtesy, the Geological Society of America)

Figure 5.12. Fine-textured topography with high drainage density formed on sedimentary rocks. Santa Fe and Los Alamos counties, N.M. (Photo by U.S. Geological Survey)

Table 5.3 Drainage density in regions with different geology and climate.

Drainage Density	Climate	Geology	Area
3-4	humid-temperate	Resistant sandstone, flat-lying	Appalachian Plateau
8-16	humid-temperate	Nonresistant, flat-lying rocks	Central-Eastern U.S.
20-30	dry summers—subtropical, seasonal	Fractured and weathered igneous and metamorphics	Southern California
50-100	semiarid	Variable lithology and structure	Rocky Mountains
200-400	arid-semiarid	Flat-lying, non-resistant sedimentary rocks	Badlands, S. Dakota
1100-1300	humid-temperate	weak clays	Northern New Jersey

Data from Strahler 1968 and Schumm 1956.

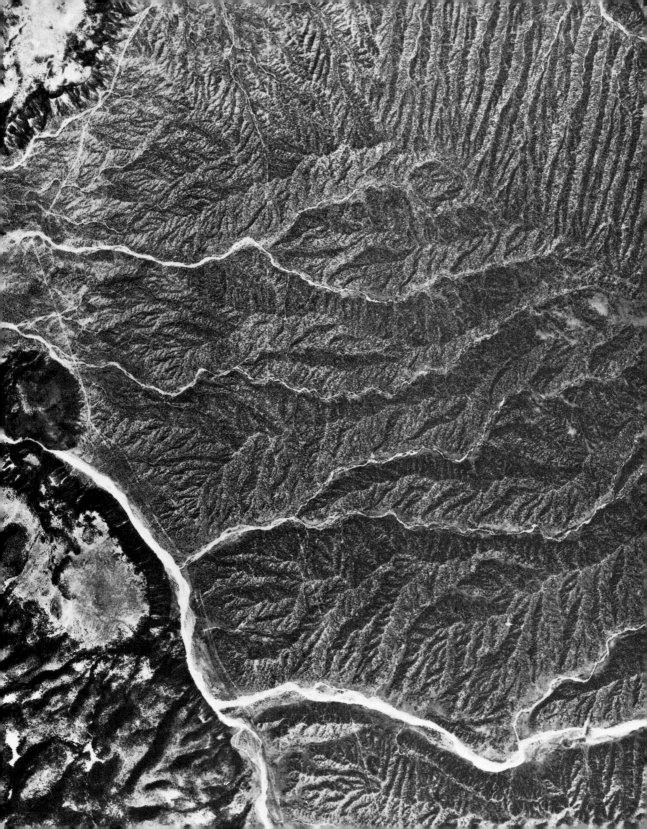

The density factor is also related to a parameter known as the *texture ratio* (T):

$$T = N/P$$

where N is the number of crenulations on the most irregular contour and P is the length of the basin perimeter. Ratios determined for many areas show the mean texture ratio to be a useful descriptive number. A region with a coarse texture has a mean value < 4, medium texture $4–10$, and fine texture > 10; figure 5.12 shows a fine-textured region where drainage density is high. Drainage density is related to the texture ratio as a simple power function

$$D = aT^b$$

where a and b are constants. The texture ratio can therefore be used as a substitute to indicate relative density values.

The drainage density not only reflects geology and climate, but it also has been used as an independent variable in the framing of other morphometric parameters. For example, the *constant of channel maintenance* and the *length of overland flow* (table 5.2) both utilize a reciprocal relationship with density to demonstrate the link between factors that control surface erosion and those that describe the drainage net (Schumm 1956). The constant of channel maintenance indicates the minimum area required for the development and maintenance of a channel; that is, the ratio represents the amount of basin area needed to maintain one linear unit of channel length. As Schumm points out (1956, p. 607) this relationship requires that drainage networks develop in an orderly way because the meter by meter growth of a drainage system is possible only if sufficient area is available to maintain the expanding channels. The ruggedness number (drainage density × basin relief) is another parameter that employs drainage density; it is useful in relating morphometry to flood peak discharge (Patton and Baker 1976).

Relief Morphometric Relationships   A third group of parameters shown in table 5.2 is used to indicate the vertical dimensions of a drainage basin; it includes factors of gradient and elevation. Like stream numbers, length, and area, the average slope of stream segments in any order approximates a geometric series in which the first term is the mean slope of first-order streams. This relationship, called the *law of stream slopes,* is reasonably valid as long as the geologic framework is homogeneous. Channel slopes and surface slopes are closely akin to the parameters for length. Horton suggested, for example, that the length of overland flow as a function of only the drainage density is at best an approximation because overland flow also depends on slope parameters.

As relief refers to elevation differences between two points, slopes that connect the points are integral factors. The choice of reference points differs, but the most useful relief parameters are the *maximum relief* (highest elevation in the basin − lowest elevation in the basin) or the *maximum basin relief* (highest elevation on the basin perimeter − the elevation at the mouth of the trunk river). The *relief ratio* (Schumm 1956), the maximum basin relief divided by the longest horizontal distance of the basin measured parallel to the major stream, indicates the overall steepness of the basin.

A different relief study is the *hypsometric analysis* (Strahler 1952b), which relates elevation and basin area. As figure 5.13 shows, the basin is assumed to have vertical sides rising from a horizontal plane passing through the basin mouth and under the entire basin. Essentially, a hypsometric analysis reveals how much of the basin is found within cross-sectional segments bounded by specified elevations. The relative height (y) is the ratio of the height (h) of a given contour above the horizontal datum plane to the total relief (H). The relative area (x) equals the ratio a/A, where a is the area of the basin above the given contour and A is the total basin area. The hypsometric curve (fig. 5.13B) represents the plot of the relationship between y and x and simply indicates the distribution of mass above the datum. The form of the curve is produced by the *hypsometric integral* (*HI*) which expresses, as a percentage, the volume of the original basin that remains. In natural basins most *HI* values range from 20 to 80 percent, the higher value indicating that large areas of the original basin have not been altered into slopes. Low integral values simply mean that much of the basin stands at low elevation relative to the area of the original upland surface. One objection to the hypsometric analysis is the tedium involved in determining the integral, but several alterna-

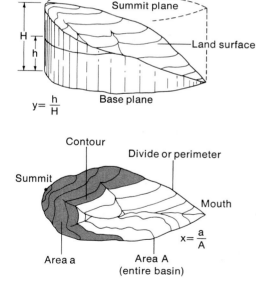

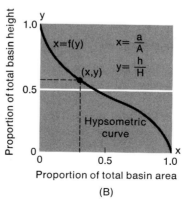

(A)

(B)

Figure 5.13. Ingredients of a hypsometric analysis. (A) Diagram showing how dimensionless parameters used in analysis are derived. (B) Plot of the parameters to produce the hypsometric curve. (From Strahler 1952. Courtesy, the Geological Society of America)

tive methods of calculation (Chorley and Morley 1959; Haan and Johnson 1966; Pike and Wilson 1971) have removed much of this difficulty.

Basin Evolution

Although morphometric values differ from basin to basin, each network still obeys the same statistical laws discussed above. Many authors have suggested that morphometry reflects an adjustment of geomorphic variables that is established under the constraints of the prevailing climate and geology (Chorley 1962; Leopold and Langbein 1962; Strahler 1964; Doornkamp and King 1971; Woldenberg 1969; and many others). Essentially this means that once a network is established, the basinal characteristics can be defined by the same quantitative terms at any time during the drainage growth. As the basins and networks evolve, an equilibrium is eventually produced by the interplay of climate and geology and maintained as a time-independent phenomenon. Once the components within a basin become balanced, any changes of climate or geology will be compensated for by adjustments of the basin parameters in such a way that the laws of drainage composition will be preserved. As originally conceived, however, these laws issued from well-developed stream systems, and the measurements needed to derive the equations were made on topographic maps of these basins. Such an approach provides no insight as to how quickly morphometric balance is attained or what changes in its character occur as the basin ages. It seems appropriate, therefore, to consider the influence of time on the morphometry of a basin.

The mode of basin evolution and its associated morphometric balance has been explained in two philosophical frameworks. Horton's infiltration approach is based on observable factors of force and resistance, which are ultimately dependent on climate and geology. It provides a rational explanation for the balance noted in the basin (Leopold et al. 1964). A projection of rill development, micropiracy, and cross-grading through time suggests that the geometric properties are the results of known hydrophysical processes.

In contrast, the same topologic equilibrium may be developed by purely random processes (Leopold and Langbein 1962; Shreve 1966a; Smart 1969; and many others), in which the distribution of geometric elements is in no way predestined but follows the laws of chance. Much of the argument for randomness is based on observations that networks of small tributary basins show too much variation from basin to basin to be rational or to predict the morphometry of larger basins formed when the tributaries join. The premise is that the consistent relationships seen in the entire basin appear only when a large number of the small component basins are used as the statistical sample, and that therefore the balance is simply a function of probability, the laws representing the most probable condition. As Howard (1971a) points out, however, the success of probability in explaining network topology does not necessarily prove an inherent randomness within the system. Factors that determine the properties of small-scale networks, such as lithology, microclimate, structure, etc., may vary in such a complex way in space and time that they may appear to have no regularity when one is predicting the properties of the entire basin network. Small-scale morphometry may fail to predict large-scale morphometry simply because we do not understand the processes or have not examined the rational elements in sufficient detail to reveal the regularity (Howard 1971a). Some studies seem to support this contention; however, equally strong arguments have been presented for the probabilistic approach (Smart 1974; Shreve 1975). In short, what morphometry is telling us about its fundamental basis is not clear. Apparently microclimatic or geologic constraints compensate for one another in a way that produces topological randomness, and so precisely what kinds of environmental controls will give rise to systematic and identifiable deviations from a random distribution is simply not known (Abrahams 1975).

Philosophy aside, some studies have provided a glimpse into the questions of how rapidly morphometry is established and what changes occur in its nature as the basin evolves. There seems to be little doubt that a quantitatively balanced drainage net forms rapidly in erodible material. This was clearly demonstrated by Schumm (1956) in the Perth Amboy (N.J.) badlands, where a statistically viable drainage was formed in less than twenty years. Although the master stream, a fifth-order segment, was only about 1 km long, the miniature network was dramatically in phase with the laws of drainage composition. Because the Perth Amboy network developed on back-filled waste from local excavations, the framework does not represent a natural undisturbed setting. Nonetheless, examples of drainage evolution in pristine areas show the same rapidity of development. Morisawa (1964) documented the growth of a drainage system on the floor of Hebgen Lake (Montana) after it was tilted and raised during the 1959 earthquake. Small basins formed on the newly exposed surface were mapped one year after the quake and remapped one year later. Although changes in channel length, gradient, and junction angles occurred, they were not significant, and in general the system development followed its original outline. Morisawa concluded that the watershed adjusted to its environmental controls quickly to produce a steady-state condition. Such a conclusion is supported by observations of the fluvial processes on a beach raised along Montague Island during the catastrophic Alaskan earthquake in 1964 (Kirkby and Kirkby 1969). In a matter of days, a newly exposed harbor floor was trenched by streams with almost uniform gradients, and the incipient drainage network was developed simultaneously. Clearly, then, the formation of drainage networks appears to be a rapid phenomenon when considered on a geological time scale. The amount of time in absolute terms probably varies according to the resistance of the material, the climate, and the initial slope angles. Unfortunately, examples of drainage establishment are documented only in areas where the least resistant materials underlie the system. Precisely how long it takes to form a balanced network in regions underlain by resistant crystalline rocks is rather conjectural.

Once basin elements attain a statistical balance, further changes in morphometry are usually revealed in the shape, drainage texture, or hypsometry of the basin. The area limits of any basin presumably are determined by the hydrophysical controls denoted by Horton and by the competition for space between adjacent basins. During the period of expansion to its peripheral limits, however, each parameter should change in such a manner as to maintain the original quantitative relationships among the factors. Hack (1957) showed that for a large number of basins, the stream length and basin area are related by the simple power function

$$L = 1.4 \, A^{0.6}$$

where L is the distance from any locality on a stream to the divide at the head of the longest segment above the given locality, and A is the basin area above the given locality. Hack noted that to preserve the original geometric balance, each variable must change at the same rate and the exponent in the equation should be 0.5. The larger exponent he observed requires that basins become more narrow and elongate as they grow.

The same tendency has been noted in arid basins (Miller 1958), and although some of the deviation from the 0.5 exponent may come from an increasing sinuosity of the main channel (Smart and Surkan 1967), the trend towards elongation is seen in enough widely divergent settings to suggest that it is an inherent property in the growth of many basins. This conlusion, however, may not be valid in all cases. Some studies indicate that as small basins develop into large ones, they reach a stage where the basin widens faster than it elongates. With time, the exponent value might decrease below 0.5 (Mueller 1972; Shreve 1974). Nonetheless, elongation is very striking in the growth of parallel drainage patterns, as exemplified in the postglacial history of the Ontonagon region of northern Michigan (Hack 1965). There the network formed as the Ontonagon Plain was exposed during the destruction of glacial Lake Duluth. Initial stream development followed parallel grooves that extended down the plain to the lake margin (fig. 5.14), and although this accentuated the basin elongation, figure 5.15 demonstrates that the lengths and areas of every order follow the 0.6 power function recognized elsewhere.

The shape of the hypsometric curve also provides some insight about the effect of time on basin morphometry. As Strahler (1952b) points out, the curve represents a continuous function relating height and area within a basin, and as such it should change its profile with time as more of the basin is consumed by erosion. Assuming the drainage system originates on a relatively flat upland surface, stages of basin evolution might generally be reflected in the shape of the hypsometric curve as the basin proceeds towards a base level condition. As figure 5.16 shows, in the early phase of drainage development, when rapid transformations are occurring within the basin, the area-height relationship is in a state of flux and the inequilibrium is usually marked by a convex hypsometric curve.

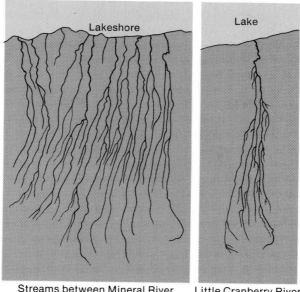

Figure 5.14. Growth of two parallel drainage patterns in the Ontonagon region, northern Michigan. (After Hack 1965)

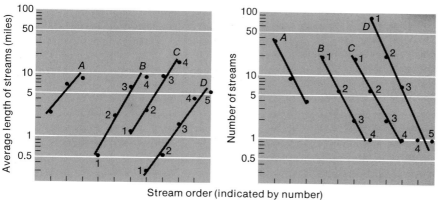

Figure 5.15. Relation of stream order to stream length and number in drainage basins of the Ontonagon Plain, Mich. A, Streams between Mineral and Cranberry rivers; data from maps; stream orders not known. B, Little Cranberry River. C, Weigel Creek. D, Mill Creek. (From Hack 1965)

As erosion proceeds, more and more components of the basin become mutually adjusted until complete equilibrium is reached. The curve gradually changes to a sigmoidal shape as the steady-state condition is attained. Equilibrium probably occurs when about 40 percent of the original basin volume is removed (Strahler 1952b). Once the character of the equilibrium curve is formed, it will be preserved as the relief is gradually lowered. In some basins, however, resistant rocks may form isolated hills, causing a distorted hypsometry with an abnormally low hypsometric integral. Strahler considered these basins to be in a "monad-

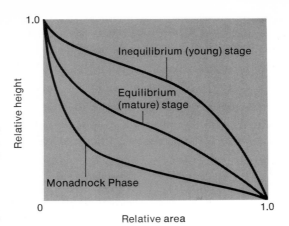

Figure 5.16. Hypsometric curves and time. (From Strahler 1968. Reprinted with permission from the *Encyclopedia of Geomorphology*, Rhodes W. Fairbridge, ed., Dowden, Hutchinson & Ross, Stroudsburg, Pa.)

nock phase,'' but he made it very clear that this phase and its hypsometric curve are transitory because erosion of the monadnock area will readjust the curve to its equilibrium shape.

Schumm (1956), analyzing hypsometry in a slightly different sense, made some salient observations concerning basin evolution when he showed that the region of most intense erosion and channel lowering migrates with time, being near the mouth during the early phase of basin development and near the headward margin in later stages. These observations demonstrate that various parts of a large basin may be evolving at different rates.

The final morphometric property that seems to depend on time is the drainage texture. As demonstrated by Ruhe (1952), both drainage density and stream frequency increase systematically in areas underlain by glacial deposits of progressively increasing age. The rate of the textural change is not constant; it probably was greatest during the first 20,000 years and then decelerated. The marked transition in the rate of textural evolution perhaps represents the time at which complete equilibrium was established and the basins were filled with as many streams as possible (Leopold et al. 1964).

It is interesting to note that throughout the period of growth in Ruhe's study, the channel lengths and numbers seem to obey Horton's geometric laws, suggesting that texture may be a surrogate for time in the analysis of basin evolution. This proposition was given added credence when Melton (1958) found stream frequency (F) and drainage density (D) to be related by a simple equation

$$F = 0.694 \, D^2.$$

In deriving this equation, which he considered to be a fundamental morphometric law, Melton used as his sample 156 basins with widely divergent geology, erosional histories and, presumably, ages. Since all stages of drainage development were thrown into the statistical pot, the significant empirical relationship between the variables most likely reflects a general law followed by basins as they grow. In fact, it was suggested that the dimensionless ratio F/D^2, called the *relative density,* should indicate how completely the stream network fills the basin (Melton 1958). High relative density values suggest that stream lengths are short and the basin outline is not yet completely filled with the stream network. As the drainage evolves and expands into each basin niche, the ratio decreases until equilibrium is established.

It has been commonly accepted that Melton's growth law is an example in which spatial parameters can be substituted for time. More recently, however, the universal applicability of the growth equation has been questioned (Abrahams 1972). Data in that study suggest that in basins with a similar environment (within-landscape relations) the values of the constants j and k in the general equation $F = jD^k$ change with the basin order. Only when disparate environments (between-landscape relations) are included in the sample do j and k assume the constant values shown in the Melton equation. This implies that the growth law is a generalization revealed only by statistically treating a group of within-landscape relations (fig. 5.17), and so space and time are interchangeable in morphometry only if the basins analyzed are environmentally similar and of the same order. Further studies, however, do not completely substantiate this premise because some within-landscape samples have been shown to approximate Melton's k value rather well (Wilcock 1975).

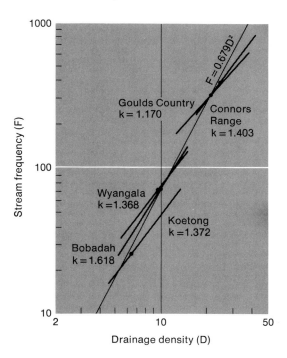

Figure 5.17. Within-landscape relations between stream frequency and drainage density and the least-squares line through the midpoints (indicated by closed circles) of the data from five landscapes; k is the exponent of D (that is, the slope coefficient) in the within-landscape equations. (From Abrahams 1972. Used with permission of the Geological Society of America)

Some of the confusion surrounding the textural relations in basin growth must stem from the fact that basin systems, which are clearly three-dimensional, are traditionally analyzed in a planimetric sense. Wilcock (1975) was able to show, however, that F/D^2 is closely related to the hypsometric integral (HI) and that much of the scatter of points in the F versus D plots is explainable as a function of hypsometry. Although the significance of the relief factor in the analysis of basin growth is not yet resolved, Wilcock's study provides a reasonable argument that all factors—linear, areal, and relief—are probably involved in the statistics of basin growth.

Even though it may be necessary to include all types of factors in the derivation of growth laws, it is clear that alterations of basin morphometry do occur with time. These changes do not violate the steady-state concept because parameters vary with one another in a systematic way. It is tempting to assume, therefore, that a well-balanced network with discernible morphometry evolved in an orderly manner from infancy to its present state. Realistically, such an assumption may be totally erroneous because basin morphometry may tell us little, if anything, about basin *history*.

A cogent example of this is the drainage system of Volney Creek, a small basin in southern Montana within the watershed of the Yellowstone River. Volney Creek heads in the piedmont region of the Beartooth Mountains, rising approximately 3 km from the mountain front. Its basin, covering an area of 60 km^2, is underlain entirely by Mesozoic and Tertiary clastic sedimentary rocks. The basin exhibits all the normal morphometric characteristics of nearby piedmont basins, and it is presumed to be in equilibrium because its morphometric parameters obey the statistical laws discussed above. The geomorphic history of the basin, however, shows that its evolution was anything but orderly (Ritter 1972). Two terraces standing well above the present level of Volney Creek are capped by gravel containing crystalline clasts that had to be derived from the Beartooth Mountains; in fact, upstream tracing of the terraces demonstrates that they cross through the present basin divide and continue mountainward in the neighboring valley of West Red Lodge Creek (see map, fig. 5.18). The significance is that prior to the early Wisconsinan (Bull Lake), the valley now occupied by Volney Creek was the drainage avenue for the master stream of the area. The crystalline-rich gravel capping the terraces indicates that the paleoriver drained from the mountains, and its basinal area must have been vastly greater than that of the modern Volney Creek. Immediately preceding the Bull Lake glaciation, a major diversion of the river placed it in the valley of West Red Lodge Creek, leaving the Volney segment abandoned until it was occupied by the very small modern river.

Piracies of the Volney Creek type are common in piedmont regions and may be important phenomena in the expansion of any drainage system, especially in the early stages (Howard 1971b). Captures, nonetheless, are catastrophic events in basin development because they drastically alter basin prop-

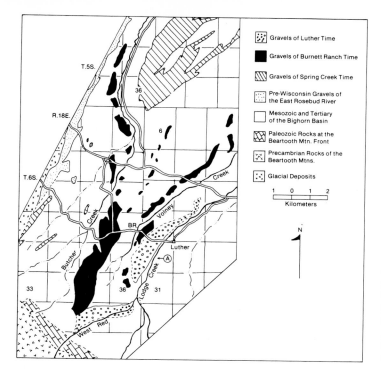

Figure 5.18. Map of terrace
deposits in a small piedmont
area of the Beartooth Mountains, Montana. Major stream
piracy occurred near (A) between deposition of Burnett
Ranch gravels and deposition
of Luther gravels. (From Ritter
1972)

Map legend:
- Gravels of Luther Time
- Gravels of Burnett Ranch Time
- Gravels of Spring Creek Time
- Pre-Wisconsin Gravels of the East Rosebud River
- Mesozoic and Tertiary of the Bighorn Basin
- Paleozoic Rocks at the Beartooth Mtn. Front
- Precambrian Rocks of the Beartooth Mtns.
- Glacial Deposits

erties such as area, relief, and stream length. The internal adjustment to changes
spurred by piracy must occur rapidly, for most basins possess a balanced morphometry even though such spasmodic events must be commonplace.

If the recognition of morphometric balance has dubious value in the historical branch of geomorphology, it is valid to ask what purpose these statistical
analyses serve. Are they merely another method by which we can describe a
basin and its drainage system in sophisticated language, or do they provide students of surficial processes with useful information? Luckily the latter is true, for
without some practical application, basin statistics become no more than
geomorphic curiosities. Perhaps the most useful attribute of morphometry is that
its elements seem to stand in a mediary position between the input and output of
basin water. On one hand, parameters are dependent on the climatic regime that
introduces water to the basin, while on the other, they indirectly control how
water is removed from the basin. In other words, water input helps determine
morphometry and morphometry helps determine river hydrology. In a hydrologic
sense, as figure 5.19 shows, certain morphometric components can exist as both
dependent and independent variables.

Morphometry and Basin Hydrology

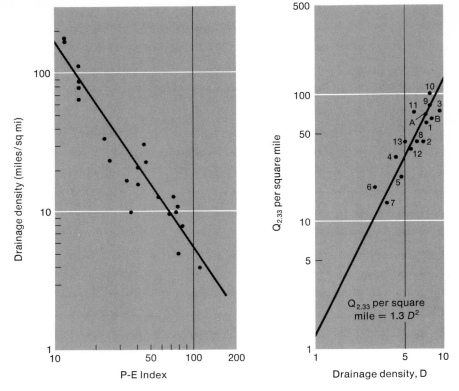

Figure 5.19. A given morphometric component may be both a dependent and an independent variable. (A) Drainage density controlled by precipitation and temperature (P-E Index). (From Melton 1957) (B) Discharge (mean annual flood, Q2.33) controlled by drainage density in 13 basins. (From Carlston 1963)

Statistical analyses have shown that valley-side slopes, drainage density, basin area, and stream length are statistically dependent on climate alone or on some combination of climate and other variables such as vegetation, infiltration capacity, and soil resistance (Melton 1957; Chorley 1957b). For example, mean monthly values of precipitation and temperature have been included in the *precipitation effectiveness index,* P-E (Thornthwaite 1931), a variable that relates in some fashion to all the morphometric factors mentioned above (see fig. 5.19). In other cases (Chorley 1957b), the P-E may be combined with other factors to create different variables such as the *index of climate and vegetation*

$$I_c = \frac{I}{P \times Q_s}$$

where I_c is the index of climate and vegetation, I is the Thornthwaite precipitation effectiveness index, P is precipitation volume, and Q_s is the intensity of precipitation. I_c has been related to several linear and area elements of basin morphometry.

A portion of the precipitation that enters most basins is ultimately transformed into stream flow. The actual percentage of the precipitation that becomes part of a river depends on the amount of water that is consumed by direct evaporation and the amount lost when vegetation transpires water vapor into the atmosphere. Presumably, if all types of water gains and losses could be measured accurately, the accounting of these would constitute the *hydrologic budget*. This can be expressed in general terms (Todd 1959) as

$$I_s + I_g + P + W_i + SS_d + SG_d = O_s + O_g + C + W_e + SS_i + SG_i$$

where I_s is surface inflow, I_g subsurface inflow, P precipitation, W_i imported water, SS_d decrease in surface storage, and SG_d decrease in ground water storage. The terms on the right side of the equation refer to losses of the same types, except for C which is consumptive loss (mainly by evapotranspiration). In reality, a balanced hydrologic budget is very difficult to demonstrate because measurements of evapotranspiration are quite unreliable. The other factors, however, are measurable, and assuming the budget equation is valid, values of evapotranspiration are usually accepted as being the difference needed to balance the budget after everything else has been considered.

A major loss of water from many drainage basins occurs as stream discharge, that is, the volume of water passing a given channel cross section during a specified time interval, or

$$Q = wdv$$

where Q is discharge in ft³/sec(cfs) or m³/sec(cms), w is width, d is depth, and v is velocity. Measurement of discharge is a relatively simple procedure whereby the total width of the channel is divided into evenly spaced segments. The depth and velocity are measured in each compartment, and total discharge is determined by summing the discharges of all the subsections. The difficult measurement to obtain is velocity, which is not equally distributed in a stream (fig. 5.20). Surface velocity is not a good estimate of the mean velocity, and so measurements must be made within the current. Many devices have been developed to measure velocity, but probably the most widely used is the Price current meter. This instrument, designed by W. G. Price in 1882, has a group of conical cups mounted on a vertical rod that is rotated by the force of the water striking the cups. As the shaft spins, it periodically closes an electrical circuit, the moment of closure being recognized by the meter operator as a clicking sound in earphones connected to the circuitry. The number of axial revolutions per unit time is easily converted into velocity values.

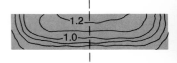

Figure 5.20. Velocity distribution in wide, shallow channel. Numbers represent velocity in meters per second.

To be of any scientific value, discharge must be measured repeatedly at the same locality, giving hydrologists a better understanding of how flow varies with time. In the United States these sampling localities, called *gaging stations*, have been in operation since the late 1800s; therefore, a wealth of flow data for a large number of streams is available from the U.S. Geological Survey and many state

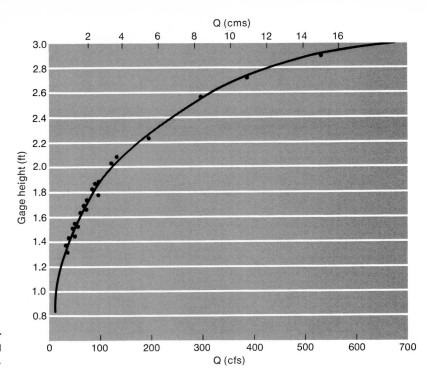

Figure 5.21. Rating curve for low flow, Rock Creek near Red Lodge, Mont.

water surveys. At most stations, discharge is measured each hour and the average of these values published as a *mean daily discharge*. The *mean annual discharge* is the average of the daily values over the entire period of record, presuming that the station has been maintained for longer than one year. Actually, hourly discharge is not measured as described above becase the procedure is too time-consuming. Instead it is estimated from a *rating curve* (fig. 5.21), which relates a wide spectrum of discharge values to the elevation of the river above some datum. Once the rating curve is constructed, the height of the river above the datum (called the *stage* or *gage height*) is the only variable observed directly, and its value is used to predict the discharge.

Geomorphologists are interested in the frequency and magnitude of flow events because each has a decided bearing on how watershed systems work. The frequency of a given discharge is compiled into a *flow duration curve* (fig. 5.22), which relates any discharge value to the percentage of time that it is equalled or exceeded. At any station, then, the lowest daily discharge in the period of record will be equalled or exceeded 100 percent of the time. The largest flow will be equalled only once out of the entire number of days in the sample, giving it a percentage value slightly greater than zero.

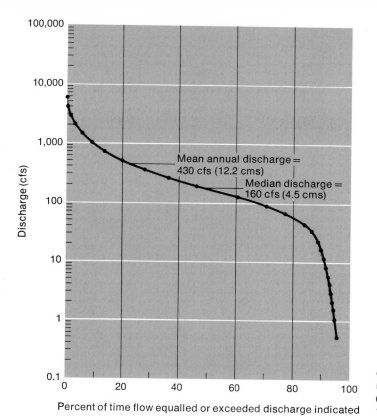

Figure 5.22. Flow duration curve for the Powder River near Arvada, Wyo., 1917-1950. (From Leopold and Maddock 1953)

Another common approach to finding the frequency-magnitude relationship is to consider only the peak discharge during each year (annual series) or only the discharges above some predetermined value (partial duration series). The statistical samples in the series approaches are much smaller than in analyses utilizing daily records; however, they are useful in studies of major flow events. The annual discharges shown in table 5.4 have been ranked according to their magnitude during the years of record, and a recurrence interval for each flow is shown. The recurrence interval, simply the average time between two flow events of equal or larger magnitude, is calculated as

$$R = \frac{n + 1}{m}$$

where R is the recurrence interval in years, n is the total number of discharge values in the sample, and m is the rank of a given flow.

Table 5.4 Annual flood series data, 1932-1963, for Rock Creek near Red Lodge, Mont. Each maximum annual flood is ranked, the highest being ranked 1. Recurrence intervals have been calculated.

| Year | Maximum Flood | | Rank Magnitude | Recurrence Interval |
	cfs	cms		
1932	935	26.5	22	1.45
1934	533	15.1	31	1.03
1935	1490	42.2	9	3.56
1936	1240	35.1	15	2.13
1937	1930	54.6	5	6.4
1938	991	28.0	21	1.52
1939	661	18.7	29	1.10
1940	673	19.0	28	1.14
1941	780	22.1	24	1.33
1942	1840	52.1	6	5.33
1943	2010	56.9	3	10.6
1944	1630	45.1	7	4.57
1945	1990	56.3	4	8
1946	774	21.9	25	1.28
1947	846	23.9	23	1.39
1948	1070	30.3	19	1.68
1949	1190	33.7	17	1.88
1950	1100	31.1	18	1.78
1951	1460	41.3	10	3.2
1952	2590	73.3	2	16.0
1953	1300	36.8	13	2.46
1954	1570	44.4	8	4.0
1955	578	16.4	30	1.06
1956	1430	40.5	11	2.90
1957	3110	88.0	1	32
1958	1250	35.4	14	2.29
1959	1200	34.0	16	2.0
1960	680	19.2	27	1.19
1961	751	21.3	26	1.23
1962	1030	29.1	20	1.60
1963	1350	38.2	12	2.67

A plot of discharge and recurrence intervals on logarithmic graph paper (fig. 5.23) allows hydrologists to estimate the magnitude of a flood *to be expected* within a specified interval of time. On Rock Creek (Mont.), as figure 5.23 shows, the flow should equal or exceed 81 cms (2860 cfs) once during each 25-year interval. As rivers unfortunately do not understand statistical theory, there is no reason to believe that 25-year floods will be evenly distributed over time. In the next fifty years there may be two 25-year floods in successive years, or floods may occur in the first and last years of the period, or a flow equalling or exceeding the discharge of a 25-year flood may not happen at all. Stated differently, a 25-year flood has a 4 percent chance of occurring in any given year; a 10-year flood has a 10 percent chance, and so on. Exactly which year it will happen is simply unpredictable.

Even though it is impossible to predict when a given flow will occur, magnitude-frequency analyses have practical value in river management, especially for lower frequency floods such as the 20-year (q_{20}), 10-year (q_{10}), or 5-year (q_5) floods. Using an annual series, the *mean annual flood* is the flow that should recur once every 2.33 years ($q_{2.33}$) and the *most probable annual flood* once every 1.58 years. The U.S. Geological Survey publishes extensive flood-frequency information on a state and regional basis. Techniques for deriving flood frequency curves from gage data or for estimating maximum probable floods at any site have been described in detail (Dalrymple 1960).

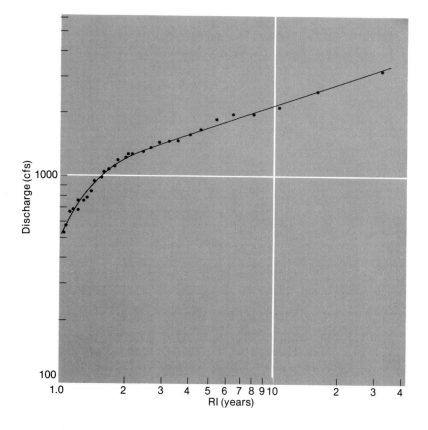

Figure 5.23. Flood frequency curve for Rock Creek near Red Lodge, Mont. Various scales and plotting techniques can be used in constructing a flood frequency curve. For details see Dalrymple 1960.

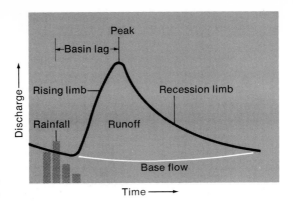

Figure 5.24. Flood hydrograph.

Where discharge is recorded continuously, a flood hydrograph depicting flow as a function of time can be plotted as in figure 5.24. Discharge appears on the hydrograph in the form of *direct runoff* (overland flow plus throughflow) or *base flow* (water infiltrated to the water table and released more slowly from the underground system). Although discharge measured at a station may contain any or all of these components, it should be clear that direct runoff occurs only in response to a storm, while in dry periods all the discharge is base flow. The maximum or peak flow usually develops soon after precipitation ends, separating the hydrograph into two distinct segments, the rising limb and the recession limb. The rising limb in general reflects the input of direct runoff, and the increase in discharge with time is relatively rapid. The recessional phase, however, is controlled more by the gradual depletion of water temporarily stored in the system, and so the decrease in discharge with time is less pronounced. Also the shape of the recession limb is not influenced by the properties of the storm causing the increased discharge but is more closely related to the physical character of the basin. Since a hydrograph represents the sum of hydrographs from all subareas of a basin, if geology and topography are fairly constant throughout, then rainfalls having similar properties should generate hydrographs with the same shape. On this premise, a type hydrograph for a basin, called a *unit hydrograph*, has been developed, in which the runoff volume is adjusted to the same unit value (one for the entire area). The unit hydrograph has been used as a connecting link in many studies attempting to relate basin morphometry to hydrology.

Some hydrologic characteristics of a basin play a significant role in determining the magnitude of the flood peak. The *basin lag,* for example, is the time needed for a unit mass of rain falling on the basin to be discharged from the basin as streamflow. It is usually estimated as the time interval between the centroid of rainfall and the peak of the hydrograph (fig. 5.24) and probably consists of two separate parts: (1) the time involved in overland flow and (2) the channel-transit time. Lag time for any basin is fairly consistent, with only minor variations in the parameter caused by the position of the storm center relative to the gaging site.

Obviously lag should increase with the size of the basin, but comparisons of basins of equal size show that lag time may be as much as three times greater in *sluggish* streams than in those with short lag times, called *flashy* streams. Lag, therefore, must be determined by more than drainage area alone, and since it is presumed to be influenced by the geomorphic framework, it should be related to morphometry in some discernible way.

There is no question that the time elements of basin hydrology have a monumental influence on the magnitude of peak discharge during floods. Take, for example, the progression of the flood crest recorded at several gage stations in the Susquehanna River basin during the flood in June 1972 (fig. 5.25). Most

Figure 5.25. Progress of flood caused by hurricane Agnes in the Susquehanna River basin, June 1972. Basins of increasing size: (A) Bald Eagle Creek; (B) Juniata River; (C) Susquehanna River.

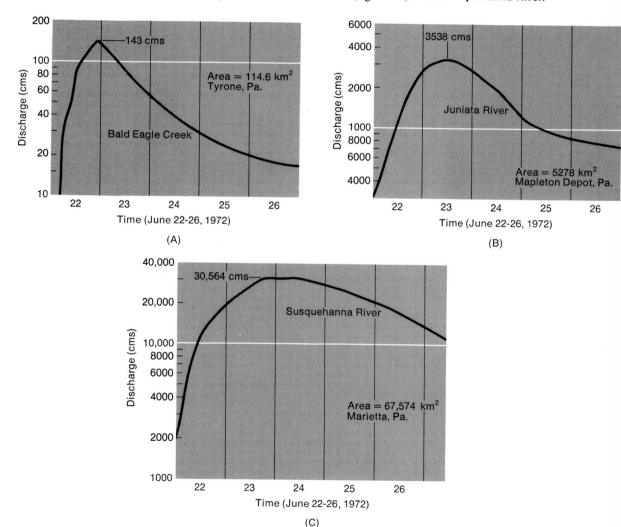

of the precipitation spawned by hurricane Agnes entered the basin during the period between June 19 and June 22. In the minor tributary Bald Eagle Creek, the flow peaked at 143 cms (5050 cfs) on the night of June 22, indicating a relatively quick response to the storm. In the larger Juniata River, the flood crested about 12 hours later with a considerably higher discharge value of 3538 cms (125,000 cfs). Far downstream on the main Susquehanna River, the flood peak did not occur until 12 hours after the Juniata peak, when discharge rose to 30,564 cms (1,080,000 cfs). The fact that discharge peaked later on the main stem, and at a significantly higher magnitude than in the tributaries, shows the importance of timing in basin hydrology.

As suggested above, lag time and peak discharge are positively related to basin size. More important, however, is the fact that simple expansion in size cannot totally explain the observed flow characteristics during a flood. Using the same data as before and replotting the peak discharge as Q/km^2, figure 5.26 reveals the interesting hydrologic property that discharge per unit area is much higher for the smallest tributary than for the massive basin of the main stem. Had the increase in discharge during the 1972 Susquehanna flood been only a function of increased basin area, the peak flow at the Marietta station should have been a simple product of the Q/km^2 value of Bald Eagle Creek times the drainage area above Marietta, or 1.25 cms/km^2 × 67,574 km^2 = 84,468 cms (2,978,454 cfs); but this is almost three times the actual measured value. Somewhere within the basin, a built-in flood control mechanism exists which not only holds down the magnitude of peak Q, but simultaneously maintains abnormally high flow over a longer period of time. Most lowering of the peak flow is due to the ability of floodplains to store or retard the movement of larger volumes of water until the

Figure 5.26. Discharge per unit area during flood caused by hurricane Agnes in the Susquehanna River basin, June 1972. Note that the main river has considerably less discharge/area than tributaries. Main stem also peaks later and discharges flood water over a longer time period, showing the effect of storing water on floodplains.

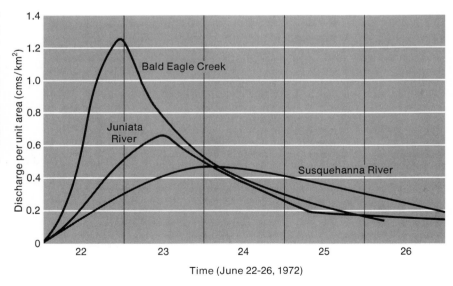

flood crest passes a given channel locality. In addition, some of the water infiltrates into the floodplain alluvium and is gradually released back to the channel as groundwater. Thus, a large portion of the runoff never adds to the increasing peak flow downstream, but shows up in the flood record after the crest has passed as part of the slow recessional phase of the flood.

Morphometric Relationships

We now know that stream hydrology, as defined by the discharge hydrograph and by time elements such as flood frequency and lag, is significantly related to many components of basin and network morphometry. The interdependence of morphometry and hydrology is statistically real but does not necessarily indicate a cause-and-effect relationship—one factor is not the cause of changes in the other. The high correlation probably exists because both factors vary in a consistent way with the same underlying climatic and geologic controls that are the bases for the hydrophysical laws examined before. In general, area and relief factors are closely related to flow magnitude, and length elements to the timing of hydrologic events. All morphometric types, however, are themselves so complexly woven together that no one factor can be isolated as a completely independent variable.

Since basin area and peak discharge are highly correlative, we should expect that many other areal parameters would be similarly related to discharge. In fact, every factor involving area differs in its success as a predictor of discharge, but one parameter, drainage density, seems to have considerable value as a gage of peak flow. In a study of 15 small basins in the southern and central Appalachians and the Interior Low Plateau region, Carlston (1963) demonstrated a very close relationship between drainage density and the mean annual flood (as was shown in fig. 5.19). Notably, the basins in his sample have wide variations in relief, valley-side and channel slopes, and precipitation characteristics, yet none of these factors seems to disrupt the flood magnitude-drainage density relationship. Carlston suggests that the general capacity of a terrain to infiltrate precipitated water and transmit it through the underground system is the prime controlling factor of the density-mean annual flood relationship in basins up to 260 km^2 in area. In larger basins, channel transit time plays the dominant role in the flow character. The rate of base flow, found to be inversely related to drainage density, is also dependent on the terrain transmissibility. Thus, as Horton suspected earlier, high transmissibility (as evidenced by infiltration capacity) spawns low drainage density, high base flow, and a resultant low-magnitude peak flood. In contrast, an impermeable surface will generate high drainage density in order to efficiently carry away the abundant runoff; base flow will be low and peak discharge high.

Linear morphometry and channel hydrology also relate in significant ways. The relative values of total stream length in basins and the length of overland flow exert some influence on lag time. This is not surprising, because both these factors are statistically related to drainage density, and high-density basins with efficient removal of runoff tend to reduce the lag. In addition, any increase in the total distance that water must travel before it is expelled from the basin will increase the lag and make the rise in peak discharge less than expected (Taylor and Schwarz 1952). Even though total discharge may increase with basin size, stream order, or the length of the master stream, the lag time should still increase. Importantly, the total area of available storage space in floodplains within a basin also increases with area, length, and order. Thus, although absolute values of discharge increase with the length parameters, they do not rise as much as they could because the increased storage holds back a large portion of the flow until the flood crest has passed.

Relief factors relate to hydrology mainly through slope angles, including both valley-side slopes and channel slopes. As we would expect, steep slopes deliver water rapidly to any gage station, thereby decreasing lag time and increasing the peak discharge at any given frequency of flow. That is, basins with high relief characteristics tend to have flashy streams and high-magnitude floods, while basins with low relief and concomitant gentle slopes have greater lag and retarded peak discharges. These general rules vary with geology and climate.

The possible interrelationships between hydrology and morphometry are seemingly infinite. Attempts have been made to derive comprehensive formulae to predict peak discharges by using multiple correlation analyses of climatic and geomorphic parameters (Potter 1953; Morisawa 1962; Benson 1962). Although these are reasonably successful within a well-defined physical setting, the variables are so complexly related that no one equation will fit all climatic or geologic situations. The approach, however, may be valid and should be considered as a possible prototype for future analyses.

Basin Denudation

In addition to being hydrologic entities, basins are also geographic compartments where sediment is manufactured, eroded, and deposited, and from which, given sufficient time, the debris will ultimately be removed. The amount of sediment leaving a basin can be readily converted into an estimate of lowering of the basin surface, called *denudation,* which is usually expressed as a time parameter or rate. Denudation seems to have no rigorous definition, but because it implies removal of basin material, it is commonly used as a synonym for erosion. The two differ, however, because denudation considers only those eroded products that are removed completely from the basin, assuming in that consideration that the sediment is derived in equal portions from all subareas of the watershed. It therefore presupposes equal surface lowering over the entire basin. Denudation rates tell us little about those erosive processes that simply redistribute sediment *within* the basin; nor do they indicate that at any given time some parts of the basin are probably aggrading rather than eroding. Denudation, then, is the long-term sum of the overall erosive process, and even though it is analogous to erosion it is not precisely the same.

Estimates of modern denudation are usually based on measurements of stream load made at gaging stations or on the volume of reservoir space lost when sediment accumulates behind a dam. All types of load (suspended, bed, and dissolved) are included in the analyses at gage stations, which require that the weight values of load be converted into volumetric terms. Once the volume of sediment leaving the basin is determined, it is divided by the area of the watershed above the gage station to provide the third (or vertical) dimension, which represents the magnitude of surface lowering (for details see Ritter 1967). Rates are commonly expressed in inches or centimeters per 1,000 years.

Factors of Denudation

Denudation is controlled by a number of interrelated geologic, hydrologic, and topographic factors which, combined in different ways, cause the magnitude of denudation to vary widely from region to region. Nonetheless, some generalizations can be made concerning those elements that seem to exert the greatest influence on denudation rates.

Precipitation and Vegetation   Intuitively we would expect the amount of sediment yielded from any basin to be related in some systematic fashion to the amount of incoming precipitation. Actually the correlation between the two is not so direct as we might hope, because precipitation is complexly interrelated with other factors that influence its erosive capability. The amount of runoff from any given precipitation, for example, varies with temperature. Vegetation, a function of both precipitation and temperature, serves as a protective screen against the erosion of surface material. Precipitation, then, cannot be considered as a completely independent variable in the realm of denudation, even though it may be a dominant factor.

In an important paper, Langbein and Schumm (1958) documented the relationship between sediment yielded from basins averaging 3900 km² in area and effective precipitation, a parameter derived by adjusting the magnitude of precipitation to values expected at a mean annual temperature of 10°C (50°F). Under those conditions, they were able to show that as precipitation rises from zero, the sediment yield increases rapidly to a maximum yield value at about 30 cm of effective precipitation (fig. 5.27). Any increase in precipitation above 30 cm promotes a decline in sediment yield because the density and type of vegetation begin to play an active role in protecting the slopes from erosion. Vegetation generally exerts a control on erosion when the ground cover is between 8 percent and 60 percent, values typical in semiarid and subhumid climates. Thus, as L. Wilson (1973) suggests, the Langbein-Schumm curve may be valid for regions with a continental climate but may not be applicable in other climatic regimes, especially nonseasonal types. In any case, the demonstration by Langbein and Schumm that the relationship between precipitation and sediment yield is nonlinear and very complex seems to be a valid geomorphic obser-

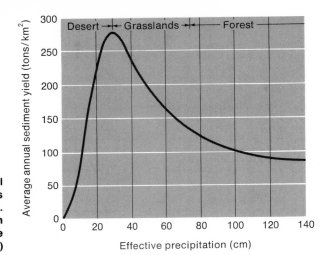

Figure 5.27. Average annual sediment yield as it varies with effective precipitation. (After Langbein and Schumm 1958, with permission of the American Geophysical Union)

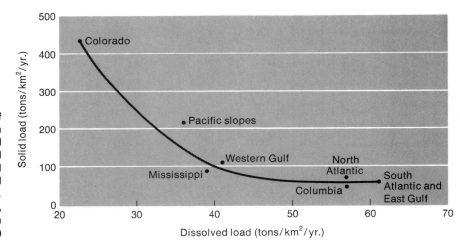

Figure 5.28. Relationship of dissolved and solid load in various parts of the United States. (After Judson and Ritter 1964, Rates of regional denudation in the United States, *Jour. Geophys. Research* 69:3395-401, copyrighted by American Geophysical Union)

vation. Precisely at what precipitation the maximum values of sediment yield will occur depends on the specific climatic setting (for example, see Fournier 1960), and total values may relate more to the seasonality of the climate than to the mean annual precipitation (L. Wilson 1973). Some studies tend to support that contention (Corbel 1959; L. Wilson 1972; Jansen and Painter 1974). It should also be noted that variations in sediment yield under different climates may be partially offset by an increase or decrease in dissolved load (fig. 5.28). Normally, dissolved load will increase regularly with precipitation, but maximum solution is probably reached at about 63 cm of annual precipitation, with little additional dissolution resulting from higher precipitation (Leopold et al. 1964).

Basin Size   All other factors being equal, sediment yield decreases as the size of the drainage basin expands (Brune 1948), as evidenced by remarkably high denudation rates in very small basins of the midwestern United States (see Schumm 1963c). The explanation for this phenomenon seems to lie in several topographic realities: (1) small basins generally have steep valley-side slopes and high-gradient stream channels which efficiently transport sediment; (2) in basins filled to capacity with streams, the drainage density always remains high near the basin divide, but it may decrease with time in the central part of the basin; (3) floodplain area increases as the basin expands, especially in the central and lower reaches of the basin (Hadley and Schumm 1961).

The integration of these factors leads us to realize that in natural basins most sediment is produced in the small headward subareas where it is quickly removed, but during its downstream transit a significant portion may be stored in the floodplain system. How long it will remain in storage depends on the rigor of the geomorphic and hydrologic processes. Presumably, sampling over a long enough time would show that sediment stored within the basin is flushed rapidly from the system during episodes of rejuvenation. Inclusion of such spasmodic bursts of high sediment yield in a long-term sample would tend to temper the variations in denudation that appear to be related to basin size.

Elevation and Relief   Mountainous terrains with excessive elevation and relief are known to produce abnormally high sediment yields, particularly where rocks are nonresistant (Corbel 1959; Hadley and Schumm 1961; Schumm 1963c; Ahnert 1970) or affected by recent or current tectonism (Li 1976). For basins at least 3900 km^2 in area, the greatest denudation rates average about 0.9 m/1,000 years and probably occur in mountain belts where relief and elevation are greatest and rocks are erodible (Schumm 1963c).

In the arid climate of the western United States, sediment yields are a function of the relief-length ratio (fig. 5.29). Utilizing this fact and holding area constant, Schumm (1963c) showed that the relationship between relief and denudation is definable in quantitative terms. It has also been shown that elevation alone produces disparate denudation rates, since low-lying areas in any climatic regime yield less sediment than higher basins of comparable size (Corbel 1959).

Significantly, relief and elevation analyses demonstrate clearly that denudation rates are not constant through time. As relief and elevation of a basin are gradually diminished during its evolution, the rate of surface lowering decreases proportionately, and each successive interval of stripping requires a longer period of time (fig. 5.30).

Rock Type   With similar climate and topography, basins underlain by clastic sedimentary rocks and low-rank metamorphics usually produce abundant suspended loads and so are characterized by higher rates of denudation than regions of crystalline rocks or highly soluble sedimentary rocks (Corbel 1964). The rela-

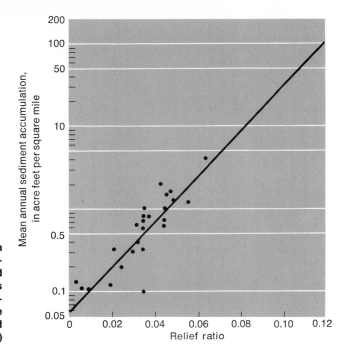

Figure 5.29. Relation between mean annual sediment accumulation in reservoirs and the relief ratio for basins located on the Fort Union formation in the upper Cheyenne River basin. (From Hadley and Schumm 1961, fig. 31)

Figure 5.30. Relation of denudation rates to relief-length ratio and drainage-basin relief. Denudation rates are adjusted to drainage areas of 1,500 square miles. Curve is based on the average maximum denudation rate of 3 feet per 1,000 years when relief-length ratio is 0.05. (From Schumm 1963)

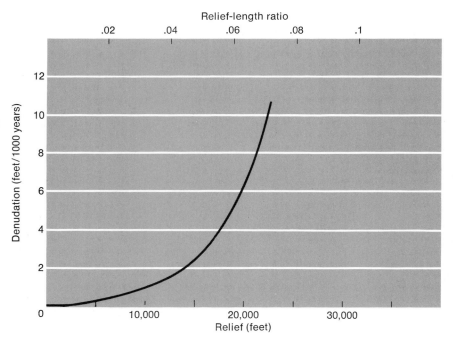

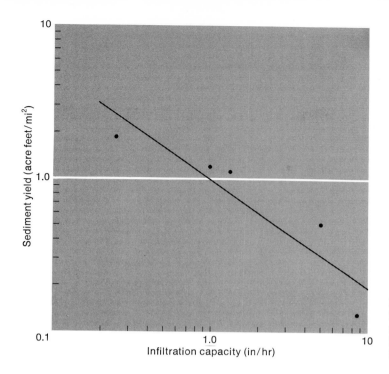

Figure 5.31. Relation between infiltration capacity and sediment yield, indicating lithologic influence. (Data from Hadley and Schumm 1961)

tionship of lithology and denudation is poorly understood in a quantitative sense and may be obscured by other rock characteristics, such as fracturing, which cause different lithologic units to behave similarly with respect to denudation. Even so, properties that are commonly a reflection of lithology, such as the infiltration capacity, seem to be systematically related to rates of denudation (fig. 5.31), indicating that the lithologic influence is real. At this time, however, geomorphologists have not been able adequately to reveal its fundamental character.

The Effect of Human Activity   Most estimates of denudation are significantly inflated where human activity disturbs the natural setting. Exactly how much humans accelerate erosion varies with the type of activity and the particular environment, but it seems reasonable to propose that human interference has the potential to alter drastically the natural denudation rate. Douglas (1967) suggests that the abnormalities induced by human occupation may be great enough to invalidate the practice of applying present rates in studies of long-term landscape evolution. Evidence also indicates that human activities may increase detrital loads by at least an order of magnitude (Judson 1968a; Meade 1969); chemical loads are expanded by pollutants introduced into streams or the atmosphere (Meade 1969). One important contributor to accelerated erosion is the replacement of mature forest cover by intensely cultivated land. Studies of this practice

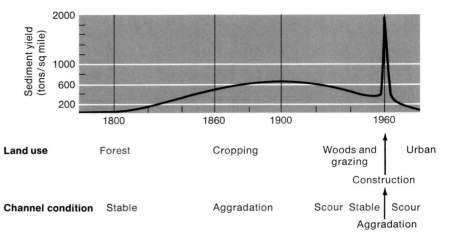

Figure 5.32. Changes in sediment yield and channel behavior in one area under various types of land use. (After Wolman 1967. Used with permission of *Geografiska Annaler*)

in the United States show an increase in sediment yield of 1 to 3 orders of magnitude when cropping is substituted for the natural vegetation (Ursic and Dendy 1965; Wolman 1967). Construction associated with urbanization causes an even more dramatic rise in the sediment yield (fig. 5.32), but after construction is completed the values decrease rapidly because much of the surface is protected from erosion by our concrete citadels (Wolman 1967). Urbanization also affects the runoff characteristics within a drainage basin (McPherson 1974); consequently, the concentration of sediment in streams, as influenced by humans, depends on variations in both hydrology and sediment yield.

In summary, rates of denudation reflect an integrated balance of many competing variables. In general, sediment yield increases with increasing runoff, altitude and relief, precipitation, temperature, and rock softness. Sediment yield decreases with increasing vegetation and area (Jansen and Painter 1974). Any of these factors could dominate in a particular situation, and all could be overwhelmed by the tremendous effect of human actions.

Present and Past Rates

Table 5.5 presents a random sample of modern denudation rates for basins of varying size. The values are imprecise and probably high because in most cases they do not include an adjustment for human impact. Nonetheless, they do indicate the general tendency for denudation rates to fall between 2.5 and 15 cm per 1,000 years when considered on a regional scale. Judson (1968b) recalculated the denudation rate for the entire continental United States by subtracting the effect of human occupancy from earlier estimates. His figure of 3 cm/1,000 yrs. agrees rather well with the denudation in very large drainage basins that are mostly unaffected by humans (Gibbs 1967) and perhaps represents a reasonable approximation for denudation on a continental scale.

Table 5.5 The influence of geology and climate on suspended-load denudation in basins of different size in the United States.

Basin	Location	Area (mi²)	Average Annual Suspended Load (tons × 10³)	Denudation (in/1000 yrs.)
Mississippi	Baton Rouge, La.	1,243,500	305,000	1.3
Colorado	Grand Canyon, Ariz.	137,800	149,000	5.6
Columbia	Pasco, Wash.	102,600	10,300	0.5
Rio Grande	San Acacia, N.M.	26,770	9,420	1.8
Sacramento	Sacramento, Cal.	27,500	2,580	0.5
Alabama	Claiborne, Ala.	22,000	2,130	0.5
Delaware	Trenton, N.J.	6,780	998	0.8
Yadkin	Yadkin College, N.C.	2,280	808	1.8
Eel	Scotia, Cal.	3,113	18,200	30.4
Rio Hondo	Roswell, N.M.	947	545	3.0
Green	Palmer, Wash.	230	71	1.6
Alameda	Niles, Cal.	633	221	1.8
Scantic	Broad Brook, Conn.	98	7	0.4
Napa	St. Helena, Cal.	81	63	4.1

Data from Judson and Ritter 1964.

Although methods other than analyses of sediment wedges have been employed (Eardly 1967; Ruxton and McDougall 1967; Clark and Jager 1969), most past rates of denudation are estimated from measurements of sediment accumulation in depositional basins. A valid estimate can be made only if (1) the volume of sediment derived by erosive processes can be accurately determined, (2) the boundaries of the source area are definable, and (3) the time interval of sediment accumulation can be ascertained within reasonable limits. It is very difficult to meet all these requirements: noneroded matter such as pelagic and volcanic rocks add to the depositional volume; material eroded from the basin as dissolved load may not be returned to the deposit by chemical precipitation; and absolute dates that bracket the time of deposition are necessarily imprecise. Nonetheless, some estimates of past rates have been made for large portions of North America (Gilluly 1949, 1955, 1964; Menard 1961). It is interesting to note that the rates found in these studies of large regions are within the same order of magnitude as those based on modern stream data. Such similarity prompted the hypothesis (Ritter 1967) that when viewed on a large enough area or over a long time interval, denudation rates will probably be about the same. Based on current evidence, the average value will probably fall somewhere between 2.5 cm and 15 cm per 1,000 years.

Summary

In this chapter we examined a remarkable statistical balance among the spatial characteristics of river networks and the watersheds that contain them. Because the parameters of this morphometry also relate in a significant way to the hydrologic and erosional properties of most watersheds, drainage basins serve as primary units for systematic analyses of geomorphology. Drainage basins and their river networks probably evolve according to fundamental hydrophysical laws, but their ultimate character is conditioned by the geological framework and the external constraints of climate. An equilibrium condition, defined in terms of mathematical balance, is probably attained early in the growth history of most basins. This does not indicate, however, that basins evolve in an orderly way with time. Geologic catastrophes that upset equilibrium tend to be filtered out in a morphometric sense because basinal parameters apparently adjust to changes rather quickly.

The quantitative description of drainage basins has great potential for applied geomorphic studies. Morphometry leads to rapid estimates of river discharge and sediment-yielding properties of the basin. Future refinements of technique and data collection should make morphometry a powerful predictive tool, singularly important in regional planning and development. In addition, as we will see in chapter 6, discharge and sediment are the major controlling factors of river behavior.

Suggested Readings

The following references provide greater detail about the concepts discussed in this chapter.

Chorley, R. J. 1969. *Water, earth and man*. London: Methuen and Co.

Doornkamp, J. C., and King, C.A.M. 1971. *Numerical analysis in geomorphology: An introduction*. London: Edward Arnold Ltd.

Horton, R. E. 1945. Erosional development of streams and their drainage basins: Hydrophysical approach to quantitative morphology. *Geol. Soc. America Bull*. 56 275-370.

Judson, S. 1968. Erosion of the land. *Am. Scientist* 56: 356-74.

Langbein, W. B., and Schumm, S. A. 1958. Yield of sediment in relation to mean annual precipitation. *Am. Geophys. Union Trans*. 39: 1076-84.

Schumm, S. A. 1956. Evolution of drainage systems and slopes in badlands at Perth Amboy, New Jersey. *Geol. Soc. America Bull*. 67: 597-646.

Strahler, A. N. 1968. Quantitative geomorphology. In *Encyclopedia of geomorphology*, edited by R. Fairbridge, pp. 898-912. New York: Reinhold Book Corp.

It seems fair to say that fluvial action is the single most important geomorphic agent. Although other surficial processes are significant, streams are so ubiquitous that their influence in geomorphology can hardly be overestimated. As discussed in chapter 1, most geomorphic analyses of stream processes in the early twentieth century were based primarily on logic and qualitative observation. The few quantitative studies of fluvial mechanics were somehow lost in the wave of geomorphology that used fluvially produced landforms to reconstruct history. Geomorphology had reached the unenviable state of having a number of basic concepts concerning stream processes and fluvial landforms that were based on a woefully incomplete understanding of how rivers actually work. Since then, there has been a rebirth of interest in the details of fluvial mechanics, and data are now accumulating faster than one can assimilate their meaning. This explosion of information is undoubtedly the key to the advancement of geomorphic thinking, for without it our basic tenets will never be critically evaluated. In this chapter we will examine briefly much of the current thinking about streams; those interested in greater detail than space allows here are referred to a number of excellent books about the fluvial realm: Leopold et al. 1964; Leliavsky 1966; Morisawa 1968; Raudkivi 1967; Chorley 1969a; Schumm 1972; Gregory and Walling 1973; and others.

Fluvial Processes 6

The River Channel

Basic Mechanics

Let us begin with what happens to a constant discharge of water flowing in a long, steeply inclined channel that has no tributaries. Gravity tends to accelerate the flow downstream continuously unless the increase in velocity is moderated by friction within the water and turbulence generated along the channel perimeter. If these resisting elements were constant and less than the gravitional force, acceleration would continue along the entire length of the channel. In natural rivers, however, the intensity of resistance is not constant but increases with the flow velocity. At any point in a river, therefore, the velocity represents the balance between the energy causing flow and the energy consumed by resistance to flow.

Resistance is spread through the water by viscous or turbulent processes and is proportional to the velocity or its square. In *laminar flow*, internal friction is the dominant resisting force, and its intensity is proportional to the velocity of flow. Because the change in velocity with depth is linear, resistance increases uniformly from the water surface to the bed, the rate of increase being determined by the molecular viscosity of the fluid. Most resistance in laminar flow results from intermolecular viscous forces as layers or particles of the fluid slide smoothly past one another (Leopold et al. 1964). The controls on viscosity are internal characteristics of the fluid such as temperature and suspended sediment concentration.

In *turbulent flow*, the water does not move in parallel layers; its velocity fluctuates continuously in all directions within the fluid. Water repeatedly interchanges between neighboring zones of flow, and shear stress is transmitted across layer boundaries in another form of viscosity, called *eddy viscosity*. Eddy viscosity greatly increases the dissipation of energy and the flow resistance. In turbulent flow, in contrast to laminar flow, resistance is proportional to the square of the velocity. Resistance and velocity, therefore, do not change uniformly with depth. In addition, because turbulence is generated along the channel boundaries, most resistance in this type of flow is due to external factors such as the channel configuration and the size of the bed material.

As depth and velocity increase, the conditions at which laminar flow changes to turbulent can be predicted by a dimensionless parameter called the *Reynolds number (Re):*

$$Re = \frac{VR\rho}{\mu}$$

where V is the mean velocity, R the hydraulic radius, ρ the density, and μ the molecular viscosity. The hydraulic radius R is determined by the relationship

$$R = \frac{A}{P}$$

where A is the cross-sectional area of the channel and P the wetted perimeter (fig. 6.1). In wide, shallow channels the hydraulic radius closely approximates the mean depth.

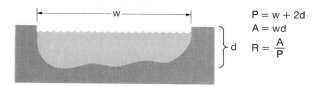

$P = w + 2d$
$A = wd$
$R = \dfrac{A}{P}$

Figure 6.1. Cross-sectional measurements of a stream channel: w = width; d = depth; A = area; R = hydraulic radius; P = distance along wetted perimeter.

Since the factor $\dfrac{\mu}{\rho}$ defines a fluid property called *kinematic viscosity* (ν), the Reynolds number is recognized as a ratio between driving and resisting forces because

$$Re = \frac{VR\rho}{\mu} = \frac{VR}{\nu} = \frac{\text{driving force}}{\text{resisting force}}.$$

In normal situations true laminar flow occurs where Re values are less than 500, and well-defined turbulent flow when Re is greater than 750.

Another dimensionless number used to describe the conditions of flow is the Froude number (F_r):

$$F_r = \frac{V}{\sqrt{Dg}}$$

where D is depth and g is gravity. The Froude number is important because it can be used to distinguish subtypes of turbulent flow called *tranquil flow* ($F_r < 1$), *critical flow* ($F_r = 1$), and *rapid flow* ($F_r > 1$).

Flow Equations and Resisting Factors

Flow and resistance have been the concern of hydraulic engineers for centuries, and a number of equations have been derived to express the relationships between the factors. Two equations of great importance to students of rivers are the Chezy equation and the Manning equation, both of which have been employed in a variety of fluvial investigations. The Chezy equation

$$V = C \sqrt{RS}$$

derived in 1769, shows that velocity is directly proportional to the square root of the RS product, where S is the slope of the channel. The Chezy coefficient (C) is a constant of proportionality that relates to resisting factors in the system.

The Manning equation originated in 1889 from an attempt by Manning to systematize the existing data into a useful form. The equation

$$V = \frac{1.49}{n} R^{2/3} S^{1/2}$$

is similar to the Chezy formula in that R and S are proportional to the velocity. In addition, the factor n, called the Manning roughness coefficient, is also a resisting element that is closely related to the Chezy coefficient because as

$$C(RS)^{1/2} = \left(\frac{1.49}{n}\right) R^{2/3} S^{1/2}$$

$$\text{then} \quad C = 1.49 \left(\frac{R^{1/6}}{n} \right)$$

Manning's n is presumed to be a constant for any particular channel framework; consequently it has been used extensively in analyses of river mechanics (table 6.1). The U.S. Geological Survey, for example, has developed a visual guide for rapid estimation of Manning's n (Barnes 1968).

Although resistance coefficients are defined by hydraulic characteristics (S, D, V), they are not independent of other factors because in alluvial channels their values vary with particle size, sediment concentration, and bottom configuration. For example, in the same reach of a river, different flow conditions (as determined by the Froude number) may mold bottom sediment into a variety of bed forms (fig. 6.2), which in turn have a decided influence on the value of Manning's n (table 6.2). Irregularity of the channel bottom due to other factors such as bars, riffles, and bends may have an equally important influence on the roughness (Simons and Richardson 1966; Leopold et al. 1964); particle size of the bed material also causes variation of roughness values. Wolman (1955) demonstrated this last fact in the Brandywine Creek (Pennsylvania), where the resistance-particle size relationship is defined by the equation

$$\frac{1}{\sqrt{F}} = 2 \log \frac{d}{D_{84}} + 1.0$$

Table 6.1 Manning roughness coefficients (n) for different boundary types.

Boundary	Manning roughness $n, ft^{1/6}$
Very smooth surfaces such as glass, plastic, or brass	0.010
Very smooth concrete and planed timber	0.011
Smooth concrete	0.012
Ordinary concrete lining	0.013
Good wood	0.014
Vitrified clay	0.015
Shot concrete, untroweled, and earth channels in best condition	0.017
Straight unlined earth canals in good condition	0.020
Rivers and earth canals in fair condition—some growth	0.025
Winding natural streams and canals in poor condition—considerable moss growth	0.035
Mountain streams with rocky beds and rivers with variable sections and some vegetation along banks	0.040–0.050

(A) Typical ripple pattern, F << 1

(E) Plane bed, F < 1 and d < 0.4mm

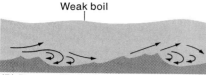
(B) Dunes with ripples superposed F << 1

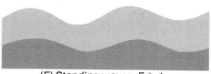
(F) Standing waves, F ≥ 1

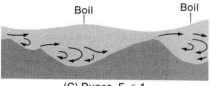
(C) Dunes, F < 1

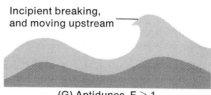
(G) Antidunes, F ≥ 1

(D) Washed-out dunes or transition F < 1

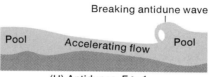
(H) Antidunes, F > 1

Figure 6.2. Bed forms in alluvial channels and their relation to flow conditions. F = Froude number, d = depth. (From Simons and Richardson 1963. Used with permission of the American Society of Civil Engineers)

Table 6.2 Variation of Manning *n* values with changes in bed form occurring under different flow conditions.

Bed Form	Manning n (ft $^{1/6}$)
Lower regime	
Ripples	0.017–0.028
Dunes	0.018–0.035
Washed-out dunes or transition	0.014–0.024
Upper regime	
Plane bed	0.011–0.015
Standing waves	0.012–0.016
Antidunes	0.012–0.020

From *Handbook of Applied Hydrology* by Ven T. Chow. Copyright © 1964 by McGraw-Hill Inc. Used with permission of McGraw-Hill Book Company.

where F is another resistance parameter called the Darcy-Weisbach resistance coefficient, d is water depth, and D_{84} is the particle diameter which is equal to or larger than 84 percent of the clasts on the channel bottom. Increased particle size should produce a correlative increase in flow resistance.

Clearly, external boundary conditions such as channel shape and particle size generate a large amount of resistance. Some of these factors produce turbulence in the form of eddies and secondary circulation. In contrast, sediment concentration (the amount of sediment per unit volume of water) affects resistance internally. This modification was first detailed by Vanoni (1941, 1946), who showed that an increase in the concentration of suspended sediment tends to lower the resistance (fig. 6.3). As the concentration increases, the turbulent effect presumably is reduced because the mixing process within the fluid is dampened. All other factors being equal, sediment-laden water should flow at a higher velocity than clear water.

The ability of a river to do geomorphic work represents a balance between driving and resisting forces. It depends on how much potential energy is provided to the flow and how much of that energy is consumed in the system by the various resisting elements. The resistance generated along the channel perimeter or within the flow is controlled by the shape of the channel, the size and concentration of sediment, and the total volume and physical properties of the water.

Our understanding of river mechanics is hampered by the difficulty of obtaining measurements in a natural channel. Much of our knowledge of fluvial mechanics has therefore been derived from studies in flumes, where slope and velocity can be varied over a wide range of values and any variable can be held constant. Unfortunately, flumes have a limited range of discharge and depth, and although the application of flume results to real systems gives an excellent first approximation, it cannot take into account the complexities of the many ever-changing and interdependent variables in natural streams (see Maddock 1969). We still need much more data about natural rivers before we can hope to utilize our knowledge in a predictive way.

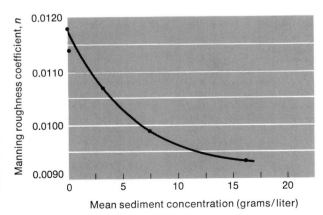

Figure 6.3. Effect of suspended load concentration on the Manning roughness coefficient n. (From Vanoni 1946, fig. 24. Used with permission of the American Society of Civil Engineers)

Although most energy in a stream is dissipated by turbulence, a small part is used in the important task of eroding and transporting sediment. These processes, often taken for granted, are extremely complex and poorly understood, yet they underlie some of our most basic geological concepts. We will briefly review the more significant ideas about the relationship between river flow and sediment.

Transportation

In general, fine-grained sediment (silt and clay) is transported suspended in the water by the supporting action of turbulence. Suspended load usually moves at a velocity slightly lower than that of the water and may travel directly from place of erosion to points far downstream without intermittent stages of deposition. Coarse particles may also travel in true suspension (Francis 1973), but they are likely to be deposited more quickly and stored temporarily or semipermanently within the channel. Except for short spasms of suspension, coarse sediment usually travels as bedload; its episodic journey may include movements close to the bed surface such as rolling, sliding, or bouncing. How long coarse debris remains stationary within a channel depends on the size of the debris and the flow characteristics of the river; such debris probably is immobile more than it is in motion. Bradley (1970) showed that gravel can be stored in channel bars long enough for weathering to drastically weaken its resistance to abrasion.

Because of fluctuating discharge, at any given time a single particle may be part of either the bedload or the suspended load. As this makes the distinction between the two load types unclear, other terms have been devised to relate sediment more appropriately to river flow. *Wash load* consists of particles so small that they are essentially absent on the stream bed. In contrast, *bed material load* is composed of particle sizes that are found in abundant amounts on the stream bed (Colby 1963). While most, if not all, bedload (sediment transported along the bottom) is bed material load, most bed material load is transported as suspended load. Besides the difficulties of measuring load and the uneven distribution of sediment in the flow (Colby 1963), constraints exerted from outside the channel further cloud our understanding of transport mechanics. For example, the relationship between wash load and discharge may seem puzzling because most streams at any flow can carry more fine-grained sediment than they actually do. In fact, the concentration of fines is a function of supply rather than transporting power; therefore, it is relatively independent of flow characteristics. Coarse sediment, on the other hand, is usually available in amounts greater than a stream can carry, and so its concentration should correlate more significantly with the parameters of flow such as depth and velocity. Theoretically then, bedload or large bed material load should relate more simply to discharge than does wash load.

Attempts to calculate the discharge of bedload and bed material load have been made, and some equations are available to predict these quantities (Meyer-Peter and Muller 1948; Einstein 1950). The equations seem to require such precision, however, that small errors in measurement cause large variations in the computed values of sediment discharge. Furthermore, the rate of sediment transport may be related to the entire size distribution rather than to the single, arbitrarily defined size that is usually employed in discharge equations. Bagnold (1967), for example, suggests the possibility that 95 percent of the total work rate in sediment transport may be attributed to moving the coarsest sediment, even though this material may constitute only a minor percentage by weight of the whole load. In view of these difficulties, it seems that bed material discharge might be more simply and accurately related to a single flow variable such as mean velocity, as is done in figure 6.4.

Entrainment

The procedure of measuring load, although complicated in itself, may be much simpler than obtaining a rational explanation for what causes the variations in sediment discharge under similar flow conditions. Part of the reason probably is

Figure 6.4. The relation of bed material discharge to mean velocity; Niobrara River near Cody, Neb. (From Colby 1963)

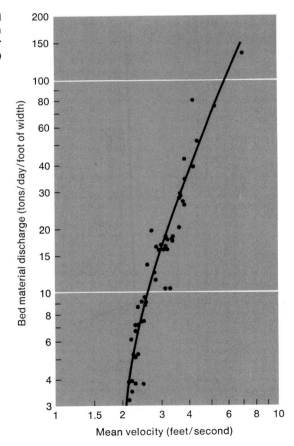

that additional sediment is being actively added to the transported load by the processes that initiate motion of bed particles, collectively known as *entrainment*. The amount of sediment entrained depends directly both on the erosive power of the flow and on the size of particles on the bed surface that are in the proper position to be eroded. Two streams with identical flow conditions may have different bed material discharge if one flows across a fine-sand bottom and the other over a cobble bed.

Generally the term *competence* refers to the size of the largest particle a stream can entrain under any given set of hydraulic conditions. It should be evident that the value of competence to geomorphology depends on how we measure the sediment being moved and, more importantly, how accurately we can determine the flow conditions. Although this may seem simple enough, in practice it is an excruciating problem for several reasons: (1) particles are entrained by a combination of fluvial forces including direct impact of the water, drag, and hydraulic lift, and each of these may be best defined by a different parameter of flow; (2) flow itself is a complex mixture of variables that are neither constant nor easily measured, especially during high discharge; and (3) sediment of the same size may be packed together differently or have shape properties that cause abnormal responses to the same flow conditions. Thus, any investigation into the mechanics of competence must settle for only partial success until we can eliminate or inhibit some of the inherent variability.

Historically, two hydraulic factors have been utilized to represent the flow condition in the competence relationship. The first, *critical bed velocity*, is used to demonstrate the obvious fact that the eroding and transporting power of a river expands rapidly with an increase in velocity at the channel bed. It has been long known (Rubey 1938) that the volume or weight of the largest particle moved in a stream varies as the sixth power of the velocity. The sixth-power law provides a sound theoretical basis for competence studies, but it is less satisfying in practice because accurate measurement of bed velocity is exasperating—if not impossible—in high-energy streams. Mean velocity, an easily determined substitute in competence work, may not be reliable unless the depth is also considered in the analysis (Vanoni et al. 1966).

The second factor, *critical shear stress* (sometimes called tractive force), signifies the downslope component of the fluid weight exerted on a bed particle (fig. 6.5). It is proportional to the depth-slope product and can be expressed by the DuBoys equation for boundary shear:

$$\tau_c = \gamma\, RS$$

where τ_c is the critical shear stress, γ the specific weight of the water, R the hydraulic radius, and S the slope. In most streams transporting coarse bedload, R is closely approximated by mean depth, and the two commonly are interchanged in the equation. The use of critical shear stress in competence studies has been criticized (Yang 1973), but the simple reality that depth and slope in a river are easier to measure than bed velocity makes it an appealing parameter.

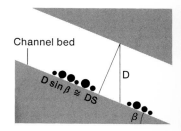

Figure 6.5. Component of flow weight exerted as shear stress on a channel bottom. The critical shear stress is the depth-slope product (DS) multiplied by the specific weight of the water and β is the angle of slope.

Precisely how the two methods correlate with one another is not completely understood, but Rubey (1938) presented evidence to suggest that in the size range between fine sand and pebbles, critical bed velocity becomes more important in the entrainment process as particle size increases (fig. 6.6). Smaller sizes move as a function of the *DS* product and seem to be relatively independent of velocity. Thus, the shear stress approach may be completely valid only for smaller sizes or low-velocity flows, and very fine-grained sediment requires higher velocities for its entrainment than the sixth-power law would predict. Wolman and Brush (1961) found similar trends in a later flume study of the movement of sand-sized debris.

The above observations fit rather nicely the curves produced by Hjulström (1939) and shown in figure 6.7, which relate current velocity, particle size, and process; as figure 6.8 shows, Rubey's size classes seem to fall along the trend of

Figure 6.6. Sediment particles of different sizes begin to move on the stream bed at different products of mean velocity and depth-slope. Smallest particles (*B*) move mainly as a function of *DS*, while largest particles (*H*) move primarily as a function of the mean velocity. (From Rubey 1938)

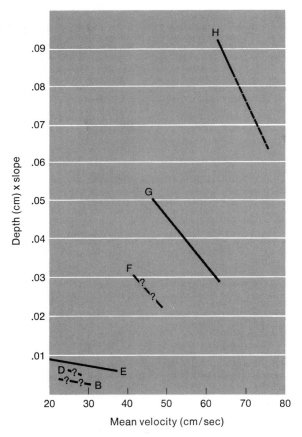

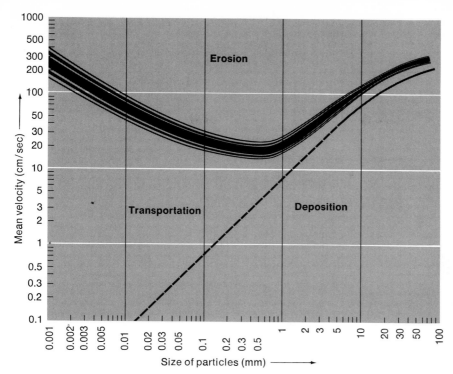

Figure 6.7. Mean velocity at which uniformly sorted particles of various size are eroded, transported, and deposited. (From Hjulström 1939. Used with permission of American Association of Petroleum Geologists)

the boundary between erosion and transportation, that is, the velocity needed to initiate motion. The velocity that produces erosion of clay-sized particles is in some cases as great as that needed to entrain larger material. This explains the commonly observed phenomenon of coarse particles being transported across stationary material of a smaller size.

Unfortunately, flumes are not useful in the study of competence when particles are larger than pebble size. Most competence investigations of coarser sediment have therefore been made in natural rivers or canals and, for reasons explained earlier, employ shear stress as the diagnostic hydraulic variable (Lane 1955; Fahnestock, 1963; Kellerhals 1967; Scott and Gravlee 1968; Church 1972). The shear stress criterion was justified mathematically by Shields (1936), but his equation also is of questionable value when particle size exceeds 7 mm, and it may be too sensitive for use in uncontrolled situations. The Shields equation has been adapted for use with larger sediment (Komar 1970); when assumptions are made about particle and water densities (Baker 1973b, 1974) it simplifies to

$$\frac{DS}{1.65d} = 0.06$$

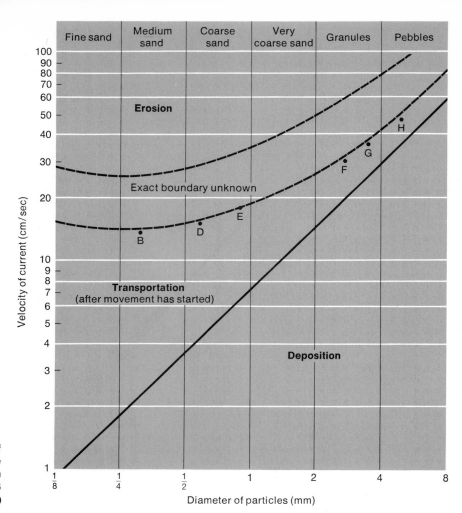

Figure 6.8. Portion of Hjulström curve with size grades *B* to *H* placed on diagram. Compare figures 6.6 and 6.7. (From Rubey 1938)

where D and S are depth and slope, and d is the particle diameter. In most studies, as figure 6.9 shows, the competence relationship is revealed by empirical rather than theoretical methods. The scatter on such plots is usually great and, as shown in the figure, the empirical regression may not agree with the line drawn by use of the Shields theory. However, there may be a hydrodynamic explanation for the disparity (Baker and Ritter 1975).

Perhaps the most appealing aspect of the shear stress approach is its applicability in geological work. With the aid of reasonable assumptions, the shear stress-particle size relationship allows educated guesses about the character of paleorivers that transported and laid down sediment in a variety of old deposits

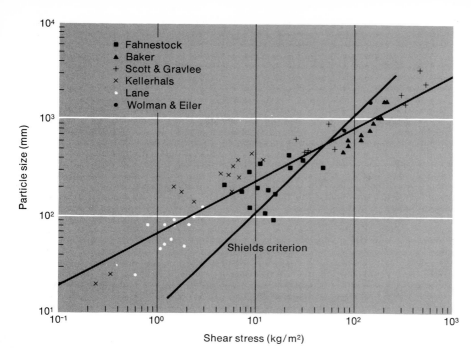

Figure 6.9. Relation of extended Shields criterion to empirical studies of critical shear stress and the particle size entrained. (From Baker and Ritter 1975. Used with permission of the Geological Society of America, Inc. Copyrighted 1975)

(Birkeland 1968; Lustig 1965; Baker 1973b, 1974). In general, such studies use the maximum particle size as a gage of the depth-slope product of the depositing river. The precision of such analyses may be questionable, but the qualitative understanding of the system that derives from the approach is probably valid.

The processes of entrainment determine the type and magnitude of erosion that occurs on the channel floor. It is incorrect, though, to assume that the only significant erosion is vertically directed. Bank erosion, which proceeds laterally, not only contributes to the sedimentary load but also, through its control on channel width, exerts a direct influence on other channel processes. Wilcock (1971) showed, for example, that how competence responds to increasing discharge depends on the rate of change in width during the high flow conditions. The rate of bank erosion in alluvial channels can be enormous or minuscule, depending mainly on the cohesiveness of the bank materials. In general, banks that are composed of fine-grained sediment or are densely vegetated (Hadley 1961; D. G. Smith 1976) have more resistance to lateral erosion than channels with sandy or gravelly banks. The actual process of erosion also differs, for clay-rich banks usually retreat by undercutting and failure of large blocks of the bank (Stanley et al. 1966; Laury 1971) while more coarse-grained banks erode by dislodgement and sloughing of individual particles. Even highly cohesive banks may erode rapidly if the dominant process is undercutting and slump.

Deposition

If the entrainment of sediment logically represents a threshold of erosion, a similar threshold must exist when sediment in transport is deposited. Suspended rock and mineral fragments tend to settle to the bottom at a rate that depends on the relative densities of the water and the particle, the fluid viscosity, and the size, shape, and density of the sediment. The distance any particle will be transported in one event depends on its fall velocity and on whether its downward settling is offset by turbulent forces in the water. Variations in these factors make the channel floor a dynamic interface where some particles are being entrained while others simultaneously are being deposited. The net balance of this activity depends on local conditions rather than on the average cross-sectional hydraulics; in fact, local effects may be different from those that occur over a long reach of the channel (Colby 1964). It is conceivable that scour and fill may be happening in the same channel reach at any given time, and a number of alternating events may coincide with fluctuating discharges. A long episode in which less sediment leaves the bed than is returned results in a distinct period of aggradation. Long-term events are caused by changes external to the channel (climatic or tectonic) that introduce into any stream reach more sediment than the available discharge can transport. These events will be considered in more detail later when we examine the relationship of time in channel mechanics.

The Frequency and Magnitude of River Work

At this stage of examining river energy, we can logically ask how and when fluvial work is accomplished. Is it the super-event of very high discharge which happens once in a millenium that allows rivers to do what they do, or is it the normal flow that occurs time and time again? The answer, I suspect, depends on how one defines geomorphic work. Wolman and Miller (1960) suggest that the work done by a river can be estimated by the amount of sediment it transports during any given flow. They concluded that in most basins 90 percent of the total sediment load (i.e., 90 percent of the work) is removed from the watershed by rather ordinary discharges that recur at least once every five or ten years. They also found that river channels form and reform within a range of flows between a lower limit set by the demands of competence and an upper limit where flow exceeds bankfull and is no longer confined to the channel. That is, channel configuration itself is probably related to high-frequency discharge, and its precise character is presumably a partial function of the river's load or work requirements.

There is ample evidence to support Wolman and Miller concerning sediment loads, since even catastrophic floods (Stewart and LaMarche 1967; Scott and Gravlee 1968) seem to transport from the basin a smaller percentage of the total load than normal flows do. In many catastrophic floods, however, tremendous erosional and depositional effects occur within the basin even though the loss of sediment is minor (Stewart and LaMarche 1967). Because material is redistributed by erosion and deposition, channels may shift their position and valley morphology may be significantly altered. It appears that great floods probably do a different type of geomorphic work which directly influences the megascopic features of a valley. This is quite unlike the work accomplished at or below bankfull stage, where the river transports sediment and modifies the channel in an apparent attempt to maintain rather than destroy the balance with its surroundings.

The normal fluvial condition has been aptly referred to as a "quasi-equilibrium" state (Leopold and Maddock 1953; Wolman 1955). It seems certain that every river strives to establish an equilibrium relationship with the prevailing discharge and load by adjusting its hydraulic variables. These flow variables, however, are mutually interdependent, meaning that a change in any single parameter requires a response by all the others. The difficulty involved in understanding rivers becomes evident when you consider that discharge and load are in continuous flux, and so all the hydraulic variables must always be adjusting. Obviously equilibrium as a steady-state condition cannot be attained in a river—thus the term quasi-equilibrium.

The Quasi-Equilibrium Condition

Hydraulic Geometry

The quasi-equilibrium condition was first demonstrated in a landmark study by Leopold and Maddock (1953). Using abundant flow records compiled at gaging stations throughout the western United States, they set out to determine the statistical relationship—the *hydraulic geometry*—between discharge and other variables of open channel flow. Because every river has wide fluctuations in discharge, any given channel cross-section must transport a range of flows that come to it from the adjacent upstream reach. Discharge, therefore, serves as an independent variable at any station, and the changes in width, depth, velocity, or other variables can be observed over a wide spectrum of discharge conditions (fig. 6.10). At a station each of the factors (*w, d, v*) increases as a power function such that

$$w = aQ^b$$
$$d = cQ^f$$
$$v = kQ^m$$

where a, c, k, b, f, and m are constants. Because discharge (Q) equals the product of width, depth, and velocity, the relationship can be expressed as

$$Q = aQ^b \times cQ^f \times kQ^m$$
$$\text{or}$$
$$Q = ackQ^{b+f+m}$$

and it follows that $(a \cdot c \cdot k)$ and $(b + f + m)$ must each equal 1. Leopold and Maddock found that the average at-a-station values of b, f, and m

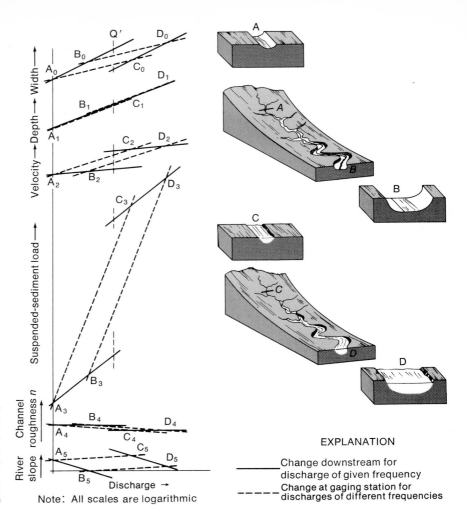

Figure 6.10. Hydraulic geometry relationships of river channels comparing variations of width, depth, velocity, suspended load, roughness, and slope to discharge at a station and downstream. (From Leopold and Maddock 1953)

Note: All scales are logarithmic

EXPLANATION

—————— Change downstream for discharge of given frequency

- - - - - - Change at gaging station for discharges of different frequencies

for a large number of midwestern and western streams were .26, .40 and .34 respectively. The exponent values, however, will differ with climate and geology, and average values will not fit any particular stream. Essentially the at-a-station exponents tell us what portion of the increase in discharge will be caused by an increase in each of the component variables.

As discussed earlier, discharge also increases with the expansion of drainage area, and so on most rivers, it must increase downstream. The question is how much of the downstream increase in volume is due to each of the variables of width, depth, and velocity. To make this analysis, care must be taken to ensure that the variables are measured during the same flow conditions. On a given day, for example, a disastrous flood with high *w, d,* and *v* values may be occurring in an upstream reach while flow conditions far downstream are normal. A comparison of the hydraulic variables in these two widely divergent frequencies of flow would be misleading. Obviously the frequency of the discharge must be considered for any observations to be valid.

Leopold and Maddock (1953) found that the mean annual discharge occurs with approximately the same frequency, usually 20–25 percent, on an amazingly large number of rivers. The mean annual discharge is probably not the flow that determines the character of the channel; its absolute value is relatively low, being equalled or exceeded one out of every four days, and the water level is well below the bankfull stage. Nonetheless, its availability in published reports makes it a useful parameter to determine downstream hydraulic geometry. In sum, at-a-station and downstream hydraulic geometry differ in that one (at-a-station) compares flows of vastly different frequencies while the other (downstream) analyzes variables at the same frequency of Q even though the absolute values in cfs units are different.

Width, depth, and velocity increase downstream with increasing mean annual discharge (fig. 6.10). The average values of b, f, and m for western streams are .5, .4, and .1 respectively, but they vary from region to region and for any particular stream within a region. In general, the rate of change in depth (f) is relatively consistent in both downstream or at-a-station geometry; values seem to range from .3 to .45. Width usually increases much more rapidly and with more consistent values downstream than at a station, probably because the b value for any cross-section depends more on bank cohesiveness than on discharge (Wolman and Brush 1961; Knighton 1974). Velocity increases more rapidly at a station than it does downstream.

Hydraulic geometry analyses have become standard in fluvial geomorphic studies for making comparisons of rivers of different climates, physical settings, and size (Wolman 1955; Leopold and Miller 1956; Myrick and Leopold 1963; Fahnestock 1963), even though the values may be affected by other geomorphic variables (Miller 1958; Hadley 1961; Wolman and Brush 1961; Thornes 1970; Knighton 1974).

The nonchalance of the statement that velocity increases downstream indicates how much fluvial geomorphology has changed since the Leopold and Maddock study. The suggestion that mean velocity, and probably bed velocity (Leopold 1953), increase downstream came as a shock to most geologists, who intuitively "knew" that small tributaries flowing on steep slopes must be traveling faster than the low-gradient trunk rivers. The surprise at this new interpretation of velocity was probably due mainly to geologists' inclination to consider slope as the major, if not the overriding, control of velocity. The importance of slope seems to be entrenched in geologic thinking and emerges in our basic concepts. Further, slope has always been involved in the interpretation of sediment transportation and deposition, and it usually is the easiest parameter to reconstruct for an ancient fluvial regime. For example, the surface of coarse-grained deposits such as terrace alluvium or fan debris is in many cases the original stream bottom, and its gradient represents the channel slope at the time of deposition. Slope, then, is a valuable tool in geologic interpretation and a prime factor of fluvial processes, but as we will see later, it cannot be singled out as the only adjustable variable in a stream channel or even the dominant one.

The possibility of a downstream increase in velocity should have been suspected because Manning's equation tells us that depth plays a greater role than slope in determining velocity. In a stream with a constant roughness, if depth increases downstream at the same rate that slope decreases, an overall increase in velocity will occur. Increased depth simply overcompensates for the loss of velocity due to a gradually decreasing slope.

All geomorphologists, however, do not accept the idea of a downstream increase in velocity. Carlston (1969), for example, suggests that on large rivers downstream velocity is probably constant, but on smaller streams it may increase or decrease according to local controls. Mackin (1963) sounds a more serious objection when he cautions against the injudicious use of empirical methods as the basis for sweeping geomorphic conclusions, citing the downstream velocity interpretation as a specific example. Indeed, some streams do decrease in velocity downstream (Brush 1961; Carlston 1969), but as Mackin points out, these individual cases may be obscured when they are placed on the same scatter diagram with measurements taken on other streams with different fundamental controls. The increase in velocity, therefore, may be the result of a particular methodology; as such it represents the exposition of a general trend rather than a conclusive rule. It is difficult to find fault with Mackin's argument, but in defense of empiricism we must recognize that streams do not lend themselves to a purely rational or scientific approach. The complex interaction in a river cannot be adequately expressed in mathematical terms, and the multitude of variables that define the system are constantly adjusting and readjusting to minor variations in flow. There may always be an element of indeterminancy in river mechanics that is simply beyond rational comprehension (Leopold and Langbein 1963). It may be better to strive for general trends and admit their fallibility in specific cases than to wait endlessly for a detailed scientific explanation.

The Role of Sediment   In most streams the amount of suspended sediment at a station increases rapidly with increasing discharge (fig. 6.11), varying as the simple power function

$$L = pQ^j$$

where L is suspended load and p and j are constants. The at-a-station value of j is normally > 1, indicating that the influx of sediment to the river is greater than the addition of water. Interestingly, the dramatic increase in sediment content is probably not caused by scouring of the channel floor; studies of floods on many rivers show that deposition occurs during the rising stages of the flood, precisely when the suspended load is increasing rapidly (fig. 6.12). Scouring takes place in the waning phase of the flood when velocities and sediment loads are lower than at the flood peak. This observation would mean that the bulk of sediment added to a river during floods derives from the valley-side slopes of the watershed. The observation also helps explain the notable variation in suspended load at any given discharge (evident in fig. 6.11) because the sediment acts as an independent variable that demands compensatory adjustments in velocity and depth. The load effect on velocity represents a dramatic case of feedback mechanics in a geomorphic system. Since the value of j is greater than unity, the sediment concentration (suspended load/unit volume H_2O) increases, thereby lowering the internal resistance to flow, as discussed earlier. The decreased n prompts an increase in velocity and enhances the river's ability to carry a greater load. In this way, a high rate of increase in suspended load seems to initiate the adjustments in hydraulic geometry that are needed to carry that load through the section.

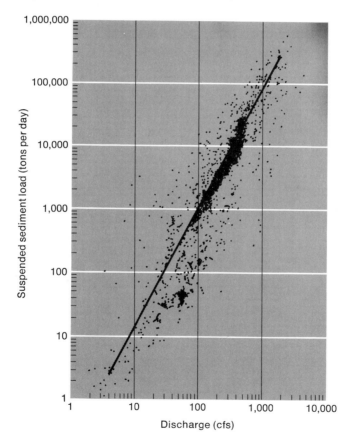

Figure 6.11. Relation of suspended load to discharge in Power River at Arvada, Wyo. (From Leopold and Maddock 1953)

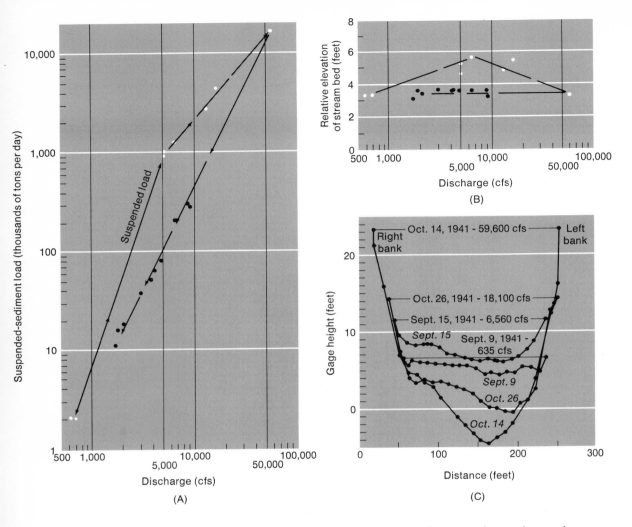

Figure 6.12. Changes in (A) suspended load, (B) stream-bed elevation, and (C) water-surface elevation with discharge during flood of September-December 1941, San Juan River near Bluff, Utah. (From Leopold and Maddock 1953)

Very few rivers have more than one sediment-gaging station, and so a consideration of downstream changes in sediment concentration is rather theoretical. However, sediment yield does decrease with increasing basin area, and discharge increases. It seems logical that sediment concentration will tend to decrease downstream as suggested by Leopold and Maddock, and the value of j will probably be less than 1 in most perennial streams.

To put this into perspective: Because suspended load seems to generate changes in velocity and depth, the ratio $\frac{m}{f}$ may be a realistic gage of changes in sediment concentration. If this is the case, for any value of b, an increase in sediment concentration ($j = > 1$) requires an increase in the ratio $\frac{m}{f}$. Leopold and Maddock (1953) suggest that where $\frac{m}{f} > .67$ the sediment concentration will probably be increasing, and at values below .67 it will be decreasing; this analysis fits their average at-a-station and downstream $\frac{m}{f}$ values (.85 and .25 respectively). Once again, these generalizations must be tempered for specific cases. For example, it was found that in arid, ephemeral streams the value of j exceeded unity downstream, indicating an increasing sediment concentration (Leopold and Miller 1956). The high j value was accompanied by an unusually high rate of increase in velocity ($m = .2$) and a retarded increase in depth ($f = .3$), producing an $\frac{m}{f}$ value of .67. The increased downstream sediment concentration is probably due to discharge being lost by infiltration into the channel bottom, a common phenomenon in arid, ephemeral streams.

The relationship between bedload and hydraulic geometry is poorly understood because so few measurements of bedload are available and no consistently reliable sampling technique has been developed, especially for bed material smaller than gravel size. Suggestions have been made, however, that the entrainment of bedload is dependent on hydraulic geometry (Wilcock 1971).

The Influence of Slope

I suggested earlier that channel slope has always been recognized as a prime adjustable property of rivers. Geologists and geographers traditionally have carefully studied the river gradient and generally have accepted the proposition that a concave-up longitudinal profile (change in elevation with increasing length) is the channel form assumed by rivers in equilibrium. There is abundant evidence to substantiate the importance of slope in a river striving to maintain balance, but whether the gradient adjustment operates to the exclusion or subservience of other variable changes is a question of considerable debate in modern geomorphology.

Observations at gage stations show that the slope of the water surface remains relatively constant during flows of different magnitudes, indicating that the adjustments to increasing discharge must be made by the other hydraulic variables. For example, we cannot call on a dramatic increase in slope to produce the high $\frac{m}{f}$-value associated with the at-a-station increase in sediment concentration. The high velocity must be generated by an increase in depth, a decrease in roughness, or both. Downstream the channel gradient does exert an influence, because in most rivers there is a notable decrease in slope. Roughness, however, usually remains fairly constant (Leopold and Maddock 1953) because of the offsetting effects of a decrease in particle size (decreases n) and a decrease in sediment concentration (increases n). As a result, any increase in velocity downstream can best be justified by the increase in depth, explaining the low m values and the retarded $\frac{m}{f}$ in that direction.

The relationships between slope and other hydraulic parameters reveal the complexities of quasi-equilibrium, but they do not explain why slope usually decreases downstream or what external factors may control the form of the longitudinal profile. As early as 1877, G. K. Gilbert concluded that slope was inversely related to discharge, and since Q increases with basin area and stream length, it is axiomatic that slope should decrease downstream. However, in most rivers particle size generally diminishes downstream, prompting many observers to suggest that channel gradient adjusts to the bed material. Actually both factors are probably involved. Rubey (1952) demonstrated that if channel shape is constant the slope will decrease with (1) a decrease in particle size, (2) a decrease in total load, or (3) an increase in discharge. He expressed the relationship as follows

$$S_G{}^a X_A = K \frac{L^b D^c}{Q^e}$$

where S_G represents the slope of the water surface, X_A is the optimum form ratio (the $\frac{D}{W}$ providing the river with the most efficient traction), L is the stream load, D is the average diameter of particles in the load, and K, a, b, c, e are constants. Rubey concludes that the channel gradient at any point along the river is a function of both sediment and discharge. If Rubey is correct, then slope is dependent, or partially so, on all hydraulic variables because they are also related to discharge.

Many studies subsequently have shown the correctness of Rubey's analysis. In one of these studies, comparing stream profiles in areas of differing geology, Hack (1957) found no consistent correlation between slope and bed-material size when all sample localities from a geologically divergent region were considered together. His plot, shown in figure 6.13, reveals a tremendous scatter; for example, streams with a median particle size of 60 mm have slopes ranging from 1.1 m/km (6 ft/mi) to 37 m/km (200 ft/mi). It was only after Hack added a third variable, drainage area, to the analysis that a significant relationship became apparent (fig. 6.14), and slope could be defined by the equation

$$S = 18 \left(\frac{M}{A} \right)^{0.6}$$

where M is the median size of the bed material in millimeters, A is area in mi^2, and S is slope in ft/mi. Since basin area can normally be used as an index of discharge (Leopold et al. 1964), Hack's study reinforces Rubey's contention that both Q and sediment are determinants of slope. It does not indicate which factor is the principal determinant; indeed, one would expect the relationship to be

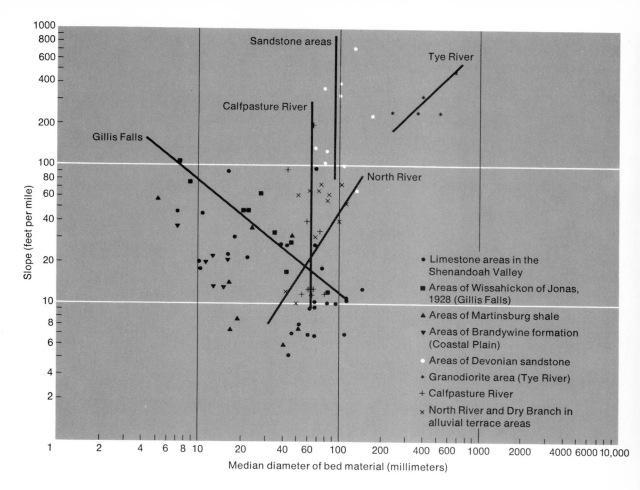

defined by different mathematical equations in different physical settings. Nonetheless, Hack demonstrated that streams flowing within a single geological unit and having the same drainage area should have similar channel slopes and particle size. Moreover, the slope at any point is logically related to linear morphometry (Hack, 1957) since

$$L = 1.4A^{0.6}$$

and therefore by substitution

$$S = 18\frac{M^{0.6}}{L/1.4} = 25\frac{M^{0.6}}{L}.$$

The relationship demonstrated by Hack may have a useful application by providing the basis for a rapid method of terrain analysis, including first approximation of bedrock character and flow conditions (Hack 1973).

Figure 6.13. Scatter diagram showing relation between channel slope and median size of bed material in selected rivers, Maryland and Virginia. (From Hack 1957)

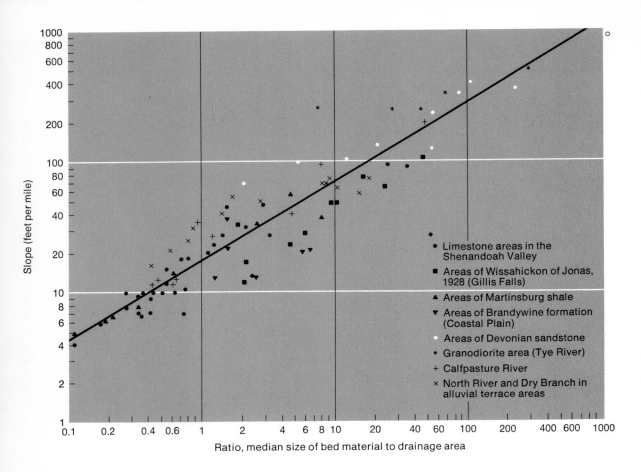

Figure 6.14. Relation between slope and the ratio of median size of bed material to drainage area in selected streams, Maryland and Virginia. See fig. 6.13 for comparison. (From Hack 1957)

Before happily assuming that slope, after all, seems to be understandable and possibly even predictable, turn once again to Rubey's equation. Note that although slope depends on both sediment and discharge, the equation also emphasizes that fluctuations in these variables do not *demand* an adjustment in the channel gradient. It is equally possible that changes may be counterbalanced by variations in the cross-sectional shape of the channel (X_A).

Channel Shape

Logic tells us that, unless velocity is completely unrestrained, rivers with a large mean annual discharge have greater cross-sectional areas than streams with smaller average flows. This fact has been verified repeatedly by casual observation and documented by the relationships exposed in hydraulic geometry. It is significant, however, that many rivers with the same mean annual discharge have different cross-sectional areas; even when the total area is the same, the width-depth ratio may vary considerably. Obviously factors other than discharge alone influence the shape of river channels, and so we look to the stream sediment for an explanation.

Schumm (1960) presented cogent arguments to suggest that channel shape, as defined by *W/D,* is determined primarily by the nature of the sediment being transported. In streams carrying a high percentage of silt and clay (particles < 0.074 mm), channels tend to be narrow and deep. In contrast, wide, shallow channels seem to be characteristic of rivers transporting coarse bedload. A wide, shallow channel shape is much more efficient in the transport of coarse bedload because higher velocities are nearer the channel floor (Lane 1937). Schumm's data, collected from semiarid and arid climate streams, show that *W/D* is related to the percent of silt-clay by the equation

$$F = 255M^{-1.08}$$

where *F* is the width-depth ratio and *M* is the percent of silt and clay in the channel perimeter. The magnitudes of the mean annual discharge or the mean annual flood do not seem to affect this relationship; in fact, only 40 percent of the variability in channel shapes can be accounted for by discharge alone (Schumm 1971).

Even though local bank and bottom conditions may create some anomalies, it seems evident that the shape of alluvial channels is mainly a function of the load being transported. This prompted Schumm (1963a) to classify rivers on the basis of their load types; more important, it demonstrates another viable method by which a river can be adjusted. A channel reach suddenly burdened with a different type of load than the one it previously carried may as easily alter its channel shape to accommodate the new load as change its gradient by deposition or erosion. The repeated changes in the shape of the Cimarron River in Kansas since 1880 are an excellent example of this phenomenon (Schumm and Lichty 1963).

Channel Patterns

This discussion of adjusting mechanics in rivers has centered primarily on the balancing factors that function in a channel cross section or a series of cross sections considered in a downstream direction. Rivers also have characteristic forms extending over long stretches of their total length which, when observed in plan view, display a distinct geometric pattern. The pattern that a river adopts is now recognized as another manifestation of channel adjustment to the prevailing discharge and load. In that sense, channel patterns are another important aspect of river mechanics.

Patterns are usually classified as straight, meandering, or braided, even though the boundaries between types are sometimes arbitrary and indistinct. The distinction between straight and meandering, for example, is based on a property called *sinuosity*. Although there is no complete agreement on how sinuosity should be determined (Leopold and Wolman 1957; Schumm 1963b; Brice 1964), we will follow Schumm's definition that it is the ratio of stream length (measured along the center of the channel) to valley length (measured along the axis of the valley). The transition between a straight and a meandering stream is usually placed at a sinuosity value of 1.5, but again this value probably has no particular mechanical significance.

The braided pattern, characterized by the division of the river into more than one channel, is more easily discerned. The designation becomes vague, however, when only part of the river's total length is multichanneled. How much of the river must consist of divided reaches to constitute a braided system is an individual decision. Another complication is that some single-channeled rivers become distinctly braided in times of high flow, requiring that stage be considered when deciding on the pattern classification. In addition, there is no reason to expect that a river will display the same pattern for its entire length. In fact, because patterns are probably a function of discharge and load, minor variations in those factors downstream may easily generate different patterns, especially in very large watersheds.

Straight Channels

Most streams do not have straight banks for any significant distance, making the straight pattern a rather uncommon one. It may seem strange, then, that straight streams display many of the same channel features as the more common meandering pattern. As figure 6.15 illustrates, straight reaches often contain accumulations of bed material called *alternate bars* that are positioned successively downriver on opposite sides of the channel. A line connecting the deepest parts of the channel, called the *thalweg,* migrates back and forth across the bottom. In rivers with a poorly sorted load, the channel floor undulates into alternating shallow zones called *riffles* and deeps called *pools.* The pools are directly opposite the alternate bars and the riffles are about midway between two successive pools. Clearly, a straight channel implies neither a uniform stream bed nor a straight thalweg, and the spacing of bars, riffles and pools is closely analogous to that in a meandering channel (Leopold et al. 1964; Dury 1969).

Meandering Channels

The most common river form by far is the meandering pattern, also shown in figure 6.15. Meandering reaches contain the same physical components observed in straight channels (pools, riffles, bars), distributed in a similar way. The thalweg also migrates back and forth across the channel, impinging against the outer bank of the meander bends and crossing to the opposite side near the riffles.

A definite lateral component is the prime distinguishing property of flow in a meander. Presumably, as water is forced against the outer bank of a meander, its slightly elevated level at that point gives the flow a circulating motion. The water moves along the surface toward the undercut bank and along the bottom toward the *point bar* (fig. 6.16). This corkscrew motion, called *helical flow,* reaches its greatest velocity slightly downstream from the axes of the meander bends at the position of the pools; the velocity decreases gradually downstream until the flow approaches or reaches the next riffle and the lateral circulation disappears. From this transition zone to the next downstream meander, the spiral

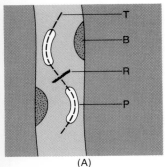

(A)

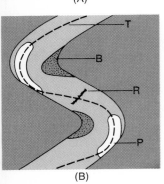

(B)

Figure 6.15. Features associated with (A) straight and (B) meandering rivers. T = thalweg; B = bar; R = riffle; P = pool.

velocity increases again, but the cellular pattern is opposite to its original direction because the orientation of the meander itself has been reversed (fig. 6.16). The magnitude of the lateral velocity varies but it can be great enough to influence the transport of sediment. Particles eroded from an undercut bank, for example, may be dragged into the center of the channel by the bottom limb of the helical cell. When the motion reverses at the transition, some of the sediment is returned to its original channel side and deposited on the next downstream point bar. This model of helical flow is probably oversimplified, and determining its precise relationship to sediment transport requires very detailed field study (Jackson 1975).

Meandering rivers shift their positions across the valley bottom by eroding on the outer banks of meander bends and simultaneously depositing point bars on the inside of the bends. Even though the location of the river varies with time, there is no compelling reason to suggest that the shape or hydraulic properties of the river stray far from average values as long as the prevailing controls of climate and tectonics remain unchanged. In fact, meanders in rivers of all sizes are dimensionally similar, with consistent geometric and hydraulic relationships as in table 6.3 and figure 6.17.

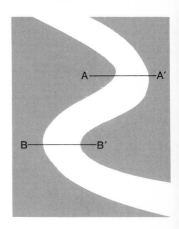

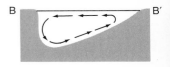

Figure 6.16. Helical flow at successive meander bends in a meandering river.

Table 6.3 Empirical relationships between parameters that define meander geometry.

Dependent Relationship	Source
Wavelength	
$\lambda = 6.6\, w^{0.99}$	Inglis (1949)
$\lambda = 10.9\, w^{1.01}$	Leopold and Wolman (1960)
$\lambda = 4.7\, r_m^{0.98}$	Leopold and Wolman (1960)
$\lambda = 30\, Q_{bf}^{0.5}$	Dury (1965)
Amplitude	
$A = 18.6\, w^{0.99}$	Inglis (1949)
$A = 10.9\, w^{1.04}$	Inglis (1949)
$A = 2.7\, w^{1.1}$	Leopold and Wolman (1960)

λ = wavelength (ft)
w = width (ft) at bankfull stage
r_m = radius of curvature (ft)

A = amplitude (ft)
Q_{bf} = bankfull discharge (cfs)

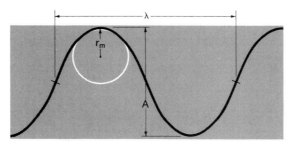

Figure 6.17. Geometric parameters of a meander:
λ = wavelength;
A = amplitude; r_m = radius of curvature.

The most revealing geometric property is the meander wavelength, which relates to many other variables including discharge, width, and the radius of curvature (r_m). The relationships are in many cases almost linear and undoubtedly reflect basic mechanical principles. For example, using the equations in table 6.3 and considering them as linear, we find that where all parameters are measured in feet,

$$\frac{r_m}{w} = \frac{\dfrac{\lambda}{4.7}}{\dfrac{\lambda}{10.9}} = \frac{10.9}{4.7} = 2.3.$$

Actual measurement of this ratio shows most rivers to have values between 2 and 3 (Leopold and Wolman 1960), suggesting that the relationship probably is a function of channel curvature exerting an influence on flow. Bagnold (1960) showed that as the radius of curvature decreases (decreasing $\frac{r_m}{w}$) the main filament of flow tends to shift toward the outer bank, causing a concomitant decrease in resistance on the inside of the bend. Greater curvature will continue to decrease resistance until a critical value of $\frac{r_m}{w}$ is attained, when flow along the inner bed becomes unstable and breaks away from the boundary (fig. 6.18). This creates eddy currents along the inside boundary, increasing the energy dissipation and so effectively establishing a minimum resistance for the flow. In most fluid systems, eddying begins when the curvature ratio is between 2 and 3, suggesting that the large number of real meanders having these values probably represents a quasi-equilibrium between flow and geometry. Hickin (1974) found the critical $\frac{r_m}{w}$ value to average 2.11 on the Beatton River in British Columbia. He demonstrated that once a developing meander attains the critical curvature, it exerts significant control on the subsequent rate and direction of lateral migration.

The Origin of Meanders   Before the recent developments in fluvial geomorphology, most geologists accepted the premise that random diversions of flow by slumped boulders or fallen trees were the ultimate causes of meanders. Such aberrations undoubtedly can start meanders, and once an initial bend develops, the sinuous nature is transmitted downstream and causes more bends to form (Friedkin 1945). Helical flow is also propagated from the initial curve. Although it may be tempting to write meandering off as the result of normal but random erosional or depositional events in a river, the explanation is unfortunately not that simple. Meanders with similar geometry occur in streams that cross sediment-free ice and in unchanneled flow such as the Gulf Stream (Leopold and Wolman 1960) where random erosional or depositional events are

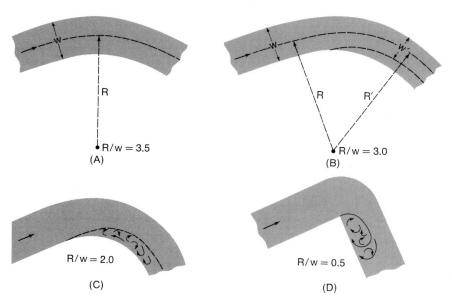

Figure 6.18. Relation between radius of curvature, width, and flow properties. Decreasing ratio _rm_/w in (B) causes flow to break away from inside of bend and create eddying zone (C and D). R = _rm_. (After Bagnold 1960)

impossible. Meanders apparently manifest a more basic aspect of fluvial mechanics, and most investigators now believe the pattern relates to the energy distribution in river flow.

In simplest terms, the _energy grade line_ is a graphic representation of the potential energy (head) possessed by the river along its longitudinal profile. The loss in potential with distance reflects the amount of energy that resisting elements consume in the system. Here we can assume that the slope of the water surface is parallel to the slope of the energy grade line. In detailed studies, however, the components that determine potential energy may vary such that the two lines are not always parallel in the downstream direction. If the same amount of river energy is dissipated over each equal unit distance along the channel, the energy grade line will be a straight, sloping line.

As diagrammed in figure 6.19, straight reaches with an alternating pool and riffle sequence have an irregular grade line, because head loss is greater over the riffles than over the pools. The construction of pools and riffles in straight reaches is probably independent of the channel pattern (Langbein and Leopold 1966). In a curved reach, however, the irregularities are removed, and the energy grade line approaches the smooth curve expected in a river that is uniformly expending energy. Pools and riffles probably are affected by this pattern. Thus, one effect of meandering is to increase resistance, and with it energy dissipation at the pools, making the grade line more uniform. This analysis is compatible with the belief that the meandering pattern approximates a condition of equilibrium, for energy is lost equally throughout the length of the river.

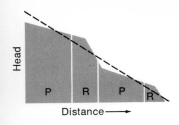

Figure 6.19. Profile of energy grade line in straight river with alternating pools and riffles. P = pool; R = riffle; head = potential energy. Head decreases with distance as energy is dissipated in flow. (From *Fluvial Processes in Geomorphology* by Luna B. Leopold, M. Gordon Wolman, and John P. Miller. W.H. Freeman and Company. Copyright © 1964)

However, the average gradient is usually steeper in a meandering reach than it is in a straight reach (Langbein and Leopold 1966; Schumm and Khan 1972), a condition that is *not* compatible with a stream's attempt to do as little work as possible (Rubey 1952). This tendency is probably satisfied because, as explained above, the curvature ratio in most meanders minimizes flow resistance and so puts work expenditure at a minimum. Langbein and Leopold (1966) suggest that depth, velocity, and slope are adjusted in meanders to provide a more uniform distribution of both bed stress and friction. As these vary differently with slope, the authors conclude that slope adjusts so as partially to modify both shear stress and friction and thereby minimize their total variance. Recent work in a stream with a pool and riffle sequence seems to verify their conclusions (Cherkauer 1973).

In a different energy analysis, Yang (1971) suggests that minimizing the time rate of energy expenditure is more important in the formation of meanders than is minimizing the total energy expenditure. He expressed this concept as

$$\frac{\Delta H}{\Delta t} = \frac{kY}{\Delta t} = \phi(Q, S_v, C_s, G \ldots) = \text{a minimum}$$

where $\frac{\Delta H}{\Delta t}$ is the rate of energy expenditure, H being head and t being time. ϕ is a function of external constraints such as discharge (Q), valley slope (S_v), sediment concentration (C_s), geology (G), etc. Y is the fall and k is a conversion factor between head and fall. Since a straight channel has the shortest length for a fixed fall, it has the largest $\frac{\Delta H}{\Delta t}$; that is, energy is dissipated in the shortest time. Meanders, in contrast, have the lowest $\frac{\Delta H}{\Delta t}$ because the fall takes place over a greater distance. Yang concludes that streams adjust their slopes to minimize the time rate of energy expenditure, and the result is a meandering pattern.

To summarize, meandering seems to be an equilibrium form that rivers strive to attain because (1) they attempt to dissipate energy in equal amounts along the length of the channel, probably by minimizing the variance between shear and friction; and (2) under the constraints of (1), streams try to minimize their total work expenditure (least work) or the rate of work expenditure by adjusting their curvature geometry or their slopes.

Which parameters best reveal the equilibrium state and which factors demand adjustments in the meandering pattern when altered are the subjects of some debate. As discussed earlier, meander geometry clearly is related to channel width, but other studies have suggested discharge (Dury 1965) or gradient (Carlston 1965) as the prime independent variables. Schumm (1967b) argues that wavelengths in rivers transporting a high percentage of sand and gravel will be higher than those in rivers with the same discharge but moving a fine-grained load.

Although many variables are related to the wavelength of meanders, it seems that if geometry reflects an equilibrium condition, discharge and sediment characteristics are once again essential ingredients. Even though width and slope are statistically related to wavelength, both these factors are dependent on discharge or sediment types; the ultimate controls, therefore, must lie with the more fundamental variables. Schumm (1963b) showed that sinuosity in semiarid streams is related to the width-depth ratio such that

$$P = 3.5F^{-2.7}$$

where sinuosity is P, and F is W/D. As discussed earlier, W/D is a function of the type of load being transported. It is not surprising that in most rivers in Schumm's study, the sinuosity also increases with an increase in the percentage of silt-clay (M) in the channel such that

$$P = .94M^{.25}.$$

In general, then, highly sinuous rivers tend to be narrow and deep and are characterized by fine-grained loads that are transported in suspension. Streams with low sinuosity tend to be wide and shallow and transport coarse material mostly as bedload. This does not mean that coarse bedload rivers never meander. Many observations show they do. It simply indicates the tendency of bedload rivers to establish less sinuous patterns.

Sinuosity does not seem to vary directly with discharge. The lower Mississippi River, which has a mean annual discharge of 14,150 cms (500,000 cfs), has a sinuosity of 2.1. Red Willow Creek in Nebraska, which has an average discharge of 1.2 cms (42 cfs), also has a sinuosity of 2.1. The point is that discharge seems to control the dimensions of meandering rather than its intensity. As discharge increases, wavelength, width, and r_m all increase, but the degree of meandering as evidenced by sinuosity seems to relate more closely to the load types.

Braided Channels

A basic part of the formation of the braided pattern is the division of a single trunk channel into a network of anastomosing branches and the growth and stabilization of intervening islands. Detailed studies of braided systems in both flumes and rivers (Leopold and Wolman 1957; Fahnestock 1963; Church 1972; Rust 1972; N. D. Smith 1970, 1974) show that divided reaches have different channel properties than adjacent undivided segments. Braided zones are usually steeper (fig. 6.20) and shallower; total width is greater although each channel may be narrower than the undivided trunk; and changes in channel positions and the total number of channels are likely to be extremely rapid (Fahnestock 1963; Church 1972; N. D. Smith 1974). Fahnestock, for example, documented lateral shifting of the channels up to 122 meters (400 ft) in eight days in the braided segment of the White River in Washington.

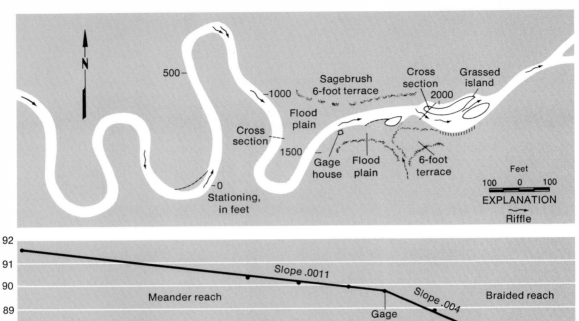

From planetable map by Leopold and Wolman, Aug. 11, 1953

Figure 6.20. Change of river pattern in Cottonwood Creek near Daniel, Wyo. Longitudinal profile of river shows that a change in slope accompanies the change from meandering to braided pattern. (From Leopold and Wolman 1957)

Any explanation of the origin of braids is necessarily oversimplified because, like all fluvial processes, it involves the simultaneous interaction of a number of factors. According to Fahnestock (1963), the most important of these are:

1. *Erodible banks.* Most investigators of channel patterns feel that bank erosion is perhaps the most necessary factor in creating a braided system. If bank erosion is prohibited by material cohesiveness or vegetation, it is unlikely that a braided pattern will develop.
2. *Sediment transport and abundant load.* Almost every braided river transports large volumes of bedload, and much of the channel shifting is prompted by temporary deposition of bars across entrances to branches of the network. It is incorrect, however, to assume that braided rivers are overloaded, since many times braids form when the channel is actively being eroded (Leopold and Wolman 1957), and at the same time, undivided segments immediately downstream from an evolving braided reach are

being actively aggraded. Even though the load exceeds the transporting capacity of the channel, no braiding occurs.

3. *Rapid and frequent variations in Q*. Fluctuations in discharge tend to produce the alternating erosion and deposition that seem to be a necessary part of braiding mechanics. However, some of the channel shifting and much of the increase in the number of channels may simply be a matter of reoccupying abandoned channels during high river stages. It is also significant that laboratory studies have produced braids under constant discharge, indicating that discharge may sustain the pattern, not cause it.

Although slope and other channel properties of braided streams are different from those of meanders, they probably are not factors in the origin of braids. More likely they represent the geometric modifications brought about by particular sediment and discharge requirements. It seems safe to say that braids do not necessarily connote instability. The pattern simply represents another condition a river may establish in response to external controls. It may be maintained for a long period of time and possibly is as close to true equilibrium as the meandering pattern.

The Origin of Braids   The development of braids follows a rather distinct sequence of events, illustrated in figure 6.21 (Leopold and Wolman 1957). During high flow a portion of the coarse load being transported is deposited because of some local channel condition. This initial accumulation becomes the locus of an incipient longitudinal bar because reentrainment of the particles requires a greater velocity than did their transportation and deposition (fig. 6.7). Continued deposition here allows the bar to grow both upward and in the downstream direction. As particles move across the reach they are deposited on the lower end of the bar where depth suddenly increases and velocity decreases. These changes occur because the width and discharge remain constant over that segment of the cross section, and since q = wdv, an increase in depth requires a decrease in velocity. Most smaller particles move easily over the growing bar, but some may be trapped in the interstices between the larger grains.

As the expanding bar begins to occupy a significant portion of the channel area, the channel is no longer wide enough to contain the total flow, and as flow is deflected around the bar, the banks are eroded. Simultaneously, the bar itself may be trimmed and the channel somewhat deepened. These processes combine effectively to enlarge the channel on both sides of the bar, allowing the water level to be lower at any equivalent discharge. Eventually the bar emerges as an island flanked by two distinct channel branches. Bars and islands do not necessarily remain fixed in position or shape since they are also susceptible to lateral erosion (N. D. Smith 1974). In documented cases, however, vegetation may spread rapidly on the islands, especially if overbank flows or wind provide silt as a capping layer; the rooting of vegetation tends to increase the resistance to erosion. The whole process from initiation to stabilization may take as little as two

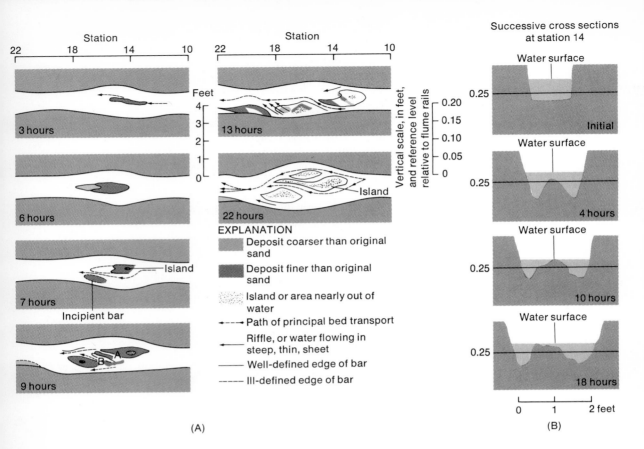

(A)

(B)

Figure 6.21. Stages in the development of a braid in a flume channel. (A) Sequential development of the pattern at various times. (B) Cross sections at one station along the flume. (From Leopold and Wolman 1957)

years. Bar growth can also begin when a river reoccupies an abandoned channel. In such an event, an erosional remnant may become a core upon which new bar growth will develop (Eynon and Walker 1974).

Other bar types also are common in braided systems. N. D. Smith (1974) recognizes transverse, point, and diagonal bars in addition to the longitudinal type. Transverse bars are tabular bodies that grow by downstream migration of foreset beds developed more or less perpendicular to the current. These bars form when sand moving along the bottom encounters a shallow depression where velocity is lowered (Jopling 1966). In response, the sand is deposited as a "delta" that builds upward until the flow velocity increases to its former level, and the top of the bar becomes the channel floor across which sand can once again be transported. At low flow, transverse bars may be exposed and dissected into a series of small channels.

Smith (1971) noted significant differences between longitudinal and transverse bars in the Platte River (fig. 6.22). Transverse bars normally are composed

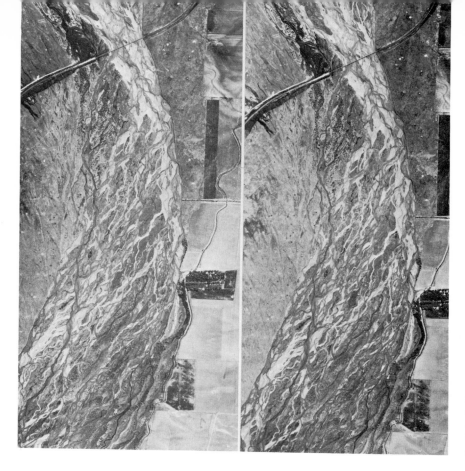

Figure 6.22. Intensely braided North Platte River in Keith County, Neb. (Photo by U.S. Dept. of Agriculture. From the University of Illinois Catalog of Aerial Photography)

of sand and are better sorted and more fine-grained than longitudinal bars. In addition, they show well-developed planar cross-beds in contrast to the crude horizontal stratification found in the coarse sediment of the longitudinal bars (Rust 1972). Assuming that most rivers have decreasing particle size downstream, the number of transverse bars should increase in that direction relative to the number of longitudinal bars. Furthermore, the surface relief on transverse bars is considerably less than that found on longitudinal bars, and the two can be differentiated by a *bed relief index* (Smith 1971). Since the contrasts between bar types are distinct and measurable, and the dominance of one type over the other may be a function of distance from the source, it seems reasonable that the characteristics of braiding should provide clues in the reconstruction of paleoriver systems. Based on these criteria, Smith (1970) has made a cogent interpretation of the Silurian rivers of the Appalachian region; however, before such reconstructions can be made, the rapidly changing properties of bars, due to a complex history of erosion and deposition, demand careful attention to details.

The Continuity of Channel Patterns

If channel patterns reflect specific adjustments to fluvial variables, it follows that boundaries between the patterns should be definable in terms of those variables. It also follows that each pattern must be stable within certain finite limits of the controlling factors, called *thresholds;* when the limits are exceeded, a viable fluvial response would be a pattern change. Patterns, then, may be ephemeral fluvial properties, especially in segments where the values of the controlling factors are critically close to the threshold condition.

As we saw earlier, straight reaches in suspended-load rivers are rare enough to be considered as unstable and probably transitional to the meandering form. This observation is supported by energy analyses and by the fact that the physical components of straight reaches are analogous to those in meanders. For example, in straight reaches the spacing of successive riffles is about five to seven times the channel width. In meanders the relationship

$$\lambda = 10.9w^{1.01}$$

where λ and w are measured in feet, suggests that the two successive riffles found in one complete wavelength are spaced about the same as those in straight reaches. Straight reaches, therefore, will probably not remain long in that form unless the banks are unerodable or a coarse-grained load requires the river to use all its energy for transportation, leaving none to be dissipated in meander bends. Even so, special cases may exist. Distributaries in deltas, for example, are straight and perhaps remain so through a combination of very low gradient and very low bedload values.

Precisely how and why straight rivers become meandering has been the subject of much recent debate (Dury 1969; Tinkler 1970, 1971; Keller 1971a, 1972, 1974; Humphreys and Hughes 1974; Lewin 1976). Generally, most in-

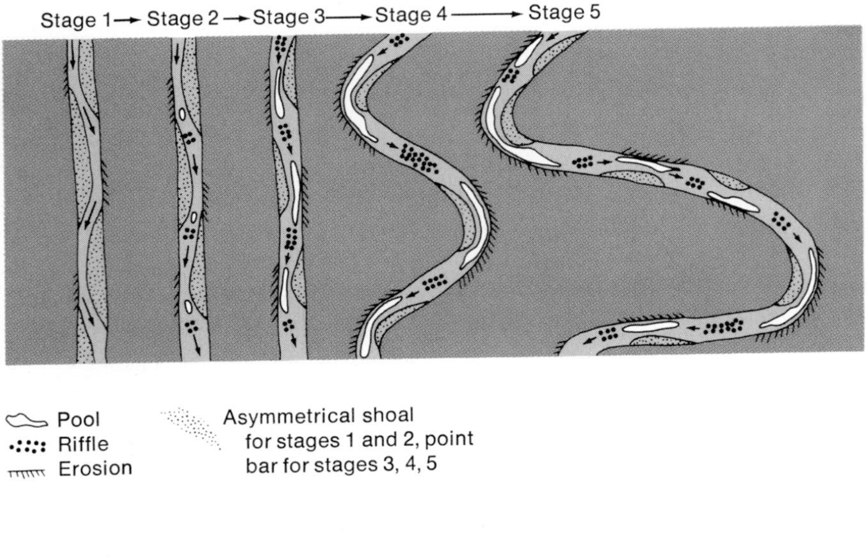
Stage 1 → Stage 2 → Stage 3 → Stage 4 → Stage 5
Pool
Riffle
Erosion
Asymmetrical shoal for stages 1 and 2, point bar for stages 3, 4, 5

Figure 6.23. Stages in the development of an alluvial stream channel as it changes from straight to meandering. (From Keller 1972. Used with permission of the Geological Society of America)

vestigators visualize the transformation as evolutionary rather than catastrophic, and most current models of the transition postulate distinct stages of development, like those shown in figure 6.23. Although geomorphologists are far from fully understanding the processes involved, some interesting observations have been made. First, the transition from straight to meandering is intimately related to the mechanics of erosion and sedimentation, especially to the formation of bars, pools, and riffles. Second, the development of meanders requires bank erosion. Initially the erosion seems to be dependent on secondary flow generated by the construction of bars of various types that modify the channel topography (Lewin 1976). Later, lateral accretion is probably the result of secondary flow rather than its cause. Third, there is no compelling reason to expect that the last developmental stage (fully meandering) in any model is forever stable. The mechanics that produces meandering also causes chutes and cutoffs that return the reach to straightness (Keller 1972; Lewin 1976), or they proceed to multilooping (Brice 1974) or other complexly woven forms (fig. 6.24).

The threshold marking the transition between meandering and braided is perhaps more precisely, although empirically, defined. Several studies have suggested a threshold boundary between the two forms based on the relationship between slope and discharge (Leopold and Wolman 1957; Ackers and Charlton 1971; Lane 1957). Although the limiting values are not consistently the same

Figure 6.24. Complex of loops and abandoned channels in strongly meandering river pattern. Rio Grande in Alamosa County, Colo. (Photo by U.S. Geological Survey. From the University of Illinois Catalog of Aerial Photography)

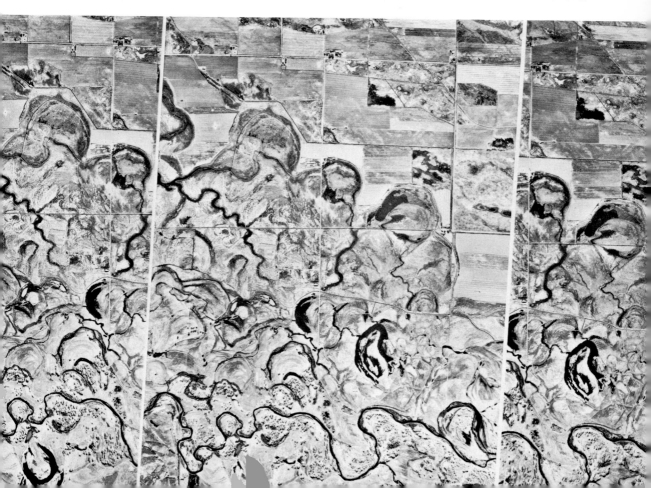

(fig. 6.25), it seems obvious that at the threshold an increase in slope at any given discharge (or an increase in discharge at any given slope) will change a meandering pattern to a braided pattern. It has also been demonstrated (Schumm and Khan 1972) that the threshold values between patterns can be defined as well, if not better, by the slope-sediment load relationship (fig. 6.26).

In sum, we can say with some assurance that the combined effects of discharge and sediment delimit the stability range for any channel pattern. Slope is probably not the inducing agent in pattern change but more likely adjusts as a dependent variable, along with the pattern, to changes in the sediment-discharge regime.

Rivers, Equilibrium, and Time

The physical operations within rivers are driven by their attempts to establish and maintain the most efficient conditions for transporting water and sediment. Because every river has a unique combination of these two factors, the parameters that define the equilibrium state must differ from river to river and

Figure 6.25. Relation of slope and discharge. Lines represent threshold slopes at various discharges as determined in different studies. (From Schumm and Khan 1972. Used with permission of the Geological Society of America)

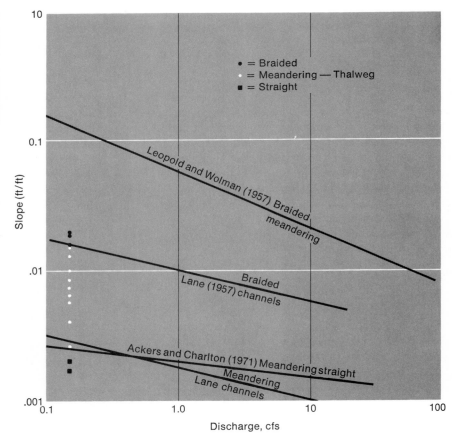

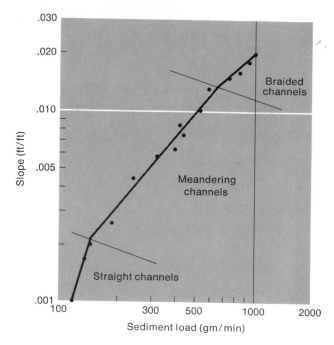

Figure 6.26. Relationship between slope and sediment load in flume study of channel patterns. (From Schumm and Khan 1972. Used with permission of the Geological Society of America)

even in various segments of the same river. Every student of rivers understands that discharge and load are not constant, and thus the element of time becomes a significant factor in any consideration of fluvial mechanics. Remember that discharge and load are not in themselves independent variables because they are ultimately a function of climate, geology, and tectonics. Furthermore, it is unreasonable to expect either discharge or load to change independently of one another since both factors are related to the same, more basic, controls. A change in climate, for example, will surely alter discharge, but it may also prompt a simultaneous change in the character of the load because vegetation and weathering will likewise adjust to the new climatic regime.

The importance of time is that the type of fluvial variable most likely to act to maintain equilibrium may depend on the time span being considered. Figure 6.27 shows a gradual increase in mean annual discharge provided to a river over a period of several hundred years. Presumably the load characteristics are also changing for the reasons stated above. Within the period, normal variations in discharge and load are accommodated by instantaneous adjustments of hydraulic geometry. Major floods may occasionally alter the valley topography or divert the river to a new position, but the channel itself will reorganize according to the average flow and load conditions. At some point, however, the gradually changing mean values of load and discharge can no longer be balanced by hydraulic variables under the prevailing channel configuration or pattern. One flow event, perhaps not even a major flood, will eventually exceed the stability limits of the

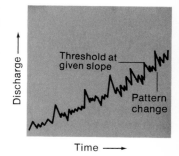

Figure 6.27. Diagram showing gradually increasing discharge with time. Discharge variations are accommodated by variables of hydraulic geometry until threshold is reached. At threshold, a major change in the river character, such as a pattern change, is required to carry the increased average discharge.

original channel morphology; a fluvial threshold is passed and a major rearrangement of the channel pattern or its configuration must take place.

During this long-time interval, the river was in a perpetual quasi-equilibrium condition if one considers only the instantaneous responses of hydraulic geometry (w, d, v, sediment concentration, roughness) to short-lived events such as floods. It is clear, however, that during the same interval, the river was *approaching* a different equilibrium condition that required a particular channel morphology or pattern to balance the new mean values of discharge and load.

We see, then, that fluvial equilibrium depends on the time scale one uses to define it. Schumm and Lichty (1965) recognized this problem by designating time spans as steady, graded, and geologic (shown earlier in table 1.4) and indicating that some variables of fluvial geomorphology are dependent or independent according to the time span being considered. Channel morphology, for example, is an independent variable during steady time and exerts a direct control on the hydraulics of flow. During graded time, however, channel morphology is a dependent variable.

Our interest here is in the adjustments a river might make to counterbalance changes in discharge and load that occur over a period of hundreds of years, the time interval known as *graded time* (Schumm and Lichty 1965). Channel morphology is the main dependent variable on this temporal scale, largely determined by mean values of the controlling factors. Rivers during this episode may appear to be quite stable, if stability is judged by hydraulic geometry. Even the morphology may show little change since its adjustment may be imperceptibly slow.

Adjustment of Gradient

A common response of channel morphology to changes extending over a graded time span is the alteration of slope. As discussed earlier, in alluvial rivers the normal downstream decrease in gradient, which promotes a concave-up longitudinal profile, is a function of increasing discharge and decreasing particle size. The concavity, however, is usually not perfectly smooth in detail but is commonly interrupted by perturbations. These can be caused by reaches where the channel is floored by bedrock, or by local zones of erosion or deposition. Local filling may be initiated by an influx of bed material load that is too coarse or too great in volume to be transported on the preexisting gradient. Coarse sediment, for example, may be introduced to a trunk channel from a tributary basin where active tectonism is occurring. The addition of the coarse load requires deposition at the confluence of the two rivers until the local gradient is increased sufficiently to allow the bedload to be transported (fig. 6.28A). As the channel floor is raised by deposition, the slope of the river upstream is effectively lowered, and a wave of filling may spread through the channel network.

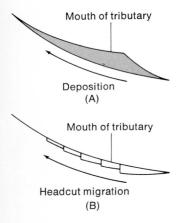

Figure 6.28. Adjustment of river gradient to changes in load. In (A) tributary delivers more coarse-grained load to main channel, inducing a wave of deposition upstream from confluence of rivers. In (B) influx of fine sediment initiates entrenchment downstream from junction and headcut migration upstream.

In contrast, a change that produces finer or less load may induce the river to entrench its channel (fig. 6.28B). This lowers the gradient downstream and creates an oversteepened segment, called a *knickpoint,* at the point of maximum entrenchment. As the knickpoint migrates in a headward direction, erosion may begin upstream. The extent of headward erosion seems to depend on the geological setting. Both flume and field data suggest that in channels formed of cohesionless material, pronounced knickpoints will be smoothed out after only a short distance of upstream migration (Brush and Wolman 1960; Morisawa 1964). Where the channel is composed of bedrock or cohesive sediments, the knickpoint may retreat for considerable distances and still preserve a vertical headcut. In this situation, material falling to the base of the headcut must be continuously removed.

It appears, then, that regardless of the change in discharge and/or load, one response available to rivers is to alter declivity. This adjustment may be accomplished by cutting or filling of the channel until the slope is regraded to an angle that can transport the new load most efficiently with the available discharge. This response is familiar to geologists as a basic part in the concept of the *graded river.* The idea of a graded condition in rivers finds its roots, once again, in the writings of G. K. Gilbert, who indicated that ''equilibrium of action'' in streams consists of a mutual adjustment between velocity, discharge, slope, and load. Later W. M. Davis (1902) refined these ideas by introducing the terms ''grade'' and ''graded slopes'' to describe the balanced fluvial condition. Although terminology left wide latitude for interpretation, the early conceptual models clearly stressed the importance of slope in the adjusting process, paving the way for the widely accepted premise that a concave longitudinal profile is the trademark of a graded river. In addition, Davis tied the graded condition to his cycle of erosion by suggesting that development of the graded profile, which is the optimum form to transport sediment, marks the beginning of the mature stage of the cycle.

The idea of a graded river did not go unchallenged, and in fact it was in answer to these challenges that Mackin (1948) put the concept into perspective and provided for the first time a clear definition of a graded river as

> . . .one in which, over a period of years, slope is delicately adjusted
> to provide, with available discharge and with prevailing channel charac-
> teristics, just the velocity required for the transportation of the load
> supplied from the drainage basin. The graded stream is a system in
> equilibrium; its diagnostic characteristic is that any change in any of the
> controlling factors will cause a displacement of the equilibrium in a
> direction that will tend to absorb the effect of the change.

The concept of grade as an equilibrium condition is valuable in understanding fluvial mechanics even though it probably overstates the role of slope. The construction of Hoover Dam, for example, caused a radical decrease of load in the Colorado River downstream from the dam. The river did not adjust to the altered load by a slope change; instead, an increase in roughness brought about the expected decrease in velocity (Leopold and Maddock 1953). Other studies have shown dramatic changes in the channel shape to be the prime factor involved in the balancing process (Schumm and Lichty 1963; Knox 1972). Thus, every altered load condition does not have to be countered with a modification of declivity alone. This has tempted geomorphologists to consider the possibility that all rivers flowing in alluvial channels are graded, except that they adjust their slope and/or other channel characteristics to transport their loads. In that sense, grade becomes analogous to quasi-equilibrium, and the graded condition represents the most probable state for the channel configuration and flow properties (Langbein and Leopold 1964).

In spite of the advantages in considering grade and quasi-equilibrium as equivalent, it is important to remember Mackin's words "over a period of years," since they may state the true distinction of a graded river. It is known, for example, that actively downcutting streams are still in quasi-equilibrium as defined by their hydraulic geometry; that is, the hydraulic variables are perfectly adjusted to flow and their measurement would not indicate that any fluvial response is occurring or, for that matter, that any change requiring a response has occurred. It is also true, however, that the very fact of progressive entrenchment indicates the river is approaching a different equilibrium, one established over a period of years. It is not, perhaps, necessary to require every graded-time adjustment to be made by a change in slope, but it is important to recognize the time distinction. Graded-time adjustments seem to be made by the variables of channel configuration and by pattern changes as alluded to by Schumm and Lichty (1965). Significantly, the initial response may appear in the form of a hydraulic variable (such as n in the Hoover Dam situation), but this may not be the ultimate response. If the inducing change is minor, the variables of hydraulic geometry may absorb it, just as they do during floods. But if the change is major, and especially if it occurs gradually, the final response may be one involving the channel configuration or pattern, and changes in those factors take time.

Adjustment of Shape and Pattern

As explained above, one of the more significant advances in recent years is the growing awareness of fluvial geomorphologists that rivers can respond to altered discharge and/or load in ways other than cutting or filling of the channel (for example, see Dury 1964). In a series of papers Schumm (1965, 1968, 1969) pieced together many of the empirical equations we have examined into a comprehensive, though qualitative, model of possible river adjustments to altered hydrology and load. The following equations are the basis for what Schumm (1969) refers to as *river metamorphosis:*

$$Q_w{}^+Q_t{}^+ \simeq \frac{w^+L^+F^+}{P^-}S^\pm d^\pm$$

$$Q_w{}^-Q_t{}^- \simeq \frac{w^-L^-F^-}{P^+}S^\pm d^\pm$$

$$Q_w{}^+Q_t{}^- \simeq \frac{d^+P^+}{S^-F^-}w^\pm L^\pm$$

$$Q_w{}^-Q_t{}^+ \simeq \frac{d^-P^-}{S^+F^+}w^\pm L^\pm$$

In these equations Q_t is the percentage of the total load transported as bed material load (sand-sized or larger), and Q_w can be either the mean annual discharge or the mean annual flood. The other variables are width (w), depth (d), slope (S), meander wavelength (L), width-depth ratio (F), and sinuosity (P). The plus or minus exponents indicate whether the variables are increasing or decreasing.

To exemplify the use of these equations, let us assume that a large area is clear-cut of its natural forest cover. We can expect an increase in Q_w because infiltration rates will be lowered and direct runoff will increase, as well as an increase in Q_t because coarse sediment normally stabilized on slopes by rooting now makes its way to the channel; the coarse sediment also will be moved more frequently because of the increased peak discharge. With both Q_w and Q_t increasing, we can expect increases in width, wavelength, and W/D, and a decrease in sinuosity. Depth and slope may vary in either direction. Slope will probably increase, however, because the channel becomes straighter, and depth will probably be constant or decrease since both w and W/D increase.

Perhaps the best geological example of river metamorphosis is found in the history of the Murrumbidgee River which flows across a large alluvial plain in New South Wales, Australia (Schumm 1968). As figure 6.29 illustrates, the present highly sinuous river flows within a floodplain containing large oxbow lakes and other features that preserve an older and larger channel of the Murrumbidgee (paleochannel 1). Evidence of a still older, low-sinuosity channel (paleochannel 2) is also present on the plain. The morphologic, sedimentologic, and hydrologic characteristics for the three channels are presented in table 6.4. Pedogenic and geomorphic data confirm that during the tenure of paleochannel 2 the climate was more arid than at present, and at the time of paleochannel 1, more humid than now. Using the present river as a norm, Schumm (1968) compared the adjustments in channel parameters that occurred in the Murrumbidgee River under changing climates. In a change toward aridity (present → paleochannel 2), the decrease in precipitation over the entire basin should decrease Q_w and increase Q_t. According to the equations of river metamorphosis, to transport the changed load with less water, the channel should become shallower and probably wider (d^-, F^+). Wavelength should probably increase (L^+) because sinuosity should decrease (P^-) and slope increase (S^+). Comparison of the Mur-

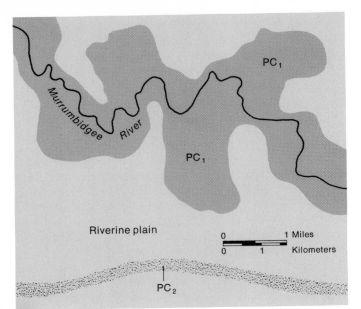

Figure 6.29. Adjustments of the Murrumbidgee River, Australia, to changes in climate and sedimentology. PC₂ (paleochannel 2) functioned at time of greater aridity; PC₁ (paleochannel 1) was sinuous and much larger than present river. Modern river is highly sinuous and flows within the channel limits of PC₁. (After Schumm 1968, fig. 9)

Figure labels: PC₁, PC₁, Riverine plain, PC₂

Table 6.4 Fluvial, geometric, and sedimentological data comparing the modern river with paleochannel 1 and paleochannel 2, Murrumbidgee River, Australia.

	Median grain size (mm)	Channel silt-clay percent (M)	Width (ft)	Depth (ft)	Width-depth ratio	Sinuosity	Gradient (ft per mi)	Bankfull velocity (ft per sec)	Bankfull discharge (cfs)	Sand discharge (tons per ft. per day)	Sand discharge (tons per day)	Meander wavelength (ft)
Murrumbidgee River near Darlington Point	0.57	25	220	21	10	2.0	0.7	3.0	10,000	9	2,000	2,800
Paleochannel 1		16	460	35	13	1.7	0.8	4.2	51,000	45	21,000	7,000
Paleochannel 2 **Northern (Kearbury pit):**												
Small	0.55	1.6	600	9	67	1.1	2.0	5.2	23,000	90	54,000	18,000
Medium			1,000	12	83			6.3	73,000	140	140,000	
Large			1,700	20	90			8.8	290,000	300	510,000	
Central (Kulki pit):												
Small	0.60	3.4	500	8	63	1.1	2.0	4.8	19,000	70	35,000	15,000
Medium			800	9	90			5.2	35,000	80	64,000	
Large			800	14	57			7.0	77,000	210	178,000	

After Schumm 1968.

rumbidgee variables in the table shows that the actual changes agree with those predicted by Schumm's equations.

An increase in precipitation (present → paleochannel 1) would increase Q_w but probably cause little change in sediment yield or size because the density of vegetation would also increase (as discussed in chapter 5, fig. 5.27). In response, the slope and W/D would show little change, but the channel would become wider and deeper, and the dimensions of meandering (wavelength and amplitude) increase (table 6.4).

One of the more significant observations made by Schumm is that changes in the Murrumbidgee gradient in response to altered controls were made without cutting or filling the channel. Most of the slope adjustment was accomplished by changes in the length of the river due to variations in sinuosity.

To summarize, our present understanding of rivers requires that we rethink the meaning of equilibrium and its relation to fluvial mechanics. The following emerge as prime points for future consideration and should be of special interest to engineers and geologists:

1. The adjustment of slope in rivers can be made by a major change in channel pattern rather than by vertical filling or trenching.
2. The initial response of a river may not be the same as its end response. In addition, regulation of a river may generate responses that do not remain local but alter channel morphology over great lengths of the river system.
3. Both discharge and sediment must be considered in the prediction of fluvial mechanics. The greater width associated with channels of low discharge (such as paleochannel 2 in the Murrumbidgee, table 6.4) is contrary to many regime equations relating width and discharge. The tendency of the river's response to a change in discharge, therefore, can be counterbalanced by a simultaneous change in the character of the load.

Summary

River action, like all geomorphic processes, behaves according to the driving and resisting forces built into the system. For example, a river will entrain, transport, or deposit sediment depending on the driving energy given to the water by velocity, depth, and slope, and on the amount of that energy consumed by the resistance to flow offered by elements such as channel configuration, particle size, and sediment concentration. The work demanded of a river—the amount of discharge and load it must handle—is determined by the geological and climatic character of the drainage basin. Each river develops a particular combination of shape, gradient, and hydraulic variables (called the hydraulic geometry) that allows it to accomplish its work most efficiently. The river will attempt to maintain its high efficiency by adjusting the above properties whenever discharge or load vary. Because discharge and load fluctuate continuously, equilibrium as a steady-state condition can never be attained, and the river variables must perpetually be adjusting. Nonetheless, the normal variations of discharge and load are accommodated by hydraulic geometry.

The type of channel pattern (straight, meandering, braided) a river displays and the longitudinal profile are other fluvial characteristics controlled by the basin environment. Each pattern originates in a specific manner, and its geometric form is designed to facilitate the work of a river, measured as the prevailing values of discharge and load. Once established, the pattern will be maintained as long as the normal variations in load and discharge can be absorbed by the mechanics of hydraulic geometry.

Major long-term changes in climate or basin tectonics may alter the average discharge and/or load to a point where adjustments of hydraulic geometry can no longer maintain the most efficient system. When those threshold values of discharge or load are reached, major fluvial responses in the form of pattern changes, degradation or aggradation, or dramatic revisions of the width-depth ratio will occur to reestablish the greatest fluvial efficiency. An excellent example of these major reactions has been detailed for the Murrumbidgee River in Australia. It is important to recognize, however, that our present knowledge does not allow us to predict which of the possible adjustments will occur in response to major changes in the fundamental controls.

Suggested Readings

The short bibliography that follows should provide you with greater detail concerning the concepts treated in this chapter.

Chorley, R. J. 1969. *Introduction to fluvial processes.* London: Methuen and Co.

Leopold, L. B., and Maddock, T., Jr. 1953. The hydraulic geometry of stream channels and some physiographic implications. U.S. Geol. Survey Prof. Paper 252.

Leopold, L. B., and Wolman, M. G. 1957. River channel patterns; braided, meandering and straight. U.S. Geol. Survey Prof. Paper 282-B.

Leopold, L. B.; Wolman, M. G.; and Miller, J. P. 1964. *Fluvial processes in geomorphology.* San Francisco: W. H. Freeman.

Mackin, J. H. 1948. Concept of the graded river. *Geol. Soc. America Bull.* 59:463-512.

———. 1963. Rational and empirical methods of investigation in geology. In *The fabric of geology,* edited by C. Albritton, pp. 135-63. Reading, Mass.: Addison-Wesley.

Maddock, T., Jr. 1969. The behavior of straight open channels with movable beds. U.S. Geol. Survey Prof. Paper 622-A.

Morisawa, M. 1968. *Streams, their dynamics and morphology.* New York: McGraw-Hill.

Schumm, S. A. 1968. River adjustment to altered hydrologic regimen, Murrumbidgee River and paleochannels, Australia. U.S. Geol. Survey Prof. Paper 598.

———. 1969. River metamorphosis. *Am. Soc. Civil Engrs. Proc., Jour. Hyd. Div.,* HY 1:255-73.

In chapter 6 we examined the basic mechanics of fluvial processes and found that the activity within a stream channel is generally related to the energy possessed by the river and to the ways that energy is utilized to carry water and sediment most efficiently. Rivers, however, are more than natural sluices; they also mold the geologic setting into discernible topographic forms. They accomplish this primarily through the erosional capability inherent in the movement of sediment-laden water, and through the deposition of debris that occurs when the transporting energy is less than the demands being made on it. Some fluvial features are purely erosional; the topographic form is clearly one of sculptured rock, and little, if any, sediment is associated with the feature. Others may be entirely depositional, and the exposed topography is formed by the burial of an underlying surface that existed before the covering sediment was introduced. In these cases, the bedrock framework may have no influence on the surface configuration. Many features spring from some combination of both erosion and deposition; the pure cases are probably end members of a continuum of possible forms.

If rivers establish or nearly establish some form of equilibrium, it seems reasonable to expect that fluvial features—the tangible results of river work—will somehow reflect the balanced processes that created them. Here we will investigate these end products of fluvial action and, wherever possible, document how the properties of features may reveal the processes involved in their origin.

Fluvial Landforms 7

Floodplains

Floodplains are perhaps the most ubiquitous of fluvial features, found in the valley of every major river and in most tributary valleys. However, a precise definition of a floodplain is more difficult than one might expect. Topographically and geologically speaking, a floodplain is the relatively flat surface occupying much of a valley bottom and is normally underlain by unconsolidated sediment. The sediments of most valley bottoms are not necessarily a function of the river occupying the valley, but may be deposited there by a variety of geomorphic processes. Nonetheless, to be considered as part of the floodplain, the surface and the sediments must somehow relate to the activity of the present river. The definition must also have a hydrologic connotation, since the floodplain is a surface subject to periodic flooding. It can easily be defined in terms of hydrology as the water level attained in some particular stage of the river. Detailed analyses (Wolman and Leopold 1957) demonstrate that most topographic floodplains are subjected to flooding nearly every year or every other year. The recurrence interval of bankfull stage, for example, averages about 1.5, indicating that most rivers leave their channels two out of every three years. If the surface flanking the river has any relief, however, part of the *topographic* floodplain will not be inundated by the annual or biannual flood that marks the *hydrologic* floodplain. Furthermore, the yearly flood may not be very significant in terms of flood damage, a concern of most engineers and basin managers. In fact, many scientists use the *flood-damage stage*—the water level where overflow begins to cause damage —to mean the flood stage. In general, the damage stage is well above both the level of bankfull and the average elevation of the topographic floodplain.

Figure 7.1. Normal sequence of floodplain stratigraphy. Lower two-thirds of deposit is composed of coarse-grained, laterally accreted, point bar gravel. Yardsticks show how gravel in lowest unit dips slightly to the left. Upper third of deposit is vertically accreted silt deposited during overbank flow. Some gravel at top of floodplain is also overbank debris. Sexton Creek, Shawnee National Forest, southern Illinois.

Regardless of how it is defined, a floodplain plays a very necessary role in the overall adjustment of a river system. It not only exerts an influence on the hydrology of a basin (lag, etc.) but also serves as a temporary storage bin for sediment eroded from the watershed. Therefore, floodplains are features that are both the products of the river environment and important functional parts of that system.

Deposits and Topography

Floodplains are composed of a variety of sediments that are created by diverse processes and accumulate in distinct subenvironments within the valley bottom. Most floodplains can be differentiated into deposits of channel fill, channel lag, splays, colluvium, lateral accretion, and vertical accretion (Happ et al. 1940; Lattman 1960). Near the valley sides, *colluvium,* resulting from unconfined wash and mass wasting, may be prominent in the floodplain sequence; toward the axis of the valley, these deposits grade into alluvial-type deposits. Coarse debris from which the fines have been winnowed are interpreted as *channel lag* deposits, in contrast to *channel fill,* which consists of a poorly sorted admixture of silt, sand, and gravel. *Splay* deposits are composed of material spread onto the floodplain surface through breaks in natural levees and usually are more coarse-grained than the overbank sediments that they cover. The most important deposits in the floodplain framework are those of *lateral accretion* and *vertical accretion,* which in some cases can be separated on the basis of particle size, the laterally accreted sediment usually being sands and gravels, more coarse-grained than the vertically accreted silts and clays (shown in the photograph in fig. 7.1). Point bars, however, the most common deposit of lateral accretion, may sometimes have the same texture as the overbank sediments (Wolman and Leopold 1957). Therefore, particle size is not an infallible criterion for distinguishing between vertical and lateral accumulation.

Although some large rivers do not have clearly organized channels (Garner 1967) and not all form normal floodplains (Tanner 1974), most valleys of large rivers are occupied by well-defined floodplains that consist primarily of laterally and vertically accreted deposits. These are usually associated with specific depositional environments. In the Mississippi River valley, for example, they have been broadly categorized (Fisk 1944, 1947) as (1) channel types, including point bars, chutes and sloughs, and sand ridges of meander scrolls; and (2) overbank types, consisting of splays, natural levees, and backswamps (fig. 7.2). The surface of the floodplain may have considerable microrelief that reflects the fluvial mechanics in the various depositional environments.

The floodplain surface is most irregular in a zone close to the river where point bars are molded into alternating ridges and swales that Leopold and his co-workers (1964) refer to as *meander scrolls.* The characteristic scroll topography may start as a longitudinal bar with a narrow trough to its rear (Sundborg 1956) or simply as a low ridge of sediment that accumulates on the inside of a

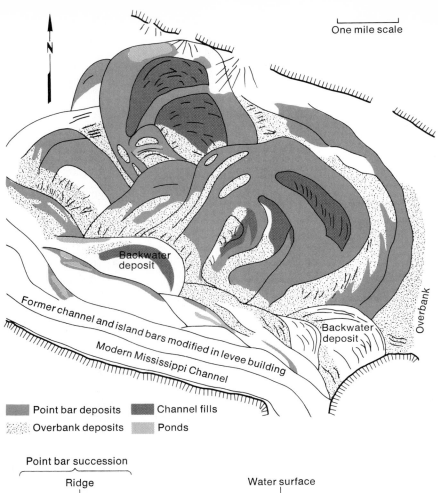

One mile scale

N

Backwater
deposit

Former channel and island bars modified in levee building

Modern Mississippi Channel

Backwater
deposit

Overbank

Figure 7.2. Map showing complex distribution of various types of deposits on a portion of the Mississippi River floodplain near Grand Tower, Ill. (Courtesy of S.E. Harris, Jr.)

Point bar deposits   Channel fills

Overbank deposits   Ponds

Figure 7.3. Meander scroll topography formed by point bar deposition in a laterally migrating river. (From Hickin 1974. Used with permission of *American Journal of Science*)

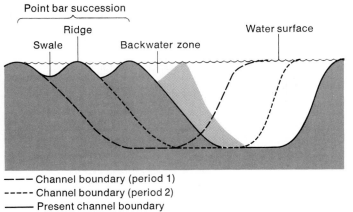

Point bar succession

Ridge

Swale   Backwater zone   Water surface

— — — Channel boundary (period 1)
- - - - Channel boundary (period 2)
——— Present channel boundary

meander bend during bankfull flow (Kolb and Van Lopik 1958; Hickin 1974). When the high discharge subsides, the ridge is exposed and rapidly vegetated, becoming, for all practical purposes, the new channel bank. The next high flow repeats the process. As the river shifts across the valley by undercutting on the outer bank, the successive ridges and intervening swales that characterize meander scroll topography develop simultaneously (fig. 7.3). Periodically, flow will break across the point bar surface, often occupying a particular swale and scouring the surface into more pronounced low channels called *chutes*. Wolman and Leopold (1957) measured velocities of up to 3 fps in chute channels, a flow that is capable of eroding the surface and transporting coarse sand. Chute erosion tends to accentuate the scroll topography, and even though the slough areas may be silted in by overbank deposition, the scalloped profile of scroll topography (fig. 7.4) can be preserved for hundreds of years (Hickin and Nanson 1975). (For greater detail about point bar mechanics and deposits, see McGowen and Garner 1970.)

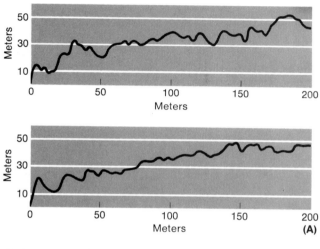

Figure 7.4. (A) Diagram of two profiles of meander scroll topography that has been preserved for considerable periods of time. Beatton River, Canada. (After Hickin and Nanson, *GSA Bulletin,* 1975, vol. 86, p. 491, fig. 6. Used with permission of the Geological Society of America) (B) Meander scroll topography on abandoned point bar of the Mississippi River in Pike County, Ill. Local relief between ridges and swales is about 3 meters. (Photo by U.S. Dept. of Agriculture. From the University of Illinois Catalog of Aerial Photography)

In addition to the irregularity caused by scrolls on the inside of meander bends, some topographic relief near the channel is due to the formation of *natural levees*. Natural levees stand along most major rivers as low ridges that commonly are broader than one might expect, often extending for hundreds of meters. Levees are usually highest near the active channel and slope gradually toward the valley sides. They owe their character to the retardation of flow velocity when rivers leave the channel, which results in the largest suspended particles being deposited adjacent to the bank. Levee deposits are therefore more coarse-grained than most other overbank sediment and are accreted more rapidly. For example, Kesel and his colleagues (1974) found a net addition of 53 cm on natural levees during the two-month overflow of the Mississippi River in 1973, while only 1.1 cm accumulated in the backswamp area during the same period.

As one proceeds away from the river, the floodplain topography becomes much more regular, and its flatness is interrupted only by oxbow channels or by splay deposits that have been debouched onto the backswamp surface. The zone away from the active river is characterized by gradual accumulation of overbank sediment that tends to subdue any existing relief. Abandoned channels represented by oxbows or oxbow lakes gradually fill with silts and clays, leaving *clay plugs* as the only evidence of the former channel. Old sloughs, chutes, or cutoffs can also fill with overbank sediment, adding to the general reduction of relief away from the channel. It is incorrect, however, to assume that all overbank sediment is fine-grained. Where banks are cohesive, coarse gravel can be transported onto the surface of the floodplain (McPherson and Rannie 1967; Costa 1974; Ritter 1975). In a short-lived flood on Sexton Creek (Illinois), gravel as large as 205 mm was deposited as lobes or fanlike splays on top of the floodplain, and particles up to 102 mm were carried there in suspension.

In general, the relationship between floodplain deposits and river processes would be straightforward if rivers would stay in the same place for extended periods of time. Actually, as the rates in table 7.1 indicate, most rivers migrate laterally across the valley bottom quite rapidly, forcing the depositional environments also to shift their location with time. A backswamp region, for instance, may include remnant deposits of a channel. The displacement of one environment by another adds to the complex maze of floodplain deposits, emphasizing the point that floodplains are dynamic rather than static fluvial features.

The Origin of Floodplains

It is now generally accepted that two dominant fluvial processes act simultaneously to develop most floodplains. As described earlier, maximum erosion in meandering rivers takes place on the outer bank just downstream from the axis of curvature. At the same time, sediment accumulates in point bars that build up along the inside of the meander bend. Detailed study of these processes has documented that bank erosion and point bar accumulation are volumetrically

Table 7.1 Rates of lateral migration of rivers in valleys.

River and location	Approximate size of drainage area (square miles)	Amount of movement (feet)	Period of measurement	Rate of movement (feet per year)
Tidal creeks in Massachusetts		0	60-75 yr.	0
Normal Brook near Terre Haute, Ind.	±1	30	1897-1910	2.3
Watts Branch near Rockville, Md.	4	0-10	1915-55	0-0.25
	4	6	1953-56	2
Rock Creek near Washington, D.C.	7-60	0-20	1915-55	0-0.50
Middle River near Bethlehem Church, near Staunton, Va.	18	25	10-15 yr.	2.5
Tributary to Minnesota River near New Ulm, Minn.	10-15	250	1910-38	9
North River, Parnassus quadrangle, Va.	50	410	1834-84	8
Seneca Creek at Dawsonville, Md.	101	0-10	50-100 yr.	0-0.20
Laramie River near Ft. Laramie, Wyo.	4,600	100	1851-1954	1
Minnesota River near New Ulm, Minn.	10,000	0	1910-38	0
Ramganga River near Shahabad, India	100,000	2,900	1795-1806	264
	100,000	1,050	1806-1883	14
	100,000	790	1883-1945	13
Colorado River near Needles, Calif.	170,600	20,000	1858-83	800
	170,600	3,000	1883-1903	150
	170,600	4,000	1903-1952	82
	170,600	100	1942-52	10
	170,600	3,800	1903-42	98
Yukon River at Kayukuk River, Alaska	320,000	5,500	170 yr.	32
Yukon River at Holy Cross, Alaska	320,000	2,400	1896-1916	120
Kosi River, North Bihar, India		369,000	150 yr.	2,460
Missouri River near Peru, Nebr.	350,000	5,000	1883-1903	250
Mississippi River near Rosedale, Miss.	1,100,000	2,380	1930-45	158
	1,100,000	9,500	1881-1913	630

From Wolman and Leopold 1957. See this work for data sources for individual rivers.

equal during any given period of lateral and downvalley migration of the meander bends (Wolman and Leopold 1957). In addition, data from the same study show that the point bars tend to increase in height until they reach the level of the older part of the floodplain. It seems clear that a meandering river can shift its position laterally without changing the channel shape or dimensions.

In channels where coarse sediment is an important part of the load, the point bars tend to collect sediment that is easily distinguished from that of overbank origin. During low flow, sediment of all sizes may be temporarily trapped in the channel floor, but at the peak of bankfull discharges this material will be removed along with any debris eroded from the undercut banks. The coarse sediment is deposited on the point bars, now submerged. As figure 7.1 showed, these deposits commonly display cross-beds dipping into the channel. Over a period of years, the point bars expand laterally, being progressively spread across the valley bottom as a thin sheet of sand or gravel (Mackin 1937; Leopold et al. 1964). If the load is not characterized by coarse-grained particles, the spreading of lateral accretion deposits proceeds in exactly the same way (fig. 7.5), but the point bar sediment may be more difficult to distinguish from the overbank materials.

The maximum thickness of laterally accreted deposits is determined by the depth to which a river can scour during recurring floods. Based on the results of empirical studies, Wolman and Leopold (1957) suggest that natural channels are probably scoured to a depth 1.75 to 2 times the depth of flow attained during a flood. This rule of thumb generally fits the scouring observed in a variety of perennial rivers (Leopold et al. 1964, p. 229), but where the channel width is prevented from expanding by bridge supports or extremely resistant banks, the scour depth may reach 3 or 4 times the water depth. The thickness of lateral accretion deposits should, therefore, increase gradually downvalley along with the normal increase in discharge and depth noted on most rivers.

The second dominant process in the origin of floodplains is overbank flow. As rivers wander across their valley bottoms, they normally leave their channel confines during periodic flooding and deposit fine-grained sediment on top of the floodplain surface. Because of this, the floodplain is vertically accreted, and in

Figure 7.5. Progressive lateral erosion and point bar deposition (cross sections) in Watts Branch near Rockville, Md. (From Wolman and Leopold 1957)

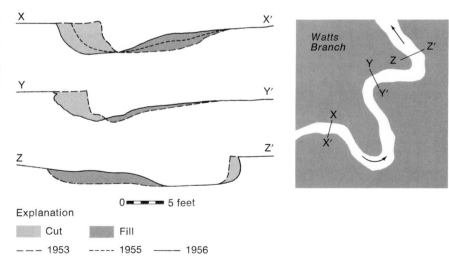

Explanation

▨ Cut   ▨ Fill

--- 1953   ----- 1955   ——— 1956

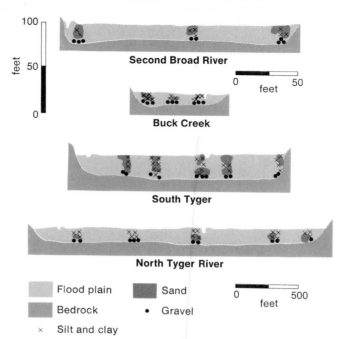

feet

100

50

0

Second Broad River

0   feet   50

Buck Creek

South Tyger

North Tyger River

Flood plain   Sand

Bedrock   • Gravel

× Silt and clay

0   feet   500

Figure 7.6. Cross sections of floodplains in North Carolina and South Carolina. (From Wolman and Leopold 1957)

most floodplain stratigraphy a thin layer of overbank silt and clay rests on the laterally accreted point bar deposits described above (figs. 7.1, 7.5).

Ideally, then, the entire floodplain sequence consists of a relatively thin accumulation of laterally and vertically accreted sediments that have been spread more or less evenly across the valley bottom (fig. 7.6). The two types of deposits can differ in their textural properties (although the process mechanics does not require it), but they will be essentially the same age.

An analogous development is associated with floodplains of braided rivers except that the systems are more dynamic and less regular. Bars and bank erosion, for example, are not restricted to one particular side of the channel, and the river can shift its position without laterally eroding the intervening material. Abandoned channels and islands gradually coalesce into a continuous floodplain surface. In a braided system the scouring depths will probably be less than those in meanders because high flow will be accommodated more easily by increasing width than by scour of the channel. In addition, the flow in a braided system is carried by a number of channels rather than one, so that depths should be lower. It seems reasonable, then, to expect floodplain sediments in a braided stream system to be less thick and more irregular. Recognition of the true floodplain sequence, however, may be complicated by the fact that braided systems are commonly associated with long-term valley aggradation. In such a case, the floodplain sequence might appear to be enormously thick. However, floodplains relate to the hydraulics of the present river only, and the sedimentary pile it produces is probably just a thin skim on top of the fill.

Relative Importance of Lateral and Vertical Processes   The model of floodplain origin just described raises the question as to which process—lateral migration or overbank flow—plays the dominant role. There is probably no universal answer to this question, because each system obeys its own unique combination of controlling factors. Nonetheless, evidence suggests that most floodplains result primarily from the processes associated with lateral migration. Perhaps the most persuasive argument for that conclusion uses a comparison of the rates involved in the two competing processes. Table 7.2 is a random sampling of sediment increments measured during floods. Although incomplete, it shows that backswamp deposition during overbank flow tends to be limited, probably because (as discussed in chapter 6) the maximum sediment concentration in a flood occurs before bankfull stage is reached. That is, most sediment is transported from the system before overbank conditions are attained.

Assuming the same vertical increment with each flood, the level of a floodplain built entirely by overbank deposition should increase at a progressively decreasing rate, as shown by the curves in figure 7.7. The initial growth would be rapid because flooding would occur frequently, and perhaps 80–90 percent of the floodplain construction would take place in the first fifty years (Wolman and Leopold 1957; Everitt 1968). However, as the surface grows higher relative to the channel floor, the stage needed to overtop the banks is also increasing. The surface is inundated less frequently, and the rate of growth is drastically retarded. The fact that most floodplains are occupied with water nearly every year argues against the importance of overbank deposition in their construction.

At reasonable increment rates for overbank deposition, a 3 m thick floodplain sequence would probably take several thousand years to accumulate. Assuming this is a valid estimate, it is instructive to note again the lateral migration rates given in table 7.1, which indicate that most large rivers migrate quite rapidly. This is especially true when their meander geometry is adjusted for efficient lateral shifting. Hickin and Nanson (1975) suggest that the rate of lateral migration in meander bends of the Beatton River (British Columbia, Can.) is

Table 7.2 Increment rates of overbank deposition in major floods.

River basin	Flood	Average thickness of deposition (feet)
Ohio River	Jan.-Feb., 1937	0.008
Connecticut River	March, 1936	.114
Connecticut River	Sept., 1938	.073
Kansas River	July, 1951	.098

From Wolman and Leopold 1957. See this work for references on individual data sources.

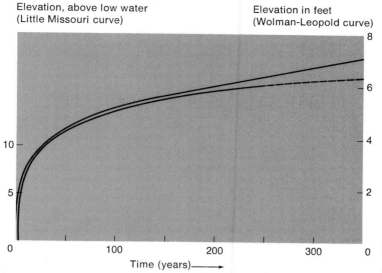

Elevation, above low water
(Little Missouri curve)

Elevation in feet
(Wolman-Leopold curve)

Time (years) ——→

Figure 7.7. Increase in elevation of floodplain with time.
Lower curve from empirical
data collected on floodplain of
the Little Missouri River;
upper curve was derived
theoretically for Brandywine
Creek (Pa.) by Wolman and
Leopold (1957). Note different
vertical scales. (From Everitt
1968. Used with permission of
American Journal of Science)

greatest when r_m/w is about 3 (see chapter 6). At values higher or lower than 3,
the rate of channel migration decreases dramatically.

It seems certain that the magnitude of vertical accumulation depends primarily on the rate at which the river migrates laterally. The total thickness will
approximate the vertical accretion that can be accomplished in the time the river
takes to migrate the entire width of the valley. For example, if a floodplain is a
kilometer wide and the river shifts laterally at a rate of 2 m a year, it will take
500 years for the river to complete one swing across the valley. At any given
locality, perhaps several meters of overbank sediment will accumulate in that
time, but the entire deposit will be reworked by lateral erosion when the river
reoccupies that position. The lateral migration rate thus becomes a controlling
and limiting factor on the thickness of overbank deposition.

The apparent preeminence of lateral processes in floodplain construction
does not mean that overbank deposition is unimportant. Schumm and Lichty
(1963) present evidence to suggest that vertical accretion probably dominated the
initial stage in floodplain development on the Cimarron River in Kansas. Everitt
(1968) also notes that vertical accretion might be more important in building
floodplains of semiarid rivers even though lateral erosion may subsequently rework the sequence. In some situations, the river might lack the widespread lateral
movement needed to rework the entire floodplain, and portions of the surface
could continue unimpeded growth by overbank deposition. A part of the Delaware River has proceeded in this manner for at least 6,000 years (Ritter et al.
1973). Similarly, portions of the lower Mississippi River seem to meander in
only a narrow zone of the valley bottom at any time. Although the position of the

meander belt shifts periodically, the restricted width of lateral migration has allowed marginal portions of the backswamp zone to grow by vertical accretion for perhaps 10,000–11,000 years (Kesel et al. 1974).

In summary, floodplains appear not only to be formed by balanced fluvial systems, but also to serve as integral parts of the system. They are constructed by simultaneous processes of lateral migration and overbank flooding. Point bars, the deposits of lateral accretion, are spread in a rather even sheet across the valley bottom, while overbank deposits accumulate over the entire floodplain surface away from the channel. The floodplain acts as a storage area for sediment that cannot be transported directly from the basin when it is eroded.

Floodplains are usually considered to be features associated with stable rivers, but there is no overriding reason why they cannot be present when a channel is undergoing long-term aggradation or degradation. In fact, the observed frequency of overbank flooding can continue during valley filling if the channel floor and the floodplain surface are raised at the same rate. Once the thickness of the valley deposits exceeds the limits of a reasonable scouring depth, however, the sediment below that depth can no longer be considered as part of the active floodplain. In a degrading channel, the floodplain becomes a terrace when channel incision prevents the river from inundating the surface annually or biannually.

Fluvial Terraces

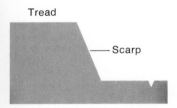

Tread

Scarp

Figure 7.8. Parts of a fluvial terrace.

Terraces are abandoned floodplains that were formed when the river flowed at a higher level than at present. The surface of the terrace is no longer related to the modern hydrology in that it is not inundated as frequently as an active floodplain. Topographically, a terrace consists of two parts: a *tread,* which is the flat surface representing the level of the former floodplain, and the *scarp,* which is the steep slope connecting the tread to any surface standing lower in the valley (fig. 7.8). The very presence of a terrace indicates an episode of downcutting; some change must occur between the conditions prevailing during formation of the tread and those producing the scarp. Usually the downcutting phase begins as a response to climatic or tectonic changes, but these are not always necessary. The tread surface normally is underlain by alluvium of variable thickness, but in a pure sense, these deposits are not part of the terrace. To avoid confusion, it is better to limit the term to the topographic form and refer to the deposits as fill, alluvium, gravel, etc.

Types and Classification

Howard and his co-authors (1968) categorize terraces as *erosional* or *depositional.* Erosional terraces are those in which the tread has been formed primarily by lateral erosion. If the lateral planation truncates bedrock, the terms *bench, strath,* or *rock-cut terrace* are commonly used. If the erosion crosses unconsolidated debris, the terms *fill-cut* or *fillstrath* (Howard 1959) have been suggested.

Depositional terraces, the second major grouping, are those terraces where the tread represents the uneroded surface of a valley fill. Figure 7.9 illustrates both types.

Erosional terraces, especially rock-cut types, are identifiable by the following, rather distinct, properties (Mackin 1937): (1) they are capped by a uniformly thick layer of alluvium in which the total thickness is controlled by the scouring depth of the river involved; and (2) the surface cut on the bedrock or older alluvium is a flat mirror image of the surface on top of the capping alluvium (fig. 7.9). In contrast, the alluvium beneath the tread of depositional terraces varies in thickness and commonly exceeds any reasonable scouring depth of the associated river. Although the tread surface may be flat, the surface beneath the fill can be very irregular (fig. 7.9).

Leopold and his colleagues (1964) suggest that terraces be classified as either *strath terraces,* which are those cut on bedrock, or *alluvial terraces,* which are topographic forms molded from the floors of alluvial valleys. At first glance this approach may seem to be the same as the one described above, but the two classifications have real and significant differences. To exemplify, figure 7.10 shows a terrace sequence formed from one alluvial fill. In the Leopold classification all the terraces are alluvial because the tread surface of each was carved from the floor of an alluvial valley. In the Howard approach, however, only the highest terrace is depositional. The lower two are erosional (fillstrath) because their treads have been developed by a laterally eroding river during quiescent periods between spasms of vertical downcutting, even though the lateral migration took place in alluvium. A series of alluvial terraces may, however, be entirely depositional if it stems from alternating episodes of cutting and filling. Figure 7.11 demonstrates that each tread in the terrace sequence is associated with a specific period of valley filling. Thus, use of the terms "erosional" or "depositional" requires some knowledge about the origin of the terrace, while the terms "alluvial" or "strath" simply refer to the material from which the terrace has been formed. In practice, recognizing the effects of lateral erosion on top of an alluvial fill is extremely difficult, and unless proof of origin is available it may be judicious to employ the less stringent terms.

Another classification scheme is based on the topographic relationship between terrace levels within a given valley, as illustrated in figure 7.12. In this method, terrace treads that stand at the same elevation on both sides of the valley are called *paired* (matched) terraces and presumably are the same age. If the levels are staggered across the valley they are said to be *unpaired* (unmatched) terraces. Most investigators interpret unpaired terraces as erosional types, formed by a stream simultaneously cutting laterally and downcutting very slowly. Levels across the valley, therefore, are not exactly equivalent in age, but differ by the amount of time needed for the river to traverse the valley bottom. Actually, unpaired terraces can also be depositional in origin if the entrenchment

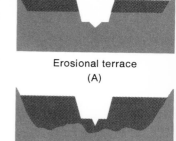

Erosional terrace
(A)

Depositional terrace
(B)

Figure 7.9. (A) Erosional terrace. Thin alluvial cover with truncation of underlying bedrock along smooth, even surface. (B) Depositional terrace. Terrace scarp underlain by alluvium that is highest level of fill deposited in valley. Note thickness of alluvium and irregular bedrock surface beneath the fill.

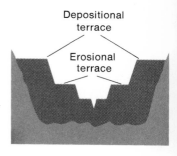

Depositional terrace

Erosional terrace

Figure 7.10. Terrace sequence in which both terraces are alluvial terraces because the scarp is underlain by alluvium. Upper terrace is depositional because it represents the level of valley fill. Lower terrace is erosional because it was formed by lateral erosion across the material of the original fill.

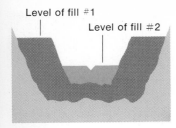

Level of fill #1
Level of fill #2

Figure 7.11. Terrace sequence with two alluvial terraces that are both depositional. Treads of both levels represent level of filling. The two episodes of filling are separated in time by a period of cutting, which formed the scarp of the high terrace and also the valley that contains the second fill (sometimes called a cut-and-fill terrace sequence).

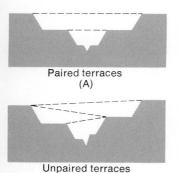

Paired terraces
(A)

Unpaired terraces
(B)

Figure 7.12. Terraces classified on basis of topographic relationships. (A) Paired terraces have treads at same level on both sides of the valley. (B) Unpaired terraces stand at different elevations on either side of the valley.

between two episodes of valley filling occurs at the valley sides rather than along the valley axis. In this situation, shown in figure 7.13, the second fill is placed adjacent to but lower than the earlier deposit; the subsequent placement of the river between the two fills gives the terraces their unpaired relationship (see Ritter 1967). Paired and unpaired terraces can be of any origin (erosional or depositional) and develop from either bedrock or alluvium (strath or alluvial). The terms, therefore, are entirely descriptive and carry no genetic connotation.

Still other terms have been used to classify and describe terraces (Howard et al. 1968), but often they seem to confuse rather than clarify the basic issue of how terraces form. For that reason, we will restrict terrace nomenclature to the few terms introduced above and pay special attention to the mechanics of terrace development.

The Origin of Terraces

Depositional Terraces   The development of a depositional terrace always requires (1) a period of valley filling and (2) subsequent entrenchment into or adjacent to the fill. This cyclic pattern is necessary because the alluvium at the tread surface takes its form from purely depositional processes. The tread, in fact, represents the highest level attained by the valley floor as it rose during aggradation. The initial entrenchment that forms the terrace scarp is primarily vertical, and so the tread surface is virtually unaffected by subsequent lateral erosion at a lower level (see fig. 7.9).

Valley filling occurs when, over an extended period, the amount of sediment produced in a basin exceeds the amount that the river system can carry away. Prolonged aggradation is usually triggered by (1) glacial outwash, (2) climate change, or (3) changes in base level, slope, or load due to rising sea level, rising local or regional base level, or an influx of coarse load because of uplift in source areas. Where tectonics are ruled out, the balance between load and discharge is determined primarily by climatic processes, although it may be driven by glaciation and may be complexly interrelated with sea level changes, etc. Although the mechanics of valley entrenchment are poorly understood, the incision, like filling, can be triggered by a variety of tectonic or climatic changes.

In regions affected by glaciation, outwash is probably the most common type of fill associated with the formation of depositional terraces. The properties of outwash and the mechanics of its deposition will be discussed in chapter 10. At this point it is enough to say that the amount and size of the load transported and released by active glaciers simply overwhelms the ability of rivers to remove it from the system.

In coastal regions, depositional terraces commonly result when alternating cutting and filling are initiated by fluctuations in sea level. During the Pleistocene, eustatic deviations of the ocean level (the base level for major rivers) resulted when ice was stored on the continents and then released. Theoretically, river entrenchment should accompany glacial expansion (when sea level is decreasing), while filling would take place during the waning phase of the glacial cycle (when sea level is rising). Fisk (1944) interpreted the depositional terrace sequence in the lower Mississippi Valley as being related directly to waxing and waning glaciations and their effects on eustatic sea level change. Although the details of aggradation and degradation were modified as more data became available, the relationship between sea level change and depositional terraces in the lower Mississippi basin is generally recognized as being real. Figure 7.14 shows these interpretations.

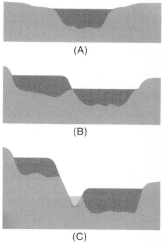

Figure 7.13. Progressive stages in development of unpaired depositional terrace sequence. Downcutting between episodes of filling occurs at or near the lateral edge of the valley. (A) First cut and fill. (B) Cutting of valley adjacent to first fill. Deposition in second valley. (C) Cutting of new valley between the earlier fills.

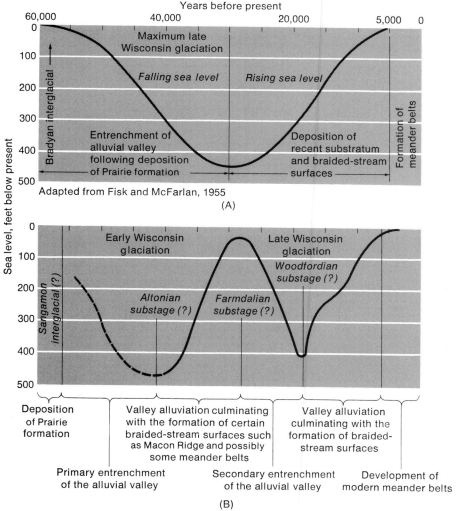

Figure 7.14. Lower Mississippi Valley chronological concepts (A) as proposed by Fisk (1944), and (B) as applied by Saucier (1968) to sea level fluctuation curve proposed by Curray (1965). (From Saucier and Fleetwood 1970. Used with permission of the Geological Society of America)

It is more difficult to explain trenching and filling in depositional terraces in regions outside direct glacial influence or far from the effects of sea level change. Most sequences of this type have also been attributed to the fluctuating climates accompanying glacial and interglacial conditions. The swing from arid to humid as glaciation begins (and vice versa) affects not only the prevailing discharge but also the amount and type of sediment delivered to the rivers. Such changes in fundamental river controls logically produce the trenching and filling. As the Holocene climatic reversals in North America are reasonably well known, it is ironic that little agreement exists as to which climate produced the filling and which caused the trenching. To complicate matters, all interpretations of the reversing sequence, so common in the nonglaciated regions of the United States, seem to be supported by field evidence (for a discussion see Flint 1971, pp. 304-307).

Part of the problem in explaining climatically induced terraces is that one type of climate change can produce a wide range of sediment yields and concentrations (Schumm 1965), depending on what local conditions prevailed prior to the alteration. Furthermore, as discussed in chapter 6, the fluvial reaction to a new sediment-discharge regime is not limited to trenching or filling. The Murrumbidgee River is a clear example. Thus, one river might develop a depositional terrace in response to a climatic shift, while another river, responding to the same impetus, might rearrange its channel configuration, pattern, or hydraulic parameters, but not undergo the trenching required to form a terrace.

Even where cutting and filling are the only responses, tributaries may not change in phase with events occurring in the master channel. Persuasive arguments have been made in support of a synchronous response of all rivers in a region or basin (Leopold et al. 1964), but equally convincing evidence suggests that in some cases tributaries may be trenching while the main channel, choked with their eroded debris, is filling (Schumm and Hadley 1957; Kottlowski et al. 1965; Haynes 1968). Obviously, in the latter case, some knowledge of a river's position in the drainage network is required before it will sensibly relate to a regional cut and fill sequence. The paradox surrounding climatically produced depositional terraces may be frustrating, but it should not be alarming. Either cutting or filling or neither may occur at the time of a climate change, depending on (1) prior fluvial and climatic conditions, (2) stream order, and (3) the particular channel character required to transport the new load and discharge most efficiently.

A less obvious and spontaneous filling and cutting may result from physical processes that have no relationship to tectonics or climate. Small streams that rise in the plains surrounding high mountains often have gentler valley gradients than do the larger rivers that head in the mountains. This trait develops best where the piedmont area is underlain by easily eroded siltstones and shales. Streams originating there adjust their gradients to the fine-grained sediment released from the weakly resistant rocks. The mountain rivers, however, must transport coarse bedload derived from the resistant rocks in the mountain core, and do so most efficiently by developing a steeper channel gradient. Because of this unique physical control, the main river stands at a higher elevation than its tributaries at an equal distance upstream from their confluence (fig. 7.15). It is well established that such a lithologic and drainage distribution leads to repeated stream captures when headwardly eroding tributaries intersect the position of the master stream (Rich 1935; Mackin 1936, 1937; Hunt et al. 1953; Hack 1960b; Denny 1965; Ritter 1967, 1972). The mountain stream is diverted into a lower tributary valley and is contained there until the process functions again.

The sudden influx of coarse load into the valley of the capturing tributary produces an untenable fluvial condition because the master stream cannot transport its oversized debris on the low valley gradient established by the tributary. The obvious result is filling of the valley until the gradient increases to an incline capable of transporting the mountain load under the prevailing discharge. Subsequent downcutting, often along the valley side, produces a depositional terrace (Ritter 1972, 1974). The enigma of depositional terraces formed in this manner is that the eroded surface beneath the gravel was formed by one river (the tributary) while the filling was caused by another (the mountain stream).

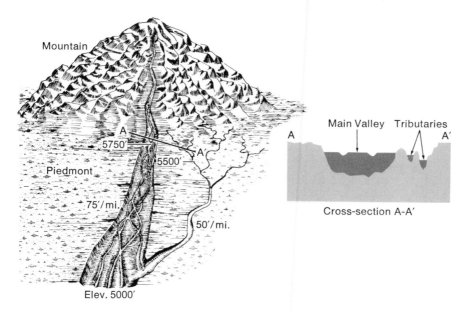

Figure 7.15. Physiographic and geologic controls of stream piracies in piedmont regions. Main river coming from mountain carries coarse-grained load on high gradient. Tributaries that head in piedmont region carry fine-grained load on gentle gradient. At an equal distance upstream from juncture of the two rivers, the tributaries stand at lower elevation and are in position to capture the poised master stream.

Erosional Terraces

One sometimes wonders if any aspect of fluvial processes escaped the genius of G. K. Gilbert. The following statement is contained in his remarkable discussion of the origin of floodplains:

> ...The deposit is of nearly uniform depth, descending no lower than the bottom of the water-channel, and it rests on a tolerably even surface of the rock or other material which is corraded by the stream. The process of carving away the rock so as to produce an even surface, and at the same time covering it with an alluvial deposit, is the process of planation. (Gilbert 1877, pp. 126-127)

Clearly Gilbert presupposed the process of lateral erosion long before any detailed understanding of its mechanics existed. Certainly he provided a theoretical base for the early analyses of fluvial terraces (Miller 1883; Davis 1902), and his thinking probably represents a cornerstone in the classic model of rock-cut terraces developed later by Mackin (1937) and illustrated in figure 7.16.

As we have said, erosional terraces are those in which lateral erosion is the dominant process in constructing the tread. Mackin (1937) presented an excellent description of the terrace origin (see that work for details of the process; the mechanics was described in the previous section). Briefly, as rivers migrate across the valley bottom, they erode one bank while simultaneously depositing point bar debris near the other. The bar sediment later becomes the capping terrace alluvium. It is usually thin and of constant thickness, and it sits on a flat surface eroded across the underlying bedrock or sediment. The buried surface is carved during floods when scouring penetrates the debris lying on the channel floor. For this to occur, the scouring depth of the river must be great enough to remove the entire pile of channel alluvium and expose the suballuvial material to short-lived erosion (fig. 7.16). Continual shifting of the channel position back and forth across the valley, combined with the occasional scouring, creates be-

Figure 7.16. Stages in the development of a rock-cut terrace. (A) During low water stages, fine sediment (f) is deposited during normal flow and coarse sediment (c) is deposited at the end of a high water event. (B) High water stage entrains all the channel sediment and scours the underlying bedrock before coarse detritus is deposited again on the channel floor. (From Mackin 1937. Used with permission of the Geological Society of America)

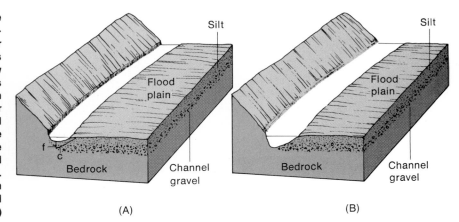

(A)   (B)

neath the alluvium the bevelled surface that is a mirror image of the plane surface on top of the deposit.

The sheet of alluvium is almost always present in an erosional terrace, but it is not a prerequisite and is certainly not the paramount characteristic of the feature. That role falls to the laterally eroded surface. Thus, it is probably acceptable to ignore the alluvium and consider the cut surface to be the terrace tread. Like any approach to terraces, this may or may not lead to difficulties in the field, depending on the particular situation.

Erosional terraces are normally thought of as the "equilibrium" model of the terrace line. It would seem that the development of a tread surface by lateral planation should require not only time, but also a long period of stability during which base level and channel functions are constant, and no vertical disruptions by filling or cutting occur. Nonetheless, even this logical rule of thumb has exceptions. For example, near Pyramid Lake (Nevada), the Truckee River has formed six erosional terraces during a period when its base level, represented by the lake, was rapidly declining (Born and Ritter 1970). The highest and oldest terrace formed sometime between 1925 (when its level was beneath the lake), and 1938, when aerial photography showed it to be a well-developed landform. It now stands approximately 10 m above the Truckee River, which is rapidly downcutting to keep pace with the declining lake level. Apparently each terrace was formed during one major flood when the river, in high flow, was able to erode laterally at a dramatic rate into its unresistant banks. The terrace levels are carved into noncohesive lake sediments recently exposed because Pyramid Lake, in hydrologic imbalance, has dropped almost 25 m in this century. Thus, where banks are easily eroded, time and stability are not essential factors in the formation of erosional terraces. In fact, it is difficult to imagine any geomorphic setting in greater disequilibrium than the Truckee River system near Pyramid Lake.

Terrace Origin and the Field Problem

Understanding local terraces and establishing a regional pattern of terrace development are not only basic in historical geomorphology, they are also useful in providing information for regional planning, land management, water supply, and locating sand and gravel for building materials. Acquiring such knowledge, however, is a painfully slow procedure requiring field study and correlation of surfaces within a valley or between valleys. Determining terrace origins is not so easy as we would like to think. Although we postulate guidelines for recognizing terraces of different origins, terraces in the real world develop in such a variety of ways that the exceptions almost become the rules. Basically, terraces are terraces are terraces—and we should probably not generalize about features that defy generalization. Each terrace must be examined according to its own geologic, climatic, and tectonic setting without preconceived ideas about its origin.

Using terraces to interpret geomorphic history is a monumental task that requires detailed analysis. Terraces are rarely preserved intact along the length of a valley, but instead are segmented into isolated and physically separated remnants, often kilometers apart. Reconstruction of the original longitudinal profile of the terrace surface requires correct correlation of the remnants, and every method used in that procedure is burdened with fundamental assumptions that may be invalid in certain situations (for details see D. W. Johnson 1944; Frye and Leonard 1954).

To elucidate the problem, it may seem to some readers that erosional terraces are really not that much different from depositional terraces. After all, both are usually covered with alluvium, and if one walked across that alluvial surface there would be nothing to indicate what type of terrace lay beneath. Nonetheless, a real and very important difference does exist—one that cannot be disregarded or minimized. When an erosional terrace forms, the capping alluvium is deposited *at the same time* that the underlying surface is eroded. In significant contrast, the surface beneath a depositional terrace was present before the influx of the alluvial fill; a finite time gap separates the deposition from the cutting of the underlying surface. Failure to recognize this subtle distinction between erosional and depositional terraces can lead to drastically different reconstructions of geomorphic history.

The differing interpretations of the terrace sequence in the Bighorn Basin of Wyoming show the problem well and demonstrate the difficulty of obtaining sufficient field data for interpretive purposes.

Mackin (1937) divided the Cenozoic history of the Bighorn Basin into two major phases: (1) a long period of basin filling throughout most of the Tertiary, followed by (2) rejuvenation and basin excavation which has continued to the present. Within the basin are a series of terrace levels standing at elevations from 330 to 6 meters above the present rivers. Mackin felt that each level represented a rock-cut bench formed when downward excavation ceased and allowed lateral erosion to become the dominant fluvial process. The evidence supporting this interpretation seemed to be clear. Where later entrenchment exposed the terrace gravels they were thin, constant in thickness, and resting on a flat, truncated bedrock surface (fig. 7.17A). All the ingredients of a rock-cut terrace were observed, and to interpret them as such was certainly reasonable.

In a later study, however, Moss and Bonini (1961) were able to gain additional information about the subsurface framework of several key terraces by running seismic profiles across the features perpendicular to the axis of the Shoshone River valley. Instead of the expected flatness, the bedrock surface beneath the alluvium showed considerable relief, and in places the gravel thickness was well beyond a reasonable scour depth for rivers of this type (fig. 7.17B). They interpreted these characteristics to mean that the surface beneath the gravel represented the valley topography that existed before it was buried by the influx of a later fill. They concluded that the terraces are depositional and that the fill was outwash from glaciers in the nearby Absaroka Range. If all the terraces had this origin, the history of the basin would change signifi-

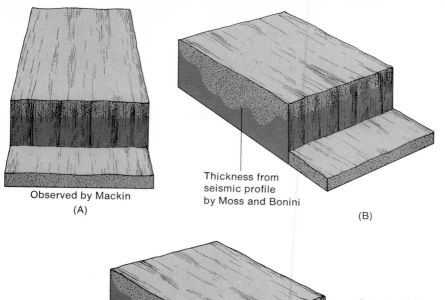

Observed by Mackin
(A)

Thickness from
seismic profile
by Moss and Bonini

(B)

Figure 7.17. Interpretations of
the Cody Terrace near Cody,
Wyo. (A) As rock-cut terrace,
based on observed alluvial
thickness (Mackin 1937).
(B) As depositional terrace,
based on seismic profiles
across the terrace (Moss and
Bonini 1961)

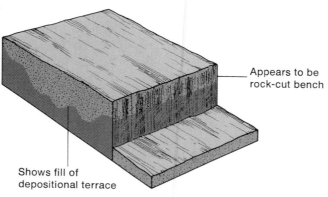

Appears to be
rock-cut bench

Shows fill of
depositional terrace

Figure 7.18. Difficulty in interpreting terrace origin from
field data: If downcutting exposes only part of fill, terrace
may appear to be rock-cut.
Data across the terrace are
needed to determine true
thickness of the fill.

cantly. The general excavation phase, in this interpretation, was periodically interrupted by glaciofluvial filling of the valleys, not by valley widening. In addition, considerable time elapsed between erosion of the underlying bedrock surfaces and creation of the terrace treads.

It is instructive to note that the Moss and Bonini interpretation was made possible by techniques and data not available to Mackin in 1937 or, for that matter, to most of us today. What Mackin observed was the edge of a fill where it intersected an eroded valley (fig. 7.18). The much-needed third dimension across the terrace could be reconstructed only with the proper field equipment and approach.

The topography of almost every region reflects an adjustment between dominant surficial processes and lithology. When the rocks have diverse resistances, geomorphic processes tend to maximize the relief between regions of greatest and least resistance. Nowhere is this more apparent than in areas where

**Piedmont
Environment: Fans
and Pediments**

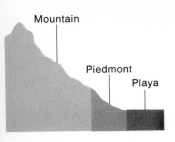

Mountain

Piedmont

Playa

Figure 7.19. Physiographic components of a mountain-basin geomorphic system.

mountains and plains adjoin, especially where the climate is arid or the region has undergone recent tectonism. Aridity serves to buffer the smoothing effects of vegetation; vertical tectonic activity accentuates relief by bringing more resistant basement rocks toward the surface, where they are commonly etched into the cores of topographic mountains.

The sloping surface that connects the mountain to the level of adjacent plains is called the *piedmont*. It extends from the mountain front to a flood plain or playa, either of which can mark the base level for geomorphic processes that function on the piedmont surface (fig. 7.19). Piedmonts consist of a number of geomorphic landforms, but most commonly they are composed of eroded bedrock plains called *pediments* and depositional features called *alluvial fans*. The relative percentage of the total piedmont area occupied by either of these features probably depends on the unique combination of local geomorphic variables. Although both features can be found in any climate, they dominate the piedmont most completely in arid regions. In humid climates, slope material is stabilized by vegetation and moves primarily by mass wasting, and sediment that could be used for fan-building is removed from the system in perennially flowing rivers. Therefore, characteristics and formative processes of fans and pediments have been investigated most extensively and in greatest detail in arid climatic zones.

Alluvial Fans

Alluvial fans are one end of an erosional-depositional system, linked by a river, in which rock debris is transferred from one portion of a watershed to another. Fans are largest and most well developed where erosion takes place in a mountain, and the river builds the fan into an adjacent basin. Deposits tend to be fan-shaped in plan view and are best described morphologically as a segment of a cone radiating away from a single point source. The point source represents the spot where the master river of the watershed emerges from the confines of the mountain; it doubles as the apex of the conical shape. The point source can also shift away from the mountain front to a position well down the original fan surface, if that surface has been entrenched at some time during its development. In those cases, the mountain stream, still occupying a confining channel, traverses a portion of the older fan material. The stream eventually emerges downfan as the point source for a still younger fan. Adjacent fans often merge at their lateral extremities; the individual cone shape is lost, and a rather nondescript deposit is formed covering the entire piedmont. These coalesced fans are commonly referred to as bajadas, alluvial aprons, or alluvial slopes.

Fan Morphology   The longitudinal slope of an alluvial fan generally decreases downfan even though its precise value at any point depends on the load-discharge characteristics of the fluvial system. Near the mountain front, slopes are commonly very steep, although they probably never exceed 10° (Cooke and Warren 1973). Fans gradually flatten to their lower extremity, called the toe, where gradients may be as low as 2 m per kilometer ($\approx$10 ft./mi.). The steepest gradients are usually associated with coarse-grained loads, low discharges, high sediment production in the source area, and transport processes other than normal streamflow (Bluck 1964; Bull 1964, 1968; Blissenbach 1954; Hooke 1965, 1967, 1968). These factors often conflict in the same region. In Fresno County, California, for example, fans derived from basins underlain by mudstones or shales are 33–75 percent steeper than fans of the same size related to sandstone basins (Bull 1964). The low gradient expected because of the small particle size is offset by a high rate of sediment production. Fan gradients may also be related to parameters of drainage basin morphometry (Melton 1965a); however, most statistical measurements probably reflect only the more basic controlling factors. Absolute slope values at any given point, then, may represent a myriad of controls in the erosional-depositional system.

Two slope characteristics deserve special attention. First, the gradient of most fans near the mountain front is approximately the same as that of the mountain river where it merges with the fan apex. Deposition on the upfan surface, therefore, is not initiated by a dramatic decrease in gradient as the master river passes from the mountain onto the fan. Second, although fans are concave-up from the apex to the toe, their longitudinal profiles are usually not a smooth exponential curve. Instead, on many fans the concavity stems from a junction of several relatively straight segments, each successive downfan link having a lower gradient (fig. 7.20).

The changes in fan slope, represented by individual segments, are genetically related to changes in the channel of the trunk river upstream from the fan apex. For example, in many fans intermittent uplifts have increased the stream gradients, and in response to each event, a new fan segment has formed, gradu-

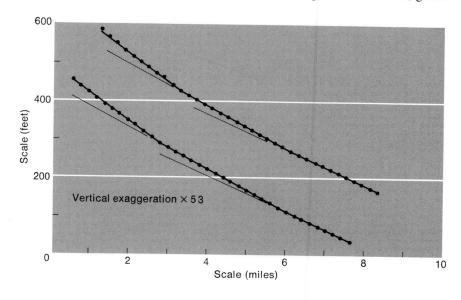

Figure 7.20. Straight segments of several radial profiles on an alluvial fan. Concave profile stems from lower gradient on each basinward segment of the fan. (After Bull 1964)

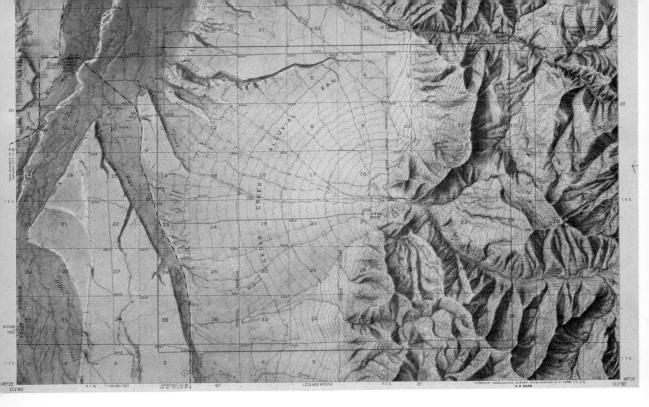

Figure 7.21. Map view of topography on typical alluvial fan. Note downslope deflection of contours indicating convex cross-fan profile. (From U.S. Geological Survey, Ennis, Mont., 15° quadrangle)

ally adjusting its slope until it approximates the newly formed steeper slope of the trunk river (Bull 1964; Hooke 1972). Under this particular control, each segment as one moves upfan is steeper and younger, and its deposits are graded to the level of the next lower segment. Segmentation may also result from climatically induced changes in the load-discharge balance, and the segments do not always become steeper toward the apex. Thus the overall longitudinal profile may be very sensitive to historical changes in the balance between the erosional and depositional parts of the system (Bull 1964, 1968).

The shape of alluvial fans is distinctly convex in cross-profile, as can be seen by the downfan deflection of topographic contours (fig. 7.21). By considering fans as segments of true circular cones, Troeh (1965) proved that fan shape can be approximated by the equation

$$Z = P + SR + LR^2$$

where Z is the elevation at any point, P the elevation at the apex, S the slope of the fan at P, R the radial distance from P to Z, and L half the rate of change of slope along a radial line. The method for deriving the numerical values for any particular fan is also given in Troeh (1965, pp. 621-622).

It is now firmly established that the area of a fan is statistically related by a simple power function to the area of the basin supplying the sediment such that

$$A_f = cA_d{}^n$$

where A_f is the area of the fan and A_d is the area of the drainage basin. The exponent n is the slope of the regression line in a full logarithmic plot of the two variables.

Bull (1968) showed that the relationship is generally very similar for a group of fans representing a variety of environments in the western United States, and the mean value of n is approximately 0.9 when A_f and A_d are measured in square miles (fig. 7.22). The coefficient c, however, seems to vary widely, reflecting the effect on fan dimensions of geomorphic factors other than drainage basin size. Chief among these are climate, source rock lithology, tectonics, and the original space available for fan growth in the collecting basin. Thus, even where drainage basins in the same region have equal areas, the areas of their fans may differ by as much as an order of magnitude if these other characteristics differ greatly.

Several specific examples demonstrate the effect of these factors and the significance of fan morphometry. Bull (1964) showed that fans derived from basins underlain by fine-grained sedimentary rocks are almost twice as large as those derived from sandstone basins of equal size. The regression lines have approximately the same slope ($n = 0.91$ and 0.98), but the effect of particle size shows up in the value of the coefficient c, which varies from 0.96 for sandstone to 2.1 in the mudstone drainage basins. The effect of tectonics is revealed in the fans of Death Valley, where eastward tilting of the valley permitted fans on the west side of the valley to grow larger while those on the east side were stunted by

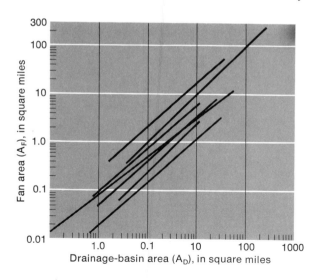

Figure 7.22. Relation of fan area to drainage-basin area for a number of fans in California and Nevada. See Bull 1968 for sources of data and equations of the regression lines. (From Bull 1968. Used with permission of National Association of Geology Teachers)

burial beneath the playa (Denny 1965). The c values are 1.05 for fans on the west side of the valley and 0.15 for those on the east side.

Processes, Deposits, and Origins   Any model that proposes to explain the geomorphic meaning of alluvial fans must be based on discernible facts concerning both fan morphology and the processes that function in the fan system. Evidence that some of the pertinent facts are still missing can be found in the unresolved controversy about the stability or instability of modern fans. Although fans in the California-Nevada region have been studied in greater detail than anywhere else, little agreement exists about their equilibrium conditions. Some authors view fans as steady-state forms, neatly adjusted and in a continuing dynamic equilibrium (Denny 1965, 1967; Hooke 1968). Others see them as actively growing (Beaty 1970) or being dissected (Hunt and Mabey 1966), processes that presumably indicate that fans may be approaching an equilibrium condition but have not sensibly attained it. Bull (1975b) cites fans as features that do not attain a steady-state condition but instead develop under the control of allometric change (see chapter 1). Still another interpretation is that fan characteristics can only be explained by cyclic changes in processes initiated by climatic fluctuations (Lustig 1965), and so every fan has vestigial properties unrelated to the present conditions. Facing these widely divergent viewpoints, it seems logical briefly to consider the evidence that led these scientists to such varied conclusions.

Deposits and Depositional Processes   The movement of sediment from source areas to depositional sites involves a variety of flow types, ranging from highly viscous debris or mud flows to normal water flow. The type of flow during any given event depends primarily on the lithology of the basin and its degree of weathering, and secondly on the magnitude of the precipitation causing the flow. The ephemeral nature of flow in arid regions results in spasmodic rather than continuous deposition, and the depositional site changes repeatedly. Only a limited portion of the fan surface is occupied by flow and undergoing deposition at any given time.

As flow leaves the confines of the trunk channel, deposition is initiated by changes in hydraulic geometry, not by a sudden decrease in gradient (Bull 1964). Generally, when the flow becomes unconfined on the fan surface, the width increases so dramatically that both depth and velocity decrease to a level where the flow can no longer transport the load. In segmented fans, however, areas close to the apex are commonly occupied by channels called *fanhead trenches* which are incised 7–12 m below the level of the fan surface (Bull 1964). These may connect to the trunk river in such a way that flow remains contained and is transmitted far downfan before it is freed to increase its width. The effect of changing hydraulic geometry is reinforced by a loss in discharge if the fan surface is permeable and water seeps into the underlying deposits. It is not unusual, for example, for the entire flow to disappear underground before it can traverse the length of the fan.

The deposits of any single flow usually form as narrow tongues, possibly up to several kilometers long, but normally only 120–700 meters wide (Bull 1968). The length of each deposit probably depends on the viscosity of the flow, the permeability of the surface, and the distance downfan that the flow is held in a distinct channel. In many cases, the flow at first follows well-established channels, but at some point along its length overtops the banks and spreads outward as diffuse flow. In the case of water flows, lateral shifting of the loci of deposition allows the braided stream system to deposit a sheet of poorly bedded sand and gravel in which individual beds can be traced laterally for only short distances. This sheetlike configuration may be interrupted by thicker deposits that represent an occasional channel entrenchment into the fan surface and subsequent backfilling. Deposits within these larger channels are generally more coarse-grained.

Debris flows or mud flows usually follow more well-defined channels because the confining limits of the channel ensure the depth of flow needed to offset the high viscosity of the fluid. During transport, however, debris flows also may overtop banks and spread out as sheets (Bull 1963). Debris flows are so dense and viscous that only the very largest particles can settle from the mass during flow. Nonetheless, they are capable of transporting extremely large boulders for considerable distances on lower gradients than normal streamflow would require. Their high viscosity, however, effectively restricts the distance of transport (in comparison with water flow), and their forward movement may simply stop even though they are still confined in a channel. Therefore, deposits from debris flow are poorly sorted with boulders embedded in a fine-grained matrix; in contrast to water-transported sediment, they are usually lobate and have well-defined margins often marked by distinct ridges.

Some fans are built almost entirely by debris flows (Beaty 1970), although their deposits may be reworked almost immediately by normal streamflow. Beaty (1963) reports debris-flow deposits being dissected within 48 hours of their deposition by streamflow that began in the same storm but continued after the debris sediment had been deposited. Beaty (1970) also found that in the Milner Creek (White Mountains, California) fan, debris was derived from the floor of the trunk channel. Accumulation of 3–7 m of sediment is needed in the channel to provide the volume found in each major debris-flow deposit on the fan. Sediment production in the drainage basin, therefore, must be rapid enough to collect sufficient amounts on the channel floor before successive high-magnitude storms remove it. If sediment production is too slow, not enough particulate matter will accumulate within the recurrence interval of major floods to produce a debris flow, and the discharge delivered to the fan surface will be in the form of normal streamflow. Some debris flows are initiated by landslides (Johnson and Rahn 1970).

In fans composed of coarse-grained deposits, large discharges may infiltrate before crossing the entire fan (Hooke 1967). Under these conditions, coarse sediment may be deposited in lobate masses called *sieve deposits*. These resemble debris-flow deposits but lack primary fine-grained material and are highly permeable. Sieve material can be deposited when the flow is confined or unconfined, but only when the surface is permeable and the flow does not contain fine-grained sediment.

Entrenchment and Location of Deposition   It is axiomatic that lateral migration is involved in developing the convex cross-profiles and the plan-view shape of alluvial fans. It is equally important to recognize that the loci of deposition also migrate along radial lines during fan development. This longitudinal shifting is accomplished by entrenchment or backfilling of the main channel that extends from the source area onto the fan. Trenched channels have lower gradients than the fan surface, and so they are deep near the fan apex and become progressively shallower downfan until they finally emerge at the surface. In segmented fans, the trenches tend to incise until they have the same gradient as the adjacent downfan segment (fig. 7.23). Entrenchment is significant in that it provides an

Figure 7.23. Fanhead trenches in segmented fans. Entrenched channels are attaining the same gradient as the adjacent downfan segments. (From Bull 1964)

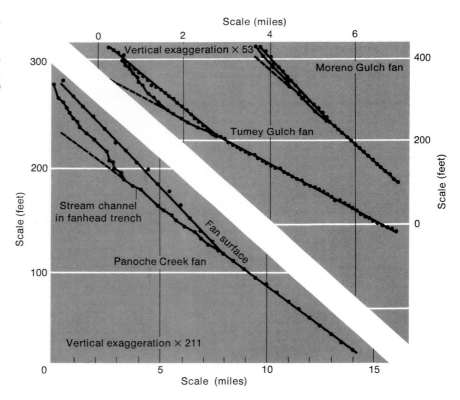

explanation for abnormal distribution of particle sizes on the fan, since even coarse sediment can be placed far downfan if entrenched channels confine the entire flow. In addition, entrenchment tends to enlarge fans because moderate flows that could not traverse the fan unconfined are transported farther in well-defined channels, resulting in deposition and fan construction in downslope areas.

Entrenchment can be either temporary or permanent, and distinguishing between the two seems to be critical in the analysis of fan origin. Many fanhead trenches appear to be rather ephemeral; that is, they have evidence of alternating episodes of trenching and filling. Temporary entrenchment sometimes occurs when debris-flow deposits plug the channel, and the flow shifts laterally to a new position where entrenchment begins again. It also may be that alternating trenching and filling are expected results of fan processes associated with changes in rainfall intensity (Bull 1964) or normal alternations of debris flows and water flows (Hooke 1967). In any case, temporary trenching seems to be a common process on most active fans and can probably be explained in terms of local conditions.

In permanent entrenchment, channels are incised to depths that cannot be easily backfilled, often being cut to levels greater than 30 m below the fan surface. In addition, the impetus for the downcutting may be outside the fan system itself and so cannot be explained by local fan processes. On the upper fan slope, permanent entrenchment may be caused by accelerated erosion, but it also can occur naturally in a uniform environment if the trunk river continues to downcut within the mountain. In either case, the fan surface standing above the trench is no longer involved with active fan processes; soils may develop on the alluvium and incipient drainage networks may be established. If the basin of deposition is open, and base level for the fan is the floodplain of a through-flowing river, downcutting of that river may initiate a wave of fan incision that is propagated upslope from the toe of the fan. In addition, lateral migration of the master river can erode the toe of the fan and rejuvenate the streams crossing the fan surface. Eventually the entire fan is dissected when the incision reaches the apex and captures the trunk river as it emerges from the mountain. This process is especially effective on small valley-side fans of glaciated valleys (fig. 7.24) and commonly relates to climatic fluctuations and the glacial cycle (Ryder 1971a, 1971b).

Equilibrium and Fan Origins   Diverse interpretations have been made of the meanings of modern fans even though all authors seem to recognize the same processes, deposits, and morphology in the fans they have investigated. What causes the divergence of opinion? A prime candidate might be the familiar nemesis—what time scale is being used to make judgments about equilibrium? We will briefly review the major interpretations.

Figure 7.24. Dissected fan in Rock Creek valley, Beartooth Mountains, Montana. Grassy areas represent original fan surface. Tree-covered zones are in portions of the fan that have been entrenched when the master river (Rock Creek) cut to lower level. (Photo by R.R. Dutcher)

Denny (1965, 1967) recognized piedmont development in the Death Valley regions as a continuous steady state in which the rate of deposition on the piedmont equals the rate at which sediment is eroded from the piedmont and moved to the playa or floodplain. For the system to remain in balance, large areas of the piedmont must be dominated by erosion, either in pediment zones or in abandoned parts of the fan. Based on field observations, Denny subdivided fans into several major zones which he categorized as *modern washes, abandoned washes,* and *desert pavements.* The two parts of figure 7.25 show an aerial photo of Death Valley fans, along with a map of their components.

On the Shadow Mountain fan, washes make up about two-thirds of the surface area, but only a few contain unweathered gravel and accommodate present-day floods. These modern washes are the primary areas of deposition on the fan surface. Most washes have scrub vegetation and gravel coated with desert varnish in their channels, indicating that they are abandoned channels and have not been flooded for considerable time, perhaps several thousand years. Desert pavements are surfaces of tightly packed gravel that armor, as well as rest on, a thin layer of silt, presumably formed by weathering of the gravel. Pavements are the primary erosional areas of a fan. They have not received sediment for a long time, as evidenced by the thick varnish coating the pebbles, the pronounced

Figure 7.25. (A) Large alluvial fans at the base of the Pana-mint Range in the north end of Death Valley. Inyo County, Cal. (Photo by U.S. Geological Survey. From the University of Illinois Catalog of Aerial Photography) (B) Components of fan in Death Valley region. (Adapted from Denny 1965)

(A)

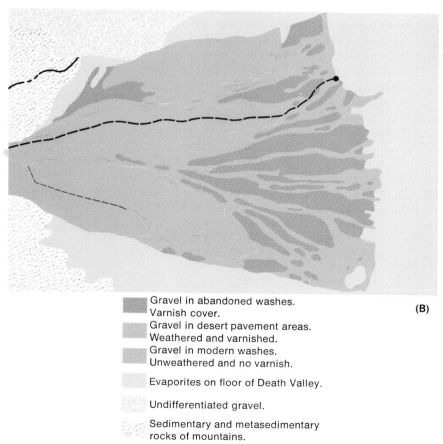

(B)

Gravel in abandoned washes. Varnish cover.

Gravel in desert pavement areas. Weathered and varnished.

Gravel in modern washes. Unweathered and no varnish.

Evaporites on floor of Death Valley.

Undifferentiated gravel.

Sedimentary and metasedimentary rocks of mountains.

weathering beneath the silt layer, and the striking smoothness of the surface, due to obliteration of the original relief by lateral downwasting into depressions. Pavements characteristically are cut by gullies that head within the pavement area and may be meters or even tens of meters deep. Because the gullies carry a locally derived fine-grained load, they often meander and, importantly, may stand at lower elevations than adjacent modern washes that head in the mountains.

The distribution of pavements and their gullies in relation to modern washes creates a geomorphic situation quite similar to that discussed earlier for the creation of terraces in a piedmont region (see fig. 7.15). Flow in the modern washes is periodically diverted into the gullies, changing part of the desert pavement area back into an active wash and shifting the position of the modern washes. Simultaneously, the lower segment of the captured wash is abandoned and, over a long period of time, imperceptibly converts to a desert pavement. Denny thus visualized fans as being in a dynamic equilibrium that is maintained by a balance between erosion and deposition. Areas of deposition are converted to areas of erosion and vice versa by repeated stream piracies that are facilitated by load and gradient differences between the modern washes and the gullies within desert pavements. This analysis requires that once equilibrium is established, both erosion and deposition must function on a fan at all times. It also means that equilibrium is maintained over a graded-time interval, and so all wash areas or all pavement areas are not of the same absolute age.

In a different perception of equilibrium, Hooke (1968) suggested that the steady-state condition is maintained by adjustments in thickness or volume of the fan. The balancing mechanism involves the relative amounts of aggradation on the fan and on the playa. If the playa is too large, its thickness will increase more slowly than that of the fan, and the fan will progressively encroach into the playa zone. The area of the playa then decreases and, with a constant supply of sediment, its vertical growth rate will increase. Such adjustments cause the playa and fan to thicken at the same rate when considered over a long period. Hooke feels that Denny's model will work only in open systems where sediment beyond the extremities of the fan itself can be continuously removed. Otherwise, the thickness of features rather than the erosion-deposition rate becomes the balancing element in the system.

In contrast to the steady-state hypotheses, Lustig (1965) proposed that the characteristics of fans are best explained by a climatically controlled sequence of events with finite episodes of fan building and fan destruction. In his model, fan building occurred under earlier, more humid, conditions when frequent widespread precipitation encouraged greater growth of vegetation on the fan surface. The combination of higher resistance to erosion and more numerous overbank sheetlike flows resulted in a general aggradation of the surface. As the climate changed toward aridity, precipitation became localized and the vegetal cover became less continuous. Debris flows became the dominant flow type, trenching the upper fans and shifting the loci of deposition far downfan.

Lustig, therefore, views fans as tending toward an equilibrium condition in each climatic phase, but not necessarily attaining it. He also would recognize the present as essentially a period of non-growth, an interpretation that agrees with others who see the modern condition as a destructive phase (Hunt and Mabey 1966; Beaumont 1972). It should be noted, however, that Lustig's climatic model requires that debris flows be the cause of entrenchment, instead of simply using incised channels as transport avenues. There is some question whether debris flows possess the necessary erosive ability to do this (Hooke 1968), and indeed, some fans seem to be constructed—rather than destroyed—almost entirely by debris flows (Beaty 1963, 1970).

Later work also suggests that caution must be used before explaining fan mechanics by climatic controls. Lattman (1973) has presented convincing arguments that abandoned pavement surfaces on fans in southern Nevada are widespread only where a strong carbonate cementation is present. The cementation, occurring in the form of petrocalcic horizons, laminar layers, or gully bed cementation, is best developed on fans composed of clasts of carbonates and basic igneous rocks. It is less well developed on fans of noncarbonate sedimentary rocks, and it is absent where the source areas are underlain by acid igneous rocks. Not only is there this lithologic control, but the cementation seems to form over a wide range of precipitation, causing Lattman to conclude that Pleistocene climatic changes probably had little effect on the distribution of cementation or, presumably, on the development of abandoned surfaces. Apparently, the creation or destruction of desert pavements can occur under any climatic conditions or at any time.

Finally, Beaty (1970) argues that the Milner Creek fan in the White Mountains of California has been growing for possibly 700,000 years and has not yet reached a steady state. If his interpretation is correct, the establishment of equilibrium requires considerably more time than most geomorphologists had previously suspected. Beaty's model is diametrically opposed to Lustig's climatic model, for the fan is actively growing today and its aggradation is primarily in the form of debris-flow deposition.

It is clear from these examples that we do not truly understand what fans tell us about their conditions of equilibrium. Since the message contained in the basic data is similar in most cases, either we are reading it differently or it is written in an unknown language. Perhaps, like many geologic features, fans can form in a number of ways under various combinations of their controlling factors. Each fan should, therefore, be considered in its own individual setting. It will probably have relict components retaining the effects of a past climate, areas of erosion and deposition, young and old parts, and a morphology that demonstrates a statistical balance. The interpretation of what all those things mean is a real challenge to any geomorphologist.

Pediments

Since Gilbert first described "hills of planation" in the Henry Mountains of Utah, geomorphologists have been intrigued with their origin, and this feature, given the name *pediment*, has been discussed endlessly during the last century. Our discussion of these interesting features will necessarily be brief, but excellent reviews of the topic are available: Tator 1952, 1953; Tuan 1959; Hadley 1967; Cooke and Warren 1973. Definitions of the term "pediment" are as numerous as the workers who have studied this feature and, like the landform itself, descriptions range from general to rather graphic and precise. For example, Denny (1967, p. 97) employs the term in reference "to the part of the piedmont that is more or less bare rock surface." On the other hand, R. U. Cooke (1970, p. 28) suggests that "pediments are composed of surfaces eroded across bedrock or alluvium, are usually discordant to structures, have longitudinal profiles usually concave upward or rectilinear, slope at less than 11°, and are thinly and discontinuously veneered with rock debris."

In order to understand better what it is that we are discussing, and in deference to clarity, it may be better to look briefly at those characteristics that are universally recognized as salient properties of pediments, rather than to adopt a formal definition:

1. Pediments are erosional surfaces that abut against and slope away from a mountain front or escarpment.
2. They are entirely erosional in their origin and commonly form in a direction that diverges from the trend of the regional structures.
3. The surfaces are usually, but not necessarily, cut on the same rocks that make up the mountain. They may truncate both bedrock and/or alluvium, but are best developed and preserved on bedrock, especially resistant types such as granite or related crystalline rocks.
4. Pediments may or may not have a thin covering of sediment which presumably represents load that is in transit. This characteristic has traditionally created problems because the question arises of how much alluvium can be tolerated before it must be recognized as a fan, younger in age than the pediment surface and therefore divorced from the processes of pedimentation. It is applaudable, then, that Cooke (1970) restricts the pediment to only that part of the eroded surface not continuously covered by alluvium (fig. 7.26). The erosional surface beneath the continuous debris cover is called the *suballuvial bench,* a term first used by Lawson (1915), and the cover itself is referred to as the *alluvial plain.* The pediment, then, is bounded upslope by the mountain front and downslope by the alluvial plain (fig. 7.26).
5. Pediments are usually found in arid regions, although most workers would not restrict the processes of pedimentation to that climate. Note that I said they are *found,* not formed, in arid climates. Recent data suggest that pediments in the Mojave Desert may be relict features that formed under a more

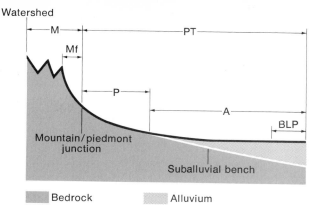

Figure 7.26. Landforms in the mountain-basin geomorphic system: M = mountain area; Mf = mountain front; P = pediment; PT = piedmont plain; A = alluvial plain; BLP = base level plain. (After Cooke 1970. Used with permission of *American Journal of Science*)

humid, Tertiary climate, and the surfaces are at present undergoing virtually no expansion under modern desert conditions (Oberlander 1972, 1974). L. C. King (1953) suggests that pedimentation may be a basic geomorphic phenomenon, present in all hillslope developments regardless of climate.

Morphology and Topography   It is ironic that in spite of the singular attention devoted to pediments, a multitude of untested hypotheses exist concerning the processes of pedimentation, but an amazingly skimpy pool of reliable data to support them. After a century of study, there is still confusion and lingering disagreement about every aspect of pedimentation. Cooke and Warren (1973, p. 188) express this succinctly in their description of the topic as "a subject dominated by almost unbridled imagination." Cooke (1970) suggests three reasons for the failure to resolve these differences of opinion: (1) we have not viewed pediments as part of an erosional-depositional system, but have studied only the pediments to the exclusion of other related forms; (2) we have not collected the precise data needed to understand the system; and (3) we have been overly concerned with general, evolutionary hypotheses of pediment formation and, in many cases, have deduced processes from the genetic model rather than employing direct measurement and observation. Regardless of these past sins, we can make some definitive remarks about pediment morphology and topography.

Size and Shape   Pediments vary in size from less than 1 square kilometer to hundreds of square kilometers, probably depending on fundamental geomorphic controls. Shape is also variable, with pronounced irregularities when the rocks cut by the pediment surface have wide differences in resistance to erosion (Hadley 1967). Generally they tend to be fan-shaped in plan view, narrowing toward the mountain front and widening downslope (D. W. Johnson 1932; Rich 1935; Gilluly 1937). Across the pediment, the shape can be either convex or concave.

Slope   The longitudinal profile of almost all pediments is slightly concave-up, although local convexities do occur. Overall longitudinal convexities have been suggested as a theoretical possibility if the suballuvial bench is also considered (Lawson 1915), but available observation and geophysical data (Langford-Smith and Dury 1964) have not demonstrated the actual presence of such a form. Slope angles on pediments range from .5° to 11° but seem to average about 2.5°.

A prevailing misconception concerning pediment slopes is that they are controlled entirely by the size of the material they are required to transport, and so it is widely accepted that they are "slopes of transportation" (Bryan 1922). Actually, detailed measurements show that the relationship between particle size and pediment slope is not so straightforward as previously supposed (Dury 1966b; Cooke and Reeves 1972). Where particle size decreases in an orderly way downslope, the rate of decline in pediment gradients seems to be greater than the rate of reduction in size. Cooke and Reeves (1972) also found that only the largest particles showed a consistent decrease with distance from the mountain front. Other statistical parameters of size varied incoherently with distance and slope decrease. They attribute these anomalous relationships to differing amounts of in situ sediment being added to the total load at any given sampling locality.

It might also be logically assumed that slope should be related in a significant way to the area of the mountain drainage basin or to the length of the pediment. It has been noted, for example, that pediments have lower gradients where they are associated with large rivers or canyons (Bryan 1922; Gilluly 1937). Although data are limited, neither of these assumptions could be substantiated in more recent morphometric studies of pediments in California and Nevada (Mammerickx 1964; Cooke 1970). Cooke (1970) did, however, find a significant correlation between pediment slope and the relief/length ratio of the pediment association (the ratio of the difference in elevation between the highest and lowest points in the association to the distance between the highest and lowest points in the association along a line of direct water flow between them). The pediment association includes the pediment, the mountain area tributary to it, and the area of the related alluvial plain. It seems likely that both the relief/length ratio of the pediment association and the slope of the pediment are directly affected by faulting. Thus, pediment slopes are probably steeper where related to tectonic activity, but more data of this type are needed to confirm this as a general rule.

The Piedmont Angle   The upper boundary of the pediment is usually marked by an abrupt change from the steep slopes of the mountain front to the low declivity of the pediment surface. In plan view the boundary is usually linear, but embayments into major valleys of the mountain front can give the trace a rounded or crenulate appearance. The angle formed by the junction of the two surfaces is called the *piedmont angle* (fig. 7.27), and its development and maintenance have traditionally been cited as evidence in theoretical models of pediment origin.

In detail, the piedmont angle can take the form of a narrow zone of intense curvature rather than a distinct angle. Twidale (1967) reports mountain front slopes of 22° changing to pediment gradients of 3° over a transition zone 100 m wide. Both the magnitude of the piedmont angle and the sharpness of the angular relationship are probably related to structural or lithologic control (Denny 1967; Twidale 1967; Cooke and Reeves 1972), but other processes including weathering and several forms of corrasion have been suggested as contributing or dominant factors. The early idea of Bryan (1922) that the angular relationship represents an adjustment of the two slopes to the size of debris they are required to transport cannot be accepted without qualification, since Melton (1965b) was unable to demonstrate a significant correlation between slope angle and the size of weathering products. At the present time, no one set of processes adequately explains the origin and development of all piedmont angles.

Surface Topography   Contrary to lay opinion, pediments are not monotonous, smooth, flat surfaces, but are dissected by incised stream channels and dotted with residual bedrock knobs, called *inselbergs*, that stand above the general level of the pediment itself. The frequency and size of both incised valleys and inselbergs seem to increase toward the mountain front, sometimes giving the topography the aspect of gently rolling hills and valleys (Gilluly 1937). Some of the channels and other depressions may be filled with alluvium up to 3 m thick (Cooke and Warren 1973), giving the false impression that the bedrock surface is smooth and perfectly planed.

Figure 7.27. View across pediment surfaces to Bear Peak, Boulder County, Colo. February 29, 1972. (Photo by H.E. Malde, U.S. Geological Survey)

Processes   Any viable model of pediment genesis must explain the morphologic and topographic elements of the pediment association. As these show considerable variation, it seems unlikely that any one combination of processes will produce all pediments or that any single evolutionary model will suffice. Over the years, a number of models utilizing a few basic processes have emerged as the prime hypotheses for pediment origin. However, pediment processes are not easy to study over large areas or during short time periods, and so most of the proposed mechanics of formation are based on intuition rather than solid observational data.

There is no doubt that water flows across pediment surfaces in several different forms. The presence of drainage patterns on undissected surfaces provides unmistakable proof that true streamflow occurs on pediments. Many authors have suggested it as a dominant process of pedimentation (Gilbert 1877; Paige 1912; D. W. Johnson 1932; Howard 1942; Rahn 1966; Warnke 1969), especially when the streams migrate laterally across the surface while simultaneously planating the rocks. Although most experts recognize the efficacy of lateral planation in developing part of the eroded surface, many believe this process is incapable of producing and maintaining the piedmont angle in areas remote from the main drainage lines. Streamflow erosion of these interfluve regions would require that rivers emerging from the mountains occasionally flow perpendicular to the pediment slope—a maneuver that defies the law of gravity, as Lustig (1969) reminds us.

Besides normal river flow, unconcentrated flows in the form of sheet and rill wash or floods also traverse pediment surfaces. The phenomenon of sheetflooding was first observed by McGee (1897), who saw the flow as the erosive mechanism in the formation of pediments. Although this concept was adopted by some later workers (Lawson 1915; Rich 1935), it was refuted by others who felt that sheetflow acting alone could not create large pediments, especially across resistant lithology, since the smooth surface itself is a necessary prerequisite for the development of unconcentrated flow. In fact, observations of storm runoffs in Arizona (Rahn 1967) indicate that large discharges on pediments occur as streamflows rather than sheetflows. It may not, in fact, be particularly important to know whether the flow at the head of the pediment is an unconfined type such as sheetwash or whether it moves in small rills or channels. What is important is that flow at the base of the mountain front appears to be a capable transporting agent (Higgins 1953; K. G. Smith 1958) and therefore plays a significant role in pedimentation by preventing the accumulation of debris there and so perpetuating the character of the piedmont angle. This fact has been documented beyond question in the development of miniature pediments where the underlying rocks are relatively nonresistant.

For example, Schumm (1962) studied the progressive changes of tiny pediments in the Badlands National Monument (South Dakota) by driving stakes into the pediment surface and the adjacent hillslope. Over a period of eight years the stakes were exposed by erosion (fig. 7.28), allowing Schumm to make some critical observations. The hillslopes were eroded by overland flow, mostly in small rills that shift their positions when deflected by irregularities in the slope surface. At the contact with the pediment, however, the flow type changes and crosses that surface in the form of sheetwash. Because of the hillslope retreat, a new strip of pediment is formed at the base of the slope. The surface of the new pediment is initially rather steep but, as the process continues, is regraded to a lower angle. The regrading is possible because the velocity of the sheetwash of the upper area of the pediment probably is the same as that of the overland flow on the steeper hillslope. Schumm explained this phenomenon by utilizing Manning's equation to show that the decrease in velocity expected on the more gentle slope of the pediment can easily be compensated for by a simultaneous decrease in roughness. Interestingly, the regrading first requires some erosion of pediment surface itself close to the hillslope.

Figure 7.28. Profiles and changes on miniature pediments in Badlands National Monument, S.D. Solid line is profile in 1961; dashed line is inferred profile in 1953. Numbers below solid line represent slope in percent; numbers above line indicate erosion depths in millimeters. (From Schumm 1962. Used with permission of the Geological Society of America)

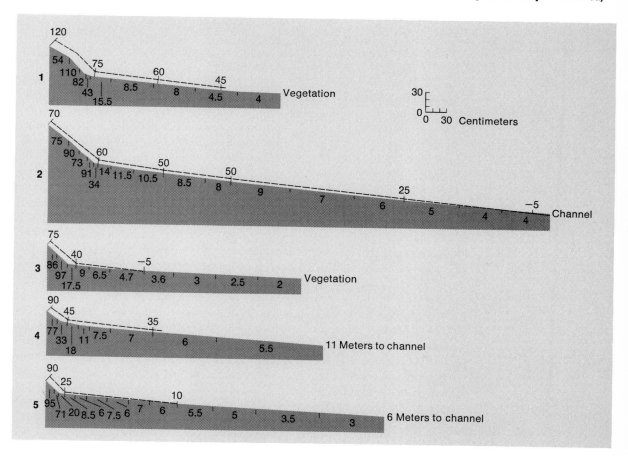

The processes of weathering also have been noted, with different degrees of emphasis, as important factors in pedimentation. The objection that sheetwash may be ineffective as an erosive agent is partly overcome if the initial strength of the pedimented rock is lowered by weathering. Weathering profiles of considerable thickness have been recognized on pedimented rocks (Mabbutt 1966; Twidale 1967; Oberlander 1972, 1974), leading to the suggestion that the rock surfaces beneath the weathered mantle, and even some suballuvial benches, may be produced by weathering rather than by water erosion. Mabbutt (1966) suggests that pediments developed on granite and related rocks may be formed primarily by continuing subsurface weathering which, because of slight differences in resistance, produces an uneven bedrock surface beneath the mantle. The surface on top of the mantle is kept flat by temporary alluviation. In other rock types, weathering may be more rapid at the surface, and the products are removed by sheetwash to maintain the relatively flat pediment surface. Some evidence exists to support the conclusion that subsurface weathering is most pronounced at the junction of the pediment and mountain front (Mabbutt 1966; Twidale 1967). Thus, headward extension of the pediment and the maintenance of the piedmont angle may be intimately related to the type of rocks involved and the efficiency of the weathering processes.

Formative Models   Cooke and Warren (1973) point out correctly that the critical boundary in piedmont areas lies between zones that are primarily depositional and those that are predominantly erosional. The position of this boundary and the presence or absence of pediments are determined by the amount of sediment produced in the mountain relative to the ability of processes to transport the material across the piedmont zone. If supply exceeds transportation, the boundary may abut against the mountain front, and no pediment will be found. In the opposite case, the boundary may be well down the piedmont slope, and pediments will be present. If equilibrium exists between rates of supply and removal, the boundary will be stable and not necessarily parallel to the mountain front. Once established, the boundary can change its position in response to climatic or tectonic alterations. Pediments may be modified after their formation by regrading, or they may be isolated from the processes that are adjusted to the new piedmont setting, as when the surface is entrenched or is buried beneath an alluvial cover. With these possibilities in mind, Cooke and Warren (1973) categorized situations under which pediments form, and their designations are presented here with some modification and addition.

In one situation, which we can call *headward pediment extension,* erosional processes allow the pediment to expand into the area of the mountain. The alluvial boundary may remain constant, or it may move toward the mountain in phase with the migrating mountain-pediment junction, depending on whether the sediment derived from headward erosion can be transported from the system. Cooke and Warren (1973) recognized three formative models as promoting headward growth: the lateral planation hypothesis, the parallel retreat hypothesis, and the drainage basin hypothesis. The *lateral planation* hypothesis was developed most fully by D. W. Johnson (1932), who believed that rock planes (pediments) are a natural consequence in arid regions. According to Johnson, rivers near the mountain front are essentially graded and so heavily loaded that they cannot cut vertically but tend to migrate and erode laterally. Extreme migration is presumed to trim back interstream bedrock spurs, forming and maintaining the piedmont angle. In response to changes imposed on the system, the rivers will perpetuate their equilibrium condition by regrading the pediment slope.

The *parallel retreat* mechanism was introduced by Lawson (1915) and championed by Rich (1935). It requires that after a mountain front achieves its diagnostic slope, weathering and erosion will maintain that declivity by making the surface retreat parallel to itself. The bedrock bench, produced as the front recedes, is swept clean by rill and sheetwash, processes that keep the angle between the mountain front and the original piedmont slope at a constant value. A slightly different mechanism of parallel retreat occurs in granitic regions where subsurface weathering may continue to level the rock surface at depth (Ruxton and Berry 1961; Mabbutt 1966). Mabbutt called this process *mantle-control planation*. It includes slight back-trimming at the hill base which stabilizes the piedmont angle and extends the pediment toward the mountain. It may also construct rectilinear profiles across the mountain front-pediment boundary (Ruxton and Berry 1961), which may also be preserved by back-wearing (fig. 7.29).

The *drainage basin* hypothesis rises from the apparent failure of either of the above models to explain the total morphology along most mountain fronts. There appears to be no dramatic change in gradients where master channels cross the mountain front onto the pediment surface, indicating the viability of lateral planation there and negating the necessity of parallel retreat. Contrariwise, the interfluve areas that do exhibit the pronounced piedmont angle are far from the main valley and, for reasons stated before, probably cannot be trimmed back by lateral corrasion. The drainage basin hypothesis, therefore, represents a compromise between the two extremes of the other theories. It recognizes that lateral planation will be dominant along the main drainage line, while interfluve areas evolve by weathering and wash. The recognition that planation, weathering, and

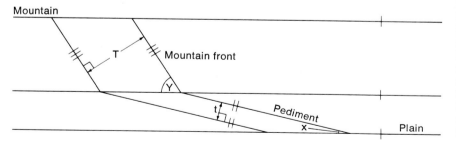

Figure 7.29. Retreat of pediment surface parallel with expanding plain. Piedmont angle at Y is maintained. Ratio of hillslope erosion (*T*) to pediment erosion (*t*) must be constant, and the rectilinear profile is perpetuated. (After Ruxton and Berry 1961. Used with permission of *Revue de Geomorphologie Dynamique*)

wash all are involved in pediment mechanics is not new. It was first espoused in early investigations (Bryan 1922, 1935; Gilluly 1937; Sharp 1940) and more recently by Lustig (1969).

A second situation may exist where there is equilibrium between the supply of debris to the piedmont and its removal to the base level zone. In this case, piedmonts consist of both depositional features and pediments that are continuously and simultaneously being formed and destroyed. The system is in a steady-state condition that persists because the amount of sediment accumulated in fans in one part of the piedmont is equal to the amount removed somewhere else by pedimentation. Stream piracies, due to the contrasting gradients between mountain and piedmont streams, repeatedly shift the zones of erosion and deposition. Gully erosion by streams heading in the piedmont area and flowing perpendicular to the mountain front forms the pediment surface. It may subsequently be buried by alluvium after piracy diverts the mountain stream onto the pedimented surface. This basic process has been employed to explain piedmont geomorphology in a variety of climatic regimes (Hunt et al. 1953; Hack 1960b; Denny 1965) and has been formalized as a widespread, pediment-producing mechanism by Denny (1967).

In the third case, erosion on the piedmont exceeds the debris supplied from the mountain, so the pediment-alluvial plain boundary is moved away from the mountain front, and new pediment zones develop farther downslope. We can call this process *basinward pediment extension*. Pediment zones already present before the boundary shift may be modified during the transition; the most common evidence cited for basinward migration of the alluvial boundary is the truncation of weathering or soil profiles developed on the pediment surface. A significant corollary to this concept is necessary. Since a basinward shift of the alluvial plain may expose older pediment surfaces by removing their sediment cover, some pediments may be truly relict features that have no sensible relationship to present conditions. Cooke and Warren (1973) refer to this as the *exhumation hypothesis* and cite Mabbutt (1966) and Tuan (1962) as its most recent proponents.

In two recent papers, Oberlander (1972, 1974) presents a thought-provoking argument that the granite pediments of the Mojave Desert were formed in their entirety before the region became as intensely arid as it is today. He concluded that boulders included in the mantles covering the pediment were originally isolated as corestones in very deep chemical weathering and have reached their present position by subsequent stripping of the weathered zone. The evidence for this conclusion is quite strong. The bouldery mantle can be traced into a well-developed weathering profile, including corestones in a grus matrix, which is preserved beneath basalts dated at >8 m.y. The pediment surface, therefore, most likely represents the weathering front formed under a semiarid Tertiary climate. The mantle cover is not material in transit but is the remaining part of a weathered profile that was progressively stripped after the region became more arid in the late Pliocene and Quaternary. Oberlander's explanation differs from others in that the pediment surface is not an exhumed, rock-cut, suballuvial bench. The pediments, in fact, were not cut in rock but were formed under a soil cover by the erosion of regolith that was being developed continuously at the weathering front. Slope retreat and maintenance of the piedmont angle were probably the results of wash processes with parallel rectilinear back-wearing similar to that described by Ruxton and Berry (1961). It is even possible that stripped pediment surfaces possess considerable relief and may be "born dissected" as was proposed earlier (Gilluly 1937; Sharp 1940).

Oberlander's proposal merits special attention as geomorphologists grope for all-inclusive models. His work demonstrates that all landforms do not necessarily yield significant relationships based on analyses of process and form. The failure of pediments to reveal any morphometric consistency may be attributed to the fact that some pediments developed under different morphogenetic conditions than those of the present. They are not equilibrium forms but, in fact, may be relicts from the distant past that are disequilibrium freaks in their modern surroundings. Oberlander states this message most clearly:

> . . .granitic landscapes in the Mojave Desert have been evolving in a sequential manner since the late Tertiary. This evolution has been triggered by climatic change, and reflects the instability of a landscape determined under conditions that no longer exist. Consequent changes in morphology will cease to be diagnostic of climate change only when all the weathered residuum inherited from the Tertiary morphogenetic regime is stripped from elevated portions of the landscape and transferred to adjacent basins of sedimentation. (Oberlander 1972, p. 19)

Although the model proposed by Oberlander applies only to granitic terrains, if correct, it demonstrates once again the irrefutable importance of climate and geology in geomorphic systems and resurrects the spectre of time in considering equilibrium. Granites in an arid climate may require imponderable time spans before their external form reflects an adjustment between processes and geology. Thus, it may be that the absolute values of graded time (Schumm and Lichty 1965) are eminently dependent on the systemic components.

Deltas

Because sediment being transported by rivers must ultimately come to rest, deposition at or near a river mouth represents a bona fide component of the fluvial system. The most important geomorphic feature produced in that environment is called a delta. Deltas are also important sedimentary entities, and geologists continue to study the depositional sequence and complex facies relationships included in the deltaic mass. Our interest in deltas here is only in the geomorphic processes that develop their form. The details of delta sedimentation will be left to the sedimentologists.

The term *delta* is usually applied to a depositional plain formed by a river at its mouth, where the sediment accumulation results in an irregular progradation of a shoreline (Coleman 1968; Scott and Fisher 1969). The feature was first named 2,500 years ago by the historian Herodotus, who noted that the land created at the mouth of the Nile River resembled the Greek letter Δ (delta). Modern deltas, however, display a great variety of sizes and shapes. At the apex of a delta, the trunk river divides into a number of radiating branches, called *distributaries,* that traverse the delta surface and deliver sediment to the delta extremities. In plan view some deltas look like alluvial fans, but they differ from fans in several important respects: (1) Deposition on deltas is due to a reduction of river velocity as the flow enters a body of standing water; the standing water can be the ocean or lakes of any size or origin. (2) Delta expansion in a vertical sense is finite, the base-level water body being the approximate limit of upward growth. (3) The gradient on the delta surface is notably flatter than that on most fans.

Sediment, Form, and Classification

Sediment deposited in deltas is usually fine sand or silt but, depending on controlling variables, may occasionally contain gravel or clay. Deltaic sedimentation and evolution were first described by G. K. Gilbert in his study of Lake Bonneville. As envisioned by Gilbert and illustrated in figure 7.30, the feature in its classic and unmodified state consists of a *deltaic plain* standing partly above and below lake level. The plain is fronted by a *delta slope* that connects its surface to the basin floor over which the delta is advancing. The basin floor, called the *prodelta* environment, is composed of fine-grained marine or lacustrine sediments that were swept in suspension beyond the delta front. The delta plain is composed of a complex of nearly flat layers (topset beds) that truncate the strata of the delta slope (foreset beds) as the delta progrades. The dip of foreset beds varies widely. In marine deltas formed by large rivers, the beds rarely dip more than 1°, making it very difficult to recognize them in the field. In small lakes, foreset beds may approach the angle of repose, and the sequence is easily discerned.

Most major rivers of the world develop deltas, and each has its own unique properties reflecting some balance between the fluvial system, the climate, tec-

Figure 7.30. Primary depositional environments and their associated layering in a classic delta.

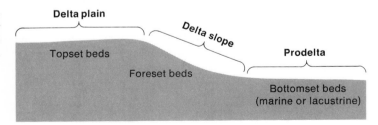

tonic stability, and the shoreline dynamics. Deltas, therefore, come in a multitude of plan-view shapes, but in general several types serve as model forms and are used for classification purposes. One or more of the controlling factors is dominant in each of the major categories of deltas. *High-constructive* deltas develop when fluvial action is the prevalent influence on the system (Scott and Fisher 1969). As figure 7.31 shows, these deltas usually occur in one of two

Figure 7.31. Classification and geomorphic characteristics of basic delta types. (From Scott and Fisher 1969. Used with permission of Texas Bureau of Economic Geology)

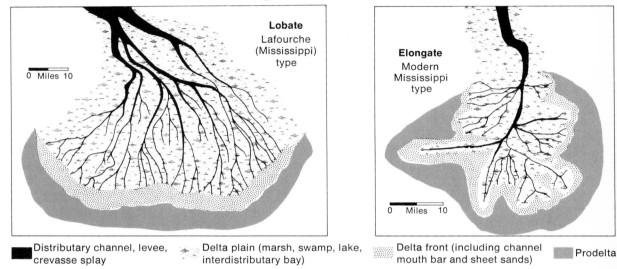

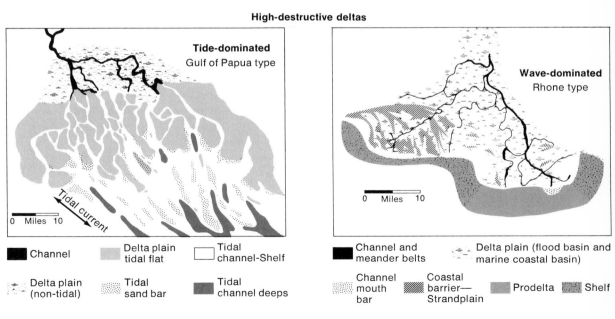

forms: an *elongate* type exemplified by the modern birdfoot delta of the Mississippi River, or a *lobate* type exemplified by the now-abandoned Holocene deltas of the Mississippi River system. Both types have high sediment input relative to the marine dynamics. Elongate deltas have a higher mud content and tend to subside rapidly when they become inactive, thereby preserving their upper sand facies. Lobate deltas sink slowly upon abandonment, and much of the sand that was prograded in the upper zones is reworked by marine processes (Scott and Fisher 1969).

High-destructive deltas originate where ocean or lake energy is high, and much of the fluvial sediment is reworked by waves, etc., before its final deposition. Figure 7.31 shows two types of these. In *wave-dominated* types, such as those of the Nile and Rhone rivers, sediment is accumulated as arcuate sand barriers near the mouth of the river. In *tide-dominated* types, tidal currents arrange the sediment into sand units that radiate linearly from the river mouth. Muds and silts accumulate inland from the segmented bars where extensive tidal flats or mangrove swamps evolve.

Dynamics and Delta Evolution

A sediment-laden river entering a body of standing water behaves much like a free jet of flow (Bates 1953). The jet flow is in the form of either an *axial jet* in which mixing is three-dimensional or a *plane jet* in which two-dimensional mixing prevails. Which flow type develops depends on the relative densities of the two water bodies. If the inflowing water is denser (hyperpycnal flow) because of its cold temperature or high sediment concentration, a plane-jet flow occurs in the form of a turbidity current moving along the basin floor. If the densities are nearly equal (homopycnal flow), axial-jet flow results, and complete mixing occurs very close to the river mouth. Almost all the river sediment is deposited immediately after entering the standing water. Homopycnal flow is most common in freshwater lake deltas, and its mechanics leads to the classic Gilbert-type construction (see Born 1972, for example). In the third possibility, where rivers flow into more dense ocean water (hypopycnal flow), mixing is rather slow and the river water spreads out laterally in a plane-jet flow.

In high-constructive deltas, distributaries develop at the mouth of the inflowing river where longitudinal bars are deposited because bedload cannot be transported when the velocity suddenly decreases. The initial bar exerts an influence on the flow and, as figure 7.32 shows, causes the river to bifurcate into two channels immediately upstream from the bar crest (for details see Welder 1959; Russell 1967b). The distributary channels are lined with natural levees that may begin beneath the surface through slow accretion of suspended load as it spreads laterally under plane-jet flow (Morgan 1970). With continued deposition, combined with minor channel scouring, the levees and bars emerge (fig. 7.32), and the distributary channels extend farther into the basin. The process may be repeated frequently, giving the basin a veinlike appearance and prograding the delta front.

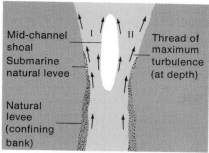

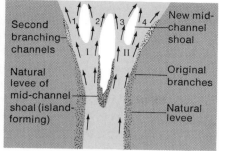

Original branching of a delta channel
(A)

Later stage of channel subdivision
(B)

Figure 7.32. Stages in bifurcation of trunk river and creation of distributary channels of a delta. (From Russell 1967b. Used with permission of the Louisiana State University Press)

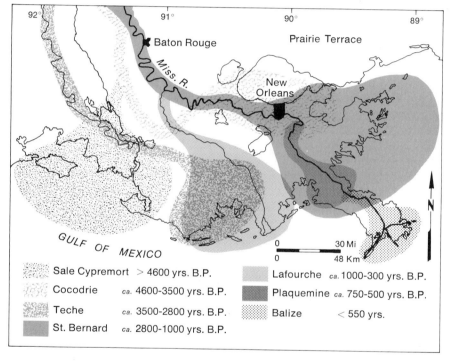

Sale Cypremort > 4600 yrs. B.P.
Cocodrie *ca.* 4600-3500 yrs. B.P.
Teche *ca.* 3500-2800 yrs. B.P.
St. Bernard *ca.* 2800-1000 yrs. B.P.
Lafourche *ca.* 1000-300 yrs. B.P.
Plaquemine *ca.* 750-500 yrs. B.P.
Balize < 550 yrs.

Figure 7.33. Sequence of the development of the subdeltas that comprise the Mississippi River deltaic plain. (From Coleman 1968. Reprinted with permission, from the *Encyclopedia of Geomorphology*, Rhodes W. Fairbridge, ed., Dowden, Hutchinson & Ross)

As the delta progrades, shorter routes to the ocean become available. These pathways often begin far inland from the delta front, usually developing when the river is diverted through a breach in the levee called a *crevasse*. The new river course shifts the locus of sedimentation and begins to form a new deltaic lobe. The Mississippi River deltaic plain, for example, actually represents the coalescence of seven major lobes that were built at different times and in various positions during the last 5,000 years (fig. 7.33). The modern birdfoot-delta growth is only a minor portion of the entire deltaic area. Abandoned lobes are no longer fed by incoming river debris and immediately become vulnerable to erosional attack by the ocean. Thus, as a new lobe develops, older lobes are being reclaimed by the ocean. For example, the Sale Cypremort lobe, the oldest lobe shown in figure 7.33, has been eroded and inundated since its abandonment about 4,600 b.p. Only a portion of its original area is now exposed above sea level.

The processes that produced the major lobes of the Mississippi deltaic plain are obscured by modifications since their abandonment. Within the modern delta, however, the genetic mechanics is known in considerable detail. Here cre-

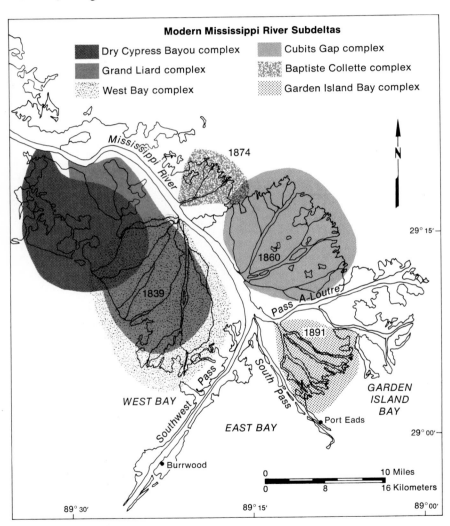

Figure 7.34. Subdeltas of the modern birdfoot delta of the Mississippi River. (From Morgan 1970. Used with permission of National Association of Geology Teachers)

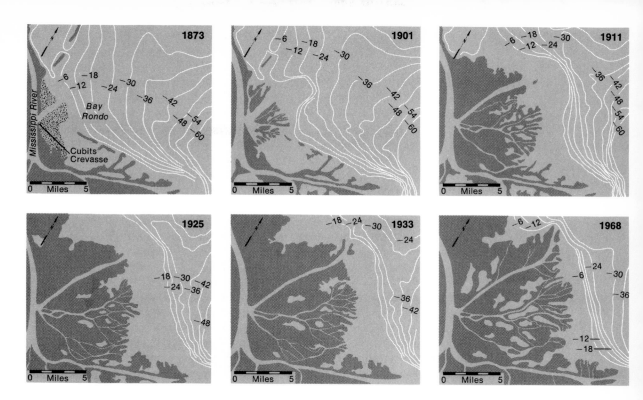

vasses in the levee system have repeatedly shifted the site of deposition. The breaks in the levee begin because of overtopping during a major flood and gradually increase in size through scouring associated with succeeding floods. Sediment diverted through a breached levee progressively builds a subdelta from deposits known as *crevasse splays*. Four such subdeltas have formed in historic times (fig. 7.34). As channels in the subdeltas bifurcate and prograde, their gradients decrease and they lose their ability to transport load. When their gradients approach that of the main trunk channel, it is no longer advantageous for the river to utilize the crevasse system. Sedimentation in the subdeltas ends, and the ocean begins to inundate the area as subsidence and compaction lower the subdelta surface. At the same time, however, a new subdelta may be developing somewhere else within the birdfoot system. Figure 7.35 illustrates the subdelta cycle.

It seems clear that delta formation can be viewed on several time scales. On a short-term basis, only a limited area (subdelta) receives any sediment, but the position of the accumulation shifts repeatedly. On a longer time scale, the entire active delta (lobe) has a periodic migration. What we observe as the Mississippi River delta is in fact a monstrous area created by the coalescence of a number of major lobes over a long period of time. Active deposition occurs on only one

Figure 7.35. Development of the Cubits Gap subdelta into Bay Rondo and postdepositional alteration. (From Morgan 1970. Used with permission of National Association of Geology Teachers)

lobe at any given time, and on only a minor portion of that lobe. Even so, the evolutionary processes on any time scale are probably similar in that they involve channel bifurcation, levee development, and crevassing.

Even though this evolutionary model is probably correct, it is valid only for deltas of the high-constructive type. The model is known in such detail only because geologists have studied the Mississippi delta for many years. Other delta types have not been so closely examined, and their mode of origin is not nearly so well known. Furthermore, we are far from understanding precisely how changes in climate, tectonics, or any of the other fundamental controls affect the mechanics of delta evolution.

Summary

The character of each major fluvial landform is due to the manner in which a particular set of geomorphic controls influences the river mechanics. The features can be predominantly erosional, predominantly depositional, or derive from the combination of both types of processes.

Floodplains originate by lateral migration of meanders and by periodic overbank flooding. The sediment composing a floodplain sequence is mainly laterally accreted point bar deposits accumulated as the river shifts its position across the valley bottom. The point bar material is usually capped by a thin layer of silt and clay deposited as vertically accreted sediment during overbank flooding of the river. The total amount of vertical accretion is probably controlled by the rate of the river's lateral migration. Because many rivers move across the valley floor rapidly, most floodplains are developed by lateral accretion, but overbank deposition may be important in the initial phase of construction.

Terraces are merely abandoned floodplains. They form when entrenchment places the river at a lower level, thereby removing the former floodplain surface from the river's hydrologic activity. The origin of a terrace usually refers to how its flat tread (the former floodplain level) was formed. The tread of an erosional terrace is produced by the lateral migration of a river and is capped by the thin point bar and overbank deposits associated with floodplains formed in that manner. In contrast, the surface of a depositional terrace represents the upper level of the sediment deposited in an episode of valley filling. The use of terraces to reconstruct geomorphic history demands some knowledge of their origins, because the salient properties of the various terrace types develop through different sequences of events.

Piedmont regions are characterized by depositional features (alluvial fans) and plains of erosion (pediments). Both features manifest some accommodation between the amount of sediment derived from a source area and the ability of the river to transport the sediment across the piedmont zone. The processes that develop fans and work on their surfaces are so complex that little agreement exists regarding their origin. Moreover, some workers believe that fans reveal a steady-state condition, while others question whether such a condition is ever attained and even whether the features are related to present conditions. Pediments also seem to defy genetic generalization. They are usually interpreted as being the result of lateral planation, weathering, and rill wash, or some combination of the various processes. As with fans, however, little agreement can be found concerning the origin of pediments; some may be relict features that are completely unrelated to modern geomorphic controls. It is clear that piedmont landforms cannot be placed into all-inclusive genetic models. Their properties vary too much to be explained by one mode of origin. Because of this, every piedmont region must be examined and interpreted according to local tectonics, climate, geology, and geomorphic history.

Deltas represent the accumulation of sediment as a transporting river enters a body of standing water. At its mouth the river bifurcates into distributaries and constructs levees. This allows debris to be transported farther into the ocean or lake basin and permits the delta to expand by prograding into the basin area. The form and size of the delta, however, depend on the balance reached between river flow and the counteracting energy of currents and waves in the ocean or lake. Detailed studies of deltas reveal a very complex growth history in which the site of active sedimentation shifts periodically through crevasses in the natural levees. Active sedimentation occurs on only a small part of the feature at any given time.

The following references will provide greater depth of information about the concepts examined in this chapter and also contain more extensive bibliographies on the various topics.

Suggested Readings

Bull, W. B. 1968. Alluvial fans. *Jour. Geol. Educ.* 16:101-06.

Cooke, R. U., and Warren, A. 1973. *Geomorphology in deserts.* London: Batsford Ltd.

Denny, C. S. 1967. Fans and pediments. *Am. Jour. Sci.* 265:81-105.

Hadley, R. F. 1967. Pediments and pediment-forming processes. *Jour. Geol. Educ.* 15:83-89.

Howard, A. D.; Fairbridge, R. W.; and Quinn, J. H. 1968. Terraces, fluvial—Introduction. In *Encyclopedia of geomorphology,* edited by R. Fairbridge, pp. 1117-23. New York: Reinhold Book Corp.

Leopold, L. B.; Wolman, M. G.; and Miller, J. P. 1964. *Fluvial processes in geomorphology.* San Francisco: W. H. Freeman.

Morgan, J. P. 1970. Deltas—a résumé. *Jour. Geol. Educ.* 18:107-17.

Oberlander, T. M. 1974. Landscape inheritance and the pediment problem in the Mojave Desert of southern California. *Am. Jour. Sci.* 274:849-75.

Wolman, M. G., and Leopold, L. B. 1957. River flood plains; some observations on their formation. U.S. Geol. Survey Prof. Paper 282-C.

Scientists have not agreed on the role of wind in geomorphology. Irrefutable proof that wind is capable of significant geomorphic work is simply not available in many instances, because definitive, quantitative data concerning eolian processes and features are woefully few. It has been described as being absolutely dominant in arid regions (Keyes 1910, 1912), in a model in which planation and mountain front retreat are attributed solely to wind erosion, and as being only a minor perturbation on features formed almost entirely by fluvial or slope processes. Disavowing these extreme views, it is probably safe to say that wind can be an effective geomorphic agent under certain physical conditions. Regions having sparse vegetation and unconsolidated sediment not tightly bound by rooting systems are most susceptible to wind attack. Extensive evidence of eolian processes is therefore found in those areas, such as modern deserts, where vegetation growth is stunted by lack of water or immature soil development. As is true of any process, however, the proficiency of wind to do geomorphic work depends on whether the driving force can exceed the resisting force of the surficial material. Thus, assigning wind processes to one particular environment alone is a gross perversion of the principles governing geomorphic processes. The emphasis placed here on deserts is simply a pedagogical tool, not a geomorphic necessity.

Detailed reviews of the physical basis of wind action and its geomorphic results are available (Bagnold 1941; Chepil and Woodruff 1963; Cooke and Warren 1973), and much of the following has been liberally excerpted from those excellent treatments.

Wind Processes and Landforms 8

The Resisting Environment

Contrary to popular thought, deserts are not the barren tracts of shifting sands depicted in Hollywood epics. In fact, only one-fourth to one-third of most desert surfaces is occupied by sand, which usually occurs in large, sandy plains called *ergs* (see I. G. Wilson 1973). Normally, deserts display a variety of erosional and depositional landforms imprinted on a diverse topography that ranges from flat plains to rugged mountains (table 8.1 and fig. 8.1). Deserts also exist in different temperature zones; polar deserts are common, although relatively unstudied. Specific processes differ in relative importance in polar deserts and hot deserts, and so the similarity of landforms that sometimes exists between the two regions is not infallible proof of an identical genetic history. In fact, the same

Figure 8.1. Map of Australian arid regions showing diverse physiography of deserts. (From Mabbutt 1971. Used with permission of the Australian National University Press)

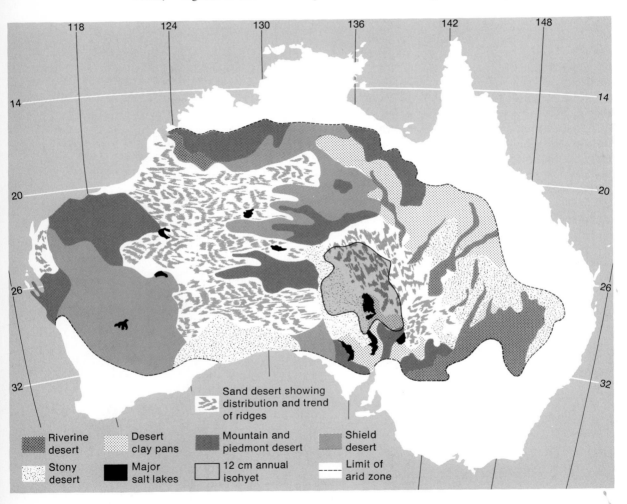

Sand desert showing
distribution and trend
of ridges

Riverine desert

Stony desert

Desert clay pans

Major salt lakes

Mountain and piedmont desert

12 cm annual isohyet

Shield desert

Limit of arid zone

Table 8.1 Area and percentage of total desert region of major components in the arid zone of Australia. (Compare with map in fig. 8.1.)

	Km³	Percentage of arid zone
Mountain and piedmont deserts	930,000	17.5
Riverine desert	210,000	4.0
Stony desert	640,000	12.0
Desert clay plains	690,000	13.0
Sand desert	1,680,000	31.0
Shield desert	1,200,000	22.5
Total	5,350,000	

From Mabbutt 1971. Used with permission of the Australian National University Press.

external form may be a function of a myriad of basic processes combined in slightly different ways. Unless stated otherwise, the discussion following refers only to hot desert conditions.

Deserts are by definition arid. They receive less than 25 cm of annual precipitation and have enormous evaporation rates, commonly 15 to 20 times greater than the precipitation (Stone 1968). Although deserts have a meager plant cover, the diversity of vegetal types is surprising, ranging from shrubs and grasses to true woodlands. This diversity is created by minor variations in soil moisture which relate to elevation. It is important because humus content and soil binding differ with particular species, and because some plants are annual and others are perennial. The type of flora influences the resisting setting; woodlands, for example, are significantly denser than other vegetation types and protect more surface area (Whitaker et al. 1968). In general, mountainous zones with stabilizing vegetation and thicker weathering mantles are less susceptible to wind attack than is the lower level of the desert environment.

Weathering and soil-forming processes are intimately involved in the character of a land surface and how it resists or succumbs to wind erosion. The angular, unaltered debris so prevalent on desert surfaces is presumed to be the residuum of rock disintegration, leading to a general concurrence that mechanical weathering predominates in the desert environment. This does not mean that mechanical weathering functions to the total exclusion of decomposition. Chemical processes, in fact, do function in the desert, as evidenced by the ubiquitous desert varnish and the common red coloration of desert soils, both of which probably require some degree of ion mobility. In addition, as explained in chapter 4, many processes of disintegration, such as exfoliation and grus formation, are abetted by chemical reactions that produce volume expansion. It should also be obvious that the relative balance between chemical and mechanical weathering varies with moisture and so with topographic position. Nonetheless, it is difficult to imagine deep weathering mantles even in mountainous areas unless an excessive vegetal cover exists, since mass wasting probably degrades the surface as rapidly as the weathering front proceeds into the rock. It is mostly on the low-angle desert surfaces that anything approaching a soil or weathering profile is preserved, and even on those stable surfaces the processes are so slow that only thin profiles develop. The situation can be complicated by the fact that weathering rates varied during the Quaternary, and some deep weathering profiles exist as relict features.

Desert soils are an important determinant of the facility of wind action. In general, A horizons are quite thin, mechanically weakened, and more permeable than the lower horizons (Cooke and Warren 1973); case hardening at the surface may in places counteract this tendency (Lattman 1973). Nevertheless, regardless of how slowly it proceeds, illuviation may cement the B and K horizons to the extent that they become nonpermeable, highly indurated zones (Gile et al. 1965, 1966; Gile 1966). The long-term result of desert processes is to create a surface that is susceptible to wind erosion and a subsurface that is eminently resistant. Stripping of the surface horizon will expose the lower, resistant zones, which are preserved as stable platforms for extended periods of time. It should be noted that stripping at one place is commonly balanced by deposition of eolian debris somewhere else in the area.

In addition to the effects of vegetation and weathering phenomena, resistance to wind action also is tangibly dependent on the size of particles strewn across the surface. The inherent diversity of the system is again clear; while some unvegetated areas are spared from intense eolian effects because the surface debris exceeds the competence of the wind, other zones in the same region are visibly ravaged because their surfaces are composed of fine-grained sediment.

Surfaces armored with a thin layer of stones that protects an underlying horizon of sand, silt, or clay are common in many environments. They are par-

Figure 8.2. Percentage of soil composed of particles less than 2 mm in diameter. Analyses of soils in the Lahontan Basin, Nevada. (From Springer 1958. Reproduced from *Soil Science Society of America Proceedings,* vol. 22, 1958, p. 65, by permission of the Soil Science Society of America)

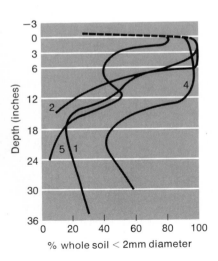
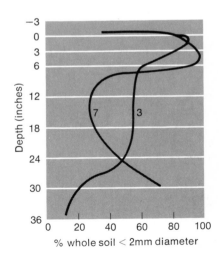

ticularly evident, however, in hot deserts where they are given a variety of names such as stone mantles, gibber, hammada, reg, and desert pavement. Desert pavement, as mentioned in chapter 7, is associated with alluvial fans and other deposits of unsorted alluvium. The surface armor is usually only one or two stones thick and consists of whole clasts or disintegrated parts of the coarse fraction found in the underlying alluvium. In most cases, a soil exists beneath the pavement, but is notably deficient in coarse particles in the upper part of its profile (fig. 8.2). This fine-grained zone has prompted the generally accepted premise that the gravelly surface layer was formed as a by-product of desert weathering. However, the surface concentration of pebbles, as well as the abruptness of the transition from fines to the pebble cap, requires some process other than those of normal soil formation.

The most commonly cited explanation for pebble concentrates on desert pavements is that the wind deflates the fine sediment from the alluvium, leaving the coarse material behind as a lag. Proof that such winnowing causes stone pavements is rare, and its universal application as the total explanation is probably unjustified. It is equally likely that surface fines are winnowed by water (Sharon 1962) or that the stone concentrations are produced when large clasts are spasmodically lifted through the alluvial matrix until they emerge at the surface. Such a phenomenon may be accomplished by cyclic freezing and thawing (Inglis 1965) or by wetting and drying (Springer 1958). In either case, the coarse particle is lifted slightly during expansion, but cannot settle back to its former position during the contractive phase of the cycle because finer material has already filled in beneath it.

In contrast to the desert pavements, some parts of the desert are covered with abundant fine-grained debris, and there deflation is a viable erosive process. Playas, for example, are notable for their silt and clay mantles; it has often been suggested that they are subject to extensive degradation by deflation (Blackwelder 1931). There is no doubt that deflation of fine sediment occurs (fig. 8.3), but

Figure 8.3. Deflation of fine-grained sediment at Pyramid Lake, Nev. Bottom sediment has been exposed by recent decline in the lake level.

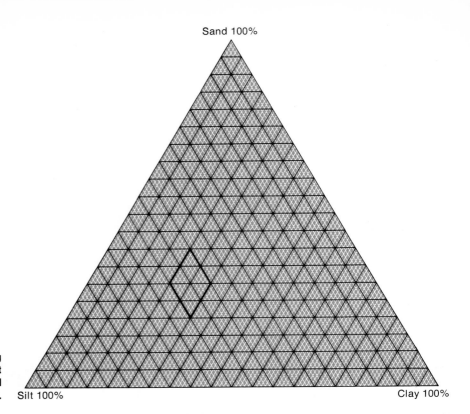

Sand 100%

Silt 100%   Clay 100%

Figure 8.4. Diagram showing mixture of particle sizes that is most resistant to wind action.

it may not be as pervasive in the desert environment as once supposed (Hooke 1968). Although most deserts do contain fine material, it is usually fabricated into a cohesive crust by a number of processes (Dury 1966a; Cooke and Warren 1973). Crustation not only gathers individual particles into aggregates larger than the wind can entrain, but it also protects unconsolidated fines immediately beneath the surface from the ravages of deflation. The greatest resistance occurs when sand, silt, and clay are mixed together in proportions of clay 20–30 percent, silt 40–50 percent, and sand 20–40 percent, as in figure 8.4 (Chepil and Woodruff 1963). But establishment of the most resistant size blend is not the only determinant of resistance. Variables of moisture, humus, and chemical composition may give considerable resistance to size blends that are usually susceptible to deflation. Calcium carbonate, on the other hand, may actually induce erodibility because it tends to inhibit the formation of aggregate clods (Cooke and Warren 1973, p. 241).

In summary, the geomorphic effect of wind action is regulated to a significant degree by the properties of the resisting framework. For wind processes to have any geomorphic consequence, the surface must be unvegetated and littered with noncohesive sediment smaller than gravel-size. Any region producing sediment of that type, and lacking the climatic, topographic, and geomorphic conditions needed to protect it, may be open to pronounced wind attack. Because hot deserts are most susceptible to wind action, they are taken as the model for the discussion of eolian processes. However, coastal regions, unvegetated semiarid or subpolar zones, and areas in front of active glaciers also often are modified by wind processes. In any situation, the amount of geomorphic work actually accomplished by the wind depends on the second prime variable—the character of the wind itself.

The Driving Force

Global wind systems are related to the large pressure differentials associated with worldwide circulation patterns. In contrast, local winds are generated by anomalies in the physical characteristics within a specific area. For example, in many cases winds reflect a difference in thermal properties of surface materials, and so we can expect the greatest wind activity in those environments where large temperature variations exist in the air layer immediately above ground level. Such conditions are common in deserts, along seacoasts, or in areas with pronounced diversity in elevation, such as the juncture of mountain ranges and the low plains fringing them.

Certain attributes of the wind—mainly its direction, velocity, and degree of turbulence—are responsible for most geomorphic effects. In areas with large thermal contrasts, *wind direction* is more or less predetermined by the temperature gradient of the near-surface air. However, temperature gradients often change from daytime to nighttime because of variations in rates of cooling and heating. In deserts, heating of the bare desert surfaces incites winds to move toward them from surrounding areas that have a vegetal cover. At night, however, the desert surface cools more rapidly than the other regions, and the winds reverse their direction. A similar diurnal direction change occurs along coasts because the ocean warms less slowly than the adjacent land, prompting on-shore winds during the day. The ocean, however, gives up stored heat less readily at night, and winds blow off-shore. The importance of wind direction is most evident in the development and preservation of eolian bed forms constructed from loose, transportable sand.

Wind velocity is important because it is the prime determinant of what material will move under wind attack and what will remain stationary. Wind velocity increases with height above the ground because it is slowed at the surface by friction. The change in velocity with height can be expressed in a number of ways exemplified by the equation (Bagnold 1941)

$$V_z = 5.75 \, V_* \log z/k$$

where V_z is the velocity at any given height z, V_* is a parameter called *drag velocity,* and k is a constant relating to surface roughness. In reality, k is the height of a thin zone immediately above the surface in which velocity is zero. The thickness of this zone depends on the surface roughness, but on a flat, granular bed it tends to be about $\frac{1}{30}$ of the diameter of the surface grains. The general relationship between height and velocity plots as a straight line on semilogarithmic paper (fig. 8.5).

As the wind blows harder and V_z increases, the drag velocity also rises, and an increasing stress is exerted on particles exposed at the surface. This follows since the magnitude of drag across any unit area of surface is related to the drag velocity such that

$$V_* = \sqrt{\tau/\rho}$$

or

$$\tau = \rho V_*{}^2$$

where τ is drag/unit area and ρ is the density of the air. Thus, for a surface with constant particle size, a stronger wind will increase the value of V_*, and therefore the height-velocity relationship is defined by a number of straight lines, each having a different V_* value. Drag velocity in this sense represents the *rate* at which velocity increases with height and can be thought of as a velocity gradient.

Figure 8.5. Relation of wind velocity and height above a surface. Lines represent drag velocities (V·), which are rates at which velocity increases with height. The value k is the height of the zone above the surface in which velocity is zero. (From Chepil 1959. Reproduced from *Soil Science Society of America Proceedings*, vol. 23, 1959, p. 425, by permission of the Soil Science Society of America)

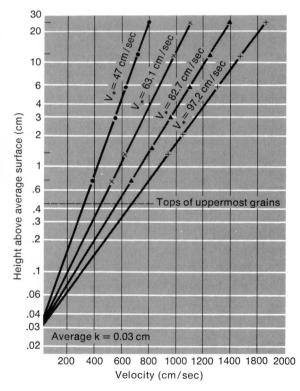

Interestingly, all lines of different V_* values merge at the ordinate where velocity is zero (fig. 8.5). This occurs simply because k is a function of surface conditions, such as particle size, rather than wind velocity.

Assisting velocity in wind attack is the factor of *turbulence*, which occurs in the form of eddy currents. Turbulence has a direct influence on the entrainment process, although it is probably capable of lifting only particles smaller than sand (Bagnold 1941; Chepil and Woodruff 1963). Sharp (1964), however, suggests that grains up to 0.125 mm may be affected by wind turbulence. Most likely, turbulence helps in the molding of desert landforms, but the types of eddy motion are so numerous and mechanically complex (Lumley and Panofski 1964) that precise measurement of turbulence is precluded, and its relationship to specific bed forms is not clear. At present little data are available to generalize about velocities and turbulence in wind processes but, in ways that are still not quantitatively clear, both factors must bear directly on the origin of wind-produced landforms.

Processes

As wind blows across a surface composed of loose sediment, a critical drag velocity exists at which particle motion begins. The ease of entrainment, however, depends not only on the size of the particles and the wind velocity but also on complicating factors such as soil moisture, packing, etc. Recognizing the inherent complexity, the velocity which initiates movement is called the *fluid threshold* (Bagnold 1941) and can be estimated by the equation

$$V_{*t} = A \sqrt{\frac{\rho_s - \rho_a}{\rho_a} gD}$$

where V_{*t} is the threshold value of drag velocity, ρ_a is the density of the air, ρ_s the density of the sediment, D is the particle diameter, g is gravity, and A is a constant of proportionality which for air is 0.1. Most desert sands have a threshold velocity of about 16 km/hr, but the precise value for any size varies with other factors such as particle shape (G. Williams 1964), sorting (Woodruff and Siddoway 1965), and surface roughness. Roughness is a function of particle size but is also controlled by vegetation. Rough surfaces reduce wind velocity; as high, dense vegetation increases roughness drastically, it lessens the erosive effect of the wind.

The influence of moisture on the critical threshold velocity was documented by Belly (1964) who suggested that a surface moistened by liquid or vapor would require a significantly higher velocity for sediment entrainment. Where soil water is present, the fluid threshold is determined by

$$V_{*t} = A (1.8 + 0.6 \log w) \sqrt{\frac{\rho_s - \rho_a}{\rho_a} gD}$$

Wind Erosion

where soil water content (w) is expressed as a percentage. If the surface is moistened by water vapor, the critical velocity is expressible in terms of relative humidity:

$$V_{*_t} = A\ (1\ +\ \tfrac{1}{2}\ h/100)\ \sqrt{\frac{\rho_s\ -\ \rho_a}{\rho_a}\ gD}$$

where h is relative humidity in percent. Calkin and Rutford (1974) found that when Antarctic dunes were moistened by summer melting or condensation, the threshold velocity needed to entrain their sands doubled.

The fluid threshold equations clearly indicate that entrainment velocity varies directly with particle diameter; nevertheless, as figure 8.6 shows, the relationship becomes invalid when size is less than 0.1 mm because of the greater interparticle cohesion and low roughness associated with finer particles (Bagnold 1941; Smalley 1970). Furthermore, grains greater than .84 mm are moved with great difficulty, and that size probably represents a logical upper limit for unaided wind entrainment. Figure 8.6 also reflects the fact that once motion begins, the surface is subjected to a continuous rain of moving particles which, on impact with stationary grains, produce entrainment at a velocity lower than the fluid threshold. This reduced threshold velocity, called the *impact threshold,* becomes progressively more significant as size increases, and particles greater than .84 mm may be entrained even when the velocity is well below the fluid threshold. It appears, therefore, that when grains begin to move at the fluid threshold, the entrainment process is self-generating because the wind speed is already above the impact threshold. Each moving grain is not only a product of the system, but also an integral component of the entrainment mechanics through the process of saltation as depicted in figure 8.7.

Figure 8.6. Relation of particle size to threshold velocity. (From Bagnold 1941. Used with permission of Chapman & Hall Ltd.)

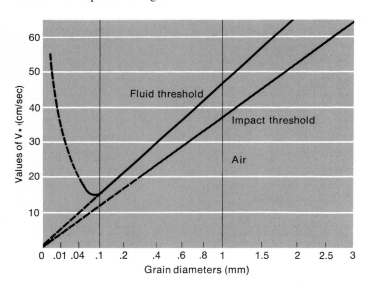

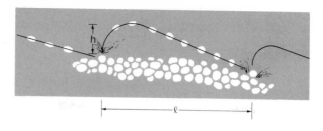

Figure 8.7. Diagram of the saltation process. Moving grain strikes the surface and dislodges a particle to elevation *h*. The particle moves downwind and repeats the process. *l* represents the distance of travel of the dislodged particle before striking the surface.

The "dust bowl" disaster in the Great Plains of the United States during the 1920s and 1930s prompted scientists to systematize data into predictive models of a potential wind-erosion hazard (Woodruff and Siddoway 1965). In general, the quantity of wind erosion is a function of the variables included in the driving and resisting forces. One expression of the relationship (Woodruff and Siddoway 1965) is

$$E = f(I', K', C', L', V')$$

where E is erosion in tons/acre/yr, I' is a soil erodibility index, K' is a roughness factor, L' the length along the prevailing wind direction, C' a wind or climatic factor, and V' a factor relating to vegetal cover. Factors may be estimated in a variety of ways. For example, Thornthwaite's precipitation effectiveness (P-E) index (discussed in chapter 5) has been employed by Yaalon and Ganor (1966) as a gage of the climatic factor C'. It is interesting to note that all factors except C' can be controlled to prevent erosion, and Woodruff and Siddoway present step-by-step applications of their general equation to show how annual soil loss may be predicted and prevented.

Features Produced by Erosive Action

Geomorphic features manufactured by wind action are due to abrasion and deflation, the two main erosive processes. Abrasion results when sand particles carried by the wind act as grinding tools to physically wear away exposed surfaces of solid rocks or rock fragments. It is most effective where relatively soft rocks, bare of vegetation, are blasted by high-velocity winds that carry an abundant amount of hard particles. Sharp (1964) suggests that maximum abrasion is determined by a unique combination of velocity and particle concentration, so that no one height above the surface has the greatest potential for abrasion. Nonetheless, it is common knowledge that most of the abrasive load travels within two meters of the ground level.

Perhaps the most frequently cited evidence of wind abrasion is the development of faceted stones called *ventifacts* (fig. 8.8). Ventifacts are produced on pebble surfaces that are oriented perpendicular to the prevailing wind. The incessant bombardment of wind-driven sand gradually cuts a smooth face into the windward portion of the original surface and forms a sharp edge between it and the leeward side of the pebble. The slope on the faceted surface is normally

Figure 8.8. Pebbles and cobbles pitted and faceted by wind-blown sand. Ventifacts were all found in Sweetwater County, Wyo. (Photo by M.R. Campbell, U.S. Geological Survey)

between 30° and 60°, but enough variability exists in these angles to suggest that some evolutionary sequence is involved in their development (Kuenen 1928; Schoewe 1932). It is equally feasible, however, that various heights of pebbles above the surface relative to various levels of maximum abrasion potential in the wind can produce a number of facet inclinations and shapes (Sharp 1964). It is also possible that some faceting is accomplished by dust-sized particles (silts and clays) carried in suspension, which would make ventifact sculpturing a function of the unique aerodynamics associated with each individual situation (Whitney and Dietrich 1973).

Facets do not form on surfaces that parallel the wind direction, and therefore the ubiquitous presence of ventifacts with more than one face and edge (fig. 8.8) has created some interpretive furor. It is tempting to regard multifaceted ventifacts as evidence of a shift in the prevailing wind direction, but it is equally possible that the stones themselves are occasionally turned over or rotated (Schoewe 1932; Sharp 1964). Overturning, of course, would result in a new portion of the original pebble being subjected to wind erosion, and any number of facets could form in a unidirectional wind.

Abrasion affects rock outcrops by shaping or grooving the surface. The most common feature etched by the wind, called a *yardang,* is an elongate ridge, normally less than 10 m high but occasionally as high as 30 m, which is aligned parallel to the prevailing wind. Yardangs are most prominently developed in regions underlain by relatively soft rocks and tend to occur in groups in which individual ridges are separated by round-bottom, wind-eroded troughs. Blackwelder (1934) emphasized the importance of eroding the intervening "yardang trough" rather than the ridge itself. He believed that most erosion occurs at the lowest level, decreasing in magnitude and efficiency toward the ridge crest. As a result the troughs are eroded more rapidly, so that sometimes the ridge sides are slightly undercut and the ridge crests are left rather ragged. In very soft materials, yardangs may be molded into a whaleback shape that is streamlined downwind. This form, however, is usually restricted to small yardangs which can be completely covered by the thin layer of sand-laden wind (Blackwelder 1934).

How much of the shape of yardangs is attributable to wind action alone is a debatable question (Twidale 1968a). Their alignment is unquestionably related to the prevailing wind direction, but possibly their form is simply a final embellishment on bedrock relief developed by other processes. No unequivocal data are available to resolve this point.

Grooving of bedrock surfaces is also a common eolian phenomenon. The abrasive processes that result in parallel furrows seem to operate at different orders of magnitude, ranging from tiny elongate flutes on ventifact surfaces measured in centimeters (Sharp 1964), to giant hollows more than a kilometer wide and up to 10 m deep. Some question lingers as to whether the mega-grooves are demonstrably erosional or merely a depressed zone between parallel ridges of deposited sand. Although the possibility of misinterpretation remains, some large erosional grooves have been reasonably well documented (D. King 1956; Shawe 1963; Stokes 1964), and they may have a distinct influence on regional drainage patterns that are established subsequently (Stokes 1964; Beaty 1975).

The process of deflation has been suggested as the prime mechanism in the genesis of many enclosed desert basins. These basins range in size from small deflation hollows to vast expanses measured in hundreds of square kilometers. Usually the small depressions are easily related to wind erosion, since they are often strung out parallel to the wind and may be associated with dunes (Flint and Bond 1968). A deflation origin for the larger areas, called *pans,* is much more difficult to prove since they have no clear orientation. Arguments about their origin, therefore, are often negative in the sense that they merely eliminate other processes and leave deflation as a viable, but untested, alternative. Very large pans, however, must be relict to some extent, since wind erosion, even if proven, functions so slowly that an enormous time span would be necessary to accomplish their formation.

Wind Transportation and Deposition

Processes

The entrainment of sediment releases particles to the processes of transportation and deposition, and the result is the array of surface forms so diagnostic in areas under wind attack. The type of transporting process is determined fundamentally by the size of the debris made available to the eolian force; silts and clays are carried in *suspension,* fine and medium sands travel by *saltation,* and coarse sands move by rolling or sliding, motions collectively referred to as *surface creep.* The most striking eolian features usually are made of sand simply because few winds can move larger grains, and sediment entrained by turbulence and carried in suspension often is so diffuse as to preclude its accumulation in explicit surface forms. In addition, entrainment of fine-grained sediment is not so easy as one might expect, because the surface is usually smooth and the small grains tend to retain moisture and have stronger interparticle chemical bonds. Grains larger than 0.1 mm probably cannot be transported in suspension (Bagnold 1941; Kuenen 1960; Sharp 1963), and particles both larger and smaller than sand often resist entrainment. For those reasons, our discussion will center on depositional features composed of sand, and on the mechanics of saltation.

As figure 8.9 shows, when entrainment occurs, the moving sand directly influences the characteristics of the wind itself because the saltating grains exert extra drag on the system. The drag velocity (V_*), when defined as $\sqrt{\tau/\rho}$ (assuming certain flow conditions), is also a measure of the velocity gradient of the wind, i.e., the rate at which velocity increases with height. The stronger the wind blows, the greater the drag on the surface and the greater the divergence of the drag velocity curves away from the vertical. The slope of each line on the figure, therefore, reflects its rate of increase in velocity with height (given as drag velocity, V_*); the higher V_* values indicate a more rapid increase in velocity with ascension above the surface.

The solid lines on figure 8.9 represent the changes impressed on the system after saltation begins. Note that under saltating conditions k rises dramatically, and it no longer represents a zero velocity but assumes a constant velocity value at the height k' where all the gradient lines merge. The wind velocity at any height (z) is then defined by the equation

$$V_z = 5.75 \ V_{*}' \ \log \ z/k' + V_t$$

where V_*' is the drag velocity under conditions of saltation, k' is the new height of the focus, and V_t is the threshold velocity (impact threshold) measured at the height k'.

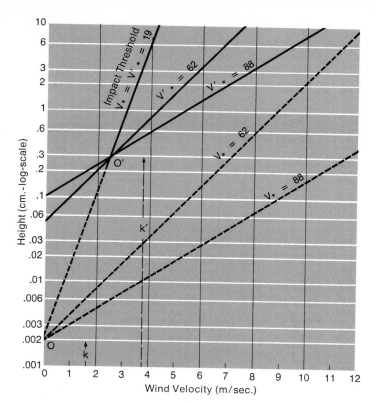

Figure 8.9. Change in drag velocity caused by saltation. Dashed lines show drag velocity before saltation begins. Solid lines represent drag velocity after saltation starts. (From Bagnold 1941. Used with permission of Chapman & Hall Ltd.)

The effect of saltation is obvious from the graph. For example, a wind with a velocity gradient $V_* = 62$ blowing across a completely resistant surface has a velocity of about 970 cm/sec at a height of 2 cm. Once motion begins, however, the velocity at that level, read along the gradient line $V_*' = 62$, has decreased to approximately 530 cm/sec. It is important to stress again that the value of k or k' is dependent on the roughness components of the surface, and so figure 8.9 defines curves that are valid for only one particle size. In addition, very few measurements have been made in the natural setting, and although estimates of the quantitative relationship between k values and particle diameter have been made (Cooke and Warren 1973, p. 259), enough variance exists in these to be cautious about their use.

Saltating grains travel in a characteristic path in which particles dislodged or bounced from a surface move at first in a predominantly vertical direction (fig. 8.7). The subsequent travel path is probably controlled by (1) the particle's initial upward velocity and (2) the velocity gradient above the surface. Bagnold (1941) suggested that most of the forward momentum is gained from the wind when a grain is near the top of its path, and from then on the actual trace of the particle movement depends on the magnitude of the wind thrust and the settling velocity of the grain. Precisely how high a grain will rise from the surface depends, of course, on the size of the particle involved and on what kind of a surface it is bouncing from. It is clear, however, that the dense curtain of saltating grains is restricted to a thin near-surface zone. Most observations (Bagnold 1941; Chepil 1945; Sharp 1964) indicate that an overwhelming percentage of the total load is carried within 2 m of the surface, although minor amounts may rise above 10 m.

Even though most sand moves by saltation, some of the load never enters the realm of airstream mechanics, but moves as surface creep. Creep results when the impact of saltating grains spasmodically shoves or rolls surface particles forward without displacing them upward. This type of motion becomes progressively more important as the grain size increases (Sharp 1964) and is probably the dominant mechanism of forward motion in very coarse sand. Nonetheless, in average wind-blown sand, ranging from 0.15 to 0.25 mm, creep seems to account for 7–25 percent of the total load.

Theoretically, it should be possible to estimate the total wind-traction load, because the discharge of sand must be related to the drag exerted on the wind by the saltating grains, and drag is dependent on V_* (drag velocity) and air density. Making some fundamental assumptions, Bagnold (1941) derived the following equation for estimating the load:

$$Q = C \sqrt{\frac{d}{D} \frac{\rho}{g} V_*'^3}$$

where Q is total saltating load, d is the particle size, D is the standard particle size (0.25 mm) used in Bagnold's experiments, and C is a coefficient relating to the surface conditions. For uniform sand, C has a value of 1.5, but this changes to 2.8 for unsorted sand and 3.5 for a pebbly surface (Cooke and Warren 1973, p. 262). Total saltation load, therefore, may be predictable on theoretical grounds and, indeed, other variations on Bagnold's approach have been developed. It is doubtful, however, that any equation will be very precise, for other uncontrolled factors such as particle shape (G. Williams 1964) and the ambiguities of grain motions in flight must introduce a finite uncertainty to any formula. This is perhaps best demonstrated by figure 8.10 where, for example, a drag velocity of 50 cm/sec may produce a significant difference in Q depending on which equation is employed.

Deposits and Features

The most striking geomorphic features associated with eolian processes are those formed by the deposition of sand. The surface features, commonly referred to as bed forms, are spaced with pronounced regularity and, like erosional features, come in a variety of sizes ranging from tiny ripples to giant forms called *draa*. (Draa apparently are restricted to the north African deserts, for reasons that will become clear later.) Intermediate in scale between ripples and draa are *dunes,* the

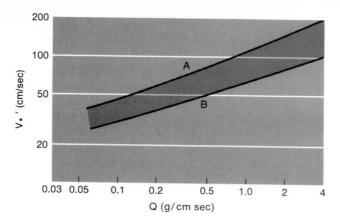

Figure 8.10. Relation of sand movement to drag velocity. Zone between curves A and B contains a family of curves derived from numerous studies. (After G. Williams 1964. Used with permission of *Sedimentology*, Blackwell Scientific Publications)

Table 8.2 Eolian bed forms and their geometry and possible origin.

Wave-length	Height	Orientation	Possible origin	Suggested name
300–5500 m	20–450 m	longitudinal or transverse	primary aero-dynamic instability	draas
3–600 m	0.1–100 m	longitudinal or transverse	primary aerodynamic instability	dunes
15–250 cm	0.2–5 cm	longitudinal or transverse	primary aerodynamic instability	aerodynamic ripples
0.5–2000 cm	0.05–100 cm	transverse	impact mechanism	impact ripples
1–3000 cm	0.05–100 cm	longitudinal	secondary vortices	secondary ripple sinuosity

From I. G. Wilson 1972. Used with permission of *Sedimentology*, Blackwell Scientific Publications.

most common depositional form in regions susceptible to wind attack. No clear dimensional boundaries separate the geometric types, which are described in table 8.2, and it is not unusual for the smaller features to be superimposed on parts of the larger ones. I. G. Wilson (1972), however, demonstrated that the dominant members of the bed form hierarchy can be differentiated by plotting their wavelength distance against a parameter of their constituent particle size (fig. 8.11). It seems clear from Wilson's work that for any given feature (ripple, dune, draa) a larger particle size is associated with a greater wavelength, and because grain size is a function of velocity, the expanded wavelength is also a reflection of a stronger formative wind. In addition, at any given particle size (and presumably wind velocity), all the types may be forming simultaneously. This strongly suggests that different process mechanics are associated with each geomorphic form. The fact that no transitional forms exist between the major varieties strengthens this supposition (Wilson 1972).

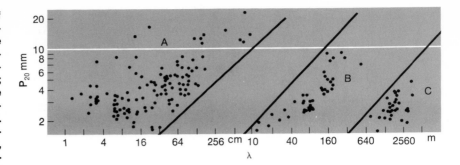

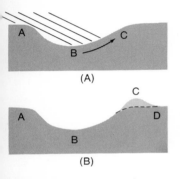

Figure 8.12. Process of creep in the development of ripples in a natural surface hollow. (A) Minor topographic irregularity increases the incidence of saltating grains on the windward face *BC*. (B) Creep builds ripple crest at *C* and a second lee slope *CD*. (After Bagnold 1941. Used with permission of Chapman & Hall Ltd.)

Ripples   Wind ripples range in amplitude from .01 cm to 100 cm and may be spaced up to 20 m apart. Their dimensions depend primarily on the wind velocity, the particle size (Sharp 1963), and the type of ripple. Some ripples are formed purely by the shear stress of the wind acting on the surface (aerodynamic ripples), but the overwhelming majority of ripples result from the surface bombardment of saltating grains and its associated creep.

Bagnold (1941) suggested that the process of forming impact ripples is closely allied to the angle at which saltating grains strike the surface. Since saltating grains descend in a nearly uniform manner, most variations in the angle of incidence are caused by minor topographic irregularities of the surface (fig. 8.12). The intensity of creep in any surface area is proportional to the number of impacts produced by saltating grains. Thus, in the natural surface hollow depicted in figure 8.12, the number of grain collisions is much greater on the windward face of the depression (*BC*) than the lee side *AB*. This results in the transport of more grains up the slope *BC* than are replenished by movement down the slope *AB* into the depression. Not only is the hollow maintained, but grains also begin to accumulate at point *C* because they are delivered there faster than they can be removed on the adjacent level surface. Eventually the accumulation at *C* produces a second lee slope *CD* which reinitiates the mechanics that operated in the original hollow. Repetition of this sequence inexorably propagates the ripple form downwind. Because of this distinct formative process, Bagnold (1941) feels that surfaces covered by unprotected sand and having no bed forms are probably unstable.

The process just described should create an ever-rising ripple crest and a gradually deepening hollow. Actually, the height of the ripple is effectively limited beause it eventually rises to a level where greater wind velocity precludes deposition. At that height, grains are transported directly over the crest and deposited in the next downwind hollow, where the wind velocity is much lower. In this way the ripples assume a consistent geometry which, for well-sorted sands between 0.19 and 0.27 mm, has a height/wavelength ratio that is normally only 1:70 and never lower than 1:30 (Bagnold 1941).

In nature, however, coarse grains are often concentrated at the ripple crest (Bagnold 1941; Sharp 1963), allowing the ripple to grow higher than expected because the large grains continue to be deposited while the smaller particles flow over the crest into the hollow. This commonly results in a height/wavelength ratio as low as 1:10, and an asymmetric ripple shape in which windward slopes are notably more gentle than lee slopes. The asymmetry, however, may be lost if a high percentage of the ripple is composed of very large grains. These "granule ripples" (Sharp 1963) have very large wavelengths and may be burdened by a high proportion of grains larger than sand size.

Dunes   Of all desert and wind phenomena, sand dunes have received the greatest scientific attention. Dunes attain a characteristic equilibrium profile which can be logically divided into three components, as in figure 8.13: the *backslope* or windward surface, the *crest*, and the *slip face* or lee slope. Measurements show that the backslope declivity, normally between 10° and 15° (McKee 1966; Sharp 1966; Inman et al. 1966), is in stark contrast to the slip face which always stands near the angle of repose for sand, between 30° and 34°. The crest, separating zones of erosion and deposition on the dune, is usually convex-up, but on very large dunes the pronounced convexity may be lost (Hastenrath 1967).

Dune height increases until it stabilizes along with the equilibrium form. Most dunes range in height from less than 3 m to 100 m, but in rare cases they have been observed to be as high as 200 m. The equilibrium height no doubt depends partly on the upward velocity gradient in the air, but most workers now seem to believe that height is mainly controlled by some poorly understood wave motion within the wind. This belief is reinforced by the fact that dunes normally occur in groups with distinctly regular spacing, rather than as randomly placed individuals. Within any given dune field the wavelength is quite consistent, although from region to region it can show considerable variation (table 8.2).

The pronounced regularity in dune spacing hints at some prevailing atmospheric motion that is capable of maintaining dune forms and their spacing in an equilibrium state. For example, formation of an initial dune possibly interferes with the airflow in such a way that it creates eddy currents or, alternatively, fixes a regularly spaced pattern of turbulence downwind from the intruding dune (Cooper 1958). Eddy currents unquestionably exist in the lee of dunes (Hoyt 1966), but considerable question remains as to whether they are modified by

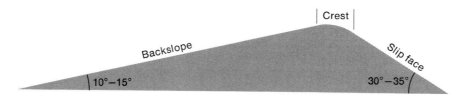

Figure 8.13. Cross-profile of normal dune showing common geomorphic components.

Backslope

Crest

Slip face

10°–15°   30°–35°

other flow mechanics and, more important, whether they possess the erosive power needed to regulate the spacing or shape of dunes (Hoyt 1966; Inman et al. 1966; Sharp 1966; Glennie 1970).

Patches of sand, which serve as the birthplace of dunes, form wherever local conditions favor deposition of sand that is moving under wind transport. This will occur when V_* (and therefore τ) is lowered, and thus deposition is commonly prompted by topographic irregularities of the surface. Once deposited, however, a sandy patch on an otherwise resistant surface becomes an integral part of the system mechanics. This is because the discharge of sand (Q) across a pebbly surface is greater than that across a sandy surface, because of the pronounced difference of the coefficient C in the discharge equation presented earlier. Thus, when an envelope of saltating sand grains traversing a pebbly surface ($C = 3.5$) enters a sandy reach ($C = 1.5–2.8$), the decrease in C across the sand requires a lower discharge, and part of the load must be deposited. The efficacy of the process is limited by the strength of the wind; Bagnold (1941) suggests that vertical accretion can be accomplished only by strong winds. In addition, accumulation requires a certain minimum length for the original sand patch, probably between 1 and 10 m. As the sand patch grows vertically, it intrudes into zones of higher wind velocity. On the upwind surface, this induces erosion of grains from the surface, since Q is a function of V_*, in addition to the deposition discussed above. In fact, the flow of sand up the windward slope must increase in intensity toward the crest because of the upward velocity gradient; conversely, it must decrease on the leeward slope.

Figure 8.14 indicates the mechanics involved in developing the equilibrium shape of a dune when the sand patch has formed. In any small zone of the surface (AE), the rate of sand removal (dQ) is given by $Q_E - Q_A$ during any time interval. The volume removed is dQ/γ, where γ is the specific weight of the sand. The volume is represented by the area of $ABDE$, which is equal in value to $ABEF$. The vertical erosion (dh) of the surface during the time interval is shown as EF, and the horizontal component (c) is AF. Thus the volume of erosion is

$$dQ/\gamma = c \ (dh)$$

Since EF/AF equals the tangent (dh/c) of the slope angle (a), $dx \tan a$ can be substituted for dh, yielding the equation

$$dQ/dx = \gamma c \tan a$$

as the rate of erosion or deposition per unit area. It follows that erosion or deposition rates at any point are proportional to the slope angle a. On the slip face, $\tan a$ is negative, indicating that dQ/dx must be negative, and therefore it represents a zone of deposition. The backslope, having a positive dQ/dx value, is an erosional zone, and the crest where $\tan a$ reaches zero can have neither deposition nor erosion.

Equilibrium in sand dunes, therefore, represents a balance between the volume of erosion and the volume of deposition occurring on the feature. This

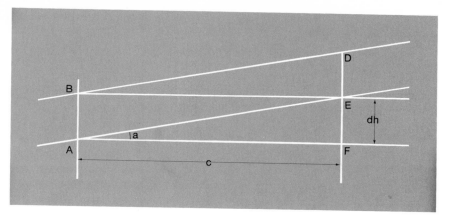

Figure 8.14. Mechanics involved in developing the equilibrium shape of a dune and the erosional-depositional system. (After Bagnold 1941. Used with permission of Chapman & Hall Ltd.)

balance is maintained through adjustments of the backslope and slip face angles which, in turn, are controlled by particle size and velocity gradients. Theoretically, the crest zone should exist as a sharp angle formed by the equilibrium surfaces of the backslope and the slip face. Actually, the wind does not respond immediately to a change in slope, and so the locale of maximum deposition will be some finite distance downwind from the crest, creating the tendency toward a convex summit. This tendency is sometimes complicated by slumping that occurs along the slip face when deposition in the crestal zone increases the lee slope beyond the angle of repose.

It is axiomatic that maintenance of the equilibrium shape requires forward movement of the entire feature, because erosion from the rear slope must be volumetrically balanced by deposition on the lee slope. There seems to be general agreement that dunes retain their original form as they advance, and that the shape components, especially the height, influence the rate of forward migration (Finkel 1959; Long and Sharp 1964; McKee 1966; Inman et al. 1966; Norris 1966; Hastenrath 1967). In fact, it has been repeatedly demonstrated that the horizontal displacement (c) that occurs in any time interval can be defined by the equation

$$c = Q/\gamma H$$

where H is the height of the dune. The absolute rates of forward migration are variable because they depend on local controlling factors, but movement measured in tens of meters per year seems to be common. This equation shows a reasonable correlation with some actual measurements (Finkel 1959), but the value of Q is subject to so many external constraints (Chepil 1959; Svasek and Terwindt 1974) that to expect precise predictions of advance rates may be too much to ask.

The cross-sectional characteristics of dunes tend to be somewhat changeable because the complexity of wind motions and the variability of surface conditions interfere with their ideal development. An even more complex matter is the attempt by students of desert landforms to categorize dunes according to their plan view, a property known as the *dune pattern*. Pattern classifications are almost as diverse and bewildering as the forms they claim to classify. Like other features, dunes have been grouped on the basis of shape, genesis, wind types involved, surface conditions, and the like, but each attempt somehow fails to account adequately for nature's incredible diversity. For our purposes, therefore, it seems best to employ a simple morphological classification system, as in figure 8.15, recognizing that it will not satisfy all possible dune patterns.

Hack (1941) suggested that in the region he studied, three basic dune forms exist, which he called *transverse, parabolic* and *longitudinal*. Transverse dunes are usually free of vegetation; they may be the most probable dune pattern if winds are unidirectional over a limited supply of sand and the sand is free to migrate. The normal transverse dune is a crescent-shaped feature, called a *barchan* dune (fig. 8.16), in which tapering edges or horns of the crescent point downwind. Transverse dunes may also stand as simple ridges oriented perpendicular to the wind, or they may form a sinuous ridge, called the *aklé* pattern by the French. The aklé ridge is made up of connected crescentic sections, for which the terms *linguoid* and *barchanoid* are employed to distinguish whether the segment is facing into or away from the wind. Although they did not employ the term aklé, the transverse pattern observed by Inman and his co-workers (1966) in the Guerrero Negro dune field of Baja California is probably a good example. Melton (1940) suggested that the type of transverse form that develops depends on the supply of sand and whether the surface beneath the dune is sandy or nonsandy. Many authors distinguish between barchans and transverse ridges or aklé by including them in separate pattern categories.

Parabolic dunes differ from transverse dunes only in that they are U or V shaped, and their horns or tapered extremities point upwind. When the lateral

Figure 8.15. Basic dune patterns.

Transverse

Dunes that form perpendicular to a prevailing wind. With limited sand and constant wind, dunes develop *barchan* form with horns pointing downwind. In areas of abundant sand, they form ridge called *transverse* dune or sinuous ridge called *aklé* pattern.

Barchan   Aklé   Transverse

Parabolic

Dunes formed perpendicular to wind but with arms fixed by vegetation. Horns point upwind; central zone migrates in direction of wind. Central segment has blowout zone. Sometimes called *U-shaped*.

Blowout

Longitudinal

Long ridge with narrow crest formed parallel to wind. May be sinuous if eddy currents develop slip faces in opposite directions. Often called *seif* dunes.

Figure 8.16. (A) Barchan dune in Sherman County, Ore., September 1899. (Photo by G.K. Gilbert, U.S. Geological Survey) (B) Asymmetric wind ripples on side of compound barchan dune. Dawson County, Mont., September 1928. (Photo by C.E. Erdmann, U.S. Geological Survey) (C) Dune forms in the Tularosa Basin, Otero County, N.M. Wind direction is from lower right. Dunes are crenulated transverse in lower right of photo. Dune forms progressively change to individual barchans and finally become U-shaped in the downward direction. (Photo by U.S. Dept. of Agriculture. From the University of Illinois Catalog of Aerial Photography)

edges of a transverse ridge become anchored by vegetation, the wind removes sand from the central zone and deposits it on the leeward slope. This, of course, allows the middle segment to advance relative to the edges and develops the characteristic parabolic shape. In addition, the dune surrounds a scoop-shaped hollow, called a *blowout,* from which the sand in the dune was derived, indicating that the pattern is partly erosional in origin.

Longitudinal dunes are narrow ridges that extend parallel to the forming wind. They are usually wider and steeper at the upwind end, gradually tapering downwind until they merge with the desert surface. In the Navajo country, longitudinal dunes are separated from one another by sand-free flats up to 100 m wide. Both the flats and the dune flanks may be vegetated, leaving only the sand of the ridge tops bare of vegetal cover and susceptible to wind transport. Locally the dunes may form in the windshadow behind an obstruction, or they may spring forth as wind-rift dunes (Melton 1940; *AGI Glossary* 1972) where blowouts provide sand that is extended downwind as a linear ridge. A special variety of longitudinal dune is called a *seif dune* (Bagnold [1941] considered it to be one of only two true dune forms, the other being the barchan). Seifs are elongate, sharp-crested ridges that often consist of a succession of oppositely oriented curved slip faces that impress a sinuous or chainlike appearance on the dune crest. In many cases seifs attain spectacular length (up to 300 km) and are really draa in terms of size. Size may be regarded as a distinguishing property of seifs, but the term has been applied to much smaller features.

In addition to the basic patterns, many other morphologic types have been reported including dome-shaped, reversing, and star (McKee 1966), and *coppice* dunes (Melton 1940) that are fixed by clumps of vegetation. It is important to understand that all dune types may be present side-by-side in a single region with the same prevailing wind, indicating that other controls are significant in pattern development. Hack (1941) visualized the transition of forms in his study area as being a function of wind strength, sand abundance, and vegetation (fig. 8.17), but it seems clear that other factors and various combination of factors will create a unique pattern assemblage in each area of dune formation. In fact, anyone perusing the literature concerning dunes must be struck with the realization that all classifications are somewhat artificial, and some may even inhibit the understanding of processes if they leave the impression that dunes are amenable to taxonomic treatment. It is increasingly clear that the variables of multiple wind directions, topography, size and abundance of sand, and vegetation are so inconstant that complex patterns are the rule, and the simple patterns of our classification the exceptions. It seems reasonable, therefore, to use the simplest ideal forms—barchans and longitudinal types—as models to demonstrate how controlling factors may influence dune patterns. This choice is predicated on the generally accepted premise that one forms perpendicular to the wind and the other parallel to it. However, dunes oriented at oblique angles to prevailing winds also are found in every region where dunes are reported. Thus, even though longitudinal and transverse dunes are most easily related to wind action,

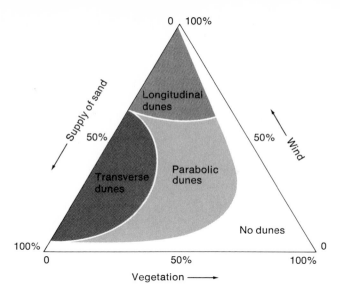

they may be less important in understanding processes than the explanations of the oblique forms.

Bagnold (1941) explained barchans as forms that develop in an area where sporadic sand patches exist within a desert pavement. It is assumed (Bagnold 1941, p. 222) that certain conditions must be met in order to mold the barchanoid form: (1) a constant, unidirectional wind; (2) a rate of sand supply that is symmetrically distributed on either side of the longitudinal axis; and (3) a slip face that is completely sheltered so that all sand crossing the crestal zone is trapped on the lee slope. Under these constraints, sand movement will be fastest near the lateral edges where the sand patch thins to meet the pebbly surface of the adjoining desert pavement. The greatest vertical growth of the sand body occurs in the middle segment, decreasing gradually to the edges. Thus, the lateral extremities will advance more rapidly than the center, and the crescentic form will develop as the equilibrium state is attained. As Cooke and Warren (1973) point out, however, Bagnold's hypothesis does not explain why barchans seem to have a regularly repeated width, nor does it account for the tremendous diversity in the length and direction of the barchan horns. Those authors suggest that divergence from the ideal form is a function of complex secondary flow patterns in the wind and of differential sand supplies. In fact, as suggested above, ideal barchans probably develop only where there is a limited supply of sand and unidirectional winds. Abundant sands lead to the aklé pattern or longitudinal dunes rather than individual barchans.

Parallel ridges of sand, either dunes or bigger draa-sized features, are prevalent in all large sand deserts. Often the longitudinal ridges bifurcate in the upwind direction, forming a Y or tuning-fork junction. Melton (1940) interpreted

Figure 8.18. Stages of seif
development from barchan
under influence of secondary
wind direction: p = primary
wind; s = secondary wind.
(After Bagnold 1941. Used
with permission of Chapman
& Hall Ltd.)

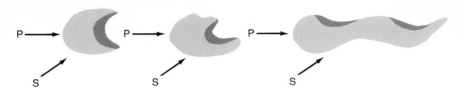

this phenomenon as the result of the wind's excavating blowouts in deep sand and depositing the freed sediment in wind-rift (*AGI Glossary* 1972) dunes. As such, the dunes are partly erosional, a hypothesis accepted by Folk (1971a, 1971b) for the longitudinal dunes in northern Australia, but rejected by Cooke and Warren (1973).

The origin of longitudinal dunes and draa has traditionally revolved around their relationship to the prevailing wind direction. Bagnold (1941) suggested that seifs develop when barchans are modified by crosswinds so that one of the horns is dramatically extended, as shown in figure 8.18. According to this idea, longitudinal dunes are not aligned parallel to the prevailing wind, but their trend represents the resultant direction of more than one wind. The secondary wind influence may arise from storm winds, diurnal reversals, or seasonal changes in wind direction. The multiwind hypothesis has received support from workers investigating dune formation in a variety of physical environments (Cooper 1958; McKee and Tibbitts 1964; McKee 1966).

Bagnold (1953) later reconsidered longitudinal dunes and suggested that they may be formed parallel to a prevailing wind that contains atmospheric, helicoidal flow. It is known that moving air when heated forms paired roller vortices whose axes are horizontal and parallel to the prevailing wind direction (fig. 8.19). The sizes of the vortices vary, but they are probably a simple function of the dimensions of the atmospheric boundary layer (Hanna 1969) and so should have some regularity for any region. Helical movement is such that any substance suspended in the air will descend into the trough areas between dunes (fig. 8.20) and rise up the dune sides along a line oriented diagonally to the dune trend. The circular component, as it impinges on the desert floor, transports sand from the troughs to the dune surfaces at the intersection of two adjacent but oppositely rotating vortical cells. The main body of the dune, therefore, is fixed parallel to the prevailing wind, but its summit zone may exhibit a sinuous course and asymmetric slip faces in response to the lateral component of the helicoidal flow (Folk 1971a).

The palatability of this hypothesis rests on the demonstration that the lateral velocity in helicoidal flow is capable of transporting sand. Few (if any) direct measurements are available to substantiate that premise, but circumstantial evidence does exist. For example, where the longitudinal ridges are of draa size, it is not uncommon for smaller seif dunes with a diagonal orientation to rest on the flanks of the draa (Glennie 1970). It is also known that upcurrent tuning-fork junctions, analogous in every way to those characterizing longitudinal dunes,

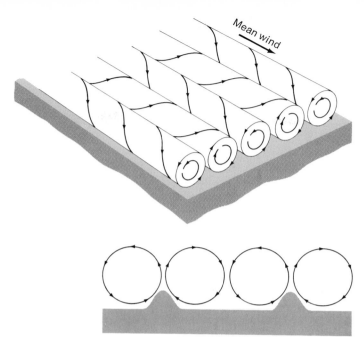

Figure 8.19. Diagrammatical view of paired roller vortices. (From Hanna, *Journal of Applied Meteorology,* 1969. Used with permission of the American Meteorological Society)

Figure 8.20. Fixing of longitudinal dunes at rising limbs of adjacent vortices having helicoidal flow. (From Hanna, *Journal of Applied Meteorology,* 1969. Used with permission of the American Meteorological Society)

have been produced in other environments where helicoidal flow of some type was involved (McLeish 1968; Allen 1969).

Accepting the helicoidal flow model explains much of the obliquity found in dune orientations. It may also account for the sinuosity of ridge crests, chains of seifs, the vagaries found in the aklé pattern, and even many features of any size that cannot be explained by simple combinations of transverse and longitudinal winds. It is too soon, however, to embrace the mechanism as the solution to all unexplained desert phenomena. In fact, the model is so appealing that its casual use and acceptance is dangerously close to becoming a theoretical crutch and may even delay the collection of hard data needed to prove its efficacy as a geomorphic agent. For example, Hanna (1969) argues convincingly that large vortices can be developed only in flat areas and only when the wind passes over the surface for a distance much greater than 20 km. If any topographical irregularity occurs before the vortices form, barchans rather than longitudinal dunes will develop. This explains why draa-sized longitudinal features are so rare in the small, fairly rugged deserts in the western United States and so common in the major deserts of Africa. However, it leaves unanswered the questions of how vortical motions on a smaller scale are generated and how obliquely oriented dunes smaller than draa size are formed.

With these questions in mind, we should return briefly to the phenomenon of crosswinds. The possibility of a resultant wind being generated by the influence of crosscurrents on the prevailing wind has never been questioned, but considerable argument remains as to how much and what type of geomorphic work it can accomplish. Substantial evidence has been reported to support the conclusion that seif chains are aligned with the resultant of several prominent wind directions (Cooper 1958; McKee and Tibbitts 1964; Brookfield 1970; Warren 1970). Equally strong data argue against a major genetic role for resultant winds. For example, seifs are often oriented in more than one direction within the same region, and some are distinctly oblique to the resultant wind direction (Brookfield 1970; Warren 1971). Such a distribution negates the argument that all longitudinal dunes form parallel to resultant currents or, for that matter, to the prevailing wind. Cooke and Warren (1973) suggest that most seifs in regions of complete sand cover probably start as oblique components of ridge deposits that were deposited by a single dominant wind. These zones are progressively accentuated as crosswinds gradually filter them out of the original pattern, and they may eventually produce a pattern with a maze of crossing elements. It seems probable, then, that crosswinds may produce resultant currents and that these may have an important modifying influence on original patterns. In most cases, however, their geomorphic effect involves minor changes in dune symmetry or orientation rather than a major role in the formative process.

Superimposed on the murky relationship between wind direction and dune orientation is the fact that other factors exert influences on the system. It is known, for example, that the well-established sorting process of wind action may produce different dune patterns (Warren 1971, 1972). Selective removal of intermediate sands may leave coarser particles as a lag that retards dune growth. It is also possible that several wind directions of different magnitudes may mold the coarse fraction into one dune type and the finer fraction into another (Warren 1970).

Vegetation, of course, is necessary for the development of parabolic dunes since the arms are fixed, allowing the central blowout to form and persist. Coppice dunes (fig. 8.21) are particularly evident as a function of vegetal control. Not only do the vegetation clumps hold the sand together, but wind crossing the brush experiences a 30-fold increase in its original k value (Olson 1958). This decreases V_z drastically and promotes deposition of sand in the windshadow in the lee of the vegetation. In a similar way, any topographic obstruction can form calm zones in its lee, deflect the flow direction, and fix the position of eddy currents. Topographic barriers can form lee dunes of considerable size, measured in kilometers, although sometimes the windshadow phenomenon produces no deposition at all. These conflicting observations sometimes arise because we forget that dunes themselves act as wind obstructions, and so the effect is sometimes the cause.

Another environmental control on dune types is the abundance or lack of sand made available to the wind process. In regions of abundant sand and unidirectional wind, the aklé pattern will form; a limited supply will result in barchans. Very large draa features require abundant sand, which perhaps explains their absence in North America and Australia. Sand supply becomes a more im-

Figure 8.21. Dunes fixed by vegetation—coppice dunes—in the Rio Grande valley near Las Cruces, N.M.

portant factor in regions with slightly greater precipitation than in the massive desert ergs. It is in those semiarid regions that we can expect an abundance of barchan and parabolic development, and probably only small longitudinal forms.

In summary, the complete understanding of depositional processes and features is veiled in the complexity of a multitude of interacting geomorphic systems. Features usually occur as ripples and dunes, but mega-scaled forms called draa are found in the great sandy ergs of North Africa. Each feature of the hierarchy occurs in patterns that are mainly a function of the wind properties. The major dune patterns are probably oriented parallel and perpendicular to the prevailing wind direction, but enough forms trend obliquely to suggest the possibility of a more complicated relationship. Helicoidal flow can adequately explain the spacing of longitudinal draa, but its mechanics is poorly understood on a smaller scale. Wind that is oriented as the resultant of several prominent winds can account for some, but not all, longitudinal dunes. In addition, vagaries of grain size and sorting, vegetation, topography, and sand supply introduce additional complications, and geomorphologists are only beginning to understand the influence of environmental changes on the genesis of dunes (Twidale 1972). Some dunes undoubtedly formed under wind and sand conditions that no longer exist, and these relict forms confuse the issue even more because their geomorphic characteristics may not be in equilibrium with the modern wind and sediment conditions. All in all, much remains to be done but, considering our knowledge fifty years ago, the scientists who have brought us this far deserve praise. The systems they study are extremely challenging, the logistics of field work a nightmare, and the environment in which they work would make most of us with fainter hearts run for the closest oasis.

Fine-grained Deposits   Our examination of wind action in geomorphology logically concludes with a look at sediment that is not normally moved near to or in contact with the surface. Strangely enough, some silt and clay accumulate in small dunes called *clay dunes*. These features usually develop near coastal lagoons (Price and Kornicker 1961) or along evaporating salt flats (Bowler 1973), where 15–25 percent clay in the residuum seems to be sufficient to make the clay aggregate or pelletize into larger grains. The sand-sized agglutinates are susceptible to the same entrainment and transport processes discussed earlier, but once they are deposited in dunes, the form stabilizes because rain impact and solution of the binding salts disperse the individual particles. Clay dunes differ from sand dunes in several important respects: (1) slopes rarely exceed 15°; (2) the steepest slopes are on the windward rather than the leeward side; and (3) the maximum height seems to be about 15 m.

Another dune type with a similar origin forms as elongate ridges about 15 m high and sometimes more than 10 km long. These ridges, called *paha,* are well preserved in northwestern Illinois where they probably formed as longitudinal dunes (Thorp and Smith 1952). They have a high silt-clay content and may have cores composed of till. Pahas most likely result when fine particles are transported in an aggregated state in much the same way as those in clay dunes.

Despite the fascination surrounding clay-rich dune forms, most wind-transported silt and clay is carried in suspension and comes to rest as blanket-like deposits of *loess* (fig. 8.22). Loess is usually characterized as homogeneous unstratified silt, up to 100 m thick, which is highly porous and has the capacity to maintain vertical or nearly vertical slopes (Lohnes and Handy 1968). It covers all surfaces regardless of their topographic position, capping drainage divides as well as valley bottoms. Most loess is moderately well sorted, with nearly 50 percent of the deposit consisting of silt grains between 0.01 and 0.05 mm in diameter (Pesci 1968). It usually contains significant amounts of clay (5–30 percent) and 5–10 percent sand. The mineral composition is fairly consistent, with quartz being dominant and feldspars, carbonates, heavy minerals, and clay minerals present in smaller amounts. Each of the minor constituents varies in percentage according to local controls. For example, the amount of calcium carbonate in loess tends to be higher in dry regions, while clay mineral content increases with greater humidity.

The hypothesis that loess originates as wind-blown dust stems from at least a century of observations. The conditions that facilitate its wind derivation are an abundant supply of loose, fine-grained sediment, moderate to strong prevailing winds, and a surface free from a continuous vegetal cover. Deserts obviously meet these requisites but, curiously, several continents with vast deserts (Africa, Australia) have almost no loess deposits, and some regions notably lacking in desert conditions have experienced major loess deposition. Chief among these is the periglacial environment, especially where glacial meltwater has spread vast outwash debris in the path of the prevailing winds (Péwé 1955). There silt is winnowed from the outwash before the surface can be fixed by vegetation. In the

Figure 8.22. Loess deposit on east side of Mississippi River valley near Chester, Ill.

midcontinental United States, for example, outwash is a logical source for many loess sheets, whose distribution shows a marked affinity to the outwash bodies occupying large valleys of the region. Presumably the loess can grow to its considerable thickness because the silt supply is continuously replenished during annual or even diurnal flooding by the proglacial rivers.

In either hot or cold loess sources, deflation, saltating grains, or other processes entrain the silt particles into the higher velocity zones, and with proper turbulence and lift, the grains can rise to elevations of several kilometers. When conditions are right, the amount of dust lifted into the atmosphere can be enormous. In a 1935 dust storm over the interior plains, about 5 million tons of dust was estimated to be in suspension over a 78 km² area near Wichita, Kansas, and at least 300 tons/km² of dust was deposited in one day of the same storm near Lincoln, Nebraska (Lugn 1962). The topographic effect of such loess deposition is different from that of most constructional geomorphic processes; it molds no important landforms but tends to form plains by filling in depressions and thereby smoothing out preexisting relief. Dust is carried as particulate matter suspended in the air, and its transport distance depends mainly on the constancy of the wind velocity, both vertical and horizontal, and the settling velocity of the grains. Loess constituents can be transported for hundreds of kilometers and deposited over vast areas.

Although the shape of grains included in loess deposits and the markings on their surfaces suggest wind transport (Millette and Higbee 1958; Krinsley and Donahue 1968), wind-blown loess is difficult to distinguish from silts deposited in other ways. In fact, much controversy has arisen in the past because sediments bearing the characteristics of loess have been deposited by processes such as mass wasting, wash, weathering, and fluvial action (R. J. Russell 1944; Fisk 1951). This has produced what some authors refer to as the "loess problem." Obviously some confusion results when loess is reworked by other transporting processes, and some of the clay and $CaCO_3$ comes from weathering of the original mass. In addition, "loess" often thickens towards topographic lows, indicating that downslope movement has been prevalent after the material's original deposition. Nonetheless, the relationship between widespread blanket deposits and the presumed prevailing winds is often so clear that an eolian origin for the original concentration of the material cannot be denied. In fact, arguments for a wind origin of the loess in the midcontinental United States are quite strong (Swineford and Frye 1951; Lugn 1962, 1968).

Some of the problem can be resolved if we restrict the term *loess* to those deposits that have the characteristics described and are definitely of wind-blown origin. Proof of an eolian genesis may be circumstantial in some workers' minds, but the evidence of field relationships and deposit properties is often more than convincing. In the western Great Plains, for example, the thickest loess (up to 70 m) occurs immediately downwind from the Sand Hills Region of Nebraska, which contains the largest accumulation of sand dunes in North America. In Illinois, as the map in figure 8.23 shows, the loess is thickest on the bluffs

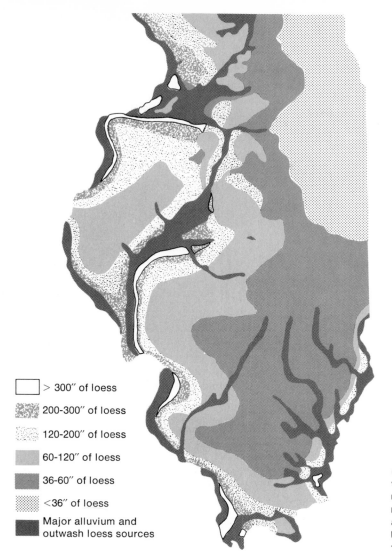

> 300″ of loess

200-300″ of loess

120-200″ of loess

60-120″ of loess

36-60″ of loess

<36″ of loess

Major alluvium and outwash loess sources

Figure 8.23. Approximate thickness of loess on level, uneroded topography in Illinois. (From Fehrenbacher et al., *The Quaternary of Illinois*. Univ. Ill. Coll. Agr. Spec. Pub. 14, 1968)

overlooking the Mississippi and Illinois valleys and thins gradually eastward into Indiana. The usual decrease in loess thickness downcurrent is so systematic it can often be expressed mathematically. For example, Hutton (1947) showed that loess in Iowa southeast of the Missouri River thins according to the equation

$$Y = 1250.5 - 528.5 \log X$$

where Y is thickness in inches and X distance in miles. Different equations fit other situations (G. D Smith 1942; Frazee et al. 1970), but all demonstrate a regular, systematic decrease in thickness with increasing distance from the source. Additionally, as might be expected, the particle size also decreases downwind, again with a demonstrable regularity (Ruhe 1969).

All the above properties indicate an eolian origin for loess and can be regarded as a function of the wind dynamics affecting the sediment, a process referred to as *eolian differentiation* (Scheidegger and Potter 1968).

Although historical geomorphology is not the prime concern of this book, perhaps the most interesting aspect of loess is its value in deciphering Quaternary stratigraphy. Individual loess deposits are widespread geographically and blanket diverse topography, so that the deposits are ideal for stratigraphic analyses. As mentioned, loess deposition is commonly related to periods of glaciation. Where several loesses are superimposed, each deposit having been formed during a specific glacial episode, the sequence provides a useful stratigraphic framework. The upper and lower boundaries of each unit within the sequence can usually be identified because each loess has distinctive properties such as soil profiles (often buried), textural or mineralogical characteristics, or fossils. These criteria may be used for regional correlation of the deposits and therefore as bases for interpreting Quaternary history.

The use of loess in regional correlation requires caution and careful field study because loess properties, like properties of any rock substance, change laterally. In addition, in glaciated areas, single loess units may divide at the position of the former glacial margin so that loess associated with the same glacial event lies both beneath and above the till of that glaciation (fig. 8.24). Beyond the glacial margin the two loess bodies are combined into one, indicating that loess was deposited continuously during the complete cycle of glacial advance and retreat. The Peoria Loess in Illinois, for example, was deposited throughout the entire Woodfordian substage and correlates with both the

Figure 8.24. Stratigraphic relationships between loess deposits and glacial deposits in the Late Wisconsinan of Illinois. (After Frye and Willman 1970)

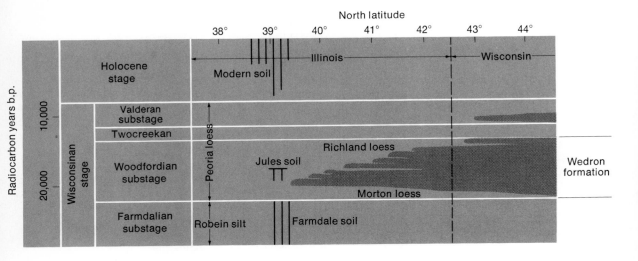

Richmond Loess and the Morton Loess, which were deposited before and after the Woodfordian till (fig. 8.24).

The fossil assemblage contained in loess also provides useful information about the environmental conditions at the time of deposition. The predominant fossils in loess are land snails. They commonly are abundant when the loess is thick, but are rare in thin loess because leaching destroys the carbonate shells. Snail ecology indicates that most loess in the North American midcontinent originated when the climate was slightly cooler and moister than today. However, considering the diverse environments in which modern loess originates, it is probably safe to say that loess by itself has little climatic significance. Climatic indicators are substances such as plant fragments or mollusks included in the wind-blown debris at the time of its deposition or shortly thereafter. As shells and wood can be dated by C^{14}, loess stratigraphy is often definable in terms of absolute time. This is especially important in the stratigraphy of the Late Wisconsinan. Older deposits may be beyond the limits of validity for the radiocarbon method.

Summary

Wind is an effective geomorphic agent in regions with sparse vegetation and an abundant supply of unconsolidated sediment. Although the most discernible evidence of wind action is found in deserts, it must be emphasized that eolian processes function in any locality having strong winds and the proper conditions of vegetation and sediment. The amount of geomorphic work actually accomplished depends also on the properties of the wind, especially its velocity and turbulence.

Particle entrainment occurs when the wind reaches a critical velocity called the fluid threshold. For any given sediment size, the value of the threshold velocity depends on a number of variables such as particle shape, sorting, soil moisture, and surface roughness. Once in motion, however, particles may strike stationary grains and cause their entrainment at wind velocities well below the fluid threshold (impact threshold). Wind-transported sediment moves in suspension, by saltation, or by surface creep. The largest portion of the load is carried within 2 m of the surface.

Geomorphic features associated with wind action are both depositional and erosional, ranging in size from microscopic to those measured in kilometers. Erosional features develop primarily from the abrasive action of wind-blown sand. The most prominent depositional features occur in a hierarchy of bed forms, including ripples, dunes, and draa, that are produced from sand-sized debris. The geometric shape and spacing of bed forms probably reflect an equilibrium condition between the wind and the characteristics of the sand. The system mechanics, however, is so complex that a precise quantitative expression of the equilibrium relationship is not yet possible. Fine-grained sediment occasionally accumulates in dune form, but most wind-blown silt and clay are deposited in sheets of loess that tend to smooth out topography rather than collect into pronounced constructional features.

Suggested Readings

The following books and articles are suggested to provide a more in-depth approach to the topics discussed in this chapter.

Bagnold, R. A. 1941. *The physics of blown sand and desert dunes*. London: Methuen and Co.

Chepil, W. S., and Woodruff, N. P. 1963. The physics of wind erosion and its control. *Advances in Agron*. 15:211-302.

Cooke, R. U., and Warren, A. 1973. *Geomorphology in deserts*. London: Batsford Ltd.

Glennie, K. W. 1970. *Desert sedimentary environments*. Amsterdam: Elsevier.

Inman, D. L.; Ewing, G. C.; and Corliss, J. B. 1966. Coastal sand dunes of Guerrero Negro; Baja California, Mexico. *Geol. Soc. America Bull*. 77:787-802.

Schultz, C. B., and Frye, J. C. 1968. *Loess and related eolian deposits of the world*. Lincoln, Neb.: Univ. Nebraska Press.

Sharp, R. P. 1963. Wind ripples. *Jour. Geology* 71:617-36.

————.1964. Wind-driven sand in Coachella Valley, California. *Geol. Soc. America Bull*. 75:785-804.

Smalley, I. J. 1970. Cohesion of soil particles and the intrinsic resistance of simple soil systems to wind erosion. *Jour. Soil Sci*. 21:154-61.

Wilson, I. G. 1972. Aeolian bedforms—their development and origins. *Sedimentology* 19:173-210.

Some of the most spectacular landscapes in the world are the results of the erosional and depositional action of glaciers, and every textbook of physical geology and geomorphology includes numerous photos and descriptions of these remarkable features. Nonetheless, to be true to the theme of this book, such features, regardless of their unique beauty, will be considered only because they manifest the processes that form them. It would also be tempting to stress the stratigraphic relationship between different glacial deposits and the effects exerted by glaciation on climate, life, and the physical setting. Obviously these subjects are of paramount interest and significance to geomorphologists because much of the modern landscape is closely associated with the physical events that occurred during the alternating glacial and interglacial stages of the Quaternary Era. While recognizing the importance of these topics, I realize pragmatically that an attempt to treat them in a general way is destined to fail. Several excellent books concerning Quaternary and Pleistocene geology examine these subjects in the depth they rightfully deserve: Charlesworth 1957; Zeuner 1959; Ericson and Wollin 1964; Wright and Frey 1965; Turekian 1971; Flint 1971. Our goal in this chapter will simply be to understand glaciers: how they form and how they move. Glacial erosion and deposition and the landforms resulting from these processes will be examined in the chapter following.

The study of glacial mechanics is an intimate part of the science of glaciology. Although this field is in its infancy, many of the techniques developed in recent years to study glaciers are extremely sophisticated and involve geophysics, remote sensing, and computer analyses well beyond the scope of an introductory discussion. Excellent reviews detail the basic concepts of the discipline (Embleton and King 1968, 1975a; Sharp 1960; Paterson 1969; Shumskii 1964), but those desiring greater depth and discussions of more recent advances in the field should refer to current periodicals, especially the *Journal of Glaciology*, which contains articles pertinent to all aspects of our discussion.

Glaciers and Glacial Mechanics 9

Glacial Origins and Types

A glacier is a body of flowing ice that has been formed on land by compaction and recrystallization of snow. Assuming this simple definition suffices, it is obvious that two critical requirements must be met before a mass of ice can be called a glacier. First, the ice must be moving, either internally or as a sliding block, and second, the mass must be due to the accumulation and metamorphism of snow. Areas in which the winter snow is entirely lost to summer melting and other forms of dissipation cannot bring forth glaciers, even if the amount of snowfall is enormous. Where a portion of the snowfall does survive the summer melt, it is buried by the next winter's accumulation; with continued annual increments, the snow pack grows in size, changes its constituent properties, and finally, with enough mass, begins to move. In any region a specific elevation exists called the *snowline* or the *firn line,* above which some snow remains on the ground perennially and permits the formation of a glacier. In polar climates, snowlines are usually near or at sea level, and they gradually increase in elevation in climatic zones with higher annual temperatures (fig. 9.1). Clearly, however, temperature is not the only determinant of snowlines since, as figure 9.1 shows, glaciers form at lower elevations at the equator than in the desert zones near 30° north. This somewhat ironic situation arises because the annual precipitation in the two regions is vastly different.

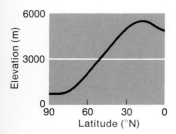

Figure 9.1. Generalized diagram illustrating the elevation of snowline at different latitudes in the northern hemisphere.

When snow accretes over a period of years, changes in the properties of the particles mark the transition of a snow pack into true glacier ice. Newly fallen snow having a very low density (usually 0.05–0.07 gm/cc) and a delicate hexagonal crystal structure is transformed into glacier ice through a series of complex but recognizable stages (fig. 9.2). In the initial phase, points of the crystalline flakes are preferentially melted, resulting in a more spherical particle shape and tighter packing due to settling of the grains. As table 9.1 shows, this tends to decrease porosity and dramatically increase the density. The time needed for this initial change varies depending on the climate and the pressure added by continuous accumulation of snow; where temperature remains near freezing or where partial melting occurs, the transition from fluffy snow to coarse granular snow may take hours or days (Embleton and King 1968). On the other hand, extremely frigid conditions retard the process such that years may be required. In temperate regions, snow lying on the surface for a complete year becomes granular and usually increases in density to about 0.55 gm/cc through the rounding and settling processes just mentioned. The material, now known as *firn,* is much denser than the original snow, but is still permeable to percolating water and is not yet true glacier ice (table 9.1). The rate of densification decreases beyond this point, and the mechanics of further transformations is different. The time needed to convert firn to ice is also variable, depending once again on temperature and the rate of increased load on individual grains. Where accumulation is rapid and water is plentiful, the transition from firn to ice probably takes place in less than 50 years, but in drier, colder regions the process may take several hundred years.

The mechanics involved in transforming firn to ice produces enlarged grains (in some cases up to 10 cm long) by recrystallization. When densities are low

Table 9.1 Increasing density of snow during transition to ice.

Material	Density (gm/cc)
New snow	0.05-0.07
Firn	0.4-0.8
Glacier ice	0.85-0.9

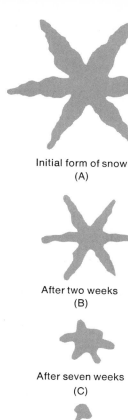

Initial form of snow
(A)

After two weeks
(B)

After seven weeks
(C)

After eight weeks
(D)

Figure 9.2. Changes in the shape of snow crystal in the transition to firn.

and porosities high, most of the stress exerted by the overlying load is in the form of vertical compression. With increased density, however, the stress pattern is hydrostatic, and crystal growth occurs in any or all directions. The enlargement process may be aided at first by a phenomenon called *sintering,* which binds the grains together (Hobbs and Mason 1964), and in zones with considerable melting, grains may cluster together by refreezing after they have been melted by pressure. Melted water may occupy air spaces lower in the firn pack, and as mentioned above, regions that produce free water will make the transition from firn to ice much more rapidly. Gradually, pore space is eliminated by crystal growth or by freezing of the downward permeating meltwater until the density is approximately 0.8–0.85 gm/cc. The only air remaining is trapped as bubbles within the crystal. For all practical purposes, the creation of ice is complete at this stage, even though the bubbles continue to be slowly expelled by compaction and the density may increase to about 0.9 gm/cc.

The time needed to transform firn to ice in temperate regions as compared to extremely cold regions is probably best demonstrated by the depth-density relationship in glaciers from the two regions (Sharp 1951; Behrendt 1965). In the Antarctic, a 0.85 gm/cc density is not achieved until at least 80 m of firn is accumulated. This is in drastic contrast to the 13 m depth needed to change snow to ice on the temperate Upper Seward Glacier in Alaska. Since accumulation rates are greater in glaciers of temperate regions, it follows that the greater depth to true ice in the Antarctic ice sheet represents the monumentally greater time needed to create glaciers in dry, cold climates.

Glacier ice is not preserved intact after its creation but is susceptible to further changes with continued increase in stress. For example, it seems clear from experimental work that the random orientation in polycrystalline ice cannot be maintained at higher stresses. The process known as recrystallization occurs when deformation of the polycrystalline mass exceeds several percent. Small, interlocking grains will develop with their basal planes preferentially oriented at 45° to the axes of compression, i.e., parallel to the direction of shear (Steinemann, 1954, 1958; Rigsby 1960). Recrystallization can also occur after stress is removed, but in this case the interlocking texture is lost and, since grain size is inversely proportional to stress, the grains get considerably larger (fig. 9.3). This explains why small, interlocked particles with preferred orientation are found in high stress zones of glaciers, while large crystals are most commonly found in stagnant or very thin ice. Some glaciers also show a gradual

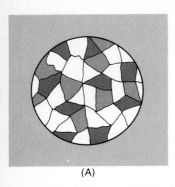

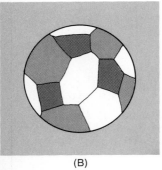

Figure 9.3. Increase in grain size caused by recrystallization process. (A) Original texture of water-soaked snow. (B) Texture after application of stress.

increase in grain size from the basal ice layers to the surface (Anderton 1974). The relationship between grain size and stress also explains how recrystallization during the transformation from firn to ice can produce crystal growth, because open spaces are still large enough to allow mineral expansion. With the final loss of space, growth is impeded by neighboring solids, and the weight of the overlying load is transferred into a shear thrust exerted throughout the entire mass.

Glacier Types

Glaciers have been classified on the basis of many salient properties but, as with any physical phenomenon, the best classification depends on the prime purpose of the groupings. The classification used here is no more astute than others; it simply serves our needs better by relating more closely to the processes of glaciation. Ahlmann (1948) suggested several bases for the classification of glaciers as *morphological, dynamic,* and *thermal.* Morphological classifications, which are based on glacier size and the environment of its growth, are most commonly employed, and the different categories, such as valley glaciers and ice sheets, are undoubtedly familiar. Although Ahlmann (1948) recognizes many morphological subdivisions, Flint (1971) suggests that glaciers can be placed in three broad categories (*cirque glaciers, valley glaciers, ice sheets*) and two intermediate types (*piedmont glaciers* and *mountain ice sheets*). In this abbreviated form the classification is quite useful, especially in a descriptive sense.

For our process orientation, however, the other two classifications are probably more useful, and we will at times utilize parts of each. The dynamic classification is based on the observed activities of glaciers and consists of three main groups: *active, passive,* and *dead* glaciers. Each type is closely related to the balance between losses and gains of ice and probably also depends somewhat on thermal properties. Active glaciers, like the one in figure 9.4, are characterized by continuous movement of ice from their accumulation zones to their edges. The movement may occur in response to normal snow accumulation, or it can be generated by avalanches or ice falls that provide the impulse for the forward motion.

A glacier is passive, as you might suspect, when it is undernourished by new snow or when slopes beneath the ice are very gentle. Although movement from accumulation zones to the fringes still occurs, its velocity is so low that it can hardly be recognized as dynamically active. The precise boundary between active and passive glaciers seems to be a matter of individual judgment, as no velocity limits for either type have been universally accepted. Passive glaciers are distinguishable from dead ice, however, because the latter no longer sustains any internal transfer of ice away from the accumulation zone. Any movement of dead ice is due purely to sliding across rather steep slopes. (The dynamic classification is particularly helpful in discussing the erosional and depositional features associated with glaciation and will therefore have greater use in the next chapter.)

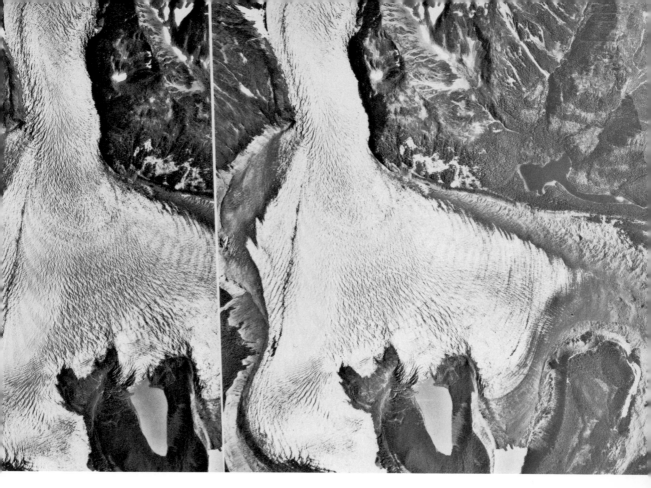

The thermal classification, based primarily on the temperature of the ice, has a direct bearing on glacial activities, but its influence is produced in a manner quite different from the balance between accumulation and dissipation. Two types of glaciers exist in this classification: *temperate glaciers* and *polar* or *cold glaciers*. In temperate glaciers the ice throughout the entire mass is at its pressure-melting point, although the upper 10 m may freeze in the winter. Meltwater seems to be present in abundant amounts within or beneath the ice mass and, in contact with the underlying rock, often facilitates slippage of the ice over the bed. This causes velocity and erosive action to be generally greater in temperate glaciers than in other types. Near the snout, meltwater may emerge as basal streams, or may be temporarily dammed within the ice; both situations promote extensive fluvial removal of debris from the terminus of the glacier.

Figure 9.4. The La Perouse Glacier in Alaska. A very active glacier that spreads onto lowland when it emerges from a narrow valley. Crevasses are distinct, and annual dark bands (ogives) are clearly visible. (Photo by U.S. Coast and Geodetic Survey. From University of Illinois Catalog of Aerial Photography)

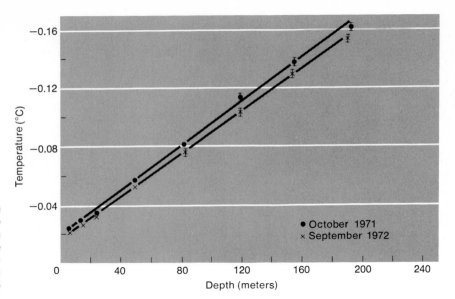

Figure 9.5. Temperature-depth relationship on a temperate glacier. (After W.D. Harrison 1975. Reproduced from the *Journal of Glaciology* by permission of the International Glaciological Society)

The suggestion that temperate ice is at its pressure-melting point is perhaps unfortunate because pressure is only one of several factors that determine the temperature at the boundary between liquid and solid. In fact, it is not uncommon for boreholes in a temperate glacier to refreeze at any or all depths (Shreve and Sharp 1970). The melting temperature of ice, therefore, may vary with impurities (Lliboutry 1971; W. D. Harrison 1972) or with stresses other than hydrostatic pressure (Paterson 1971). Thus, most "temperate glaciers" have a temperature distribution that disagrees with the pressure distribution predicted by the thickness of the ice (W. D. Harrison 1975), and even though the temperature may vary linearly with depth (fig. 9.5) and most temperatures will be within 1° C of their predicted values, it is now accepted that the simple model of pressure-melting is a good approximation but not a precise truth.

Polar glaciers (fig. 9.6) were subdivided by Ahlmann into *subpolar* and *high-polar* types. In subpolar glaciers the accumulation zone is characterized by a thin layer of firn, perhaps 20 m thick, which contains some water in the summer if temperatures are warm enough to melt the surface ice. The surface of a high-polar glacier remains below freezing at all times, resulting in a completely water-free ice mass and a thick firn zone extending to at least 75 m before true ice is encountered. The absence of meltwater within polar glaciers is an extremely significant difference between polar and temperate glaciers, and its geomorphic importance cannot be overemphasized. It requires that ice at the base of polar glaciers be below its pressure-melting point and, for all practical purposes, be solidly frozen to the underlying bedrock. Since slippage cannot occur over the bed, ice movement is totally internal, and the glacier's erosive action is greatly diminished.

It would be helpful if the dynamic or thermal classifications could be equated with the morphological types of glaciers. Such a correlation would allow a rapid first approximation of internal mechanics from simple geometric form. There is some suggestion that most valley glaciers are active and temperate, and most modern ice sheets are to a great extent polar and rather passive. Unfortunately, enough exceptions have been recognized to make anything more than gross generalizations unwarranted. The major problem seems to be that temperature distributions within glaciers are not systematic but subject to irregularities brought on by a number of factors in addition to those discussed above. Most prevalent among these are temperature fluctuations at the surface, heat generated by shearing within the ice, ice thickness, and geothermal heat flow (Embleton and King 1968). The result is that any single morphological glacier type may not have the same thermal characteristics throughout the entire ice mass. This is especially true for ice sheets or ice caps, which may be polar glaciers at one location and temperate glaciers at others (Robin 1955; Schytt 1964; Loewe 1966). Evidence also suggests that many valley glaciers have temperate ice only in their accumulation zones, while portions of the ablation zone are characterized by cold ice (Loewe 1966; Schytt 1968, 1969; Paterson 1972).

The Glacial Budget

The mechanical behavior of glaciers and the geomorphic work they accomplish are intimately related to their mass balance, sometimes called the *glacial budget*. The mass balance is essentially an accounting or budgeting of the gains and losses of snow that occur on a glacier during a specific time interval. The water equivalent of ice and snow added to a glacier during the period in question is called *accumulation* and may result from a variety of processes including snowfall, rain and other water that freezes on the surface, and avalanches. Processes that remove snow or ice, collectively known as *ablation,* commonly include melting, evaporation, wind erosion, sublimation (Beaty 1975), or the breaking off of large blocks into bodies of standing water, a process called *calving* (discussed in Holdsworth 1973). Losses by melting within or beneath the glacier are usually minor compared with surface volumes and are therefore neglected in budget studies.

Figure 9.6. The Greenland ice sheet near Thule. Much of the ice sheet consists of polar-type ice. (Photograph by S. McGee. Courtesy of Wm. C. Brown Company Publishers)

The time interval used in most balance analyses is the budget year (or balance year), which is ordinarily taken as the time between two successive stages when ablation has attained its maximum yearly value. Usually these values are achieved at the end of the summer season, but the two ablation maxima may not occur on the same day, and so the budget year is not necessarily 365 days long.

On a glacier surface two values of accumulation and ablation can be determined: (1) a gross annual accumulation or ablation, representing the total volume of water equivalent added to or lost from the glacier during the budget year; and (2) a net annual accumulation or ablation, representing the difference between the gross values and indicating whether there was an actual gain or loss of mass during the year. The latter value is simply the algebraic sum of accumulation and ablation, and for the budget year at any single measurement locality is called the *net specific budget*. If net specific budgets are determined for a network of points distributed over the entire glacier surface, their values can be integrated into a *total mass budget*, or the mass balance. Any glacier may have a positive total budget, meaning that more accumulation has occurred during the year than ablation, a negative budget indicating an excess of ablation, or a mass balance of zero if the volumes of accumulation and of ablation have been precisely the same.

It should be obvious that even though both accumulation and ablation occur on all parts of the glacier surface, the higher elevations of the glacier will usually receive more accumulation and experience less ablation than the lower reaches of the glacier, where the opposite is true. Thus, large areas with either positive or negative net specific budgets can be identified on most glaciers, as diagrammed in figure 9.7. The two areas are separated by a line called the *equilibrium line*, along which the annual volumes of accumulation and of ablation are equal. The equilibrium line should not be confused with the firn line or snowline because the two may not occur at the same place. On many glaciers, the firn line is clearly marked as a contact between snow above the line and dense blue ice below the line. On others, some of the dense ice downglacier from the snowline may have been formed by refreezing of meltwater, and as such represents a net accumulation of mass (fig. 9.7). This *superimposed ice* is found in enough cases to warrant the distinction between the firn line and the equilibrium line, and it sometimes creates a rather complex transitional zone between the accumulation area and the ablation area (Müller 1962). Nevertheless, equilibrium lines and firn lines are usually close enough to approximate one another.

The measurements needed to determine the values described above are painstakingly difficult to obtain (see Meier 1962) although new and easier techniques based on satellite imagery are being developed (Østrem 1975). Actually, unless one wishes to live on a glacier and physically measure increments of snowfall, the determination of gross accumulation is nearly impossible. Net accumulation is more easily measured by digging pits or taking cores at selective positions in the accumulation area. The net accumulation within any specific time interval is easily determined from the depositional sequence exposed in the pit, provided that the boundary of successive budget years can be established.

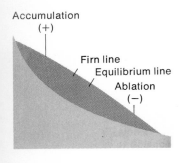

Accumulation
(+)

Firn line
Equilibrium line
Ablation
(−)

Figure 9.7. Zones of accumulation and ablation on a glacier as determined by budget analyses.

Usually a thin layer of dirt marks the end of the budget year, a horizon sometimes referred to as the *summer surface* (Paterson 1969), but this layer is not always present. The boundary may also be marked by a harder zone of ice caused by refreezing, a change in density, or an increase in grain size. Density must be determined for each annual increment in order to convert the ice thickness into water equivalents. This all sounds reasonably simple, but where accumulation rates are high the pit must be extended to a considerable depth to ascertain the net accumulation for more than one budget year. It is not uncommon for annual accumulation on some glaciers to exceed 2 m, requiring a pit at least 4 m deep to resolve the net accumulation for only two budget years.

Measurement of net ablation is normally much easier. In the ablation area, stakes are driven into the ice, and the change in the distance between the ice surface and the top of the stake represents the magnitude of ablation during the time interval between observations. On some glaciers, however, ablation may be so intense during the summer that stakes may have to be redrilled repeatedly. This is especially true near the snouts of temperate glaciers where 12 m of ice loss in one ablation season is not uncommon. Loss by calving may also be extremely large; rather precise surveying techniques are needed to determine its magnitude.

The regime for the entire glacier (total mass budget) can be estimated if the areas of individual net specific budgets are taken into account. The average net accumulation or net ablation can be taken as a simple arithmetic mean of all pit or stake sites located between two successive elevation contours, and that value multiplied by the area between the contours. The sum of these products gives the total net accumulation and the total net ablation, and the difference between these two values represents the mass balance of the glacier.

What does all this have to do with glacial mechanics and the resulting processes of erosion and deposition? If glaciers are viewed as open systems, both the mass balance and the absolute total amounts of accumulation and ablation are important factors in the character of glacial movement. When the net budgets are perfectly balanced (total mass balance equals zero), no expansion or shrinkage of the glacier occurs, and the glacial extremities remain stationary. This equilibrium condition, however, is seldom maintained for a long period of time, and so, the glacial front and sides usually fluctuate constantly. Glaciers with a positive mass balance actively advance and characteristically maintain a relatively steep or vertical front. The details of the forward motion may involve a kinematic wave action associated with extending or compressing flow (which will be discussed later). In contrast, an overall negative budget induces recession of the glacial front, and the snout will be gently sloping and often partially buried by debris released from the ice. The total mass budget, therefore, is closely related to the position and type of morainal system constructed by the ice.

The absolute amount of snow and ice relative to the area of a glacier (gross accumulation and ablation) directly affects its internal activity. Large budget values on small glaciers promote very rapid flow from the accumulation area to

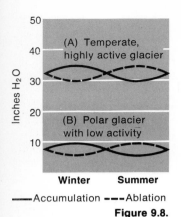

(A) Temperate, highly active glacier

(B) Polar glacier with low activity

Inches H_2O

Winter   Summer

——Accumulation ---Ablation

Figure 9.8.

Figure 9.8. Annual budgets on two glaciers in which the net mass budget of each equals zero. Active, temperate glacier (A) has greater total amounts of ablation and accumulation than the low-activity polar glacier (B). Even though both are in equilibrium and their fronts are stationary, (A) does more work because it must transfer more ice during the year from the accumulation zone to the snout.

the ablation area; in most cases temperate glaciers are likely to have large accumulation and ablation values and, consequently, high flow velocities. In contrast, polar glaciers, which tend to be rather passive, usually have small values of accumulation and ablation and low internal flow velocities. It is very important to recognize that large and small gross budgets can have the same net balance and that, in fact, if they are in an equilibrium condition, no advance or recession of the glacier front is occurring even though the internal transfer of ice may be enormous. Figure 9.8 is a hypothetical diagram demonstrating a polar glacier with low activity and a highly active temperate glacier, both of which have total mass budgets equal to zero. The snouts and lateral edges are stationary in both cases, but the temperate glacier, having a large value for accumulation and ablation, is moving at a higher rate internally. This dynamic activity causes pronounced erosion and rapid transportation of debris through the system. With the proper budgets, large moraines may result at the terminal and lateral boundaries. The polar glacier depicted in figure 9.8 will have little if any internal motion of its ice, and as a result, will probably not form any significant depositional features. Every glacier's depositional and erosional character, therefore, is fundamentally determined by the characteristics of the snow and ice added or lost from its surface.

The Movement of Glaciers

Internal Motion

Several hundred years ago, through direct observations, Alpine residents realized that glaciers move, and measurements of the rate of glacier flow were made as far back as the early eighteenth century. It is now generally accepted that glaciers move by two mutually independent processes: (1) internal deformation of the ice, called *creep*, and (2) *sliding* of the glacier along its base and sides.

The first real attempt to explain the physical dynamics of englacial ice motion seems to have been in the work of J. D. Forbes, who in 1843 suggested that glaciers respond to stress much like a plastic substance. By the beginning of the twentieth century, numerous models based on a plastic flow mechanism had been developed, although different ideas were also being proposed. They included flow caused by differential shearing along numerous, closely spaced planes (Phillip 1920; Chamberlain 1928); and the widely accepted notion that ice behaves as a viscous liquid that deforms in linear proportion to stress (Lagally 1934). In addition, a popular alternative explanation of glacial movement, called *extrusion flow,* was formulated and most fully developed by Demorest (1938, 1942). According to this theory, differential pressure beneath accumulation zones would force the lower layers of ice to move outward at a greater velocity than the surface ice. The concept was attractive because it explained not only englacial motion but also many of the large structural features associated with ice masses. Many geologists in this country perhaps established their first glacial romance with the extrusion model firmly established in their minds. Unfortunately, direct measurements of englacial flow velocities, which were made after the theory was postulated, have never revealed a subsurface zone of increased velocity that might have substantiated the extrusion mechanism, and the model has been abandoned by most glaciologists.

The advent of more precise measurements and new analytical techniques fostered the realization that the presumed plastic flow was closely associated with ice crystals lying in a preferred orientation (Perutz 1940). This close relationship between crystal form and flow prompted increased efforts to understand ice itself, especially how it responds to stress, and laboratory studies have become intimately coordinated with field observations.

It is perhaps instructive at this point briefly to consider the differences between the models of ice deformation that have been proposed during the last 100 years because, after all, it is the physical behavior of ice that determines the internal motion of a glacier. Fundamental to any understanding of glacier flow is the relationship between ice deformation and the stress that produces it. The resulting relationship defines the basic flow laws of ice. For the purposes of demonstration let us consider the two simplest possibilities: (1) that ice behaves as a viscous fluid and (2) that ice deforms as a perfectly plastic substance.

If ice behaves as a Newtonian viscous material, and we assume a constant viscosity, the application of stress should result in a linear relationship between the stress value and the strain rate (fig. 9.9). In addition, deformation will begin in the ice as soon as stress is applied and will maintain the linear proportionality regardless of the changes occurring in the stress. In contrast, as the diagram shows, a plastic substance shows no immediate response to stress but is capable of supporting a certain amount of stress without sustaining any deformation. Thus, at low stresses the strain of a plastic will be zero. As stress is increased, however, it eventually attains a value, called the *yield stress*, where the ice will experience limitless and permanent deformation. An entire glacier would behave plastically only if the shear stress along its base were equal to the yield stress (Kamb 1964).

It is now known that neither of the simple cases discussed above prevails in the mechanics of ice motion, although in many instances glacial properties predicted according to pure plasticity closely approximate the observed characteristics (Nye 1952b). Laboratory studies since the late 1940s have shown that single ice crystals under stress deform by recrystallization and by slip within the grains, a process involving a dislocation of planes of atoms inside the crystal (see Glen 1958; Weertman and Weertman 1964). The dislocation process works most efficiently through gliding on the basal plane of the ice crystal, but other, nonbasal, gliding surfaces can occur, even though deformation in these abnormal directions requires a much greater stress (Muguruma et al. 1966). The deformation of ice begins as soon as stress is applied, and the motion is more like creep than true plastic flow. Under any given stress the strain will increase rapidly at first (Glen 1952, 1955), but within a short time (tens of hours) it approaches an almost steady value that plots as a nearly straight line on a graph (fig. 9.10). This con-

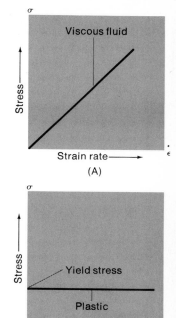

Figure 9.9. The relationship between stress and deformation. (A) Newtonian viscous fluid. Relationship is normally plotted as a function of stress and the rate of strain. (B) Perfect plastic deformation. The material experiences no deformation until stress is increased to the value of the yield stress. The material will then continue to deform as long as the stress is applied. Normally plotted as a function of stress and strain.

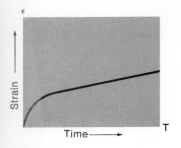

Figure 9.10. A typical curve representing the deformation of ice by creep. Stress is constant during the time interval. Creep curves are plotted as a function of strain and time. (After Glen 1955. Used with permission of the Royal Society of London)

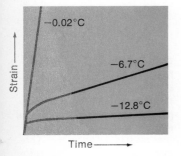

Figure 9.11. The effect of temperature on the creep deformation curve. All curves represent deformation under a stress of 6 bars. (After Glen 1955. Used with permission of the Royal Society of London)

tinuous deformation with no increased stress is the creep process that allows glacial ice to flow steadily under its own weight.

The rate of strain during the creep process is related to varying stress values by the following equation derived by Glen (1952, 1955) and now commonly referred to as the *power flow law*:

$$\dot{\epsilon} = k\tau^n$$

where ϵ is the strain rate, τ is stress, and k and n are constants. The values of n, determined by a number of investigators, seem to vary from approximately 2 to 4 for individual crystals. In polycrystalline ice they range from 1.9 to 4.5, with the mean value being close to 3. In any case, the n values associated with the power flow law are significantly greater than 1, the value required for a linear viscous flow.

It is significant to note that glacier ice is always polycrystalline. The flow law, although based on single crystal deformation, still seems to predict the response of glaciers, although minor modifications of the equation may be needed. The value of k in glacier flow might be reduced by interference of adjacent grains and recrystallization. In addition, as shown in the diagram in figure 9.11, cold ice deforms less readily than temperate ice, mainly because the constant k is also dependent on temperature. In Glen's experiments the value of k decreased by two orders of magnitude (0.17–0.0017) when the temperature was lowered from 0° C to −13° C. In fact, Paterson (1969) suggests that the strain produced in ice at −22° C by any stress is only 10 percent of its value when the ice is at 0° C.

Thus, ice clearly does not behave like a viscous fluid, although it may approximate a viscous response under low stress when the strain rate is still in its transient phase. At stress <1 bar in temperate ice, for example, n values as low as 1.3 have been measured (Colbeck and Evans 1973). However, when the nearly steady strain is attained or the ice is under high stress, intragranular slip probably dominates the motion and the deformation becomes more plastic. Although this condition approximates plasticity, there is no distinct yield stress associated with the creep process, indicating that ice is not a perfectly plastic material.

The power flow law derived in Glen's laboratory seems to fit the actual observed motions of glaciers. However, as mentioned above, it would be too much to expect that polycrystalline ice would respond in precise agreement with the flow law; modifications of the equation are generally required as the situation demands. For example, Meier (1960) found that ice movement in the Saskatchewan Glacier is best defined by the equation

$$\dot{\epsilon} = k_1\tau + k_2\tau^n$$

and the simple power law may not apply at all when stresses are below 0.7 bars. Recent creep tests conducted at the base of Blue Glacier (Olympic Mountains, Wash.) in ice at its pressure-melting point (Colbeck and Evans 1973) suggest that the ice deforms according to the equation

$$\dot{\epsilon} = k\tau + k_1\tau^3 + k_2\tau^5.$$

In addition, where stress systems are complex because of irregular valley floors, cross-sectional shapes, or non-isothermal ice, the flow law will have to be generalized or modified (Nye 1957, 1965). Nonetheless, in spite of inherent difficulties, Glen's power flow law seems to provide an excellent first approximation between theoretical and observed flow data and, with proper modification, probably best describes the internal mechanics of glaciers.

Sliding

Part of the difficulty in applying the power flow law directly to glaciers is that their motion is due partly to sliding over the underlying bedrock, in addition to deformation of the englacial ice. The processes of sliding in actual glaciers are poorly understood simply because to observe them in action one must tunnel through the ice mass to the bedrock floor, an endeavor seldom tried and even more rarely accomplished. Nonetheless, it is generally accepted that in temperate glaciers, sliding accounts for a significant portion of the total glacier movement, especially when the ice is thin and rests on steep slopes. In contrast, polar glaciers probably slide little or not at all, because the ice is frozen to the bedrock at its base (Goldthwait 1956, 1960; Haefeli 1963). This unique difference between polar and temperate glaciers has significant geomorphic consequences because debris-laden ice slipping over bedrock accomplishes much greater erosion than ice that cannot slide.

The premise that lubrication of the bedrock surface will initiate slippage is probably easy to accept, but the question does arise as to how ice can move over an irregular surface by this mechanism alone. A glacier sliding up a slope on a film of water stretches credibility. How, then, does sliding work?

Sliding consists of two main components: *regelation slip* and *enhanced creep*. The process of regelation slip involves the melting and refreezing of ice because of fluctuating pressure conditions and usually results in a texturally distinct layer of ice, only a few centimeters thick, that rests in contact with the bedrock floor (Kamb and LaChapelle 1964).

The mechanics of regelation was first revealed by Bottomley (1872), who demonstrated that a thin wire under tension could be passed through a block of ice without splitting it apart. The pressure exerted by the wire melts the ice beneath it, and the water thus released flows in a thin layer to the upper surface of the wire where it refreezes. The temperature of the ice is lower beneath the wire than above it because the pressure depresses the melting point. The heat of fusion released by refreezing above the wire is transferred through the wire to provide the heat needed for the pressure-melting along the leading edge, and so the speed at which the wire moves through the ice block is partly dependent on the rate at which the wire conducts heat. Other objects such as cubes, spheres, and discs have been forced through ice (Kamb and LaChapelle 1964; Barnes and Tabor 1966; Townsend and Vickery 1967; Morris 1976), and wires of various size and composition have been utilized to study the regelation process (Nunn and Rowell 1967; Drake and Shreve 1973). Each experiment has reinforced the correctness of Bottomley's original interpretation of the process. Any object, it seems, can be passed through ice without severing the mass and without changing the properties of the ice except along the path of transport. Along that path, the ice develops a new texture, similar in all aspects to that observed in the thin basal layers of glaciers.

In glaciers the process of regelation slip, shown in figure 9.12, allows ice to circumvent minor irregularities of the bedrock surface over which the glacier is moving. The basal ice melts at the upstream edge of the bedrock knobs where pressure is the greatest. The meltwater thus released flows around the obstacle and refreezes in the region of least pressure, which is along the downstream edge of the barrier. The mechanism seems to be quite effective when the obstacles are small, but becomes less important when protuberances become larger (Weertman, 1957, 1964). This, of course, begs the question as to what magnitude of bedrock irregularity will render the regelation process inoperative and how thick the layer of regelating ice can be.

The problem is complicated by the fact that ice near the glacier base also may deform according to the flow laws explained earlier. In that case, ice at the upglacier edge of the obstacle, where pressure is greatest, will have a higher strain rate than that at the downstream edge. Larger obstacles will augment the stress and cause higher flow velocities; i.e., the velocity will be directly proportional to the size of the bedrock knob. Because of these complications, Weertman (1957, 1964) suggested that sliding over an irregular bed is produced by components of the two processes. One (regelation slip) is the predominant control on sliding when the barriers are small, and the other (enhanced creep) when the obstacles are large. It follows that some intermediate size, called the *controlling obstacle size,* determines the sliding velocity of a glacier; at this size, the regelation and creep velocities will be equal. In the subglacial bedrock topography, ice moving over or around protuberances smaller than the controlling obstacle size will do so mainly by regelation; where surface bulges are larger, creep will prob-

Figure 9.12. The process of regelation as it functions beneath a glacier. Melting occurs on upstream flank of obstruction where pressure is greatest. Refreezing occurs downstream of obstruction where pressure is least, although some cavitation may form.

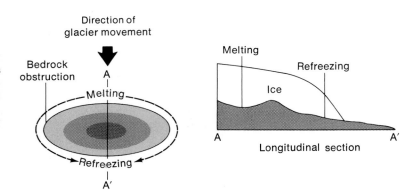

ably dominate. Figure 9.13 shows the relationship between sliding velocity, shear stress, controlling obstacle size, and the roughness factor (r), which is simply the ratio between the distance separating obstacles and their height (d/h).

The complete understanding of sliding is probably years away and will require many direct observations of the actual process to test the theoretical models we now have. It seems certain from many studies that the real mechanism will be infinitely more complex than we presently assume (for details see Kamb and LaChapelle 1964; Weertman 1964; Nye 1969, 1970; Kamb 1970; Drake and Shreve 1973; Morris 1976).

While we understand regelation and enhanced creep at least in a qualitative way, the effect of free water at the base of a glacier is a real enigma. We know it is there, but as yet we have no clear understanding of its role in sliding. Intuitively we suspect that free water should accelerate sliding. Weertman's (1964) data, for example, suggest that sliding velocity would increase by as much as 20 percent if a water layer only 1/10 the thickness of the controlling obstacle size existed along the glacier floor. Since the controlling obstacle size is probably less than 10 cm, a water film only millimeters thick could significantly influence sliding velocities. In fact, formation of a water layer equal in thickness to the controlling size may result in catastrophic increases in velocity.

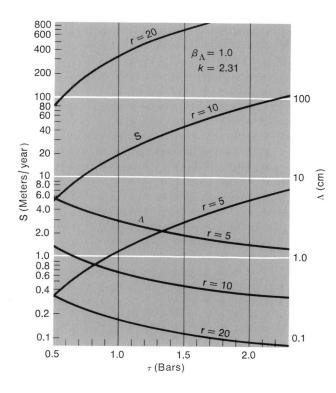

Figure 9.13. Relationship of sliding velocity (S) and controlling obstacle size (λ) with shear stress (τ). Lines have various values of roughness (r). (From Weertman 1964. Reproduced from the *Journal of Glaciology* by permission of the International Glaciological Society)

The critical unresolved questions are whether water can exist as a continuous sheet beneath a glacier and, if not, how much of the ice-rock contact must be occupied by water to produce an influence on sliding. We know that water-filled cavities exist along the basal contact (Haefeli and Brentani 1955; J. E. Fisher 1963; Savage and Paterson 1963; Vivian and Bouquet 1973), and although they are usually isolated pockets, they might possibly merge by enlargement or coalesce as they migrate along with the moving ice. Lliboutry (1964, 1968) developed a sliding model based on the hypothesis that ice will separate from its bedrock floor downstream from a bedrock obstruction, and that the cavities thus formed will fill with subglacial water under pressure. Assuming a washboardlike bedrock topography, water-filled cavities may nearly submerge the obstacles, thereby reducing the area of contact between ice and rock. The reduction of overall friction thus produced will allow a drastic increase in sliding velocity.

Lliboutry's model has been questioned on the grounds that the value of water pressure cannot be accurately determined and that his view of a sine-curve topography for bedrock surface is unrealistic (Weertman 1967; Nye 1969). Nonetheless, the effect of water as a lubricant and as a buoyant force should not be totally disregarded until it is tested by more realistic data. For example, it is rash to assume that all water freed at the base of a glacier is trapped by impermeable rock. The subsurface drainage of meltwater may reduce the velocity of ice moving over a rough surface, as laboratory experiments have demonstrated (Chadbourne et al. 1975).

The problems of subglacial water are far from being solved. Studies suggest that a water layer beneath a glacier would normally be unstable (Shreve 1972) and would probably not even form where a bed slopes downhill (Röthlisberger 1972). It may, however, develop on a horizontal bed and exert a buoyant force on the ice, but only if the exponent n in the power flow law is small (Röthlisberger 1972). Recent attempts to directly determine water pressure have been successful (Hodge 1976), and additional data of this type should provide a more sound basis for future interpretation of the water component. Until more data become available, all aspects of the sliding mechanism will be an intriguing playground where, unfortunately, our imaginations will not be restricted by the facts.

Velocity and Flow

We can now ask whether measured glacial velocities agree at all with the mechanics of ice motion discussed earlier. For the purpose of developing a theoretical model, let us first assume that an active glacier remains unchanged in its total and local dimensions over an entire budget year; i.e., the total accumulation equals the total ablation, and all cross sections of the glacier are equal in area and remain constant in area during the year. To maintain its area, each successive downglacier cross section must transport from its lower boundary exactly the amount of ice and snow delivered to its upstream boundary. As figure 9.14 shows, in the accumulation zone, the cross section at the highest elevation has

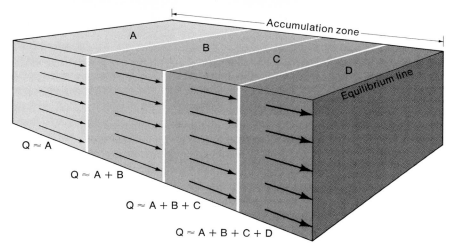

Figure 9.14. Hypothetical diagram showing why the velocity of glacier flow reaches a maximum at the equilibrium line. If no change in cross-section area (width x thickness) is allowed, increasing discharge down-ice requires the increase in velocity shown by arrows.

only a small area of accumulation above it and so must discharge only a small volume of ice, equivalent to the snow accumulated in that restricted surface area. Each section farther down-ice, however, must transfer a progressively larger volume of ice, since it moves not only the ice-equivalent of accumulation on its surface but also the cumulative volumes of all the higher sections. Without a change in cross-sectional area, the velocity of flow must increase to a maximum at the equilibrium line. This follows because discharge is increasing to that level and because glacier discharge, like river discharge, is equal to the area times the velocity. In the ablation area, we can similarly expect a gradual decrease in velocity from the equilibrium line to the terminus.

The model assumes that glaciers strive for and maintain some type of equilibrium, and that the ice movement obeys the power flow law in some form. Mathematical treatment of even this simple glacial model is quite complicated unless we further assume, as did Nye (1952a), that the motion is two-dimensional, plastic, and laminar such that the lines of flow are parallel to the bed and surface at all places (fig. 9.15). Accepting these conditions, the shear stress at the base of a glacier, measured along the central longitudinal axis and perpendicular to the surface, is given as

$$\tau_b = \rho g h \sin \alpha$$

where τ_b is the shear stress at the glacier base, ρ the ice density, g the acceleration of gravity, h the ice thickness, and α the slope of the surface.

By assimilating Glen's flow law into the above equation, the velocity at any depth along the central axis (xy axis) can be estimated by assuming that shear stress is proportional to depth, and that the strain rate directly relates to that stress. Theoretically, then, the internal velocity profile of a glacier in a longitudinal section should show a decreasing rate of flow from the surface to the bedrock floor. Even though the strain rate is greatest where shear stress is the highest (at

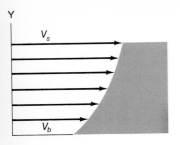

Y

V_s

V_b

Figure 9.15. Velocity distribution in longitudinal section of glacier. Surface velocity (V_s) is sum total of flow rates at every level within the ice. Strain rate due to shear stress is highest at base of ice and is shown as the basal velocity (V_b).

maximum thickness), the velocity increases from the base to the surface because each internal layer not only moves in response to the shear stress generated at that level, but is also moving forward on top of, and at the speed of, the adjacent lower layer. Thus, the surface velocity (v_s) is the sum total of strain rates for all the layers within the ice mass (fig. 9.15). The equation also shows us that internal velocity is proportional to the product of surface slope and ice thickness. The product seems to be fairly consistent, meaning that where the ice is thin the surface gradient will be steep, and vice versa.

Using the same basic approach, Nye (1957, 1965) predicted the velocity distribution for an entire cross section, using different cross-sectional shapes as models, and showed that velocity should decrease from the center of the ice to the lateral boundaries both at the surface and at depth (fig. 9.16). Thus, the simplest ideal model of a glacier, assuming continuity of discharge and shape, and laminar flow, shows velocity decreasing with depth and distance from the central axis. In the downstream direction, velocity should increase toward the equilibrium line and decrease away from it.

The velocity of ice at a glacier surface is obtained by marking the position of stakes driven into the ice relative to some nearby fixed point. This can be done by normal surveying techniques or by photography repeated at some specified time interval. Actual measurements of surface velocities substantiate in a general way the theoretical predictions (Meier 1960; Meier et al. 1974), although few measurements have been made in accumulation zones. In most temperate valley glaciers, surface velocities range from 10 to 200 m a year, but vary locally above or below these values. Outlet glaciers and ice streams associated with ice sheets may also attain similar and, in some cases, even higher velocities (Embleton and King 1968; Flint 1971). For example, the ice lobes around the periphery of the Pleistocene ice sheets of North America also probably advanced at rates between 10 and 200 m a year (Ruhe, 1975, p. 192). As predicted, velocities tend to increase from the glacier head to a maximum near the equilibrium line (Meier and Tangborn 1965) and to decrease from there to the snout (Meier 1960; Meier et al. 1974). The transverse velocities are usually greatest along the central axis, decreasing to the lateral margins. In both patterns, however, enough variations occur to recognize that the flow model is infinitely more complex than our original assumptions allow. In fact, even though flow may show a relatively simple relationship to ice thickness and surface slope as predicted, in detail it may not do so on a local scale; other factors may be involved in ways that are not yet understood (Meier et al. 1974).

Figure 9.16. Velocity distribution in glacier flowing in a channel with a parabolic shape. Diagram represents one-half of the channel cross section. Numbers represent velocity expressed in any common velocity units. (After Nye 1965. Reproduced from the *Journal of Glaciology* by permission of the International Glaciological Society)

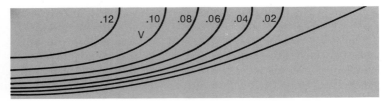

Measurements of englacial velocity require that a borehole be drilled through the ice and cased to prevent its closure. The differential movement with depth can then be calculated from an *inclinometer,* which measures the angle between the axis of the borehole and the vertical. Almost every borehole that has penetrated to a glacier floor or to great depth shows a velocity profile similar to that predicted by theory (fig. 9.17). Velocity does not change significantly with depth in the upper zones of most glaciers, although Meier (1960) has demon-

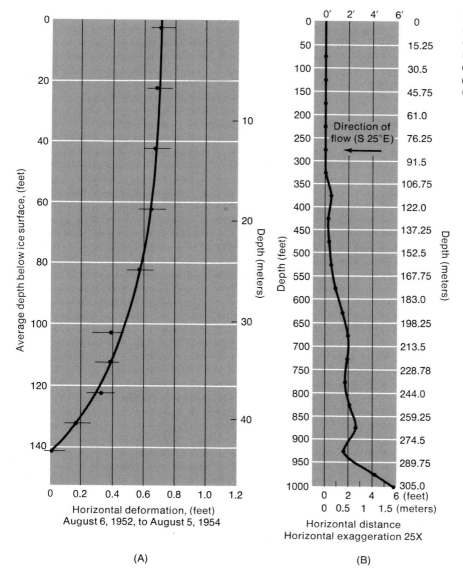

Figure 9.17. Internal velocity of two glaciers. Horizontal displacement of boreholes indicates the velocity. (A) Saskatchewan Glacier (Meier 1960); (B) Malispina Glacier (Sharp 1953).

strated that some differential movement can occur even in the upper meters of a glacier. In the lower half of most glaciers, the velocity decreases more rapidly with depth.

Complications of the Simple Model   Real glaciers seem to possess the general traits that were predicted in idealized flow, at least in the distribution of velocity. It is too much to expect, however, that glacier characteristics will be predictable in detail. The initial assumptions are invalid on a local scale and, in fact, velocity itself is controlled by several factors that are external to the system defined by the flow laws; we know also that thickness and surface slope are free to change. We will briefly examine some of the phenomena that may cloud the relationship between predicted and observed ice motion.

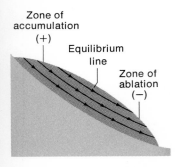

Zone of accumulation (+)

Equilibrium line

Zone of ablation (−)

Figure 9.18. Longitudinal flow lines in hypothetical glacier.

Extending and Compressive Flow   Returning to our ideal model, we see that allowing no change in cross-sectional area requires that the elevation of each point along the surface of a balanced glacier be maintained. If this is to happen, then the flow cannot consist of laminar horizons moving parallel to the surface, because the ice must move slightly downward in the accumulation zone and slightly upward in the ablation zone. (Accumulation tends to elevate the surface and ablation to lower it unless these responses are counterbalanced by motion of the ice.) Thus, the pattern of flow should be as shown in figure 9.18 and not as laminar sheets.

Actually ice will tend to thicken in some places and to thin in others, processes called *compressive* and *extending* flow. Nye (1952a) first investigated these flow types and suggested that rates of accumulation and ablation, as well as changes in the slope of the underlying bedrock, will determine which type will prevail. If R represents the radius of bed curvature in a longitudinal section, α the surface slope of the ice, and q the ice discharge, the expression

$$dq/dx + q/R \cot \alpha$$

will determine whether the flow is extending or compressive. When the expression is positive, the flow is extending (stretching), and when it is negative, the flow is compressive (thickening). Thus, where R is negative, meaning the bed is concave-up, or when dq/dx (rate of addition of ice over distance x) is negative, we can expect compressive flow (table 9.2). The opposite conditions will produce extending flow. At a constant bedrock slope, therefore, accumulation zones should be characterized by extending flow and ablation zones by compressive flow.

This analysis generally fits the velocity distribution that we surmised in our balanced glacier and observed in real glaciers. In ablation zones, with the pervasive compressive flow, the velocity should decrease downglacier because the ice is being compressed (Nye 1952a), and in accumulation zones, with extending flow, the velocity should increase downglacier. We must ask, however, whether

Table 9.2 Conditions and effects of extending and compressive flow.

	Compressive flow	Extending flow
Effects	Upper layer in compression No crevasses Thrust planes	Upper layer in tension Transverse crevasses (Other shear faults?)
Conditions	$\frac{dq}{dx}$ negative. Ablation area R negative. Concave bed $\left(\dfrac{dq}{dx} + \dfrac{q}{R}\cot \alpha \right)$ negative	$\frac{dq}{dx}$ positive. Accumulation area R positive. Convex bed $\left(\dfrac{dq}{dx} + \dfrac{q}{R}\cot \alpha \right)$ positive

R = convexity of the bed;
$\frac{dq}{dx}$ = rate of addition of ice to upper surface of glacier;
q = rate of discharge.

After Nye 1952a. Reproduced from the *Journal of Glaciology* by permission of the International Glaciological Society.

it fits the direction of flow predicted for a balanced glacier. Nye (1952a) showed that extending and compressive flow should each generate a pattern of potential slip planes as shown in figure 9.19. The family of planes are such that their resolution at any point will give the two directions of maximum shear; thus, they will be perpendicular and parallel to the bed of the glacier and form 45° angles with the surface of the ice. The potential slip planes show that in zones of extending flow (accumulation zones), the predominant downglacier slip direction will be downward at the surface; while in compressive zones, such as ablation areas, the low-angle downglacier slip will be upward.

This seems to coincide with the predicted mode of flow in our ideal glacier (fig. 9.18), but in real glaciers the pattern is much more complicated. An irregular bedrock profile will cause reversals of flow type in the ice (fig. 9.20), and where measurements have been made for an entire glacier (Meier and Tangborn 1965) the flow pattern (fig. 9.21) is not the simple one of our ideal case. Furthermore, more than one velocity maximum may occur, and the equilibrium line does not necessarily mark a zone of maximum flow velocity. In fact, on the South Cascade Glacier (Washington), the equilibrium line is usually near a zone of lower surface velocity, but the thickness is so great that the total discharge through the section is still very high.

Slip   A second complication arises because theoretical estimates of velocity do not consider the added component of basal sliding, the magnitude of which is represented by the displacement of boreholes at the bed (as was shown in

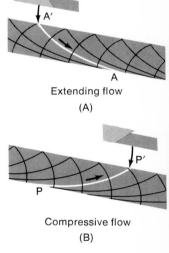

Extending flow
(A)

Compressive flow
(B)

Figure 9.19. Potential slip planes and possible faults under (A) extending and (B) compressive flow. (From Nye 1952a. Reproduced from the *Journal of Glaciology* by permission of the International Glaciological Society)

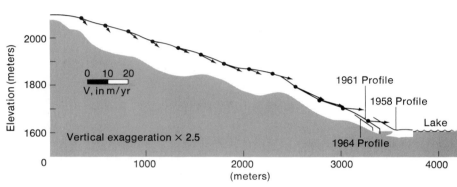

Figure 9.20. Longitudinal sections of hypothetical glacier. Sign of *R*, indicating changes in the bedrock profile, and type of flow are given at bottom. The lower section shows the slip field and the upper section shows the velocity distribution. (From Nye 1952a. Reproduced from the *Journal of Glaciology* by permission of the International Glaciological Society)

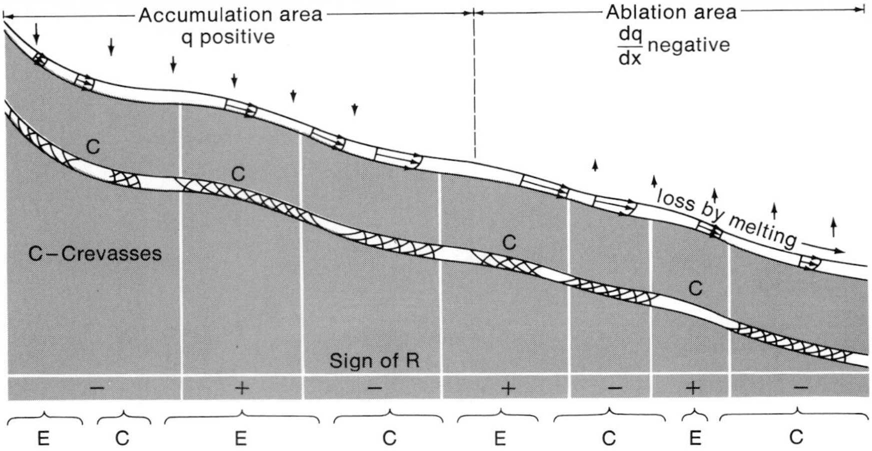

Figure 9.21. Longitudinal section showing average calculated bedrock profile and surface velocity vectors on the South Cascade Glacier, Washington. (From Meier and Tangborn 1965. Reproduced from the *Journal of Glaciology* by permission of the International Glaciological Society)

fig. 9.17). Although few direct measurements are available, they show that sliding may account for as little as 10 percent of the surface velocity in some glaciers and as much as 90 percent in others (McCall 1952; Kamb and LaChapelle 1964). Even worse, sliding velocities may be extremely variable within different portions of the same glacier (Savage and Paterson 1963). Although, as discussed earlier, the mechanics is not understood, it is clear that the influence of basal sliding must complicate the simple relationship between flow laws and velocity, at least in temperate glaciers. Slip also can occur along the lateral margins of glaciers, and abnormally high velocities are sometimes encountered at the glacier sides (Meier et al. 1974).

Variations with Time   The third factor complicating the relationship between flow theory and observed flow rates is that velocity is not constant in any given glacier. On most glaciers it varies significantly with time; to add to the confusion, the interval over which velocity variations occur is different from one glacier to another and even in specific locations on the same glacier. For example, many investigators have noted sudden, jerky motions that increase surface velocity for hours, days, or weeks (Meier 1960; Glen and Lewis 1961; Goldthwait 1973). These spastic movements, which may involve a change in velocity of as much as several hundred percent, seem to be a surface-ice phenomenon (Goldthwait 1973) and tend to be restricted in extent. They are normally explained by local controlling factors such as weather conditions, fault slips, or a sudden release of ice that has been retarded in its flow by some obstruction.

Seasonal or yearly variations in flow also are common. Velocities in the ablation zone usually increase at or near the end of the summer. These fluctuations most logically reflect accelerated basal and side slip which is facilitated by the abundance of free water at that time (Meier 1960; Elliston 1963; Paterson 1964; Hodge 1974). Seasonal velocities may deviate by as much as 80 percent from the mean annual values. They are not restricted to temperate glaciers, but also affect the movement of subpolar glaciers (Friese-Greene and Pert 1965). Yearly variations probably represent internal adjustment to accumulation rates, but they also can be influenced by basal meltwater lubrication (Meier et al. 1974).

Short-term variations in velocity are usually recognized because the span of most investigations is long enough to demonstrate their presence. However, imagine the sampling program needed to observe and document long-term fluctuations of velocity that are produced by changes in the glacial regime. Any positive change in a budget should increase the discharge of flow, and according to theory, the mechanical response to the increased accumulation will probably involve the propagation of a *kinematic wave* moving down the glacier 2–5 times faster than the actual particles of ice. The wave will reach the glacier snout long before the new ice formed from the flux in accumulation could possibly be transported that far. Without going into details, it appears that the wave motion will have little effect on ice that is extending; but in the ablation zone, where compressive flow dominates and velocity decreases down-ice, the oncoming wave accentuates the compression and the glacier becomes very unstable there. As the wave approaches any point in the ablation zone, the ice will thicken and the surface will rise dramatically. As the wave passes, it will quickly thin and subside to its former level. On some glaciers the ice surface may rise and fall more than 100 m in the sequence of events.

If, as Nye (1960) suggests, kinematic waves represent the mechanism by which glaciers respond to changes in mass balance, it becomes important to recognize that the time needed for the wave transmission varies from glacier to glacier. For example, the time needed to reestablish a steady state (i.e., the response time) in single, temperate glaciers varies from 3 to 30 years. The significance of this fact can be seen in a hypothetical case. Suppose that during the years 1954–1958, abnormally high amounts of snow accumulated on two adjacent valley glaciers. In one with a short response time, a kinematic wave rapidly traveled the length of the glacier, and the snout advanced dramatically in

1960. The second glacier, having a longer response time, showed no visible effects of the accumulation by 1960; the kinematic wave did not reach the terminal zone until 1975, when the terminus suddenly advanced. The two glaciers were completely out of phase, although both advanced in response to the same event. In addition to problems of mechanics, measurements made on a portion of the ice experiencing the wave action are not documenting the flow of ice but rather the movement of waves. In ice sheets, where response times are as long as 5,000 years, the difficulty of relating flow to changing regimes is obviously magnified.

Kinematic wave velocities may attain catastrophic values, reaching hundreds of meters a day (Meier and Johnson 1962; A. E. Harrison 1964); they have often been cited as the mechanics governing *surging glaciers*. However, in many cases glaciers surge even though there is no discernible net accumulation of mass to the glacier or any other external stimuli for the movement (Post 1960, 1969; Meier and Post 1969). It is probably best to separate surging from kinematic wave transfer even though the two movements have many of the same characteristics (Palmer 1972).

A surging glacier, then, is one in which sudden, brief, large-scale ice displacements periodically occur. The ice moves 10–100 times faster than its flow rate in the quiescent periods between surges. The periodicity seems to range from 15 to more than 100 years (Post 1969) and probably results from unique conditions that create a cyclic instability within the glacier. Some surges have no distinct periodicity and may be generated by outside events such as earthquakes or episodes of abnormally high geothermal heat.

Glacial surges are now recognized as fairly common phenomena, with 204 surging glaciers identified in western North America alone by 1969 (Post 1969). In fact, Budd (1975) has suggested that they may represent a completely separate type of glacier. However, they have no distinct size, shape, or activity, and they can contain temperate or subpolar ice. Their only unifying characteristics seem to be that surges are initiated in the ablation zone slightly down-ice from the equilibrium line, and a long and pronounced stagnation of ice occurs in the terminal zone in the interval between surges (Post 1960, 1966). During the surge the glacier surface is broken chaotically, and medial moraines and ice bands are intensely contorted.

Apparently, for some unknown reason, a reservoir of ice collects in the upper portion of the ablation zone, where ice is still active, while the terminal area is physically lowered by ablation. Eventually a critical disequilibrium condition is reached, and the response is a sudden burst of glacial movement into, and often overriding, the stagnated terminal zone. The key to understanding cyclic surging thus seems to lie in two factors: (1) how thickening of ice, with a concomitant increase in basal stress, can occur in one region of the glacier while stagnation occurs in another; and (2) what triggers the sudden release of the ice reservoir.

Several models have been proposed to explain the internal controls of periodically surging glaciers. Budd (1975) suggests that certain glaciers with the proper sliding velocity gradually develop a lubricating factor which in turn lowers the basal stress enough to induce a sudden movement. Surging glaciers, in his model, do not have sufficient mass flux to continue this rapid flow, and so they periodically oscillate from the slow flow of an ordinary glacier to sudden bursts of activity.

Another interesting model (Robin and Weertman 1973) proposes that surges end when the bed shear stress reaches a low value, a conclusion postulated first by Meier and Post (1969). During the quiet period following the surge, the glacier develops zones of vastly different basal shear stress. In the stagnating terminal area, basal shear will be relatively low. Further up-ice, however, the stress will progressively increase with time along with the thickening of the ice. Between the two regions is an area called the *trigger zone,* where the *gradient* of basal shear stress is continuously increasing and eventually reaches a critical, unstable condition. Robin and Weertman also suggest that a distinct pressure distribution, related to the basal stress, is promoted in the water at the bed. The water-pressure gradient is inversely related to the shear-stress gradient, and water is dammed in the trigger zone. The next surge begins when the trapped water lowers frictional resistance so much that the sliding velocity drastically increases.

One objection to this model, clearly recognized by Robin and Weertman, is that the basal water to be dammed must be prevented from draining downglacier through interconnected channels. This objection might be partially overcome if the glacier snout is frozen to the bedrock, a possibility suggested for subpolar surging glaciers (Jarvis and Clarke 1975). Even in temperate glaciers, patches of cold ice may exist at the glacier floor (Robin 1976).

To summarize, it has been shown theoretically that surges can be triggered by lowering basal shear stress over a period of time (Campbell and Rasmussen 1969); presumably, the lubricating effect of meltwater can accomplish the necessary decrease in stress. If that is true, as the proposed models suggest, the key to surging lies in the mechanics of basal sliding. Unfortunately, as we saw earlier, our understanding of sliding, both of regelation and the influence of water, is primitive, and so to bind ourselves at this time to any single explanation of glacial surging would be premature. Furthermore, with all the complications involved, we cannot hope to understand fully the mechanics of glacier movement until long-term, intelligently conceived sampling programs have been completed.

Ice Structures

Glaciers usually display a variety of structures that develop during growth of the glacial mass—which we can call primary structures—as well as secondary structures that develop in response to glacial movement. Primary structures are indicative of accumulation and ablation characteristics during the glacier's history, and secondary structures relate closely to the glacier's mode of flow and stress field.

Stratification

Primary structures in glaciers appear as discernible layers or bands within the ice. The layering results from processes that reflect an annual cycle of snow accumulation and ablation above the firn line. During the winter a thick pile of new snow is added to the glacier and, with time, proceeds through the phases of metamorphism that culminate in a layer of clear white ice. In the ablation season, however, the upper portion of the winter accumulation is subjected to melting and refreezing and develops a texture different from that forming lower in the snow pack. In addition, sediment and organic debris may collect in the partially ablated upper zone, making it slightly darker in color. The alternating white and "dirty" layers give the ice a stratified appearance and allow glaciologists to estimate the annual growth in the glacier thickness.

The layers tend to be tilted and deformed as the glacier moves. At the surface of the ablation zone, the bands look like truncated beds of plunging folds (fig. 9.22) with the layers dipping into the glacier in an up-ice direction. The layers exposed at the surface ablate at different rates. Debris-laden bands release their encased sediment and downwaste more rapidly, sometimes causing the adjacent up-ice layer of clear ice to protrude slightly above the level of dirty ice.

Secondary Features

Foliation   A secondary type of layering, called *foliation,* is produced by shear during ice motion; it is sometimes difficult to distinguish from primary stratification because both types may display similar grain size or textures. Foliation usually appears as alternating bands of clear blue ice and white bubble-rich ice. The layers, which dip at all angles, are most prevalent near the ice margins and commonly offset or wrinkle the primary stratification (figs. 9.23, 9.24). The

Figure 9.22. Saskatchewan Glacier—outcrop pattern of stratification in Castleguard sector. View upglacier from cliff on south margin below Castleguard Pass. Splaying and en echelon crevasses are also visible. Province of Alberta, Canada. (Photo from Meier 1960, plate 7-C)

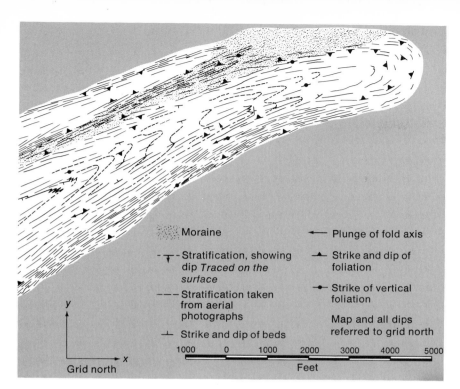

Figure 9.23. Map of a portion of Saskatchewan Glacier showing foliation and other structural features. Values of dips and plunges are not shown. (After Meier 1960)

Moraine

Plunge of fold axis

Stratification, showing dip *Traced on the surface*

Strike and dip of foliation

Stratification taken from aerial photographs

Strike of vertical foliation

Strike and dip of beds

Map and all dips referred to grid north

1000   0   1000   2000   3000   4000   5000

Feet

y

x

Grid north

Figure 9.24. Saskatchewan Glacier—gently dipping stratification, wrinkled and intersected by nearly vertical foliation. Exposed on east wall of a crevasse, 4.5 km below firn limit in midglacier. Province of Alberta, Canada. (Photo from Meier 1960, plate 8-B)

origin of foliation is poorly understood. Some evidence suggests that it originates where shear stress is greatest, but Meier (1960) argues that it is neither formed nor preserved at great depth, where shear stress would presumably be at a maximum.

Crevasses   Crevasses are cracks in the ice surface that range in size from miniature fractures to gaps several meters wide. The fractures are important in that they provide avenues for surface meltwater to penetrate the interior of the glacier, although the openings are rarely deeper than 30 m. It is generally assumed that crevasses develop perpendicular to the direction of maximum elongation of the ice. On some glaciers, however, the crack directions do not precisely coincide with the measured surface strain rates (Meier 1960). Nonetheless, crevasse types (fig. 9.25) do reflect a local tensional stress environment. *Splay crevasses* or *radial crevasses* form near the flow centerline where spreading exerts a component of lateral extension. In contrast, near the ice margins the shear stress parallels the valley walls, and crevasses develop diagonally to those sides. Here the crevasses are either *chevron* or *en echelon* types (fig. 9.26).

Transverse crevasses develop where the ice extends in a longitudinal direction. In temperate glaciers, these cracks occur most commonly in ice falls where the glacier cascades over convex irregularities in the underlying bedrock slope. In the summer, large ice crystals may develop in the transverse cracks when water trapped there repeatedly melts and refreezes. Sediment may also accumulate in these openings. At the base of the ice fall the crevasses are closed by compression, and a band of dirt-stained ice forms. During the winter, however, ice descending the fall reconstitutes into clear, bubbly ice (King and Lewis 1961). The annual downvalley flow, therefore, tends to produce a distinct series of alternating white and dark bands, called *ogives,* that are prominent at the foot of ice falls (fig. 9.27). They are different from primary layering in that they do not dip as strata into the glacier, but are purely a surface phenomenon.

Figure 9.25. Types of crevasses in valley glaciers: (A) marginal—(1) old rotated crevasses, (2) newly formed crevasses; (B) transverse; (C) splaying; (D) radial splaying. Arrows show flow direction. (From Sharp 1960. Used with permission of the University of Oregon Press)

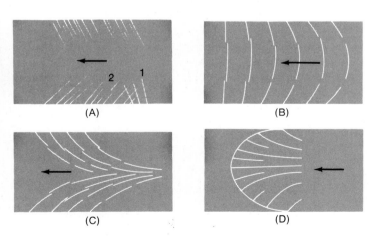

(A)

(B)

(C)

(D)

Figure 9.26. Saskatchewan Glacier—en echelon crevasse belts near south margin, 2.6 km below firn limit, viewed northwest. Province of Alberta, Canada. (Photo from Meier 1960, plate 9-C)

Figure 9.27. Merging of several individual glaciers into the Grand Plateau Glacier, Alaska. Strong medial moraines separate the ice streams. Arcuate ogives are plainly visible. (Photo by U.S. Coast and Geodetic Survey. From University of Illinois Catalog of Aerial Photography)

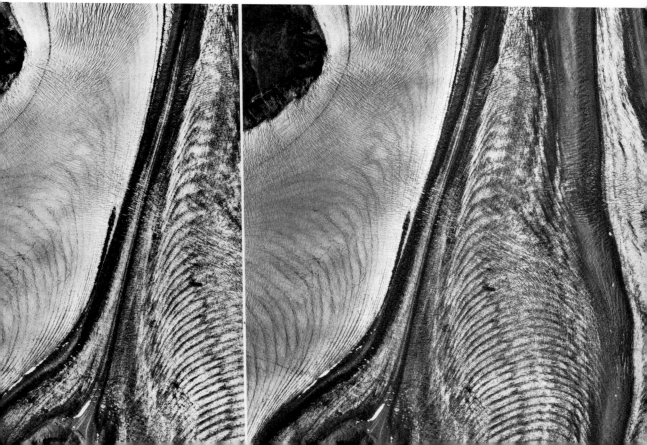

Summary   In this chapter we examined the origin and movement of glaciers, as well as the factors that might explain the properties associated with the various glacier types. Glaciers can be classified on the basis of their morphology or geographic position, their activity, or their thermal characteristics. Any glacier develops when snow accumulated over a period of years is transformed into ice by compaction, recrystallization, and melting and refreezing. These processes progressively increase the density of snow and firn as the space within the original mass is removed by pressure, crystal growth, and orientation of the grains. When the ice reaches a critical thickness, it is capable of movement.

The amount and type of geomorphic work a glacier can accomplish depends on its mass balance and on the total volume of ice added and lost during a period of time. The mass balance, or budget, is the net difference between accumulation and ablation during the time period in question. A positive mass blance indicates that more ice has been added to the glacier than has been lost; and the transfer of ice from the accumulation zone to the ablation zone causes the glacier front to advance. A negative budget results in a retreat of the front because the amount of ice transferred from the accumulation zone to the ablation zone is not sufficient to replenish the volume of ice lost by ablation. The total volumes of accumulation and of ablation determine the level of glacial acitivity. Two glaciers may have identical mass balances, but the one having the higher total volumes of addition and loss will be more active and do more geomorphic work.

Glaciers move by internal deformation of the ice and by sliding along the bedrock floor at the base of the glacier. The internal movement occurs as a type of creep that is mechanically different from viscous flow or pure plastic flow. In most glaciers the deformation caused by flow can be estimated by an equation called the power flow law or some modification of it. The mechanics of sliding is poorly understood but probably consists of the combined effects of regelation slip, lubrication by water under pressure, and enhanced creep. In regelation, ice melts on the upstream side of obstructions, where pressure is greatest, and refreezes downstream from the obstruction, where pressure is least.

Velocity of flow relates well with modified versions of the power flow law, but many characteristics of glaciers complicate a direct relationship. For example, ice does not flow in parallel laminar sheets, but thickens and thins along the length of the glacier. This develops the compressive and extending flow that causes ice to move downward in the accumulation zone and surfaceward in the ablation zone. In addition, flow velocity varies with time. The phenomenon of surging is difficult to fit into an all-inclusive flow model.

Primary and secondary structures in glaciers are related to the processes of accumulation and ablation, and to the stress field generated during the ice movement. The major structural features are stratification, foliation, crevasses, and ogives.

The following references are suggested to provide you with greater detail about the topics discussed in this chapter.

Embleton, C., and King, C.A.M. 1968. *Glacial and periglacial geomorphology*. Edinburgh: Edward Arnold Ltd.

Glen, J. W. 1955. The creep of polycrystalline ice. *Proc. Royal Soc. London*, ser. A, 228:519-38.

Kamb, B. 1964. Glacier mechanics. *Science* 146:353-65.

Kamb, B., and LaChapelle, E. 1964. Direct observation of the mechanism of glacier sliding over bedrock. *Jour. Glaciol.* 5:159-72.

Meier, M. F. 1960. Mode of flow of Saskatchewan glacier, Alberta, Canada. U.S. Geol. Survey Prof. Paper 351.

Meier, M. F., and Post, A. S. 1969. What are glacier surges? *Can. Jour. Earth Sci.* 6:807-17.

Nye, J. F. 1965. The flow of a glacier in a channel of rectangular, elliptic, or parabolic cross-section. *Jour. Glaciol.* 5:661-90.

Paterson, W.S.B. 1969. *The physics of glaciers*. Oxford: Pergamon Press.

Sharp, R. P. 1960. *Glaciers*. Eugene, Ore.: Univ. of Oregon Press.

Shumskii, P. A. 1964. *Principles of structural glaciology*. New York: Dover.

We are now ready to consider the geomorphic significance of the glacier mechanics we examined in chapter 9. It would be ideal if we could directly correlate geomorphic features with specific ice processes, but we seem to be far from such sophistication. The reasons for this involve several main factors: (1) Our conceptual models of glacier mechanics are greatly oversimplified and based on theory rather than extensive observation. (2) Much erosion and deposition take place at the base of glaciers where it is rarely feasible to document the actual process. (3) We are just beginning to understand the feedback mechanics that function between the basal ice and the geologic framework. For example, we know that glaciers have the ability to erode because we can see the results of that erosion after the ice disappears. What we don't understand is precisely how the erosive process itself is modified by remolding of the subglacial framework. Does removal of certain obstructions by erosion decrease the subsurface roughness enough to facilitate rapid sliding velocities, as suggested by Kamb (1970), and might this in turn accelerate further erosion?

Until we can resolve these deficiencies in our understanding, the process-feature relationship will be arrived at by traveling a one-way street in the wrong direction—we are interpreting the process by the character of its results. Nonetheless, glaciers have left us a variety of deposits and a myriad of features that demand our attention. It is not completely satisfying to reconstruct their origin without a wealth of solid observations, but this type of deductive reasoning has been a part of geology for a long time and is not necessarily incorrect. Furthermore, investigations of features and deposits can provide tangible clues about their origins that become invaluable in deciding how and where to study glaciers in the future. It seems inevitable, however, that new techniques for investigating glacier mechanics will demand continuous modification and testing of our present interpretations of features and deposits.

Glacial Erosion, Deposition, and Landforms 10

Erosional Processes and Features

Minor Subglacial Features

Glacial erosion is accomplished primarily by two processes, a scraping action called *abrasion* and a dislodgement or lifting action called *quarrying* or *plucking*. Since ice is not a hard mineral (1.5 on Moh's scale at 0°C), it cannot abrade most solid rock material unless it utilizes as grinding tools the fragments of rock carried in its load. Therefore, the efficiency of abrasion and the features it produces depend on the character and concentration of the debris being dragged along the base of the ice and, of course, on the properties of the bedrock being overridden. Abrasion is also influenced in a complex way by the nature of the subglacial topography and the velocity and direction of ice flow. The rate of abrasion, estimated by various methods, usually ranges from 0.06 to 5 mm a year, but it may be considerably higher under thick, high-velocity glaciers. Boulton (1974) reports that the abrasion of a marble plate inserted beneath the Glacière d' Argentiere in France proceeded at a rate up to 36 mm a year where the ice was 100 m thick and moving at 250 m a year.

The apparent dependence of abrasion on ice thickness does not agree with McCall's (1960) theoretical analysis that the amount of abrasion is independent of thickness once ice is more than 22 m thick. He states (1960, p. 57): "In other words, given the same bed conditions and ice temperatures and the same type of cutting tools, an extremely deep glacier would not cause any more abrasion than would a thin glacier whose depth was greater than 22 m." Presumably the yield stress for ice (2 kg/cm² at 0°C) is such that any depth greater than 22 m would cause the ice to flow around and under the rock fragments without increasing its ability to abrade the underlying bedrock surface. Boulton (1974), however, reports observations of extensive contact between loose particles and the bed beneath ice 150 m thick; cavities at the interface are usually occupied by rubble, not ice. He therefore rejects McCall's hypothesis and suggests that at any given ice velocity, assuming a constant concentration of basal debris, the abrasion rate will increase with increasing normal pressure until it attains a peak value, as plotted in figure 10.1. Further increase in normal stress will cause the rate to decrease rapidly to zero, where abrasion ceases and any debris in transport will be deposited beneath the ice.

Small erosional features produced by abrasion are usually in the form of linear scratches or crescentic marks that show the relationship between the size and composition of the abrading particles and the resistance of the underlying bed. Very fine particles in sufficient abundance produce a smoothly polished surface composed of microscopic scratches; as the grain size of the load increases, the scratches become larger in a transitional sequence from polish to striations to grooves or furrows.

Striations like those shown in figure 10.2 have been noted in every glaciated region of the world. They are best preserved on fine-grained rocks that have not been deeply weathered and on smooth bedding planes that dip gently away from the direction of ice movement. Striations and larger linear features also can form

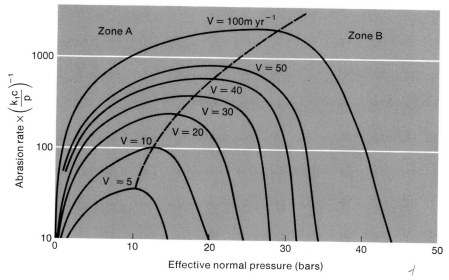

Figure 10.1. Theoretical abrasion rates plotted against effective normal pressure for different ice velocities. The rate increases in zone A and decreases in zone B as pressure increases. At any given velocity, a pressure increase greater than that at the higher X-axis intercept causes deposition of lodgement till. *k, c,* and *p* are constants depending on relative hardness, debris concentration, and penetration hardness of the ice. (After Boulton 1974. Used with permission of Donald R. Coates)

Figure 10.2. Striations on outcrop of limestone in south central New York State.

in unconsolidated material, such as till or loess, if it is highly compacted (Westgate 1968). Striae are only millimeters deep and are most likely eroded by sand grains or by jutting edges of larger particles carried in the basal ice. Where sand is the abrasive tool, the grains probably must be overlain by a large clast, since ice cannot exert enough pressure on the small surface area of an individual sand particle to produce a scratch. Striations tend to be continuous for only relatively short distances, probably because the sliding clast itself produces a carpet of plowed debris at its forward edge. When sufficient debris accumulates, the scratching particle will ride over the material, thereby interrupting its contact with the solid bedrock until it reaches a fresh surface downstream from the debris cover (Boulton 1974).

Increasing grain size in the load produces grooves and furrows rather than striations. Normally grooves are up to 1–2 m deep and 50–100 m long, but under the proper controlling factors they may achieve giant proportions. For example, H. T. U. Smith (1948) describes grooves in the Mackenzie River valley of Canada that are 30 m deep, 100 m wide, and several kilometers long. These are not the product of a single boulder but possibly represent gouging by a pocket of boulders solidly frozen together (Embleton and King 1975a). There may, in fact, be a limit to the size of boulder that can act in the grooving process, because large fragments, having too much surface contact with the underlying rock, will force the ice to flow over and around the boulder instead of carrying it as part of the basal load.

In addition to scratches of all types, a group of small features, generally referred to as *crescentic marks* or *friction cracks* (Harris 1943), are formed by chipping of the underlying rock surface (fig. 10.3). The marks are usually lunate in form, 10–12 cm long and 10–25 mm deep, and are perpendicular to the direction of ice flow as determined by other criteria. Three types of marks (*crescentic gouges, lunate fractures, crescentic fractures*) are associated with distinctly oriented fractures in the bedrock that apparently form during the chipping process. A fourth mark, called a *chattermark,* has no fractures deeper than the chip that was removed, and therefore Harris (1943) did not consider it to be a member of the friction crack family.

The features of abrasion are thought to reflect the direction of ice movement. Remember, however, that other processes such as mudflows or snowslides can form striations, and even floating ice blocks can cause them in nonresistant materials (Dionne 1974). Ice, and the striations it forms, will also diverge and converge over an irregular bedrock topography. Thus, striae have somewhat limited value as directional indicators unless a large number of measurements are obtained and treated for statistical significance. In addition, fractures associated with friction cracks dip both in the direction of glacier movement and opposite to it (Harris 1943; Dreimanis 1953; Flint 1955b); the limbs of the lunate shape also can point in either direction. Thus, the use of these features to postulate the direction of ice motion is valid only if the type of mark can be positively identified.

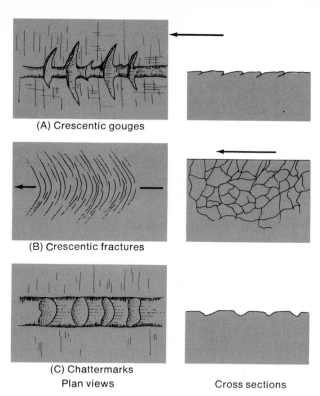

(A) Crescentic gouges

(B) Crescentic fractures

(C) Chattermarks

Plan views   Cross sections

Figure 10.3. Plan views and cross sections of several common minor chip marks made by ice moving across a bedrock surface. (After Harris 1943, fig. 3. Used with permission of The University of Chicago Press)

Quarrying differs from abrasion in that the functional success of the process depends less on the type of load being transported than on the properties of the underlying rock. In fact, fractures must exist in the bedrock if plucking is to operate at all. Intense shattering of rock in preparation for plucking probably requires some form of pressure release (Lewis 1954; Glen and Lewis 1961), crushing (Boulton 1974), or cyclic freezing and thawing. These processes, which weaken the internal cohesion of the bedrock, may occur subglacially or in association with periglacial conditions prior to the arrival of the glacier.

In detail, plucking has two basic requirements: first, the ice must exert a shear force on the loosened particles, and second, this force must exceed the resistance caused by friction when the particle is dragged over the residual bedrock (Boulton 1974). It is not enough simply to dislodge the particle from its original position, but the driving force must overcome any frictional resistance generated in the system. Boulton suggests that where ice flows over unconsolidated sediments, plucking is a relatively straightforward process related totally to the shear stress, which is expressed as

$$\tau_g = \tau_o + (\rho g h - w_p) \tan \phi$$

where τ_g is the shear stress at the glacier bed, τ_o is the cohesive strength, h is the ice thickness, w_p is the pressure in any water at the interface, ρ is density, g is gravity, and $\tan \phi$ is the coefficient of internal friction.

In tightly lithified materials, the mechanics is more complex. The process is directly influenced by periodic opening and closing of cavities beneath the ice, which are caused by obstructions and fluctuations of ice thickness or velocity. When cavities are open, free water may freeze to the fragmented particles or within the shattered mass. If the cavity is subsequently closed, the glacier incorporates this new ice into its basal layer; its forward motion will pluck some rock fragments away from the surface. Thus, plastic flow may also be involved in the encasement of shattered material and its subsequent removal (Boulton 1974). It is important to recognize that frictional resistance to plucking increases rapidly as the cavity is closed because it is directly proportional to the normal pressure. As Boulton suggests, plucking is probably most effective when the normal pressure is sufficient to incorporate the loose particles into the ice but not great enough to inhibit their forward movement by increasing the frictional resistance. Exactly when this condition prevails will vary from glacier to glacier, and other factors may complicate the dynamics. For example, rock fragments projecting downward from the ice may dislodge particles from the rock floor even when the interface is slightly open. These do not necessarily become encased in the ice but are shoved spasmodically along the bedrock surface until they encounter the downglacier end of the cavity.

It is difficult to characterize the glacier type that will produce the greatest quarrying action. Glaciers that tend to have fluctuating temperature-pressure conditions and concomitant melting and freezing certainly should produce intense regelation. There seems to be no question that rock debris is entrained by ice at the pressure-melting point; Kamb and LaChapelle (1964) observed the regelation layer to be heavily laden with such debris. But it is also true that subpolar glaciers and even some polar glaciers contain large loads eroded from the bedrock floor (Boulton 1970a). This apparent contradiction stems from the fact that in any single glacier different zones may contain ice with a variety of temperature-pressure conditions, and therefore, plucking may be part of any glacier system.

The evidence of plucking action in glaciated landscapes is usually found in erosional features somewhat larger than those produced by abrasion, but commonly the two erosive processes are closely associated in the same landforms. Where abrasion is dominant, the landscape may be indented with smoothly curved elongate surfaces whose long axes are subparallel to the direction of ice flow. Some of these surfaces are distinctly higher at one end, and they taper laterally and longitudinally until they blend into the surrounding ground level, producing a unique teardrop or raindrop shape which Flint (1971) describes as a "whaleback form." Some whaleback forms may be related to streamlined depositional features, such as drumlins, in that their shape represents the minimum resistance to flowing ice. The composition of streamlined forms

seems to be of little consequence, however, as they can be entirely bedrock, entirely sediment, or any combination of the two. Whalebacks, therefore, may be merely a transitional form in a range of streamlined features from pure bedrock to pure drift (Flint 1971).

When plucking is a significant factor, whalebacks develop a pronounced asymmetry, having a gently sloping upstream surface and a steep rock face on the down-ice side of the feature. Such a form, commonly called *roche moutonnée*, is the result of abrasion on the upstream slope and intense quarrying at the position of the steep, downstream face. Its development is due to irregular spacings of fractures within the bedrock. Where joints are widely spaced, abrasion is the dominant process, while closely spaced jointing facilitates plucking and more rapid erosion (fig. 10.4). Flint (1971) objects to the term *roche moutonnée* because of its wide misuse and suggests that *stoss and lee topography* better describes the forms developed by the combination of abrasion and plucking.

All subglacial erosion is not accomplished by the action of ice alone. Some glaciated regions display a multitude of surface features, called *p-forms* (plastically-sculptured forms), that are molded from the underlying rock in some complex way. Some of these are curved troughs and associated potholes that probably reflect the work of meltwater rushing through subglacial tunnels (Dahl 1965). Rapid drainage of englacial water through these restricted channels can produce high flow velocities that can extensively modify the surface. Some of these features may be produced by the mobilization of water-soaked till (Gjessing 1967), although there is no direct evidence to substantiate such a mechanism (R. J. Price 1973). Furthermore, Boulton (1974) cautions that the presence of some fluviatile features has little bearing on the origin of p-forms, contending that most of these features can be developed by normal abrasion mechanics of glacier ice.

Cirques

The striking landscapes found in the uplands of glaciated mountains have been sculptured primarily by the erosive action of ice contained in cirques. The term *cirque* was first used in the early 1800s to describe the collecting basins for valley glaciers in the Pyrenees, and locally the feature has been given a variety of names including cwm, corrie, kar, and botn. A cirque by any name is still a deep erosional recess with steep and shattered walls that is usually located at the head of a mountain valley. It is normally semicircular in plan view, often being de-

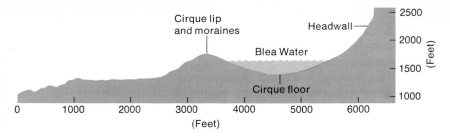

Figure 10.5. Long profile through Blea Water corrie, a cirque. (Adapted from Lewis 1960. Used with permission of The Royal Geographical Society)

scribed as an amphitheater, and it is floored by a distinct rock basin where the surface has been smoothed by abrasion. As figure 10.5 shows, the bowl-shaped rock basins commonly contain lakes, called *tarns,* that are dammed in by a convex-up rock lip which stands as a threshold boundary between the cirque floor and the downstream part of the valley. The cirque lip is often capped by small moraines that contribute to the damming effect. The rock basins can be of spectacular dimensions. For example, the rock basin floor of Blea Water corrie in England (shown in the figure) is 96 m (316 ft) lower than the rock lip (Lewis 1960).

Cirques range in size from shallow depressions to monstrous cavities that are kilometers wide and several thousand meters high along the rear wall. Their dimensions and their geomorphic form depend not only on the rocks into which they are cut, being larger and more perfectly developed in igneous or high-rank metamorphic rocks, but also on the rock structures, the preglacial relief, and the time span of the formative glaciation. Most maturely developed cirques seem to possess a reasonably consistent geometry when their length to height ratios are compared, indicating that cirques probably attain some equilibrium form related to the processes of their formation.

Cirques are often preferentially oriented according to the direction of solar radiation and the prevailing winds. For example, in the mid-latitudes of the northern hemisphere a large percentage of cirques face toward the northeast quadrant, presumably because the shade produced across their floors in that direction allows more of the winter snow to escape the summer radiation. The winds blowing from the west aid in the process by protecting the loose snow deposited in the windshadow of the mountains from deflation. The orientation also determines how long glaciers will survive in the present climate, because snow and ice are protected in cirques with the proper form and direction (Graf 1976). The elevation of cirques seems to be related to and controlled by the position of the snowline at the time of formation (see Porter 1977); many workers believe that cirque floors provide a first approximation of Pleistocene snowlines. Thus, although most cirques originate in the headward reaches of stream valleys, any hollow, regardless of its origin, that stands at the proper elevation and has the ideal orientation may progressively accumulate snow and finally become a maturely developed cirque like those in the photograph in figure 10.6.

The significance of cirque processes in the development of Alpine scenery is that cirque expansion by continued erosion gradually eliminates the preglacial upland surface. As a number of cirques grow headwardly and laterally, they progressively consume much of the intervening upland region and leave as its only vestiges spectacular *horns* and *arêtes,* the features so indicative of mountain glaciation (figs. 10.6, 10.7). With prolonged headward erosion, adjacent cirques may merge, forming *col* depressions in the knife-edged arêtes. The depressions are eventually lowered enough to allow ice to spill from one cirque into the other. The process ultimately may result in the coalescence of many cirque glaciers into an ice cap where only residual peaks of rock, called *nunataks,* stand above the ice level as reminders of the original upland surface. The erosive work in cirques lends itself to the suggestion that Alpine topography can be classified according to stages of its development, and such hypotheses are not uncommon (see Embleton and King 1975a). Our purpose here, however, is to understand the processes that create the Alpine topography, not the sequence of its development. Therefore, we must examine the origin of cirques and the erosive mechanics that functions within their boundaries. Of utmost importance are the processes that scour the basin in the cirque floor and those that cause the recession of the cirque walls.

Figure 10.6. Cirques at the head of Carbon Glacier, Mount Rainier National Park, Washington. Note well-developed arête between the two cirques. (Photo from Fiske et al. 1963, fig. 48)

Figure 10.7. Schwan Glacier in the Chugach Mountains of Alaska. Note dark lobes on glacier surface where rock-slide avalanches have moved onto the ice. Medial moraines are displayed as long linear dark bands. Horns and arêtes are shown in mountain uplands. (Photo by Austin Post. From Post 1967, fig. 11)

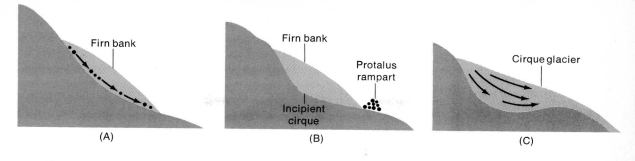

(A) (B) (C)

Figure 10.8. Stages of cirque development. (A) Nivation beneath firn bank. (B) Nivation cirque. (C) Cirque with fully developed cirque glacier.

Cirques result from two separate groups of processes: (1) mechanical weathering and mass wasting, and (2) erosion by cirque glaciers. The development of a cirque, diagrammed in stages in figure 10.8, begins in a patch of firn that fills a small depression and stands near the regional snowline. In the ablation season, meltwater released during the day percolates into fractures of the bedrock beneath the firn bank and refreezes there at night. The repeated pressure associated with freezing and thawing presumably wedges out particles of rock that are then moved slowly downslope by creep and by water flowing at the base of the firn. The combined processes are commonly referred to as *nivation* (Matthes 1900). Gradually the shape of the original depression is deepened and widened, and eventually it approaches a semicircular form which can logically be called a nivation cirque. The nivation hypothesis also contends that with continued accumulation, the firn changes into true glacier ice, and glacier erosion rather than nivation becomes the dominant process in further development of the cirque. Exactly when this transition occurs is not clear, but it is virtually impossible for rock basins cut into cirque floors to be formed by the nivation process alone, since it cannot carry particles upslope to the cirque lip.

Cirque Glaciers   Observations made in tunnels excavated into cirque glaciers indicate that such glaciers move by a process known as rotational sliding, in which ice slides over the arcuate bedrock floor, rotating at the same time around a horizontal axis. The ice exposed in the tunnels displays recognizable yearly accumulation layers that are separated by marked ablation surfaces, giving the entire glacier a banded stratigraphy (Grove 1960). Some of the ablation zones are laden with debris that fell onto the ice surface near the cirque headwall. As figure 10.9 shows, these layers originally dip downglacier at the angle of the ice surface, and with time each layer is incorporated into the ice mass as more snow accumulates in the headwall area. As the ice moves, however, the layers are reoriented so that (as in the figure) near the equilibrium line they are almost horizontal; while close to the terminus they dip steeply upglacier. The deformation of the layers, combined with changes in the position of stakes in the tunnels and on the surface (McCall 1952, 1960), makes it clear that flow lines are moving downward near the headwall, parallel to the surface at the firn line, and upward near the terminus.

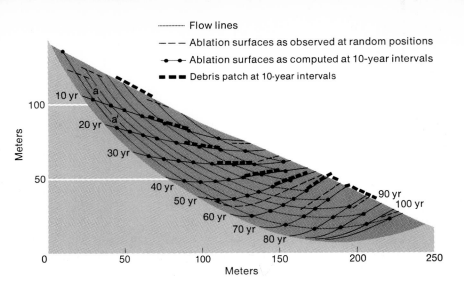

Figure 10.9. Long section through a cirque glacier in Norway, showing ablation surfaces and debris patches at 10-year intervals. Note rotation of ablation surfaces in the down-ice direction. (After McCall 1960. Used with the permission of The Royal Geographical Society)

Rotational sliding is an appealing mechanism to explain the scouring of the bowl-shaped depression in the cirque floor, for this process should be capable of carrying the products of abrasion or frost wedging upslope and over the cirque lip. We should emphasize, however, that not all cirque glaciers exhibit rotational sliding as the dominant flow mechanism. As the size of the cirque increases, the flow becomes more complex and much of the motion is the result of the normal creep process discussed earlier.

Headwall Erosion   In addition to erosion beneath the ice, the creation of Alpine topography requires expansion of the cirque dimensions. Such growth is related mainly to the removal of large chunks of bedrock from the head and side walls rather than the millimeter-by-millimeter abrasion occurring on the cirque floor. The surfaces of the cirque walls are assumed to be prepared for erosion by severe frost action associated with climates in high elevations. In addition to the shattering caused by freezing and thawing, joints often develop parallel to the wall faces when pressure is released by the removal of outer layers of rock (Glen and Lewis 1961). These *dilatation joints* aid in the fracturing process by providing avenues for percolating water and by isolating rock material into discrete units.

The actual processes of frost shattering are not so simple as they first appear and, like many glacial processes, seem to be accepted more on faith than on solid evidence. Since W. D. Johnson descended into a deep *bergschrund* in 1904, many workers have investigated the role played by bergschrunds in the mechanics of headwall retreat. A bergschrund is a crevasselike opening near the headwall that separates actively moving ice of the glacier from nonactive ice frozen to the headwall. The gap is usually open at the surface but is sometimes covered by a thin roof of ice; it narrows downward and, depending on its depth, may or may not intersect the rock of the headwall at some lower level. Based on his observations, Johnson (1904) suggested that surface meltwater gained access to the base of the headwall by percolating down the bergschrund. Since the gap was usually open to the atmosphere, air temperatures during the summer were assumed to cross the freezing point repeatedly and thus produce extensive frost wedging with water that had permeated into the rock fractures. The rubble accumulating in the riving process was removed when the bergschrund became filled with snow and congealed with the actively moving glacier.

Although the bergschrund hypothesis is attractive, it suffers because the assumed diurnal freezing and thawing at the base of the opening cannot be proved. Laboratory experiments (Battle 1960) suggest that maximum shattering by frost action occurs only when the temperature falls rapidly to between $-5°C$ and $-10°C$. A slow decrease in temperature will have little if any effect. In addition, freezing of water in a crack must proceed from the top of the opening downward to the bottom. An ice plug must form first at the top of the fracture in order to produce the closed system that will allow pressure from ice growth to exceed the tensile strength of most rocks. Actual measurements of temperature fluctuations in bergschrunds (Battle 1960) demonstrate that summer temperature ranges are only $0°C$ to $-2°C$, and in the spring temperature decreases only to $-4°C$. Battle, therefore, concludes that frost shatter in deep, closed bergschrunds is probably a minor phenomenon.

Small, shallow gaps that expose the headwall at the rear of a glacier surface (*randklufts*) might endure the shattering process and accelerate erosion along the upper zones of the headwall. This still leaves us with the problem of how cirques maintain such steep headwalls, since a rapidly retreating upper segment and a slowly receding base should produce a gentle slope on the rear wall. The situation is complicated by the fact that the lower segments of some headwalls show no evidence of shattering, while others have extremely jagged surfaces that obviously have been rived in some way.

In light of the above, we must entertain the possibility that we have misinterpreted the driving mechanics. S. E. White (1976a) suggests that hydration may exert almost as much pressure as that produced by ice expansion at $-22°C$. As minerals adsorb water, expansion and contraction of this nonfreeable water are produced by temperature fluctuations in the range of freezing and thawing. The shattering seen at the base of the headwall might then take place beneath a glacier without the stringent conditions necessary to generate forces by ice growth. The hydration-shattering proposal is thought-provoking and deserves careful investigation. The process not only can explain shattering where temperature fluctuations are minimal, but it also dispels the problems associated with the bergschrund hypothesis. Hydration also provides a viable alternative explanation in situations where rock shattering is apparent but frost action cannot be substantiated.

Glacial Troughs

Glacial erosion is not limited to the cirque environment, for ice passing over the cirque lip can also remold the preglacial valley topography into a characteristic glaciated form. The ability of ice to remove rock protuberances tends to produce valleys with steep, nearly vertical, sides and relatively wide, flat bottoms (fig. 10.10). These U-shaped valleys are essentially the channels of the ice streams, although the surface of the ice seldom accords with the bedrock configuration beneath the ice. Glacial troughs normally have cross-sectional profiles adjusted into a parabolic shape (Svensson 1959; Graf 1970), presumably because that form exerts the minimum resistance to glacier flow (Flint 1971). The precise width-depth dimensions, however, probably depend on the intensity of the glacial dynamics (Graf 1970) and the properties of the geological framework. The typical U-shape of glaciated valleys seems to be primarily a modification of initial relief; little evidence exists to suggest that glaciers themselves cut deep valleys in areas of high relief and resistant rocks. Where geomorphic conditions are right, however, ice sheets covering areas of low relief may be able to reorganize pre-glacial drainage systems by cutting new valleys beneath differentially flowing basal ice. The Finger Lakes of New York may represent such erosion (Clayton 1965), and there is little question that giant grooving is possible beneath ice sheets (Smith 1948). Furthermore, ice streaming (zones of rapid flow) appears to be a reasonably common phenomenon in ice-sheet mechanics (Sugden 1968), suggesting that some trough development may be possible if erosion is selectively concentrated along these flow lines. In mountains, however, the re-formation of the cross-sectional shape occurs by both lateral and vertical erosion of the preexisting valley. Whether the parabolic form is derived predominantly by widening or predominantly by deepening depends on the properties of the rocks and the ice, as well as the extent of preglacial weathering and the amount of load available within the ice to be utilized as cutting tools. In any case, the erosion leads to the truncation of rock spurs jutting into the valley and the formation of *hanging valleys*. These occur particularly where trunk valleys carry more ice and are more extensively eroded than their tributaries.

Glaciated troughs are also characterized by uniquely irregular longitudinal profiles that essentially represent a series of interconnected basins and steps; the diagram in figure 10.10 shows this typical profile in Yosemite National Park. The basins may contain lakes, a series of which are sometimes called *paternoster lakes*. Immediately downvalley from the basins, holding in the lakes, are rock bars or steps that commonly show the effects of intense abrasional smoothing on their relatively flat surfaces. At the distal end of the step, a sharp break in slope occurs where plucking has produced a steep scarp that connects the rock bar to the next lower basin or step.

The staircase profile has intrigued geologists for years and has been the subject of considerable speculation. Hypotheses for its origin include (1) variation of rock structures, especially spacing of joints, causing differential erosion,

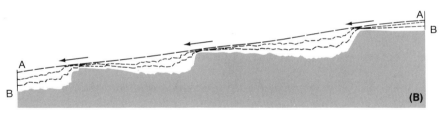

Figure 10.10. (A) U-shaped cross-profile of glaciated valley and a hanging valley— view up Yosemite Valley from vicinity of Artist Point. El Capitan at left, the Cathedral Rocks and the Bridalveil Falls at right. Yosemite National Park, Mariposa County, Cal. (Photo by J.I. Boysen. From Matthes 1930, plate 3) (B) Longitudinal section of glacial valley in Yosemite National Park illustrating staircase profile. AA is preglacial valley floor; BB is present valley floor. Broken lines represent intermediate stages of development. (After Matthes 1930)

(2) preparation of weak zones in the rock by preglacial weathering, (3) irregularities in the preexisting valley topography, and (4) increased erosive power at the confluence of tributary valleys with the main valley (see Bakker 1965). It is possible that rock bars producing a steplike profile require no special conditions for their formation other than the normal slip-plane orientation noted in glacier flow (Nye and Martin 1967). Theoretically, rock bars can form where there is no obviously harder rock, and conversely, zones of resistant rocks do not necessarily evolve into rock bars under glacial erosion.

The abrasion prevalent on the treadlike steps and the quarrying at the position of the steep scarps led Lewis (1947) to note that each step resembles a large roche moutonnée and so probably has a similar origin. He concluded that the mechanics probably worked best beneath thin glaciers where meltwater could easily penetrate to the glacier floor and shatter the rock by refreezing. Although freeze-thaw mechanics might be a reasonable accomplice in the formation of staircase profiles, we cannot ignore the fact that development of the associated basins probably required some differential ice movement, capable of flowing uphill and carrying load in the process. Such a requirement does not encourage the belief that the ice was thin; in fact, it is common knowledge that many valley glaciers have thickened to the point of overtopping their divides and spreading into adjacent valleys. In addition, as discussed earlier, the freeze-thaw component in rock shattering is somewhat suspect because we do not completely understand the thermal conditions at the base of glaciers. What we are left with, then, is the problem that while plucking is necessary to form the scarps, the rock shattering required for this process to function is probably not a result of frost action. Furthermore, some scarps form where no regional joint system is present to facilitate the shattering process.

Boulton (1974) suggests an alternative hypothesis of shattering that deserves our consideration because it not only resolves the above problem, but also relates directly to glacier mechanics. When a glacier moves across a horizontal bedrock surface, the effective normal pressure will be $\rho gh - w_p$ (where ρ, g, h and w_p are the same as defined earlier). The dashed line in figure 10.11 represents the normal pressure on the horizontal bed and is estimated by $\rho_i gh$ (where $\rho_i = \rho$). ΔP in the figure represents the increase or decrease of normal pressure produced by a bedrock obstruction; the total normal pressure at any point becoming $\rho_i gh + \Delta P$. Therefore, if the bed is irregular the normal pressure will fluctuate so that it will be higher than average on the upglacier side of a bedrock obstruction and lower than average on the lee side. Boulton shows that the *shear stresses* included in the bedrock will be greatest where the normal pressure is lowest, i.e., down-ice from the crest of the obstruction. The exact position and absolute magnitude of the maximum shear stress depend on whether cavitation occurs downglacier from the bedrock knob. Furthermore, Boulton points out that the shear strength of the rocks will be least in the downglacier position where the shear stress is greatest, creating the ideal situation for rock failure at that locale. Assuming this analysis

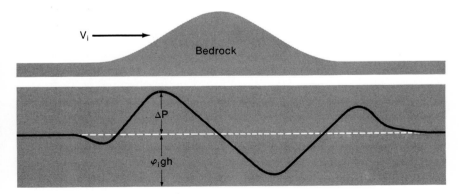

Figure 10.11. Schematic view of normal pressure distribution at glacier bed as ice flows over a bedrock obstruction. ΔP represents the increase or decrease of normal pressure produced by a bedrock obstruction. The normal pressure where no obstruction exists is represented by the dashed horizontal line and has a value equal to ρ igh. (After Boulton 1974. Used with permission of Donald R. Coates)

is correct, the inevitable conclusion is that with the proper bedrock configuration, even hard unfractured rocks may be crushed beneath the ice, and rocks already jointed can be intensely shattered. Preglacial irregularities in the long profile, structural weaknesses, and lithologic variations all will be accentuated by this process because shattering will prepare those zones for the plucking action that produces the scarps in the staircase profiles.

Fiords A special type of glacial trough exists mainly in high-latitude coastal regions that are underlain by resistant rocks, so that the general land surface stands at considerable elevation above the nearby ocean. These troughs are called *fiords,* and they differ from other types only because they are partially submerged by the ocean. The inundated bottoms of fiords have the same variety of topographic elements, both erosional and depositional, that exist in a normal continental glaciated valley (Holtedahl 1967). Their history may include components of both glacial and fluvial processes, and they may be partly controlled by tectonic and lithologic factors. For these reasons, it is unwise to make sweeping generalizations about their origin.

Perhaps the most salient property of fiords is that part of their development took place when the ice was physically beneath the ocean (Crary 1966). Flint (1971) reminds us that a glacier 1000 m thick with a density of 0.9 will remain in contact with its bed and be fully capable of erosion at water depths up to 900 m. Even at greater depths, when the snout begins to float, high topographic irregularities of the valley might still be eroded (Crary 1966). The water depths over fiords, several hundred meters in many and greater than 1000 m in some, are well beyond that which can be attributed to a postglacial rise in sea level (see Flint 1971), adding credence to the suggestion that much fiord erosion was accomplished in a submarine environment.

Deposits and Depositional Features

Before glaciers were recognized as viable geomorphic agents, deposits containing boulders that obviously came from a distant source were called *drift*. This term arose because elimination of known processes led to the belief that the anomalous boulders reached their site of deposition by riding on top of floating ice. After glaciers were recognized as the transporting vehicles, the term *drift,* or *glacial drift,* was retained and expanded to include all deposits associated with glaciation. It is estimated that drift covers 8 percent of the Earth's surface above sea level and almost 25 percent of the North American continent. The thickness of this cover varies greatly. In the United States, for example, only a thin layer of drift (<20m) covers upland areas in most of New England, although drift may be several hundred meters thick in buried valleys. The drift in the central United States is generally from 10 to 60 m thick, but once again these are average values. In some places the drift is merely a thin mantle on top of bedrock, and in other regions, such as parts of Michigan, it exceeds 200 m in thickness. The exact volume of drift deposited depends on the time span of glacier activity, but with high velocities and loads, as much as 30 m of drift can be accumulated in less than 10 years (Flint 1971).

Through the years geologists have been intrigued with the amazing variety of glacial drift and the complex interrelationships that exist between the different types. This complexity arises because (1) drift may be deposited from mediums that contain vastly different amounts of water; (2) deposition occurs beneath, within, or on top of ice, at the glacier margins, in bodies of standing water, or in fluvial settings far from the glacier, the debris being transported there by streams rising in the ice mass itself; (3) the depositional sites and environments and the drift composition all change with time because glaciers themselves are not constant in their properties nor fixed in their position; and (4) the glacier may be active or stagnant. Because of these complicating realities, any discussion of glacial drift and depositional features is difficult to organize in a way that is entirely satisfactory. In this approach, I will first briefly examine the varieties of glacial drift and then discuss features according to the depositional environment in which they originate. An attempt will be made to relate the morphology of features and their sedimentary properties to the dynamics of the system. Throughout this discussion it is important to recognize the distinction between the sedimentological character of drift and the morphology of features that result from its deposition. Morphological terms such as moraines and kames are not to be interpreted as implying a particular drift type. Many features with similar morphology are composed of a number of drift varieties, especially where the environment of deposition is subject to repeated change.

Drift Types

Over the years glacial deposits have normally been divided into two categories based on their sedimentary characteristics. Particularly important are the presence or absence of layers and the degree of sorting in the deposit. In our discussion we will separate drift into stratified and nonstratified types.

Nonstratified Drift   Sediment originating directly from glacial ice characteristically has no discernible stratification. The material is usually called *till* and typically is a nonstratified mass of unsorted debris that contains angular particles composed of a wide variety of rock types. In addition, the term *till* usually connotes material that has been transported and deposited by the ice itself, a process often indicated by striations or microscopic fractures on the grains (Krinsley and Donahue 1968). Many examples justify this description of till; absence of layering and poor sorting (especially an almost universal bimodal size distribution) seem to be the most reliably consistent properties. The bimodality observed in most tills probably is due to the differences in grains produced by abrasion and those derived by plucking. In an excellent review, Goldthwait (1971, p. 4) points out that till is probably more variable than any other sediment that is described by a single name.

Any of the identifying criteria in our definition may be missing at a particular till locality as a result of varying transporting and depositing mechanics and of heterogeneity in the rocks over which the ice has passed. For example, many clasts in till have a subangular pentagonal or triangular shape, but these forms may be significantly altered by rounding during their transportation. The extent of angularity, therefore, depends on the transport distance and the lithology of the particle, as well as its shape when it was first picked up. Some sedimentary rocks become rounded with increasing transport (Holmes 1960) while other, softer varieties may be crushed during movement, halting the rounding process (Vagners 1966). The rounding of crystalline rock particles probably depends on the original shape of the clast when it becomes part of the glacier load (Drake 1968). Shapes capable of being rounded will lose their sharp edges and corners within a very short distance, sometimes less than one mile (Crosby 1902). In addition, glaciers that override older stream deposits may incorporate in their load boulders that have already been rounded, resulting in a till that is notably less angular than one would expect. Figure 10.12 shows till that contains rounded boulders.

Overall particle size tends to be reduced by attrition during glacial transport, each particular lithologic type proceeding toward a specific terminal grade size. Granites and high-rank metamorphic rocks terminate in sands, carbonates reduce to silts, and shales proceed to clays (Dreimanis and Vagners 1971). The overall angularity and texture of till thus depend on the lithologic heterogeneity of the source rocks as well as the distance of transport from the location of their outcrop. This has led to some confusion concerning how far the material contained in till has been transported. A number of studies have shown that most till particles

Figure 10.12. Late Wisconsinan (Pinedale) till in Rock Creek valley, Beartooth Mountains, Mont. Note rounded boulders in till. (Photo by R.R. Dutcher)

come from rock sources within 50 miles of the depositional site (for example, Holmes 1952), and the accepted conclusion has been that the bulk of till is locally derived. It is now clear that this conclusion is not necessarily true for all till (Anderson 1957; W. Harrison 1960) but only indicates the source for part of the total size distribution. It seems reasonable that the lithologic composition of coarse fractions of the till will have a greater affinity to the rocks near the site of deposition (Drake 1971; Gross and Moran 1971; Dreimanis and Vagners 1971). When all sizes in the till are considered, a rather high proportion of long-traveled particles may be found, but they will be in the matrix of the deposit rather than in the larger clasts. Dreimanis and Vagners (1971) suggest that a predominance of terminal grade size over clast sizes indicates a long distance of transport during which the process of diminution presumably could proceed to its ultimate stage.

The unpredictability of till increases still more because the material is carried in different parts of the glacier where the ice is characterized by diverse mechanics. In ice sheets, debris is carried in layers 1–100 m above the base of the ice, as indicated by the fact that the surface is free of any load except near the margin (Goldthwait 1971). In valley glaciers, on the other hand, considerable debris may be shed onto the surface from the valley sides. Internal debris layers become thicker and more concentrated with sediment near the base of the glacier and toward its terminus, but in any case, they usually contain only minor percentages of the sediment weight, rarely attaining more than 50 percent. It seems certain that most till is transported as englacial debris and that it changes its position during deposition. It can be deposited directly on the bedrock floor as subglacial matter, and a considerable thickness of the glacier will remain above it at the moment of deposition. It can also rise within the ice along shear planes or along the normal upward flow lines in the terminal zone. This material finally emerges as supraglacial debris when ablation of the surface ice releases the contained particles. In this case the till will be emplaced on the bedrock surface as the ice beneath it dissipates, and no ice will remain above the till at the time of its final deposition. Till laid down in the subglacial environment under pressure of the overlying ice is usually referred to as basal till or *lodgement till*. In contrast, *ablation till* occurs where the debris is concentrated at or near the surface and is gradually let down onto the bedrock as the ice disappears.

Ablation till forms as one of two types, *flow-till* (Hartshorn 1958; Boulton 1968) and *melt-out till* (Boulton 1970b). When englacial debris is released at the surface, the till tends to be highly mobile and likely to move downslope as a quasi-mudflow, by creep, or as a semiplastic slide. However, when the thickness of the surface cover reaches about 3 cm, the ablation process is severely retarded, and if the surface layer exceeds one or two meters, ablation ceases. Further increase in the thickness of the supraglacial cover can occur only if more debris moves to the site from higher levels of the ice surface, or if the ice melts beneath the layer already concentrated at the surface. Boulton (1971, 1972b) asserts that supraglacial till may have two components: an upper zone accumulated from debris transported from a higher source (flow-till), and a lower zone accumulating in situ and never actually exposed at the surface (melt-out till). Melt-out tills rarely exceed 3 m in thickness, but the total supraglacial layer can be much greater if flow-till continues to accumulate on the surface. Little is known about the rate of accumulation of ablation tills, but Mickelson (1973) has calculated that the accumulation rate of basal melt-out till of the Burroughs Glacier in Alaska ranges from 0.5 to 2.8 cm a year.

The mechanics of the deposition of lodgement till is highly speculative since once again we are dealing with a phenomenon operating beneath an ice mass where direct observation is rarely possible (for a discussion, see Goldthwait 1971). Possibly lodgement till accumulates particle by particle when individual clasts strike some obstructing element on the bedrock floor and are dropped from the ice because of the frictional resistance. Material might also be deposited en masse, when near-bottom layers of englacial debris melt from the basal ice. The heat required to produce melting is generated by friction during sliding, or by geothermal heat that cannot be conducted away (Nobles and Weertman 1971). Still another possibility is that resistance along the bottom becomes so great that the ice shears above the basal layer. In this case the till becomes a lag, concentrated when the basal ice melts away. Once shearing occurs, the material is no longer moving along with the glacier but is being overridden by ice above the sheared zone. It is equally feasible, however, that lodgement till results when the ice simply stops moving and the lower, debris-laden ice zones are let down onto the underlying surface as the basal ice dissipates.

Presumably the type of till and its mode of emplacement can be determined from the texture of the deposit and the arrangement of the particles, a property known as the *fabric*. Along with a compact nature and the absence of extensive washing, a strong orientation of elongate clasts has generally been accepted as evidence of basal ice motion and is presumed to typify lodgement till. Many workers believe that clast orientation represents the direction of the glacier movement because in many cases it parallels striae and other directional indicators. This should not be an irrevocable conclusion, however, because particle shape also influences the orientation (Drake 1974). In addition, the fabric can be reorganized after deposition if the till is saturated and the overlying ice exerts considerable pressure (MacClintock and Dreimanis 1964; Evenson 1971; Ramsden and Westgate 1971). In till that has not been squeezed after its deposition, the clasts apparently take on their dominant orientation while still being transported as englacial load, and the orientation therefore should reflect the direction of glacier movement. Deposition only partially changes the englacially derived fabric, the degree of reorganization depending on many factors (Lindsay 1970; Clague 1975).

Ablation till is usually somewhat washed, with the fines being winnowed from the mass. It thus tends to be more coarse-grained, but generalizations are dangerous because some ablation tills contain substantial amounts of silt and clay. The material in ablation till is usually less compact than in lodgement till, and the fabric tends to be less well oriented. A series of flow tills may also be contorted or even stand at steeply dipping angles produced when the material slides or flows downslope.

Stratified Drift   The second major category of glacial drift is distinguished by the fact that the sediment was transported by moving water before its final deposition, thereby acquiring a degree of stratification not normally seen in tills. Such drift is often referred to as *fluvioglacial* because running water is involved in its origin, even though the water may not always be confined in discrete channels. Fluvioglacial deposits are also distinguished from till in that they are usually sorted and the clasts contained in the mass are more rounded. However, the demarcation between some types of fluvioglacial deposits and thoroughly washed ablation tills is a matter of degree rather than substance. Exactly where the line between the two is drawn becomes somewhat arbitrary. Highly saturated flow-till, for example, might move in a nearly fluvial manner.

The layering and sorting in a fluvioglacial deposit depends on precisely where it is formed with respect to the ice that provides the transporting meltwater. Sorting is also partly a function of the energy possessed by the meltwater, the distance of transport, and the continuity of the sorting process. Because the discharge of meltwater is notoriously inconstant, varying drastically with time of day, local climate, and the characteristics of the ice, significant differences in the sedimentology of fluvioglacial deposits can be noted over short distances. These are especially evident where deposits are formed in contact with the ice and the free circulation of meltwater is restricted (see Shaw 1972 for sedimentary characteristics). If debris is transported away from the glacier terminus, the sedimentary characteristics tend to vary more regularly.

We should stress again that the prime requisite of stratified drift is transport by water, much of which is released from melting ice. However, this puts no constraints on the environment in which the sediment is deposited. Fluvioglacial debris can come to rest in stream channels, floodplains, lake or ocean floors, deltaic plains, in contact with ice, or in any other place where sediment-laden running water loses its transporting energy. In some cases sedimentologic properties in the deposit, such as the degree of roundness and mean grain size, can help identify the depositional environment (King and Buckley 1968), but normally there are too many variables in the system to rely on these criteria alone (R. J. Price 1973). Detailed field study is almost always necessary to reach a firm conclusion about environments of deposition.

Deposits that originate in contact with the ice often contain interbedded bodies of ablation till, and their particles tend to be less well rounded and sorted because of the limited distance of transport. The characteristics of such a deposit often vary from the bottom of the sequence to the top because the environment of deposition in contact with the ice repeatedly changes with time, especially if the ice is stagnating. In any deposit, then, a particular layer may be superseded

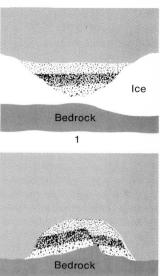

Figure 10.13. (A) Gravel pit cut in kame, south central New York, shows stratification in kame deposits. (Photo by J.H. Moss) (B) Structures and deformation of strata in ice-contact stratified drift develop as ice melts and debris collapses and is lowered onto bedrock floor.

vertically by one with different sedimentary properties that reflect a new depositional environment. The stratigraphy is complicated by the fact that much of the drift is physically supported by ice during its deposition, and when the ice dissipates, the support is removed and the sediment collapses. Such a process, as figure 10.13 shows, leads to flexures in some of the layers, minor faults, and beds dipping at angles well beyond the angle of repose for such material. Overall, the ice-contact setting at places produces an interconnected maze of stratified and nonstratified drift in which every conceivable process and environment is possible and probably has been present at some time during the depositional history. Moreover, the manner in which the ice ablates influences the resulting deposit. Drift produced while the ice is still active may be quite different, especially in distribution, from that derived from a large mass of stagnant ice which is simply downwasting as it melts and is not influenced by internal glacial movement.

Sediment deposited beyond the terminal margin of the ice is formed in the *proglacial* environment and is often referred to as *outwash*. Outwash is usually well sorted and normally consists of rounded sands and gravel representing bedload carried and deposited in stream channels. Silt and clay are usually carried as suspended load and are commonly removed from the system unless, as in the lower Mississippi River valley, the transport distance is so great that some of the outwash is silty in texture. It is important to understand that streams transporting outwash do not usually head at the glacier terminus but begin on top of or within the ice, well upglacier from the margin. Proglacial features and deposits often can be traced into and through the maze of ice-contact deposits, increasing the complexity of the depositional sequence developed near the ice margin. Outwash is less well sorted, more indistinctly stratified, and much thicker near the ice terminus than it is downstream. Away from the ice margin, outwash may be delivered into a variety of sedimentary environments. Drift deposited in bodies of standing water or in deltaic systems has all the characteristics of normal fluvial sediment accumulated in those settings.

The Depositional Framework

Before considering what geomorphic features might be produced during deposition, we must first establish a realistic framework within which we can give some order to the subtle variations of depositional features associated with glaciation. This is done in table 10.1. From our brief look at the nature of drift, it is obvious that both stratified and nonstratified deposits can be formed in contact with the ice. They differ only in whether the ice alone was the primary transporting and depositing agent or whether a stage of water transport intervened between the release of particles from the ice and their final deposition. As a result, the *ice-contact environment* must be considered as one of the major settings of our depositional framework. The ice-contact environment can be subdivided geographically into *marginal* and *interior* zones depending on where the drift was originally deposited (table 10.1).

Features formed in the ice-contact environment can be composed of either stratified or nonstratified drift (till or fluvioglacial sediment) or a combination of both types. The region beyond the terminal edge of the glacier is classified as the *proglacial* environment; in contrast to the ice-contact setting, features formed there are composed entirely of fluvioglacial sediments (outwash).

The distinction of marginal and interior zones in the ice-contact environment is problematical because glacier margins migrate forward and backward

Table 10.1 The depositional framework and associated features

Setting	Features	Type of drift
Ice contact		
Marginal	End moraines	Till and fluvioglacial
	Kames and kame terraces	Fluvioglacial
	Kettle holes	Fluvioglacial
	Eskers	Fluvioglacial
Interior	Medial and interlobate moraines	Till
	Ground moraines	Lodgement till
	Fluted surfaces	Lodgement till
	Drumlins	Lodgement till
Proglacial environment	Sandar	Fluvioglacial (outwash)
	Kettled sandar*	Fluvioglacial (outwash)

*May merge with marginal environment.

with time according to the mass budget. An active glacier with a negative mass balance should have a terminal margin that is progressively receding toward its source; marginal features will be formed in regions that were interior when the ice was at its greatest extent, and deposits of the two zones may be complexly intertwined. The marginal zone is therefore somewhat difficult to define because various types of glaciers ablate in different ways. Some stagnate over wide areas while others dissipate in a narrow zone surrounding a well-defined, active, though receding, ice front. In this discussion the ice margin consists of the stagnant ice standing in front of the active glacier zone, but also includes the area within the active zone where end moraines are developed.

The utility of the classification is shown in the map in figure 10.14. In this region the terminal moraine, marking the ice-contact marginal zone, is charac-

Figure 10.14. Map view of the depositional framework after ice has disappeared. Part of the Whitewater, Wisconsin, guadrangle (U.S.G.S. 15′). Marginal zone contains terminal moraine and ice-contact stratified features, and interior zone is characterized by ground moraine. Proglacial zone is a large outwash plain or sandur.

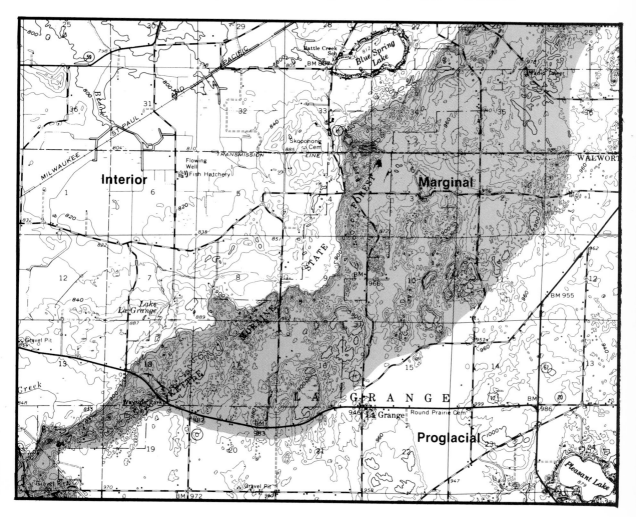

terized by a maze of small hills and depressions (kames and kettles) that distinguish deposits formed near the boundary between active and stagnating ice. North of the moraine the relatively low, flat area is underlain by material that was deposited beneath the active glacier (ground moraine); it represents the ice-contact interior zone. South of the moraine, the proglacial zone is marked as a plain composed of debris (outwash) transported and deposited by meltwater streams heading within or on top of the ice.

The framework classification fails to accommodate those features that originate in more than one environment or pass from one setting into another without changing form. Eskers, for example, may originate in tunnels beneath or within the ice or in channels on the ice surface. In stagnating ice masses, all these passageways might be interconnected in a glacial "karst topography" that includes sinkholes, sinking streams, blind valleys, moulins (nearly vertical cylindrical holes), and the like (Clayton 1964). As the ice finally disappears and the deposits are gradually let down onto the ground surface, they may form a continuous or partly segmented ridge that provides little evidence about the exact position of the material when it was originally accumulated. Nonetheless, a geographic classification related to the ice terminus still seems to satisfy our needs best.

Marginal Ice-Contact Features

Moraines   The term *moraine* originated several hundred years ago as a local name for ridges of debris found at the edges of glaciers in the French Alps. Since then many definitions have appeared, but for our purposes we can think of a moraine as a depositional feature whose form is independent of the subjacent topography and which is constructed by the accumulation of drift, most of which is ice-deposited (Flint 1971). A precise morphological definition is not possible since, as table 10.2 shows, moraines take many different forms and have a variety of dimensions. The term "moraine" is not synonymous with till, as many have suggested, but in reality refers to a suite of topographic forms on which the only restriction is that they must be composed of drift (and even that rule has been violated in some cases).

Figure 10.15. (A) Terminal moraine of Pinedale glaciation, East Rosebud valley near Roscoe, Mont. Low ridge above road in left center of photo is lateral moraine merging with terminal zone. (Photo by R.R. Dutcher) (B) Recently formed lateral moraine in Rocky Mountains of Canada. (Photo by J.H. Moss)

The most spectacular moraines develop at or near the edges of active glaciers and so are designated as *end moraines*. The end moraine constructed at the downstream edge of the ice at the farthest point of advance is called a *terminal moraine*. In valley glacier systems, it merges imperceptibly into *lateral moraines* on both sides of the valley because ablating ice deposits debris along the glacier's lateral extremities as well as at its terminus (fig. 10.15). Ice sheets form terminal moraines that tend to be long (often hundreds of kilometers), linear, topographic highs marking the forward boundary of the ice, but lateral moraines are absent because there is no side to the ice sheet. Where ice sheet movement was distinctly lobate, however, end moraines called *interlobate moraines* (Flint 1971) may develop along the junction of two lobes. These are

Table 10.2 Moraine types

End Moraines	Moraines produced at front or sides of an actively flowing glacier.[1]
Terminal moraines	Mark the farthest advance of an important glacial episode.
Lateral moraines	Deposited at or near the side margin of a mountain glacier.
Recessional moraines	Formed at glacier front during temporary halt or readvance of ice in a period of general recession.
Ground Moraine	Gently rolling surface formed of debris released from beneath the ice.
Interior and Minor Varieties	
Washboard moraines	Small, parallel ridges oriented transverse to direction of ice movement. Also called moraine ridges or cross-valley moraines.
Interlobate moraines	Formed at junction of two ice lobes.
Medial moraines	Elongate ridge developed at junction of two coalescing valley glaciers.

1. Lateral moraines may be excluded by some geomorphologists because end moraines are commonly considered only as topographic features developed at the front of a glacier.

somewhat similar in origin to *medial moraines,* which form when two valley glaciers coalesce.

Ideally, end moraines assume a rather narrow, ridgelike shape, but actually the form and size depend directly on the amount of glacial load, the mass budget, and the volume of meltwater circulating in the system. Temperate valley glaciers tend to build higher and more massive terminal moraines (some reaching 300 m in height) because these glaciers have higher flow velocities, loads, and total budgets. Ice sheets seem to generate moraines that are less dramatic in size, seldom exceeding 50 m in height. The position of a glacier front depends on the nature of the net budget, however, and so a ridgelike terminal moraine can form only when a zero net budget obtains for some time, allowing the ice front to remain stationary during the accumulation of drift. In most glaciations the ice front migrates over a wide zone as it responds to fluctuations in the mass balance. The receding and advancing ice front changes the moraine configuration from a ridge to a belt of complex topography, often several kilometers wide, which represents the limits of the frontal migration.

Within the morainal belts the locus of deposition shifts periodically, giving rise to ridges, mounds, and depressions of varying size. The ridges in the sequence, usually small and composed of till, commonly parallel the orientation of the ice front. They have been given many names, such as *cross-valley moraines* (Andrews 1963; Andrews and Smithson 1966) and *washboard moraines* or *moraine ridges* (R. J. Price 1970), and several theories have been suggested for their origin (Elson 1968). The ridges are usually segmented or interwoven with many deposits of stratified drift, giving the entire topography a chaotic expression rather than a regular undulation of ridge crests and intervening sags.

Several processes are responsible for the construction of end moraines and the minor ridges that are commonly included within the morainal belts. R. J. Price (1973) distinguishes these processes as dumping, squeezing, and pushing, but much of the difference is closely allied to the vertical distribution of the drift when it is released from the ice. The mechanics of the *dumping* process were first described in detail by Goldthwait (1951), who observed that debris in the Barnes ice cap is contained in a series of shear planes near the terminus of the ice (fig. 10.16). The sheared zone extends only approximately 150 m from the snout, and the attitude of the planes suggest they cannot penetrate the ice to depths greater than 75 m. This indicates that, if the debris is brought surfaceward from the basal ice layers, the process occurs only in the outer 500 m of the glacier.

Figure 10.16 shows that as the ice fringe thins by ablation, dirt contained in the shear cracks is released, creating a film of sediment on the ice surface which Goldthwait refers to as "black ice." Some of the debris moves downslope, as described earlier, where it blankets the lower ice up to a thickness of a meter, thereby retarding the ablation process. Black ice with only a thin dirt cover ablates more rapidly than the white ice upglacier from the sheared zone, and a trough is produced parallel to the ice front that separates the active black ice zone from ice-cored debris at the front edge of the glacier. The final hummocky surface, so characteristic of moraines, results from debris sliding off the ice-cored ridge, debris collapsing as the underlying ice melts, and from fluvioglacial action.

The shear-plane hypothesis is attractive when the outer fringe of the ice is stagnant. Bishop (1957) suggested that in the Greenland ice sheet near Thule, active ice flowing from upglacier was physically impeded when it encountered the stagnant ice zone, producing a series of shears that thrust basal debris to the surface (fig. 10.17). Ablation would result in the surface cover as observed by Goldthwait. Weertman (1961), however, contested the shearing mechanism as the manner by which drift is lifted to the surface in the Greenland ice. His greatest concern was that some of the debris layers, presumed to be encased in shear cracks, were more than one meter thick and consisted of thoroughly disseminated dirt. The dispersal of particles within the debris layers does not fit the

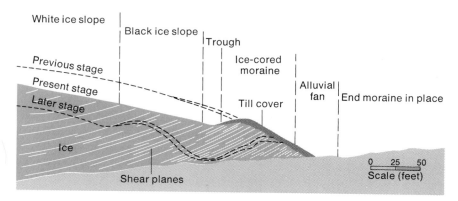

Figure 10.16. Diagram showing retreating margin of Barnes ice cap. Material moves up inclined shear planes and accumulates as till cover on core of ice. (After Goldthwait 1951, fig. 2. Used with permission of The University of Chicago Press)

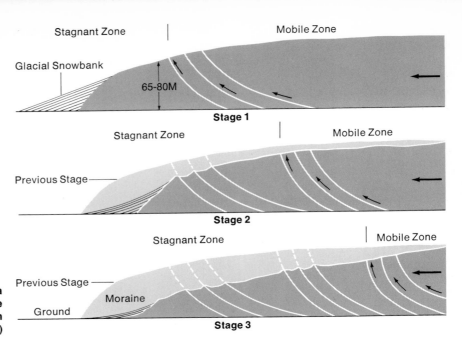

Stagnant Zone | Mobile Zone

Glacial Snowbank

65-80M

Stage 1

Stagnant Zone | Mobile Zone

Previous Stage

Stage 2

Stagnant Zone | Mobile Zone

Previous Stage

Moraine

Ground

Stage 3

Figure 10.17. Formation of a series of moraines by the shear hypothesis. (From Bishop 1957)

concept that the material was picked up by subglacial shearing, nor does it encourage one to believe that the debris layers are located in well-defined shear fractures. Weertman concluded that the disseminated nature of the dirty layers could only be due to freezing of water at the base of the ice. Boulton (1967) accepted the possibility that both shearing and freezing were responsible for creating debris bands in the glaciers of Vestspitsbergen in the Arctic Ocean and that the layers could be brought surfaceward along the shear fractures or along normal flow lines that consistently intersect the ice surface near the terminus.

Regardless of the precise mechanics by which load rises to a glacier surface, the evolution of topography in a morainal belt proceeds in a complex manner depending on (1) the character of the ice, (2) the spacing of the debris bands as they emerge at the surface, and (3) the thickness of the supraglacial cover. These factors produce differential ablation rates and widespread zones of drift that are held up by a core of solid ice. As these ice cores melt at different rates, the surface and near-surface till is progressively shifted into an uneven distribution, resulting in the typical chaotic configuration noted in morainic zones.

The development of ice-cored drift seems to be an integral component in the evolution of morainic belts, and many workers accept ice-cored ridges as a separate type of end moraine. Østrem (1964), however, found that *ice-cored moraines* do not necessarily contain glacier ice in their cores, but that many consist of firn banks. The firn banks were packed in against the glacier snout or stagnant ice and later overridden by an advancing glacier and buried by its drift.

Thus, even though they are a type of end moraine, some of these features develop on ice that was not truly part of the glacier system.

In Arctic glaciers, the englacial debris bands commonly are distributed in zones parallel to the ice front and separated by zones that are relatively free of sediment. Such a spacing of debris leads to the formation of a series of ice-cored ridges and intervening valleys, which are called "controlled" disintegration forms (Boulton 1971, 1972b) because they reflect structures within the ice. If the debris is spread unevenly through the ice, the thickness of supraglacial till becomes extremely uneven, creating a hummocky surface topography that Boulton refers to as "uncontrolled."

The rate at which ice cores melt is another extremely important determinant of the final morainic topography. Slow melting allows till to flow into and fill hollows and trenches interspersed between the ice-cored ridges and mounds. When the ice finally disappears, the topography is inverted so that the former depressions, now filled with considerable thickness of drift, stand as hummocks or ridges (fig. 10.18). Rapid melting, on the other hand, creates extremely fluid supraglacial till that flows down even the gentlest slope and spreads easily. It produces a surface of low relief with few irregularities and a subdued morainic topography. Finally, dumped moraines may have a complicated intermixing of till and fluvioglacial deposits, especially if the original ice-cored ridges restrict the outward flow of meltwater. In that case, lakes may form within the morainal belt, streams assume sinuous courses around ice-cored barriers, and the till flowing off the ridges mixes with any number of fluvioglacial deposits (Boulton 1972b).

Figure 10.18. Hummocky topography in terminal moraine. East Rosebud valley at front of Beartooth Mountains near Roscoe, Mont. (Photo by R.R. Dutcher)

In contrast to moraines formed by the dumping process, in which till is formed within or on top of the ice, moraine ridges are developed by *squeezing* drift originally deposited beneath the ice. Ridges generated by squeezing are usually smaller than dumped moraines, normally standing less than 10 m high, although they can be higher where the subglacial till is highly saturated. Theoretically, the process involves the response of water-soaked lodgement till to pressure exerted by the weight of the overlying glacier. The till will move from under the ice and emerge along the ice front, or it will be squeezed into zones of low pressure within the ice, represented by crevasse openings (R. J. Price 1970; Mickelson and Berkson 1974). In either case the ridges seem to be typified by (1) steeper distal slopes than proximal slopes, (2) a plan-view outline consisting of linked arcuate segments that parallel the ice front, and (3) a till fabric with pebbles oriented perpendicular (or nearly so) to the ridge crests, though occasionally oriented parallel to the trend of the ridge (Mickelson and Berkson 1974).

R. J. Price (1970) examined a series of small morainal ridges in Iceland that formed during the last 70 years as the ice retreated from the main, and larger, terminal moraine. The ridges apparently rise quickly (in one area 15 ridges formed in 16 years), prompting Price to suggest that they develop annually when summer meltwater descends to the glacier base and saturates the subglacial till (a proposal made earlier by Andrews and Smithson 1966). Squeezed ridges also seem to originate beneath bodies of standing water (Mickelson and Berkson 1974); since saturation of the debris would be rather complete in such an environment, this adds credence to the proposed mechanism. Hydrologic processes, however, may be involved in the origin of sublacustrine or submarine ridge development (Barnett and Holdsworth 1974).

A third mechanism capable of forming a moraine ridge is the collision of advancing ice with older deposits, which deform them into a ridgelike feature. Such end moraines, called *push moraines,* are only partly composed of sediment carried by the ice and may in fact include blocks of older rocks of a nonglacial origin. Kaye (1964b), for example, describes a Late Illinoian moraine on Martha's Vineyard (Massachusetts) as being composed of Cretaceous and Miocene deposits as well as pre-Illinoian drift. Moraines of this type can be of dramatic size, sometimes more than 100 m high, several kilometers wide, and 30 km long (Okko 1955).

Push moraines usually have a steep distal slope, and they often display a sharp break in slope at the contact between the moraine and the proglacial deposits (Embleton and King 1975a). They are more distinctive, however, in their internal structure, as in figure 10.19. Here individual layers, sheared away from the ground lying in front of the ice, are intensely deformed into large (sometimes overturned) folds and faults of all kinds (Mills and Wells 1974). Thrusting, with plates as thick as 30 m, is not uncommon and may occur as imbricate slices or even as an underthrust phenomenon (Kaye 1964a,b). Thrusting may also bring

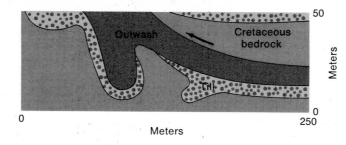

Figure 10.19. Simplified cross-section of folding and faulting in glacial deposits on Long Island, N.Y., caused by ice-shove deformation. (Adapted from Mills and Wells 1974. Used with permission of the Geological Society of America)

basal lodgement till from behind the moraine and place it on top of the moraine itself (Okko 1955). The exact style of internal deformation depends on the rigidity of the bedrock being shoved by the ice front, or on how solidly unconsolidated material is held together by freezing around the ice perimeter.

Stratified Marginal Features   As pointed out earlier, many glaciers are characterized by marginal zones of extensive areas of thin, stagnating ice. Within these zones a suite of genetically related ice-contact features is developed, composed predominantly of stratified drift and morphologically distinct from the moraines just described. These features form by deposition of drift (1) where water flows through openings in and beneath the ice, or in ice-surface channels, (2) in spaces between the ice and the bedrock of the valley sides, and (3) where disseminated sediment is passively concentrated by melting of the encasing ice.

Many of the channelways for the flowing water originate when stagnating ice breaks into individual segments along planes of structural weakness in the ice, and so the features are genetically related to many of the ice-disintegration forms described by Gravenor and Kupsch (1959), Stalker (1960), Clayton (1967), Parizek (1969), and many others. Our discussion, however, is restricted to those features that consist primarily of fluvioglacial drift, even though ablation till may be a minor ingredient. Their morphology is entirely constructional, although they can be affected by slumping as the ice walls that supported the drift during its deposition are melted away.

Kames and Kettles   *Kames* are moundlike hills of layered sand and gravel that vary in size from minor swells to conical protuberances standing up to 50 m high and extending 400 m along their base. Kame material can accumulate at the ice-substratum interface, and also in cavities located within stagnating ice or on its surface. Englacial or supraglacial debris can be lowered onto the ground surface as the ice dissipates (Cooke 1946b; Holmes 1947). The moundlike shape results only if the accumulate was originally isolated and nonlinear (fig. 10.13), because sediment deposited in linear openings is ultimately converted to ridges rather than mounds. In addition, some kames form when debris collects as fans or small deltas built against the ice or outward from the ice, with the apex resting at the stagnant margin. In either case, melting of the supporting ice allows the drift to collapse into a kame, a process evidenced by slumped strata within the deposit.

Kames are only one of many forms with essentially the same origin. They are transitional into eskers or minor ridge and circular forms that Parizek (1969) calls ice-contact rings and ridges, or into other features whose names utilize the term *kame* as a descriptive adjective, i.e., kame delta, kame moraine, etc. Perhaps the most common of the latter type is a *kame terrace*. These originate from drift deposited in narrow lakes or stream channels between the valley side and the lateral edge of the stagnating ice. When the supportive ice disappears, the inner edge of the deposit collapses into the terrace scarp. Kame terraces differ from normal river terraces in that they are restricted in their longitudinal extent and are usually narrow; the tread may slope gently into the valley, and the surface may be dimpled by kettle holes (McKenzie 1969).

Kettle holes are circular depressions that are formed in a variety of ways (Fuller 1914), most commonly by the burial of isolated blocks of ice by stratified drift. The gradual ablation of the ice leads to a gentle downward flexing of the layers as they settle over the dissipating mass. Some kettles are almost 50 m deep and up to 13 km in diameter (Flint 1971), but these giants are exceptions to the normal kettle size of less than 8 m deep and 2 km wide. Kettles usually form in association with kames and other related features, producing an irregular surface described as "kame-and-kettle topography," but similar surfaces can develop in the absence of either or both of these features.

Eskers   The term *esker,* evidently stemming from the Gaelic word for "crooked" or "winding," has been applied to a wide variety of ridged ice-contact features. Eskers range in shape from the single, narrow, sinuous ridge that is the classic form (fig. 10.20) to a complexly intertwined maze of branching and joining ridges. In some cases the ridges may be looped or circular and possibly could be cored by ice-pressed till (for a discussion see Parizek 1969). Here we refer only to those ridges composed of fluvioglacial drift. The type eskers seem to form most commonly in stagnating margins of large ice sheets where the underlying surface is broad and relatively flat. The ridges are not necessarily continuous but may consist of crudely connected linear segments. In broad valleys the eskers usually parallel the slope of the valley floor, but this is not always the case. Some ridges, for example, ascend the valley sides, transect the divide and descend the flanks of the adjacent valley.

Eskers show dramatic inconsistencies in dimension. They range from 2 m to more than 200 m in height, from several meters to as much as 3 km in width, and from tens of meters up to 500 km in length (if gaps are considered in the total distance). In cross-profile, they usually have rather sharp crests and steeply inclined sides (up to 30°), but broad eskers can maintain rather flat upper surfaces

Figure 10.20. Esker in Cheboygan County, Mich. Narrow, curving ridge (crossing area from upper right to lower left) is steep-sided esker approximately 10-20 m high. (Photo by the U.S. Geological Survey)

or may be pitted by kettle development. R. J. Price (1973) suggests that the height and width of eskers may be directly related to their overall length, longer ridges being proportionately higher and wider than shorter ones.

Eskers are constructed with stratified, often cross-bedded, sands and gravels. As in all ice-contact features composed of fluvioglacial drift, slumping of the layers is prevalent along the sides where the supporting ice melted. Along the edges, therefore, the beds may dip parallel to the angle of the side slopes, giving the esker a pseudo-anticlinal structure when viewed in cross section. Gravel included in the drift is usually derived from local sources, rarely being transported more than 10 km (Lee 1965; Flint 1971; Price 1973). The local source is not bedrock but any material susceptible to reworking by meltwater, such as widely spread till (Hellaakoski 1931; Price 1966) or even surface debris sapped from medial moraines (Price 1969).

It seems certain that eskers result from sediment accumulation in a variety of openings, such as (1) ice channel fillings (crevasse fillings, fillings between stagnant blocks, etc.), (2) tunnels beneath or within the ice, (3) supraglacial channels or even, in rare cases, (4) narrow longitudinal embayments of the ice front. The subglacial origin is made possible by meltwater that descends from the ice surface to the base through moulins and crevasses. Along the subglacial surface a network of interconnected tunnels passes water and sediment toward the terminus, and normal fluvial deposition fills or partially fills the openings. Filling probably takes place during the last stage of deglaciation when the ice is stagnant and thin, for basal tunnels could not remain open if the glacier were still moving or if the ice were thick enough to cause pressure flow (Flint 1971). It is difficult to imagine, however, a thin, stagnant margin that extends for hundreds of kilometers at any given moment, and so perhaps very long eskers are time-dependent features. In that sense, the downstream end of the esker is formed first, and the site of deposition migrates up-ice as the stagnant margin progressively recedes during deglaciation.

Researchers also have repeatedly suggested that esker sediment can accumulate in englacial tunnels, in supraglacial channels, or in crevasses. The fact that eskers do not always conform to the underlying topography was recognized years ago (Crosby 1902) and used as evidence that englacial or supraglacial deposits were passively let down on the underlying ground surface to produce the eskerine form. Flint (1971) has argued that such an interpretation is not demanded because (1) where eskers cross valley divides they pass through the lowest part of the divide, and (2) the preservation of the internal layering and the external form is not logical, considering the magnitude of the settling process, unless special conditions prevail. It should also be emphasized that subglacial water may be under enough hydrostatic pressure to flow uphill, thereby negating the requirement that eskers formed subglacially must follow the subjacent topography.

Notwithstanding the above arguments, field studies and photographic analyses of esker formation by the Casement Glacier in Alaska (Price 1966; Petrie and Price 1966) and the Breidamerkurjökull Glacier in Iceland (Price 1969) provide rather conclusive evidence that englacial and supraglacial deposits can be transformed into eskers. In those areas, some of the eskers are definitely ice-cored and were measurably lowered in elevation during the period 1948–1963 by wastage of the buried ice. Presumably the final melting of the ice will rest the esker on the original subglacial floor.

Finally, some eskers or eskerlike ridges can originate in the marginal environment (Price 1973). Along the ice margin, crevasse openings perpendicular to the front may be occupied by meltwater and filled with debris coming from the ice. Subsequent melting of the supporting ice leaves an esker connected directly to the outwash in front of the ice margin. It is also possible for ridgelike topography to develop in the proglacial environment when linearly oriented ice blocks are buried by outwash and subsequently ablated.

Interior Ice-Contact Features

Behind the marginal zone in the interior portion of a glacier, the predominant features are deposited at the base of the ice. There pressure from the overlying ice either spreads till rather evenly across the ground surface or molds water-soaked till already in that position into distinct morphologic shapes. Supraglacial drift is somewhat rare in the interior zone, but where two glaciers join, the debris dragged along their lateral edges may coalesce into medial moraines. Although these moraines appear as striking linear belts on the surface of the ice (fig. 10.7), they are superficial in that the deposits are shallow. Medial moraines are rarely preserved on the ground surface because they are let down in the middle portions of valleys where meltwater streams are likely to destroy them. The dominant forms are those deposited from beneath the ice such as ground moraines, drumlins, and fluted surfaces.

Ground Moraine   In contrast to end moraines, *ground moraine* is distinguished by its apparent lack of topographic expression. It is accepted as a moraine despite its low relief and a complete absence of transverse ridges (Flint 1971) because its surface expression is independent of the topography it covers. Ground moraine occupies much of the surface in North America and Europe that was covered by major Pleistocene ice sheets. The moraines usually exist as smoothly undulating plains, like that in figure 10.21, seldom exceeding 10 m in total relief; they range in size from small areas interspersed among younger marginal features to regions covering thousands of square kilometers behind the terminal moraine.

Although in most cases the primary building material of ground moraine is lodgement till, it is a mistake to consider ground moraine and lodgement till as

Figure 10.21. Flat till and
loess plain in southern
Illinois. Gently undulating
topography on ground
moraine has been smoothed
by influx of younger loess.
(Photo by C. William Horrell)

synonymous terms. The deposits from which ground moraine is constructed also include ablation till and interbeds of fluvioglacial deposits that originate in the same glacial advance (R. P. Kirkby 1969; Boulton 1972a,b) or in more than one episode of glaciation.

Fluted Surfaces and Drumlins   The monotony of ground moraine topography is sometimes broken by ridges or elongate hills that manifest the mechanics functioning at the base of glaciers, especially in temperate ice sheets. The most common features developed in the subglacial milieu, *fluted surfaces* and *drumlins,* appear to be oriented by moving ice; Flint (1971) refers to them as streamlined molded forms. The two forms, however, may develop in different ways, and although they might be gradational (Gravenor 1953; Lemke 1958), they are related only by their common subglacial origin and their orientation parallel to the direction of ice movement.

Figure 10.22. Fluted surface east of Havre, Mont. (Photo by U.S. Geological Survey)

Fluted surfaces (fig. 10.22) consist of narrow, regularly spaced, parallel ridges and grooves. The ridges are normally less than 5 m high and several hundred meters long, although individual grooves or ridges may be considerably

larger. Lemke (1958) reports ridges in North Dakota that are 10 m high and 10 km long, although he suggests they may be narrow, linear drumlins. Small ridges are usually composed of till, and their origin may be related to pressure-squeezing of saturated debris into longitudinal cavities at the base of the ice, a process similar to the formation of the minor moraine ridges discussed earlier. There is no question that till is mobilized and arranged in ridges where cavities open on the lee side of large boulders (Hoppe and Schytt 1953), but it seems unlikely that all fluted surfaces develop in this manner. The spacing between ridges is too regular to be dependent on the coincidence of equidistant boulders beneath the ice, and extremely long ridges would require that the cavities remain open far down-ice from the obstruction, an unlikely happening under thick ice.

Some larger ridges are composed of material other than till (Lemke 1958; Gravenor and Meneley 1958), and the surfaces between adjacent ridges are often noticeably grooved. These characteristics have led many workers to believe that fluted surfaces are both erosional and depositional in origin. Gravenor and Meneley (1958) suggest that alternating high and low pressure zones at the base of the ice produce the unique fluted surface. In their model (fig. 10.23), grooving occurs in the material beneath the high pressure zones, and debris eroded from there is moved not only downglacier but also upward into regions of low pressure. The ice between adjacent grooves becomes overloaded with debris, and till is deposited on a core of older and undisturbed sediment or rock. The model proposed by Gravenor and Meneley explains many field observations that cannot be accounted for by deposition alone, but no data as yet substantiate alternating high and low pressure zones in ice, especially with such a regular and closely spaced distribution.

Drumlins also are elongated parallel to the direction of ice flow, their long axes deviating only slightly from the average trend of the glacier movement (fig. 10.24). These distinctive forms have received attention from a large number of geomorphologists during the last century, but as yet their origin has not been fully explained (see Muller 1974 for an excellent review). Drumlins have been described as having a plan-view shape that is similar to a lemniscate loop (Chorley 1959b) or an ellipsoid (Reed et al. 1962) and in long profile a form that Flint

Figure 10.23. Hypothetical cross section at base of a glacier showing pressure distribution in the origin of fluted surfaces. (After Gravenor and Meneley 1958. Used with permission of the *American Journal of Science*)

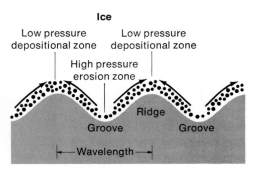

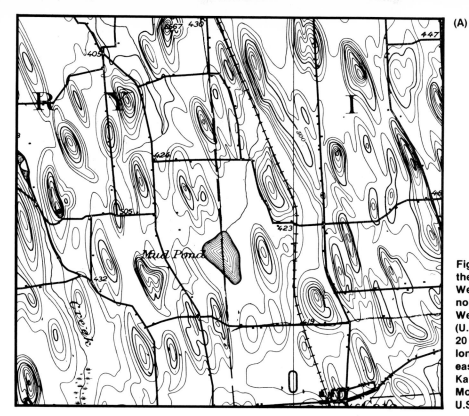

Figure 10.24. (A) A portion of the drumlin field located near Weedsport, N.Y. From the northeast quarter of the Weedsport, N.Y., quadrangle (U.S.G.S. 15'). Contour interval 20 feet. (B) Drumlin shown in longitudinal profile. Looking east, four miles northwest of Kalispell, Flathead County, Mont. (Photo by W.C. Alden, U.S. Geological Survey)

(B)

(1971) likens to an inverted bowl of a spoon (fig. 10.24). The exact shape, however, is probably variable enough that no particular model will fit all drumlins. Trenhaile (1971), for example, found that the ellipsoidal shape fit only about one-third of the drumlins he studied. The lemniscate loop seems to be the shape that provides the least resistance to flow (Chorley 1959b) and therefore probably has some theoretical advantage if drumlins are to be considered as streamlined forms (Muller 1974). In any case, it can be stated with some assurance that drumlins are higher and wider near their rear edges and that they narrow and thin downstream until they merge imperceptibly with the surrounding surface. Drumlins average in size from 1 to 2 km in length and from 400 to 600 m in width, and stand anywhere from 5 to 50 m high; individuals can be smaller or larger. Their length-width ratio seems to be reasonably consistent, ranging from 2 to 3.5 (Vernon 1966; Reed et al. 1962; Trenhaile 1971).

Drumlins display a variable internal composition; many are fabricated entirely from clay-rich till, but others have obvious cores of solid rock or preexisting drift which may or may not be stratified. Where lodgement till is present, pebble orientations tend to parallel the long axis of the feature (Wright 1957; Hoppe 1959), but the degree of the parallelism depends on where the fabric is measured. Andrews and King (1968) found that fabrics within a single drumlin can vary by as much as 90°. In the basal zones of the till the pebbles are most closely aligned with the orientation of the drumlin, but closer to the surface the fabric deviates more from the axis. This suggests that drumlin growth is accompanied by increased deflection of the ice around a basal obstruction, and studies in New York seem to substantiate that interpretation (Muller 1974). There is also evidence to suggest that pebble fabrics in drumlins differ from those in normal ground moraine deposits. Both Wright (1957) and Hill (1971) found that elongate pebbles plunge at higher average angles in drumlins and that the percentage of pebbles plunging at <10° was notably less in drumlins than in ground moraine deposits.

In addition to shape, orientation, and internal characteristics, certain distributional properties of drumlins are critical in any hypothesis concerning their origin. First, drumlins rarely exist as individuals but instead cluster together in fields that are commonly wider than most morainal belts (Gravenor 1953). The density of drumlins within a field seems to be inversely related to their sizes; that is, very dense clusters are composed of relatively small drumlins (Doornkamp and King 1971). Second, many studies show a strong probability that drumlins within any field are spaced in a nonrandom manner; that is, spacing between neighboring individuals is somewhat regular (Reed, et al. 1962; Vernon 1966; Smalley and Unwin 1968; Trenhaile 1971). Third, most glaciated areas have no drumlin development. Fourth, drumlin fields seem to be located in zones close to but behind the terminal moraines that mark the limit of a particular glaciation.

The models used to explain drumlin genesis fall into two main groups: (1) drumlins are erosional features developed when moving ice streamlines preexisting drift or rock, or (2) drumlins are depositional features formed when a moving glacier deposits till and molds the material as the ice continues its forward motion. There is reasonable evidence to support both theories. For example, when the internal composition of drumlins is stratified drift or bedrock that pre-dates the ice advance in which the drumlins were formed, it is rather difficult to ignore erosional processes in the drumlin origin (Gravenor 1953; Kupsch 1955; Lemke 1958; Whittecar and Mickelson 1976). It is true that some of the cored drumlins are thinly covered by till and perhaps the drumlinoid form is related to a final depositional event, but such an explanation cannot account for all such instances. Vernon (1966) suggests that irregular topography on older drift or bedrock might produce obstacles to flow which cause deposition on top of the older material. Such a mechanism is attractive because it explains the role of cores in drumlin formation, but it requires a selective use of available obstacles to account for the regular spacing in drumlin fields. In addition, many drumlins have no cores of older material but consist entirely of lodgement till of the same age as the surrounding drift. They often display an internal concentric banding that suggests some process of gradual accretion (Hill 1971). Drumlins such as these can hardly be doubted as being depositional, even though, as Muller (1974) points out, uncored drumlins pose a problem as to what might initiate the deposition.

In light of all this, we should probably accept a multiple origin for drumlins, as Embleton and King (1975a) suggest, and not pay undue attention to hypotheses that utilize one process to the exclusion of all others. Yet it is bothersome that drumlins have such unique spatial properties. It seems, somehow, that any explanation of drumlin development, whether by erosion or deposition, should have its roots in the mechanics of glacier flow and the character of the glacial load. One attempt to integrate all drumlins into a single genetic model (Smalley 1966; Smalley and Unwin 1968) is based on the fact that most tills have *dilatant* characteristics. In simplest terms, dilatant materials increase in volume under stress. As stress increases on till at rest, the debris will resist deformation until the stress produces the loose packing attained in the dilatant condition. As the curve in figure 10.25 shows, when dilatancy prevails, deformation proceeds more easily and will continue even if stress is decreasing, until finally the stress becomes too low to maintain the dilatant property. Conceivably then, dilatancy occurs only within certain well-defined upper and lower stress limits.

Smalley and Unwin (1968) suggest that the proper stress conditions for dilatancy are related to the thickness of the ice mass that exerts pressure on the underlying till. Where the glacier is thick, the till is highly mobile because the weight of the ice exceeds the load limit needed to initiate dilatance, and all the basal drift will be carried forward. Near the ice margin, however, the glacier thins to a point where the stress exerted on the basal load is less than the minimum required to maintain dilatance, and the material will revert to a more tightly packed texture, thereby becoming stable and resistant to further deformation. Drumlins, therefore, develop in the zone intervening between the thin mar-

Figure 10.25. Load-deformation curve for glacial till. Dilatant condition prevails in stress range shown in zone C. (After Smalley and Unwin 1968. Reproduced from the *Journal of Glaciology* by permission of the International Glaciological Society)

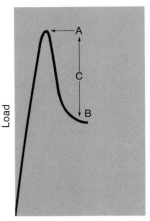

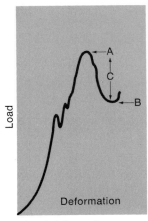

(A) Simple device for demonstrating dilatancy

(B) Load-deformation curve for dry sand

Load-deformation curve for glacial till

ginal ice and the thick interior ice. There the loading stress is less than that required to initiate dilatancy but greater than the pressure at which the till returns to a stable form (fig. 10.26). Some material collapses into stable cores in this zone because of its particle size distribution, but the till that remains dilatant is accreted and streamlined over the cores.

Some glaciers, however, do not form drumlins, indicating that all tills are not necessarily dilatant. Smalley and Unwin (1968) observe that till must contain a certain minimum concentration of boulders to attain dilatance, and so the formation of drumlins depends on both ice loading and the textural character of the till. In addition, although drumlins are depositional features in the dilatant model, older drift with low boulder concentrations or weathered bedrock may be streamlined in the same zone by erosional processes (Smalley and Unwin 1968). It should also be clear that drumlins can be formed only by temperate glaciers because the moving ice must be in contact with the subglacial debris, a phenomenon not possible in polar glaciers where the basal ice is frozen to the surface.

Finally, it is probably reasonable to accept the premise that fluted surfaces and drumlins are simply members of a genetically related group of forms ranging from those molded entirely from bedrock to those composed entirely of till (Flint 1971; Muller 1974). These forms, ranging from roches moutonnées to drumlins, can be both erosional and depositional. They are kindred in the sense that each has been oriented by moving ice and elongated parallel to the direction of glacier movement.

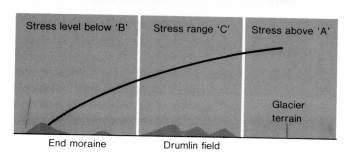

Stress level below 'B'   Stress range 'C'   Stress above 'A'

Glacier terrain

End moraine   Drumlin field

Figure 10.26. Cross section at edge of ice sheet showing most likely stress conditions for formation of drumlins (see figure 10.25 for load levels A, B, C). (After Smalley and Unwin 1968. Reproduced from the *Journal of Glaciology* by permission of the International Glaciological Society)

Proglacial Features

The large volume of water released from glaciers carries with it a tremendous quantity of sediment that is deposited in a number of environments beyond the margin of the ice. This debris usually accumulates in stream channels and associated floodplains which, because of their continuous lateral shifting, spread the sediment into a large plain called a *sandur* (from Icelandic; plural *sandar* or *sandurs*). Downstream from the sandar, the meltwater streams may empty into bodies of standing water and construct deltas, beaches, and other geomorphic features from the fluvioglacial sediment. These forms do not differ appreciably from their counterparts developed in normal fluvial, lacustrine, or marine environments and so will not be discussed here. Sandar are somewhat analogous to alluvial fans, but they are unique in that the hydrology of streams that form them is controlled by intense seasonal variations in the melting of ice. In addition, because each glacial advance develops its own related sandur, the distribution and age of the alluvial surfaces are extremely helpful in unraveling Pleistocene history. Thus, the origin of sandar and the geomorphic criteria for recognizing their form in ancient settings deserve our attention.

The study of sandar began in the nineteenth century in both Europe and North America. They are now recognized as consisting of two primary types (Krigstrom 1962). A *valley sandur* (*dalsandur*), which originates within well-defined valleys, is created by one main river and its anabranches; the entire system rarely occupies the total valley bottom at any given time. In the United States, valley sandar have been called *valley trains* and are usually associated with individual mountain glaciers (fig. 10.27). The second type of sandur, called a *plain sandur* (*slattlandssandur*), differs in that it develops with no lateral constraints, but represents the coalescence of many braided rivers that spread debris across wide areas in the form of a massive plain. In North America, these are often referred to as *outwash plains* and are usually associated with large ice sheets.

Most sandar are composed of gravel, although the deposits may include lenses of sand. A general decrease in particle size is sometimes, although not always, apparent in the downstream direction (Fahnestock 1963; Church 1972).

The deflation of very fine material on active sandar that are unvegetated helps to produce a coarse-grained surface.

In general, the long profiles of sandar are similar to river profiles, being concave-up in form and expressible as a simple exponential function (Church 1972). The concavity, however, may not be perfect because of the presence of linear segments similar to those characterizing alluvial fans. Cross-profiles tend to be convex-up, but the exact form may be irregular or may slope continuously in one direction. In addition, the cross-profile shape seems to depend on where it is measured in relation to the ice margin.

Krigstrom (1962) has recognized on sandar three distinct zones relative to the ice front, called the proximal, intermediate, and distal zones, each of which has different surficial characteristics. The *proximal zone,* closest to the ice, is usually traversed by only a few main rivers that flow in well-defined entrenched channels. These rivers and their deposits may pass continuously onto the ice mass itself, where the drift sometimes buries extensive areas of stagnating ice. As the ice subsequently melts, kettle holes form and the proximal sandur takes on a rough, pitted configuration. These kettled sandar (or pitted outwash plains) are difficult to place in our framework classification because they form in the marginal ice-contact environment, but they are continuous with the surface of the proglacial sandur and develop with the same original mechanics (described in Price 1973). In addition, on many sandar the proximal surface stands well above the elevation of rivers emerging from the ice. Several hypotheses have been suggested to explain why rivers in the proximal zone are so entrenched in their own deposits: (1) the rivers may be regrading in response to hydrologic and load

Figure 10.27. Valley sandur in front of the Sherman Glacier, Alaskan Gulf region. (Photo by Austin Post. From Post 1967, fig. 7)

characteristics (Fahnestock 1969); (2) the sediment is supraglacially deposited and left elevated as the ice front recedes and rivers emerge at a continuously lowered level; or (3) modest incision may be normal in the proximal zone, with the sandur surface simply representing the high flow level. It is conceivable that each of these interpretations is correct.

In the *intermediate zone,* the channels become wide and shallow and distinctly braided, and the entire depositional network shifts its position rapidly from side to side. This active lateral migration leaves a maze of abandoned channels with a relief of one or two meters impressed on the surface topography of the plain. Commonly the main channel is aggraded to a higher level than the smaller channels, facilitating rapid changes in the position of the river. Downstream the system changes gradually into the *distal zone,* where channels become so shallow that the rivers may merge into a single sheet of water during high flow. The flow here commonly feeds deltaic growth when the river enters a body of standing water. However, the sandur may extend itself downstream by growing over the rear portion of the delta, while the deltaic front simultaneously progrades (Church 1972).

Sandar originate from the combined effects of a large sediment supply and the high floods associated with melting ice. Most of the abundant load is derived from older drift, morainal deposits, and the continual delivery of new debris to the ablation zone and its release from the ablating ice. The greatest fluvioglacial work occurs near the ice margin where floods are produced by summer melting or as *jökulhlaups* (sudden release of lake water dammed within the ice; an Icelandic word pronounced "yokel-lawp"). These floods are characterized by rapid and drastic increases in discharge. Church (1972) reports the release of almost 5 million m^3 of water in a 30-hour period during a break out of an ice-dammed lake on Baffin Island. In that region, 25–75 percent of the total annual sediment transport and deposition occurs in four or five peak flows. Normal flows, however, do much of the total fluvial work on the sandur by redistributing the sediment across and down the plain.

The bulk of aggradation on a sandur takes place during high flow events as channel fills, sandur levee deposits, and overbank sedimentation. Overbank deposits are more prevalent in the intermediate zone where channels are shallower and interchannel reaches are covered more frequently by floods. Although high flow does initiate pronounced channel scouring, the amount of aggradation during the peak and waning stages of a flood simply obliterates the scour channels. Thus, aggradation may be rather rapid. For example, Fahnestock (1963) measured a net elevation gain of 0.36 m in a two-year period on the sandur produced by the Emmons Glacier (Washington).

In general, then, sandar can be considered as transport surfaces that aggrade during high flows but are probably eroded and changed in form when discharge and load are at normal volumes. The seasonal variations in load and discharge may also be accompanied by changes in the river pattern (Fahnestock 1963).

Therefore, the ultimate size and properties of a sandur are probably related to a quasi-equilibrium condition established by the balance between meltwater volumes and the quantity and size of the sediment made available for transportation. The surface will always be a montage of flood sediments, but the exact topography will change incessantly.

Summary

In this chapter we examined the landforms developed by the process of glacial erosion and deposition. Erosional features range in size from minor embellishments of exposed bedrock to major forms that dominate the landscape. Minor features such as striations, grooves, roches moutonnées, and friction cracks are a function of the subglacial mechanics that control abrasion and plucking. Major erosional landforms develop in two environments. In mountain uplands, the expansion of cirques results in the creation of features such as arêtes and horns that give glaciated mountains their characteristically rugged appearance. Cirques are created by nivation and rotational sliding of cirque glaciers. Cirques increase in size by erosional retreat of their walls facilitated by repeated freezing and thawing or possibly by hydration shattering. Glaciated valleys are created by large-scale abrasion and plucking that cause the dominant staircase longitudinal profile and the U-shaped cross-profile.

Deposits associated with glaciation, called drift, consist of either stratified or nonstratified material. Stratification requires that some sediment be transported by meltwater after the debris is released from the ice. Nonstratified drift is deposited directly by the ice. Certain types of depositional landforms tend to accumulate in particular geographic positions with respect to the ice front. In the marginal zone, moraines and stagnant ice features are most common. In the interior zone behind the ice margin, ground moraine, fluted surfaces, and drumlins are most conspicuous. Downstream from the glacial margin, in the proglacial zone, all drift is fluvioglacial and usually accumulates in the form of large plains called sandar. Processes operating in each of these depositional environments have been discussed with regard to how they might generate the landforms developed in the specific regions.

Suggested Readings

The following references are suggested for greater detail concerning the topics discussed in this chapter.

Boulton, G. S. 1974. Processes and patterns of glacial erosion. In *Glacial geomorphology,* edited by D. R. Coates, pp. 48-87. *Proc. 5th Ann. Symposium,* S.U.N.Y., Binghamton.

Church, M. 1972. Baffin Island sandurs: A study of arctic fluvial processes. *Canada Geol. Survey Bull.* 216.

Embleton, C., and King, C. A. M. 1975. *Glacial geomorphology.* New York: Halsted Press.

Flint, R. F. 1971. *Glacial and Quaternary geology.* New York: John Wiley and Sons.

Goldthwait, R. P. ed. 1971. *Till: A symposium.* Columbus: Ohio State Univ. Press.

Lewis, W. V., ed. 1960. Norwegian cirque glaciers. *Royal Geogr. Soc. Res.,* Ser. 4.

Muller, E. H. 1974. Origins of drumlins. In *Glacial geomorphology,* edited by D. R. Coates, pp. 187-204. *Proc. 5th Ann. Symposium,* S.U.N.Y., Binghamton.

Price, R. J. 1973. *Glacial and fluvioglacial landforms.* New York: Hafner.

A group of processes and features called *periglacial* characterize regions having extremely cold climates. The term *periglacial* was first used (Lozinski 1912) to describe the processes operating and the features developed in zones adjacent to ancient or modern ice sheets. Since then, the original connotation of "near-glacial" has been expanded, and most workers now accept the term as encompassing all nonglacial phenomena that function in cold climates, even if glaciers are not present (Dylik 1964; Butzer 1964; Washburn 1973). The periglacial environment is difficult to define in terms of precise temperature and precipitation values, although several attempts have been made to do so. For example, Peltier (1950) suggested that average annual temperatures from $-15°C$ to $-1°C$ ($5°F–30°F$) and an average annual rainfall between 127 mm and 1,397 mm would constitute a periglacial morphogenetic region. L. Wilson (1968, 1969) suggests somewhat different physical limits of $-12°C$ to $2°C$ ($10°F–35°F$) temperature, and 50 mm to 1,250 mm for the precipitation values.

Another question as to what constitutes a periglacial regime arises because some scientists believe that permanently frozen ground, called *permafrost,* is an essential ingredient (if not a prerequisite) for periglacial conditions (Tricart 1967; Péwé 1969). Certainly some periglacial features occur in close association with frozen ground, and a complete understanding of periglacial systems therefore requires some fundamental perception of permafrost. However, periglacial processes also function where no permafrost is known to exist. Most students of periglacial phenomena, then, would not view permafrost as a requirement of a periglacial system. The two points that all investigators seem to agree on are that (1) nearby glaciers are not necessary for the processes to function, and (2) the fundamental controlling factors of periglaciation are intense frost action and a ground surface that is free of snow cover for part of the year. Many regions meet these requirements, ranging from polar to subpolar lowlands to high-elevation mountains that may rise in any regional climate including temperate or even tropical zones.

It is not new for humans to occupy the high latitude regions of the world; Eskimos and other nomadic groups have wandered these areas for millenia. These inhabitants, however, lived in rather simple social organizations, and their life styles brought them into little conflict with their natural surroundings. In more recent times, the emplacement of defense installations in periglacial regions, the discovery of oil in the Arctic Circle, and the increased exploration for

Periglacial Processes and Landforms

11

mineral wealth brought with them the realization that developing a highly technical society in these regions is an engineering and scientific nightmare. For inhabitants of temperate zones, shattered highways and frozen or broken water pipes are inconveniences of a rigorous winter and the subsequent spring thaw. Multiply these problems by a thousand; add disruption of services, construction difficulties, and complications of all the trappings associated with a modern civilization; and you will begin to understand the meaning of life in a periglacial environment.

Many governmental agencies are beginning to direct research efforts to problems of cold climates, and symposia and reports in scientific journals increasingly treat such topics. Nonetheless, detailed understanding of the processes involved is lacking and, faced with an inevitable human migration into cold regions, it is imperative that geomorphologists devote additional efforts to this discipline. We have the basics to do so because the processes responsible for periglacial effects are not greatly different from those we already know. In fact, frost action and mass movements form the core of periglacial processes. The main difference between periglaciation and temperate phenomena seems to lie in the magnitude of the processes and the manner in which they are combined in the systemic operations. Thus, as L. W. Price (1972a) stresses, if we are ever completely to understand the periglacial environment and provide viable information for future planning, we must cast aside our provincial, mid-latitude approaches to the study of this system.

Our treatment of periglaciation will be necessarily brief, examining only the major processes and features. For greater detail, several excellent texts and reviews are available: Embleton and King 1968, 1975b; Price 1972a; Washburn 1973; Tricart 1969; Péwé 1969.

Definition and Thermal Characteristics

Although permafrost is not a requirement for the functioning of periglacial processes, the most troublesome engineering problems occur in regions underlain by zones of perennially frozen ground, and many periglacial features are related to abundant ground ice. Permafrost was originally defined in terms of temperature only (Muller 1947) as being soil or rock that remained below 0°C continuously for more than two years. By this definition water is not a necessary component of permafrost and, in fact, "dry" permafrost has been recognized. In most cases, however, ice is so important in the mechanics of periglacial processes that some workers (Stearns 1966) believe moisture must be present before any material can be considered as true permafrost. Other investigators have placed the temperature criteria for permafrost below 0°C (Brown 1967; Ferrians 1965). In a geomorphic sense the presence of ice, regardless of the temperature value, seems to be critical. The ice may exist as a cement between soil particles or as larger masses of pure ice. Where distinct ice masses are prevalent, they usually occur as horizontal lenses, but they may also fill cracks in the parent material and stand as vertical veins or wedges (fig. 11.1).

Figure 11.1. Ice wedge (ground ice) in permafrost exposed by placer mining near Livengood, about 50 miles northwest of Fairbanks, Alaska. Tolovana district, Yukon region, Sept. 1949. (Photo by T.L. Péwé. From Ferrians et al. 1969, fig. 9)

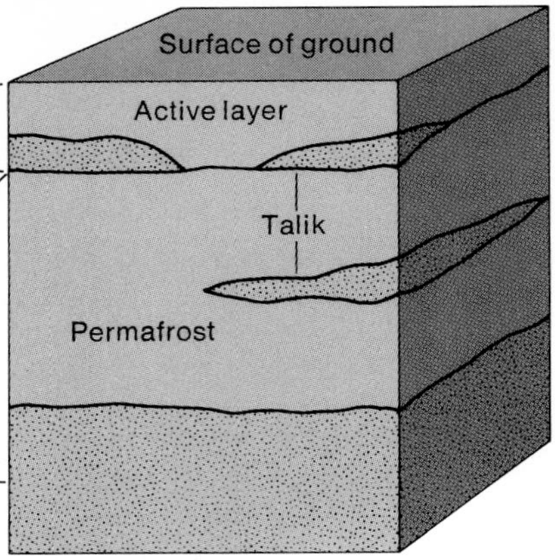

Suprapermafrost layer (ground above permafrost table including taliks, if present)

Permafrost table (upper surface of permafrost)

Surface of ground

Active layer

Talik

Permafrost

Talik

Figure 11.2. Profile in a region underlain by permafrost. Taliks are zones of unfrozen ground within or beneath the permafrost or between the permafrost table and the base of the active layer. (After Ferrians et al. 1969)

Figure 11.2 shows that permafrost is usually a subsurface phenomenon existing as a zone of permanently frozen ground that can extend downward to incredible depths. The upper surface of the permafrost, called the *permafrost table,* is overlain by a thin layer (0–3 m thick) of material that freezes and thaws on a seasonal basis. This uppermost layer, called the *active layer,* is thickest in the subarctic region (Price 1972a), becoming shallower toward both the pole and the mid-latitudes. The mechanics in the active layer is similar in most respects to the normal seasonal freezing and thawing in temperate climate zones. In permafrost regions, however, water released by thawing cannot percolate into the solidly frozen substrate, and this tends to accentuate the effects of frost action and mass movements.

Even within the permafrost itself, the temperature fluctuates with the seasons. Temperature variations, however, become less radical with depth until some level is reached, generally at 20 to 30 m deep, where the temperature never changes (fig. 11.3). This level is known as the *zero annual amplitude.*

The thermal operations in permafrost are so complicated that they are understood only in general terms. Spring thawing allows heat to penetrate into the active layer, and melting proceeds rather evenly from the surface downward. In the autumn, however, the decrease in temperature does not spread downward at a uniform rate, and the ground moisture freezes unequally. Thus, unfrozen lenses may develop between the permafrost table and the solidly frozen ground surface. This phenomenon causes unusual pressures in the soil water and depresses its freezing point. In addition, heat of fusion is generated when the zone around the ice-free lenses freezes. The result is that pockets of free water may exist in the ground for considerable time after the surface is frozen, possibly for several months.

Further complications in the distribution of frozen ground occur because the permafrost table tends to mirror the surface topography, rising beneath hills and lowering under valleys. In addition, the permafrost table is influenced in various

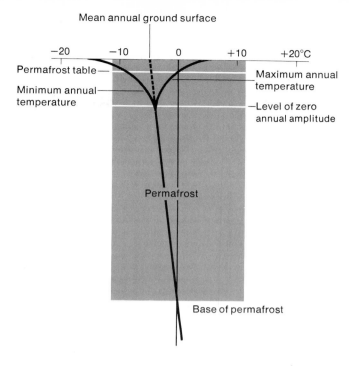

Figure 11.3. Ground temperature change with depth in permafrost regions. (After J.R. Williams 1970)

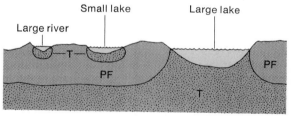

Figure 11.4. Schematic cross section showing the effect of surface water on the distribution of permafrost. (PF is permafrost; T are taliks.)

ways by the position of surface water (fig. 11.4), depending on whether the rivers or lakes freeze over in the winter (Ferrians et al. 1969).

Distribution, Thickness, and Origin

Where annual temperatures average zero or below zero C, ground freezing during the winter will penetrate deeper than the depth of summer thawing. Each passing year will produce another increment of perpetually frozen ground, and the permafrost zone will gradually thicken. The process, however, cannot go on indefinitely because temperature within the Earth rises according to the geothermal gradient, that is, at a mean ratio of 1°C per each 30 m increase in depth. Thus, the total thickness of permafrost depends on the depth at which internal heat balances the wave of cold temperature filtering down from the surface.

Table 11.1 Selected northern hemisphere permafrost thicknesses.

Location	Mean Annual Air Temperature	Thickness of Permafrost
Alaska		
Prudhoe Bay (70°N, 148°W)	−7 to 0° C (20 to 32° F)	609 m (2,000 ft.)
Barrow (71°N, 157°W)	−12 to 7° C (10 to 20° F)	405 m* (1,330 ft.) 16 km (10 miles) inland
Umiat (69°N, 152°W)	−12 to −7° C (10 to 20° F)	322 m (1.055 ft.) 235 m (770 ft.) under Colville River
Cape Thompson (68°N, 166°W)	−12 to −7° C (10 to 20° F)	306 m* (1,000 ft.)
Bethel (60°N, 161°W)	−7 to 0° C (20 to 32° F)	184 m (603 ft.) 13 m (42 ft.) under Kuskokwim River
Ft. Yukon (66°N, 145°W)	−7 to 0° C (20 to 32° F)	119 m (390 ft.) 5.5 m (18 ft.) under Yukon River
Fairbanks (64°N, 147°W)	−7 to 0° C (20 to 32° F)	81 m (265 ft.)
Kotzebue (67°N, 162°W)	−7 to 0° C (20 to 32° F)	73 m (238 ft.)
Nome (64°N, 165°W)	−12 to −7° C (10 to 20° F)	37 m (120 ft.)
McKinley Natl. Park-East Side (64°N, 149°W)	extremely variable	30 m (100 ft.)
Canada		
Melville Island, N.W.T. (75°N, 111°W)	————————	548 m (1,800 ft.) near coast, probably thicker interiorward
Resolute, N.W.T. (75°N, 95°W)	−16.2° C (2.8° F)	396 m (1,300 ft.)
Port Radium, N.W.T. (66°N, 118°W)	−7.1° C (19.2° F)	106 m (350 ft.)
Ft. Simpson, N.W.T. (61°N, 121°W)	−3.9° C (25.0° F)	91 m (300 ft.)
Yellowknife, N.W.T. (62°N, 114°W)	−5.4° C (22.2° F)	61-91 m (200-300 ft.)
Schefferville, P.Q. (54°N, 67°W)	−4.5° C (23.9° F)	76 m (250 ft.)
Dawson, Y.T. (64°N, 139°W)	−4.6° C (23.6° F)	61 m (200 ft.)
Norman Wells, N.W.T. (65°N, 127°W)	−6.2° C (20.8° F)	46-61 m (150-200 ft.)
Churchill, Man. (58°N, 94°W)	−7.1° C (19.2° F)	30-61 m (100-200 ft.)

Table 11.1 Continued.

Location	Mean Annual Air Temperature	Thickness of Permafrost
U.S.S.R.		
Upper Reaches of Markha River (66°N, 111°E)	———————	1500 m (4,920 ft.)
Udokan (57°N, 120°E)	−12°C (10.4°F)	900 m* (2,950 ft.)
Bakhynay (66°N, 124°E)	−12°C (10.4°F)	650 m (2,130 ft.)
Isksi (71°N, 129°E)	−14°C (6.8°F)	630 m (2,070 ft.)
Mirnyy (63°N, 114°E)	−9°C (15.8°F)	550 m (1,805 ft.)
Ust'-Port (69°N, 84°E)	−11°C (12.2°F)	425 m (1,395 ft.)
Salekhard (67°N, 67°E)	−7°C (19.4°F)	350 m (1,150 ft.)
Noril'sk (69°N, 88°E)	−8°C (17.6°F)	325 m (1,070 ft.)
Yakutsk (62°N, 129°E)	———————	195-250 m (650-820 ft.)
Vorkuta (67°N, 64°E)	———————	130 m (430 ft.)

*A calculated depth not actually measured.

From Price 1972. Various sources, but chiefly the following: Brown 1967, map; Ferrians 1965, map; Yefimov and Dukhin 1968. Used with permission of the Association of American Geographers, Commission on College Geography. Resource paper # 14, 1972, L. W. Price.

Permafrost today may extend to formidable depths (table 11.1). It averages between 245 and 356 m in North America and tends to be slightly thicker in Eurasia. The thickest known permafrost is in Siberia, where a depth of approximately 1500 m has been reported. The great thicknesses of permafrost probably reflect both past and present climatic conditions. Some permafrost must be very old, for normal accumulation rates are usually only centimeters per year. Some of the thickness values given in table 11.1 would necessitate a growth history involving thousands or perhaps tens of thousands of years.

Some present-day climates are probably cold enough to accumulate modern permafrost and certainly to maintain ancient permafrost. The destruction, protection, or growth of permafrost, however, varies with factors other than climate, such as the type of surface cover. The presence of ice or dense vegetation has a decided influence on permafrost thickness. In Antarctica, for example, some areas beneath the ice sheet may have no permafrost (Ueta and Garfield 1968), while zones that are glacier free have permafrost up to 150 m thick (J. R. Williams 1970).

Permafrost underlies 26 percent of the world's land surface (Black 1954), and so is not, as we may tend to think, an unusual phenomenon. As the map in figure 11.5 shows, most of the known permafrost exists in the polar regions of the northern hemisphere, extending to a southernmost limit at a latitude of approximately 55°N in both North America and Eurasia. Permafrost in these regions is usually divided into *continuous* and *discontinuous* types (Ray 1951). Continuous permafrost usually consists of thick layers of perennially frozen

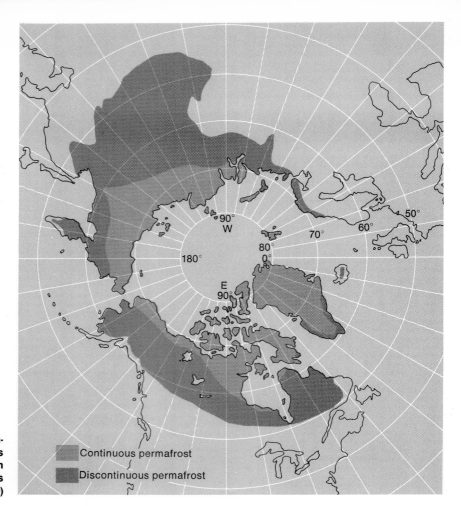

Figure 11.5. Extent of continuous and discontinuous permafrost in the northern hemisphere. (After Ferrians et al. 1969)

Continuous permafrost

Discontinuous permafrost

ground that are spread rather evenly under a wide areal surface. The continuous nature of the permafrost alters only where it thins under deep, wide lakes or rivers. In contrast, discontinuous permafrost is shallower and contains unfrozen zones within the frozen ground or wide gaps that remain unfrozen (fig. 11.6). These unfrozen areas, called *taliks,* increase in size and number southward until true permafrost exists only as isolated patches. Taliks appear as islands exposed at the surface, lenses or layers within the permafrost, or unfrozen ground beneath the permafrost (figs. 11.2, 11.6). The southern limit of continuous permafrost coincides in a general way with the −6°C annual isotherm, and the discontinuous boundary with the −1°C isotherm (Brown 1970).

It should not be assumed from the above that permafrost exists only in high latitudes. Isolated zones of frozen ground, called *sporadic permafrost* (see fig. 11.6), can be found far south of latitude 55°N and are probably relicts of a once colder climate. Pockets of permafrost persist also where elevations are sufficiently high. In the United States, for example, permafrost is found near the summit of Mt. Washington in New Hampshire (Goldthwait 1969) and in the Rocky Mountains above the tree line (Ives and Fahey 1971). Permafrost has even been reported near the summit of Mauna Kea, Hawaii (Woodcock et al. 1970; Woodcock 1974), at an elevation of 4140 m. There a 10 m thick layer of permafrost survives, even though the average air temperature is 3.6°C, because cool air is trapped locally and a low angle of sunlight combined with unusual dryness maintains the frozen condition. Such examples indicate that permafrost can be preserved for long periods in disequilibrium with the prevailing regional climate if microclimatic conditions meet the necessary requirements.

The general but imprecise relationship between average air temperature and permafrost boundaries suggests that modern climate determines the distribution of frozen ground even though most of the included ice formed at an earlier time. This conclusion fits the interpretation that the recent retreat of the discontinuous permafrost boundary in Manitoba (Canada) is related to a climatic shift that began only about 120 years ago (Thie 1974). All other things being equal, the climate most amenable to preservation of permafrost is one with cold, long winters followed by cool, short summers, and low precipitation in all seasons (Muller 1947). Other variables, such as vegetation type and density, composition of surface materials, topography, and surface water, exert an influence on the spatial character of permafrost. For example, most permafrost lies beneath the northern boreal forests or, north of the tree line, under a tundra vegetation dominated by low sedges, grasses, and mosses. The boundary between continuous and discontinuous permafrost often parallels the southern limit of the tundra. Because the thermal properties of vegetation control how efficiently temperature can penetrate the ground (Corte 1969; Price 1972a), it is not clear whether the climate or the vegetation produced by the climate is the determining factor of permafrost distribution. Probably all factors exert some control, and therefore we cannot expect a precise correlation between air temperature and permafrost boundaries.

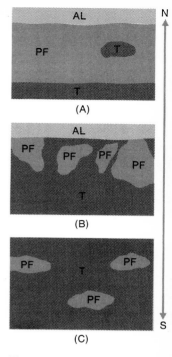

Figure 11.6. Types of permafrost zones: (A) continuous, (B) discontinuous, (C) sporadic. AL = active layer; PF = permafrost; T = talik.

Periglacial Processes

Among students of periglacial phenomena, the consensus is that the regime is dominated by the processes of frost action and mass movement. Although these were discussed in an earlier chapter, an examination of their prowess in the periglacial setting can give a different perception and a greater appreciation of their geomorphic significance. Frost action encompasses a group of processes—wedging, heaving, thrusting, cracking—that all serve to prepare bedrock or soil for erosion. Mass movements transport the loosened debris. The two groups of processes function in periglacial environments as they do in temperate zones. As suggested earlier, however, major differences arise because frost action is considerably more severe in periglacial zones, and because mass movements may be intensified during thawing because the material is saturated with excessive water that cannot drain downward through the system. The combination of these factors results in geomorphic features that are unique to periglacial regions.

Frost Action

The driving force in all processes included in the realm of frost action is the growth of ice within a soil or rock. Intuitively we might expect the explanation of freezing in a porous substance to be relatively simple, but in fact, the thermodynamics of the process are very complex (see Miller 1966; Anderson 1968, 1970). In addition, other factors within the system influence the final geomorphic effect of frost action because they respond to freezing in different ways. For example, segregation of ice into discrete lenses generates a stress field that differs from that produced by freezing of disseminated pore water, and the resulting processes also are dissimilar. Whether ice lenses actually form depends mainly on (1) the rate of freezing, (2) the ability of water to be drawn to a central freezing plane or point, and (3) the size of the pore spaces.

Generally lenses of ice are smaller and less common as depth increases because the greater pressure lowers the freezing point. However, there are exceptions to even this generalization when the grain size is conducive to ice-lens development. The water in fine-grained sediment (small pore spaces) freezes at lower temperatures and also has a greater propensity for moving to a central freezing plane, a property called the *suction potential*. Because of these factors, combined with the tendency for fines to hold more water, segregated ice masses are most common in fine-grained sediment. In contrast, coarse sediment usually contains interstitial ice. On the other hand, extremely fine-grained sediment may be impermeable. Therefore, ice lenses form preferentially where suction potential and permeability are most advantageously combined, and this usually occurs in deposits of uncemented silt (discussed in Washburn 1973).

The expansion associated with freezing exerts pressures that can produce results as variable as shattering of solid rock or physical lifting of the ground surface. Precisely what event will transpire during freezing and subsequent thawing is determined by a multitude of complex interdependent variables that operate within the system.

Frost Wedging
Frost wedging is the prying apart of solid material by ice (Washburn 1973). It is synonymous with splitting, riving, and shattering. In a perfectly closed system near the ground surface, the maximum pressure exerted by a phase change from water to ice is approximately 2100 kg/cm². At such a pressure the freezing temperature is $-22°C$. It is doubtful that such a condition can exist in a natural setting because even if a perfectly closed system could be created (a dubious possibility requiring very special circumstances), the tensile

strength of most rocks is well below the 2100 kg/cm² value. Granites, for example, have a tensile strength of only 70 kg/cm², and according to Grawe (1936), the maximum tensile strength of any rock is about 250 kg/cm². This means that rocks should begin to shatter and release the stress well before the maximum freezing pressure is attained.

In porous substances that are saturated, frost wedging is facilitated by rapid freezing of water in the near-surface pores. This tends to close the system, allowing the build-up of some extra pressure which is needed to cause shattering. Slow freezing seems to be an effective wedging process only if water is free to migrate to a freezing plane where excess pressure is generated by growth of large ice crystals. As discussed in chapter 10, wedging in cracks takes place only if freezing proceeds from the surface downward (Battle 1960).

Other factors influence frost wedging besides the rate of freezing and the closure of the system. It is known, for example, that a high water content increases the strain on rocks (Mellor 1970) and that most rocks shatter more readily if they are immersed in the fluid (Potts 1970). This probably explains why wedging is often more pronounced at the base of sheer cliffs where groundwater is available than at the top, and also why porous rocks shatter more easily than nonporous rocks. In addition, the efficacy of frost wedging seems to be linked to the number of freeze-thaw cycles. Some caution is required, however, in correlating freeze-thaw frequency and shattering because fluctuations of air temperature across the freezing point do not always coincide with those beneath the surface. Furthermore, diurnal freeze-thaw events may affect only a thin upper fringe of the soil (Fahey 1973). Thus, prolonged freezing with extremely low temperature may have more geomorphic significance than frequent freeze-thaw cycles (Rapp 1960; Washburn 1973).

The products of frost wedging are notably angular and range in size from blocks as large as buildings to fine-grained debris. It has been traditionally accepted that the terminal size of frost shattering is silt (Hopkins and Sigafoos 1951; Taber 1953). Although this is probably true in most situations, some workers now believe that clay-sized particles may be formed under certain conditions (McDowall 1960) or that any terminal size is only rarely attained (Potts 1970). In any case, coarse angular debris is common in periglacial regions, accounting for the prevalence of talus rubble at the base of alpine slopes. Wedging also breaks particles loose from bedrock covered by a soil, and these progressively make their way to the surface, creating additional problems for farmers working the land.

Frost Heaving and Thrusting   *Frost heaving* refers to vertical displacement of matter in response to freezing, while *frost thrusting* connotes horizontal movement (Eakin 1916). In the natural setting the two processes are virtually indistinguishable, and their designation as separate processes is probably more imagined than real. Heaving is directly responsible for several phenomena that are commonplace and accentuated in periglacial environments. First, heaving causes the ground surface to move vertically as ice formation expands the ground material. The extent of this displacement depends on the physical variables within the system, and the surfaces of adjacent local areas may be lifted at disparate velocities and amounts. Such differential heaving causes building foundations and other types of construction to crack. Second, heaving has the ability to sort heterogeneous debris by forcing larger particles to migrate surfaceward relative to their finer counterparts. The particles may move upward as much as 5 cm a year (Price 1972a). It is well to recognize that any object inserted into the ground or dispersed within the near-surface mass is subject to the mechanics of heaving—not only stones of various size that emerge in farmlands or on slopes, but also fenceposts, telephone poles, pilings, or anything else that people might conceivably shove into the ground.

The heaving process also functions in indurated solids and is capable of displacing joint-bounded blocks of bedrock, even though these are held rather tightly by the surrounding mass (Price 1972b). These upheaved blocks may project up to 1.5 m above the ground and, where soil is absent, give the surface a jagged appearance.

The Heaving Mechanism   In several classic studies of heaving mechanics, Taber (1929, 1930) showed that heaving in a closed system is limited to that produced by the 9 percent volume increase caused when water freezes. In open systems, heaving is much greater because additional force is gained from crystal growth along the freezing plane. The amount of extra heaving depends on how much water can be drawn to the point of freezing from the surrounding material. Taber also showed that heaving stress due to crystal growth is exerted normal to the freezing isotherm and not in the direction of least resistance. Larger particles of varying compositions will have different thermal conductivities and thus may locally influence the orientation of the cooling surface. This certainly introduces lateral movements in the expansion that are not perpendicular to the freezing isotherm. Some workers also disagree with Taber's conclusion that differences in resistance to expansion do not influence the direction of the heave (Beskow 1947).

Any theory attempting to explain why large particles are moved surfaceward in a nonsorted soil mass must consider not only the forces that lift the clast but also the reasons it does not return to its original position during contraction. Details of the two plausible theories concerning heaving mechanics, called *frost pull* and *frost push,* are discussed by Washburn (1973). Briefly, the frost-pull hypothesis suggests that stones are lifted vertically along with the fines when the ground expands during freezing. The fine sediment, being more cohesive, is brought downward quickly upon thawing and collapses around the larger clasts while the bases of the stones are still frozen. Cavities formed when a large particle is heaved (fig. 11.7) may also be partly closed by thrusting into this zone of lowered resistance.

The frost-push hypothesis is based on the fact that individual stones are better conductors of heat than porous soil. As a result, stones will cool more quickly, and the first ice to form along the freezing plane will be adjacent to and at the base of the stones, thereby pushing them upward. The phenomenon of ice forming preferentially near stones embedded in a silty soil has been observed and is not conjecture; nor is it conjecture that thawing occurs first around the stones. Presumably, however, material thawed adjacent to the top of the stone will collapse against the upper part of the clast while the base remains frozen and thus prevent its return to the original position. The shape of the stone may become important in this procedure; for instance, wedge-shaped particles with the narrow edge projecting downward will have greater difficulty returning to their original level.

Evidence exists to support both the frost-pull and frost-push theories, and probably both processes function simultaneously in a heaving environment. Washburn feels, however, that rapid heaving or movement that breaks a vegetal cover probably requires that frost push be dominant. Slow ejection of clasts can probably be accomplished by frost pull with little necessity for rapid ice build-up beneath the stones.

Other factors affect the magnitude of frost heave and its continuation in the upper part of the soil. Experiments conducted in Greenland by Washburn (1967) showed that moisture content, vegetation, and depth are all critical variables in the heaving process. The greatest heave occurred in zones that had abundant moisture, and deeper objects were displaced farther than shallow ones. Heaving is generally less where the vegetation cover is dense, presumably because of the insulation provided by the vegetal mat. At or near the surface another form of ice, called *needle ice* (or piprake), aids in the lifting process. Needle ice consists of groups of slender ice crystals that are usually 1–3 cm long, although they have been observed to attain lengths up to 40 cm (Troll 1958). The ice clusters are capable of lifting stones up to cobble size. Needle ice, which forms best in moist, loamy soils with no vegetation, also is very important as an aid in mass movement.

Frost Cracking   Frost cracking is the development of fractures at very low temperatures. The process seems to function best in permafrost regions, although it has been reported from other environments. Frost cracking is usually considered as a frost action phenomenon, but Price (1972a) points out that in fact it is due to thermal contraction rather than the expansion associated with freezing.

At very low temperatures frozen ground often evolves into a polygonal network of contraction fractures, but cracking may depend more on the rate of cooling than on the absolute temperature value at the moment of fracture (Lachenbruch 1966; Black 1963, 1969). Initial cracking at the surface may extend to a depth of 3 m and can become progressively wider and deeper. The results of frost cracking will be discussed more in a later section when we examine polygonal ground features.

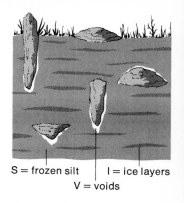

S = frozen silt   I = ice layers
V = voids

Figure 11.7. Displacement of stones during the freezing process. (After Taber 1943. Courtesy of the Geological Society of America)

Mass Movements

Any variety of mass movement (discussed in chapter 4) can occur in a periglacial environment, but there is little doubt that two types, *frost creep* and *solifluction,* dominate the cold regime. The movement of debris occurs simultaneously with the frost action processes, and it is questionable how effective the erosion would be if it operated alone. Mass movements, however, are responsible for a variety of landforms that are unique to periglacial regions.

Frost Creep   Frost creep, like any form of near-surface creep, is the downslope movement of particles in response to expansion and contraction and under the influence of gravity. It is unique only because freezing and thawing generate the cycle of expanding and contracting. As explained earlier, the freezing front in the soil usually parallels the ground surface. Individual soil particles are heaved perpendicular to the surface, but during thawing they are affected only by gravity and should contract in a vertical direction (shown earlier in fig. 4.20). Actually, a component of upslope migration has been observed in the contracting phase (Washburn 1967; Benedict 1970). This motion, called *retrograde movement,* probably stems from the attraction of fine-grained particles for one another and the cohesive strength developed between them. On the other hand, some downslope flowage may occur during the thaw, and so the overall route followed by any soil particle involved in frost creep probably resembles the path shown in figure 11.8.

Detailed studies in northeast Greenland (Washburn 1967) showed frost creep to be the dominant process where soils are silty and slopes stand between 10° and 14°. It was noted, however, that in any given year other processes could dominate the system. Therefore, it may not be safe to generalize about the conditions that promote frost creep because soil texture, moisture content, vegetation, and freeze-thaw frequency all have some bearing on the efficacy of the process.

Figure 11.8. Route of downslope movement followed by a particle under combined frost creep and solifluction.

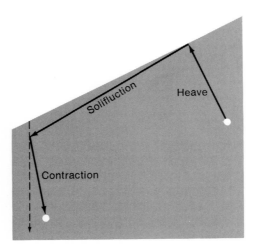

Solifluction (Gelifluction) The term *solifluction* was first proposed to describe the action of slow flowage in saturated soils. The original meaning carried with it no climatic restrictions, but over the years the term has generally been used to connote a process that functions in periglacial regions. We will follow this practice here, although you should realize that it is technically incorrect because solifluction can occur anywhere. The term *gelifluction* (Baulig 1957) refers to soil flowage associated with frozen ground and as such is a specific type of solifluction.

Solifluction is most dramatic in permafrost zones because water released in the active layer during thawing cannot penetrate below the permafrost table. Soils in the summer are often saturated, and the concomitant loss of friction and cohesion causes them to behave like viscous fluids. Such materials can flow on slopes as gentle as 1°, but maximum displacement seems to occur where the gradient is between 5° and 20°. Soils on slopes steeper than 20° tend to drain easily, and water escapes as surface runoff.

It is clear that abundant moisture is the overriding factor in the solifluction process (Washburn 1967; Chambers 1970; Price 1972a). Many students of periglacial geomorphology feel that significant flow can occur only when the moisture content equals or exceeds the liquid limit, but this interpretation cannot be accepted unequivocally because solifluction has been observed at lower moisture contents (Fitze 1971). Obviously, other factors such as soil texture, slope angle, and vegetation play a modifying role on a local basis. For example, gravel and coarse sands are so permeable and easily drained that they virtually never flow. Clays, on the other hand, are extremely cohesive. Thus, silty soils, especially those with a bimodal size distribution, are most susceptible to solifluction. Nonetheless, the overwhelming importance of water content was clearly demonstrated in Greenland (Washburn 1967), where the highest rates of movement commonly occur on the most densely vegetated areas of the slopes and at gradients that are lower than the dry portions of the slopes. This indicates that moisture can overcome both the binding effect of vegetation and the apparent deficiency of a transporting gradient.

The rates of combined frost creep and gelifluction in northeast Greenland and other regions are shown in table 11.2. It is difficult to ascertain how much of the total movement is accomplished by frost creep and how much can be attributed solely to flow.

Landforms Associated with Permafrost

Some periglacial features are so closely allied to the distribution of permafrost that a genetic relationship can hardly be denied, and observation of these features preserved in temperate regions affords the best evidence known for reconstructing the areal extent of ancient permafrost boundaries. The basic processes involved in the development of these features are no different from those that form other periglacial landforms; that is, they rely on frost action and/or mass move-

Periglacial Landforms

Table 11.2 Representative values of combined frost creep and gelifluction. [1]

Rate (cm/yr)	Slope	Source
2.0	15°	Rapp 1960
0.9-3.7	10° -14°	Washburn 1967
1.0-3.0	3° - 4°	Jahn 1960
5.0-12.0	7° -15°	Jahn 1960
10.0	20°	P. J. Williams 1966[2]
2.5	6° -13°	Benedict 1970
2.6	2.5°- 4°	French 1974

1. List is intended as random sampling and is not complete. Great variability exists depending on differences in ground moisture, vegetation, depth, slope face direction, grain size and soil texture, etc.
2. Pebbles on bare slope surface.

ment. The permafrost layer, however, places a distinctive imprint on the system and allows us to consider the geomorphic features as a separate group.

Ice Wedges and Ice-Wedge Polygons   When thermal tension exceeds the strength of the surface materials, frost cracking begins and fractures penetrate into the active layer and the upper portion of the permafrost. At first the cracks are only millimeters wide, but they may extend vertically downward to depths of several meters. The actual moment of cracking seems to occur sometime between January and March when temperatures in the ground reach their annual minima (Mackay 1974). By spring and early summer the crack may be partially or totally filled with snow that has filtered down from the surface and with water released by the onset of thawing in the active layer. As the diagrams in figure 11.9 show, this mixture freezes below the permafrost table and occupies the initial crack as an ice veinlet. Because the initial fracture now represents a zone of weakness within the permafrost, subsequent cold winters produce cracking along the trace of the original opening, and new ice is added as before. Over the years, this accumulation of relatively pure and vertically oriented ice takes the form of a downward-tapering wedge, called an *ice wedge,* shown in the diagrams (fig. 11.9) and pictured in figure 11.1. In dry permafrost, the material filling the contraction fissures may be particles of the parent soil, and the resulting feature is called a *sand wedge*.

Ice wedges range from 1 cm to 3 m wide and from 1 to 10 m deep, with the depth to top-width ratio normally ranging from 3:1 to 6:1. They are best developed in fine-grained soils with a high ice content (Black 1976). The host sediments are usually upturned within 3 m of the wedge, in response to the horizontal compression generated as the wedge grows. Ice in wedges contains oriented air bubbles and a complicated fabric of ice crystals (Black 1974).

The growth of an ice wedge does not necessarily proceed on an annual basis. Mackay (1974, 1975) observed that only 40 percent of the ice wedges on

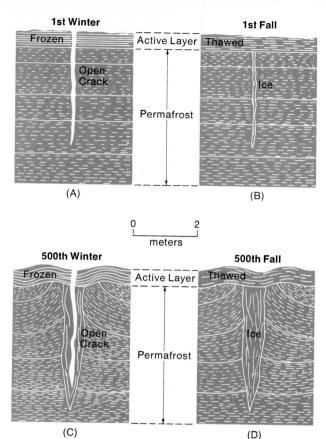

1st Winter

Frozen

Active Layer

Open
Crack

Permafrost

(A)

1st Fall

Thawed

Ice

(B)

0   2

meters

500th Winter

Frozen

Active Layer

Open
Crack

Permafrost

(C)

500th Fall

Thawed

Ice

(D)

Figure 11.9. Evolution of an ice wedge by thermal contraction. (After Lachenbruch 1966)

Garry Island (Northwest Territories, Canada) cracked in any given year, the frequency of cracking apparently controlled by the depth of snow cover. Even if cracking does occur, the openings often narrow before ice-veinlet growth begins. The size of a veinlet thus is dependent on the timing between closure of the winter cracks and the appearance of meltwater (Mackay 1975). In most cases the thickness of an annual increment (usually 1–20 cm) is considerably less than the size of the thermal crack.

Ice wedges are commonly connected in a polygonal pattern that is similar in most respects to the more familiar mud cracks. They differ mainly in that *ice-wedge polygons* are much larger, having diameters that range from several meters to more than 100 meters (fig. 11.10), and they contract and expand according to the laws that govern temperature effects on ice (Black 1974). The perimeters of some polygons are accentuated as minor ridges, so that the boundary is elevated relative to the center of the feature (*low-center ice-wedge polygons*). Others (*high-center polygons*) are depressed along their edges by melting and erosion of

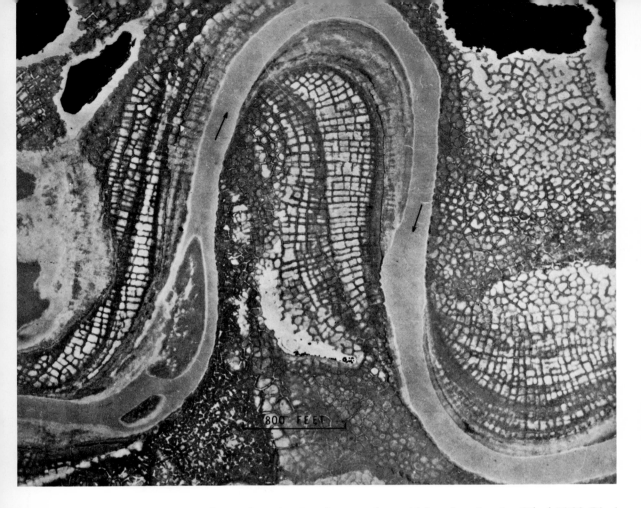

800 FEET

Figure 11.10. Polygonal markings on ground surface caused by ice-wedge polygons in the vicinity of Meade River, about 35 miles southeast of Barrow, Alaska. Barrow district, Northern Alaska region, July 1949. (Photo by U.S. Navy. From Ferrians et al. 1969, fig. 8)

the ice wedges, leaving the central core higher than the rim (Péwé 1966; Black 1974, 1976).

Actively growing ice wedges are restricted to areas of continuous permafrost, and their southern boundary seems to be the −6°C to −8°C isotherms (Péwé 1973). Farther south in the discontinuous permafrost zone, they become inactive and disappear. As ice wedges melt, the surrounding sediment may collapse into the space opened by the disappearance of the ice, thereby creating an ice-wedge cast, a feature that is best preserved in gravels (Black 1976). Although ice-wedge casts are probably the best proof of an earlier permafrost condition, other wedgelike features unrelated to permafrost look very similar. As a result, some caution is required in the identification and interpretation of ice-wedge casts (see Black 1976).

Pingos   The term *pingo* (Eskimo for "mound" or "hill") was first suggested in 1938 by Porslid for large, ice-cored, domelike features that exist only in permafrost regions. They are most perfectly developed in areas of continuous permafrost, although they do form in the discontinuous zone (Holmes et al. 1968). Active pingos rise from a few meters up to heights greater than 100 m and have basal diameters up to 1200 m (Flemal 1976). They are oval or circular in plan view, as in the photograph in figure 11.11. The ice core in a pingo is typically massive and is often exposed by tension fractures that develop at the summit of the mound as it rises. Sometimes the exposed ice melts, and small freshwater lakes occupy the craters bounded by the tensional cracks.

Many varieties of pingos are known (Pissart 1970; Flemal 1976) but most can be placed in two major genetic categories. *Closed-system* pingos (Mackenzie type) develop in level, poorly drained, shallow lake basins. They are most distinctly preserved in the Mackenzie delta (Canada) where a clutch of 1,450 pingos gives evidence that pingos tend to form in fields rather than as single individuals. In general, the draining of a lake allows the permafrost table to rise to the level of the former lake floor. As the table rises, water is trapped in the saturated soil. The cryostatic pressure (pressure from ice formation) displaces the water upward until it also freezes and becomes the core of the pingo (see Flemal 1976). The size and shape of the pingo often reflect that of the original residual pond; the pingo may grow in distinct stages (Mackay 1973).

Open-system pingos (East Greenland type) develop most readily on slopes where free water under artesian pressure is injected into the site of the pingo. As the water approaches the surface, it freezes. The continuing introduction of water

Figure 11.11. The Discovery pingo. A dense stand of large birch trees on the pingo contrasts sharply with the open stand of small black spruce in the surrounding muskeg. Goodpaster district, Yukon region, Alaska, Aug. 1960. (Photo by H.L. Foster. From G.W. Holmes et al. 1968, fig. 14)

from below under hydrostatic pressure builds up the ice mass and domes the surface material into the pingo shape (see Müller 1963). The initial growth of some pingos may involve ice lenses above the water table, but their continued growth requires a proper combination of hydrology and soil texture (Ryckborst 1975). Open-system pingos are much more common in discontinuous permafrost where taliks and freely circulating water are not unusual.

The rate of pingo growth is variable, ranging from few centimeters to more than a meter a year (Washburn 1973; Mackay 1973). This indicates that many large pingos are quite old. In fact, C^{14} dates have shown two closed-system pingos to be 4,000 and 7,000–10,000 years old (Müller 1962). On the other hand, some pingos are growing today (Mackay 1973), suggesting that pingo formation did not occur in a unique or specific time in postglacial history but is probably a continuing process.

Fossil pingos (Wayne 1967; Mullenders and Gullentops 1969; Flemal et al. 1976) provide good evidence of an extinct permafrost environment in regions that are now temperate. Proving that features are fossil pingos, however, requires detailed field and laboratory analyses. The DeKalb mounds (northern Illinois), for example, consist of approximately 500 circular or elliptical mounds shaped from late Pleistocene sediments. Good evidence for their pingo origin is the large cluster of features and the fact that their centers consist of lacustrine silts and clays. Topographically, however, most of the mounds are less than 5 m higher than the surrounding ground level, and many are up to 300 m in diameter. Their recognition as fossil pingos could not have been suggested without careful field examination.

Thermokarst   Thermokarst features are a variety of topographic depressions that result from thawing of ground ice. They are more abundant where ice wedges are present and where the thermal equilibrium is somehow disturbed. The destruction of ground ice is due either to lateral degradation (back-wearing) where lateral erosion of surface water exposes the ice, or to vertical degradation (down-wearing) when the surface thermal properties are altered (discussed in Czudek and Demek 1970). These features are similar in many respects to those found in normal karst regions, but the fundamental process is melting rather than solution of rock material. The major landforms developed and brief descriptions are given in table 11.3.

Many of the processes that culminate in thermokarst features are initiated by broad climatic changes, but minor events such as fires, clearing of forest vegetation for farming, shifting stream channels, etc., probably also can upset the thermal balance enough to be the impetus. Mackay (1970), for example, reports a remarkable case in which trampling of vegetation by a dog led to subsidence of 18–23 cm in a small area within two years. Observations such as this demonstrate the remarkably fragile equilibrium of the periglacial system.

Table 11.3 Common thermokarst features.

Type	Formation	Features
Back-wearing	By lateral erosion, by streams, lakes, or ocean	Gullies Thermocirques Thaw lakes Thermokarst mounds
Down-wearing	By melting downward of ground ice. Usually develop as flat, shallow depressions in level areas.	Alases Thermokarst valleys

Patterned Ground

Periglacial regions are often characterized by a peculiar arrangement of surface materials into distinct geometric shapes. The features, collectively known as *patterned ground,* include such diverse shapes as polygons, circles, and stripes. They are common in permafrost regions, but perennially frozen ground is not a prerequisite for their development. They can be found anywhere within the periglacial regime or even in other morphogenetic regimes. Since patterned ground was first described in the late 1800s, almost every worker in periglacial areas has noted its striking appearance, and theories concerning its origin are almost as numerous as the features themselves. Perhaps the most extensive review of patterned ground and its origin was provided by A. L. Washburn in 1956, and that excellent work still stands as the cornerstone of our knowledge about these forms.

Classification   According to Washburn (1956), patterned ground features can be placed in a descriptive classification on the basis of two primary criteria: (1) the geometric shape (circle, polygon, etc.), and (2) whether the material composing the feature has been sorted (table 11.4). Sorting separates the larger-size fractions from the fine particles and usually generates a feature rimmed by stones and centered with fines (for a review see Goldthwait 1976). The most common forms seem to be circles, polygons, and stripes, with steps and nets as minor transitional types.

Polygons   Polygons are best developed on flat, nearly horizontal, surfaces. The sorted variety is bounded by straight segments composed of stones that surround a central core of finer material. They range in diameter from a few centimeters to over 10 m and always occur in groups rather than individuals. The stones marking the boundary are often oriented parallel to the direction of the rim. The stones increase in size with the dimension of the polygon and decrease in size with depth.

Table 11.4 Simplified classification of patterned ground features.

Types	Subtypes	Processes
Circles	Sorted Nonsorted	Features subdivided on basis of necessity of cracking (frost cracking, permafrost cracking, dilation, joint control). Where cracking not essential, heave and mass displacement important.
Polygons	Sorted Nonsorted	Includes ice and sand wedges. Cracking of all types. Heave, mass displacement, and thaw processes important in non-cracked types.
Nets	Sorted Nonsorted	Includes earth hummocks. Cracking of all types except joint controlled types. Heave and mass displacement in noncracked types. Thaw also important in sorted varieties.
Steps	Sorted Nonsorted	Cracking unimportant. Heave, mass wasting, and displacement in nonsorted. Frost sorting and thaw also important in sorted varieties.
Stripes	Sorted Nonsorted	All types of cracking important. Heaving, mass wasting, and displacement in nonsorted. Frost sorting and thaw also important in sorted types.

After Washburn 1956, 1970.

Nonsorted polygons differ in several ways from the sorted type: (1) the polygonal borders are devoid of stones, the geometric form being marked by furrows, and (2) they can be considerably larger than the sorted variety, often reaching 100 m in diameter. Nonsorted polygons are associated with permafrost as are the ice-wedge polygons discussed earlier, but they can also form by dessication and even in solid rock.

Circles   Sorted circles are stone-rimmed circular forms that range from a few centimeters to several meters in diameter (fig. 11.12). As with polygons, the stone size correlates with the dimension of the circle and decreases with depth. Circles, unlike polygons, can exist singly or in groups. Nonsorted circles lack the coarse, bouldery rim and are bounded instead by vegetation surrounding a central core of bare soil. In some cases the barren core develops because scouring and winnowing action by the wind preferentially erodes and initiates the circular form (Fahey 1975).

Stripes   Stripes are linear alignments of stones, vegetation, or soil on slopes. The sorted stripes are elongated strips of stones separated by intervening zones of fine sediment or vegetation. In general, the pattern changes as the slope increases, from polygons through sorted nets or steps into stripes. Stripes usually range in width from several centimeters to several meters. Their length varies but is some-

times greater than 100 m. As in other sorted features, larger stones are associated with bigger stripes, and the largest particles are found at the surface of the accumulation. Nonsorted stripes consist of vegetation or soil with intervening zones of bare ground. They are usually not as long as the sorted variety and are sometimes discontinuous.

Figure 11.12. Sorted stone circles caused by frost action, in Alaska Range, south central Alaska. June 1968. (Photo by T.W. Péwé. From Ferrians et al. 1969, fig. 12)

Origin   Patterned ground probably originates in a variety of ways; in fact, Washburn (1956) discusses 19 major hypotheses and concludes that patterned ground is polygenetic and that some forms result from a special combination of processes. Despite the conflicting opinions, the properties of the major features do imply that patterned ground exists in a range of transitional types. The characteristics of the different types reflect variations in a few controlling factors rather than completely different origins. Although details are still missing, it does seem certain that several processes are basic to the formation of patterned ground: (1) cracking or dessication, (2) heaving and its associated sorting, and (3) gravity as expressed by the slope inclination. Cracking is instrumental in developing polygonal geometry and is primary to both sorted and nonsorted types. Heaving separates the coarse and fine sediment fractions. In the usual case, repeated freezing and thawing move the stones upward and outward and simultaneously create a fine-grained nucleus as the smaller grains with greater cohesion contract farther inward and downward during the thaw phase. The process continues until adjacent cells coalesce, and the coarse stones are concentrated as polygonal or circular rims along the lines of interference. This model is accepted by most periglacial students although some evidence exists to suggest that texturally unsorted debris does not always respond to frost action in the same way (Ballard 1973). Slope angle aids gravity in elongating the forms, tending to form stripes rather than polygons or circles as the incline increases. The movement of stones during the formation of some sorted stripes, however, is greatly influenced by needle ice, perhaps even more than by heaving (Mackay and Matthews 1974).

Landforms Associated with Mass Movement

In contrast to patterned ground features, which are shaped primarily by frost action, certain periglacial forms result when the dominant mechanism is transportation. This does not mean that frost action is absent. On the contrary, frost action is a necessary prelude to the mass movement, and probably the two processes function simultaneously to fabricate the ultimate landform. Nonetheless, the final geometry of these features reflects the motion of rock and soil particles rather than a static, in situ disintegration of the parent material (for reviews see Benedict 1976 and S. E. White 1976b).

Solifluction Lobes   In regions covered by arctic or alpine tundra, large, tonguelike masses of surface debris, called *solifluction lobes,* may be prevalent. A lobe may occur as a single tongue, usually 30–50 m wide, or as one of many individuals that are joined laterally into a much broader field. In long profile an individual lobe is marked by a pronounced scarp at its leading edge that ranges from 1 to 6 m high. Commonly, however, a succession of lobes develops on a slope in which each upstream lobe overlaps the rear of the next downslope lobe, giving the entire sequence a staircase profile. Internally the lobes consist of angular unsorted debris, with particle sizes varying from silt to coarse boulders. The long axes of the boulders are usually oriented parallel to the direction of movement. Various types of solifluction lobes and related forms (turf-banked lobes and terraces, stone-banked lobes and terraces) are discussed in detail by Benedict (1970).

Solifluction lobes, as you might expect, move primarily by solifluction, but other processes are undoubtedly involved. Flow mechanics is indicated by the fact that the greatest movement occurs near the surface, and essentially no movement is perceived below a depth of 25 cm (Price 1972a). As the material moves downslope, it encounters zones of high resistance to flowage. These retard the forward motion and cause the soliflucting layer immediately upslope from the resistant zone to bulge slightly, producing the scarplike front. As Price (1972a) points out, however, the front is often stabilized, and only a thin surface layer slides to and cascades down the stationary front. The advance of the lobe in this manner is much slower than one might expect, probably averaging 1–3 cm a year, but the rate varies according to where the measurement is made. Maximum velocities are found along the longitudinal axis and decrease progressively toward the lateral margins (Washburn 1973), but rates are significantly affected by moisture content and gradient (Benedict 1970).

Blockfields   Blockfields (often known as *felsenmeer*) are usually thought of as broad, relatively level areas covered by moderate to large angular blocks of rock (Sharpe 1938). Similar accumulations of blocky debris also are found on slopes and are variously referred to as *block slopes, block streams, rubble sheets,* and *rubble streams* (see Washburn 1973). All these terms have been used interchangeably, and some semantic confusion exists. Our concern here is with block accumulations that occur on slopes, either concentrated in valleys or disseminated across a wide area. Although we may employ the term *blockfield* in reference to these accumulations (as have many others), some workers would consider this technically incorrect. True blockfields as originally described are not the product of mass movement but result in situ from frost wedging and heaving of the underlying bedrock. In this discussion we will actually be treating block streams or block slopes, but the term *blockfield* has been so widely applied to these features that we will employ it to mean any accumulation of blocky debris, including those found on slopes and subjected to mass movement.

Blockfields on slopes tend to be elongate with widths averaging between 60 m and 120 m and lengths from 350 m to 1.3 km. Maximum and minimum values vary drastically depending on the block size, the slope angle, and the distribution of the bedrock source. The gradient on the accumulation ranges widely, from 1° to 20°, although it most commonly is between 3° and 12°. Some workers feel that an appropriate upper limit for these features is 15° (Caine 1968). Blocky slope accumulations normally display an internal fabric indicative of movement; i.e., long axes of individual blocks are oriented downslope. The orientation may become transverse to the slope near the distal end of the deposit (Caine 1968), or it may diverge locally from the norm. Accumulations seem to be 4–20 m thick, but very few observations of the total thickness have been made. One deposit which was totally exposed displayed an internal layering characterized by an open-textured surface zone and a general increase in matrix with depth (Caine 1968). This type of fabric distribution, however, cannot be accepted as a general trait until more observations become available. Individual

Figure 11.13. River of Rocks blockfield (block stream), Berks County, Pa. Slope is 5°. Blocks in foreground are as much as 5 m long. (Photo by R. Craig Kochel)

blocks contained in boulder masses are striking in their enormity, most being 1–3 m in intermediate diameter and some as large as 6 m (fig. 11.13).

A number of mass-movement processes—ranging from avalanches or landslides to catastrophic floods—have been proposed for the development of block slopes or streams. Precisely which mechanism or combination of processes leads to the features is mostly conjecture, and the distinct probability remains that block accumulations have more than one mode of origin. The most commonly cited transporting mechanism is solifluction. Caine (1968), for example, concluded that in Tasmania the block slope material moved when the entire deposit contained abundant matrix. The open texture in the surface zone (up to 3 m deep) could have been created after movement ceased when the matrix was removed by ground or surface water (Caine 1968; Potter and Moss 1968). The solifluction model will work efficiently only if the lower portions of the accumulation contain a high content of fine-grained matrix. Such a layer was observed by Caine (1968) and suggested by Denny (1956) and Sevon (1969). However, it has not been proven as a universal property of these features, and until more conclusive data are found, the solifluction mechanism can be only a viable hypothesis. Nonetheless, it would explain the terrace and lobe development noted on some block streams (Denny 1956; Potter and Moss 1968). Further, the orientation see of blocks is similar to that characteristically produced by solifluction transport.

Some investigators have suggested that slope blockfields are emplaced by movement as *rock glaciers,* i.e., an accumulation containing interstitial ice rather than a soil matrix (Patton 1910; Kesseli 1941; Blagborough and Farkas 1968). Such a mechanism has the advantage of easily accounting for the open texture in the blocky debris, since no matrix sediment must be removed. The ice simply melts away as the rock glacier becomes inactive. A rock glacier transport, however, fails to explain why slope blockfields lack such normal characteristics of rock glacier deposits as steep fronts and pronounced transverse ridges and furrows. In addition, a greater shear stress is required in moving a rock glacier than in solifluction of a fine soil, and so, to attain the proper mobility, rock glacier accumulation should be considerably thicker than blockfield deposits (Wahrhaftig and Cox 1959; Potter and Moss 1968).

In comparison with inactive deposits, active blockfields conspicuously lack lichen cover, and the clasts show no signs of weathering. Because of the wide distribution of active block features in modern periglacial regions, inactive deposits in temperate zones or at altitudes well below functioning periglaciation have been widely interpreted as evidence of earlier periglacial conditions. For example, many blockfields have been described in the middle Appalachian Mountains (H.T.U. Smith 1953; Denny 1956; Potter and Moss 1968) where present climates are patently not periglacial.

Rock Glaciers   Rock glaciers have been described as features that are composed of angular rock particles and usually head in cirques or in high, steep-walled recesses. In plan view their shapes are similar to a true glacier (Sharpe 1938), often spreading in lobate or tonguelike forms (see White 1976b). The surface characteristically is bouldery, containing blocks up to 3 m in diameter, but this texture may be somewhat misleading because many contain finer-grained debris at depth.

Rock glaciers are several hundred meters to more than a kilometer long and vary in width, narrowing toward their source where the valley sides are more constricting (fig. 11.14). The deposits contain ice several meters below the surface. According to Potter (1972), the ice exists either as an interstitial cement (ice-cemented type) or as an extensive mass that contains widely dispersed rock fragments (ice-cored type). In either case the features may represent a transitional phenomenon between a nonglacial process and a glacier with a thick mantle of surface debris or, alternatively, an ice-cored moraine (for discussion see Barsch 1971; Østrem 1971). Topographically, rock glaciers may have furrows, cre-

Figure 11.14. Inactive rock glaciers in cirques on the north side of the Continental Divide, Sawatch Range, Colo. Arrows point to rock glaciers. (Photo by U.S. Geological Survey)

vasses, and lobate or transverse ridges similar to those found on true glaciers. Their front is normally near the angle of repose and therefore quite steep.

Modern, active rock glaciers are known in polar or subpolar regions and in the high mountains of the mid-latitudes. They vary in thickness from 15 to 50 m and move continuously at rates ranging from 2 cm to 158 cm a year, depending on local conditions. In some cases they are known to be advancing in areas where glaciers are presently retreating, perhaps because the surface rubble insulates the underlying ice and slows its response to climatic change (Osburn 1975). Inactive or fossil rock glaciers are widely distributed in regions that now have temperate climates. Whether they have paleoclimatic significance depends on precisely how such features originate. Although a variety of origins have been suggested, only those processes that produce a continuous forward motion are acceptable. The main controversy surrounding the genesis of rock glaciers revolves around the question of whether or not a true glacier is a required precursor (Outcalt and Benedict 1965). In other words, is the ice contained within the rock glacier necessarily glacier ice, or could it have accreted in a periglacial setting? The fact that some fossil rock glaciers in the southwestern United States developed in regions having no evidence of glaciation (Blagborough and Farkas 1968; Barsch and Updike 1971) seems to support a completely periglacial origin for at least some of the features.

The exact process of rock glacier transport is still not clear. Frost creep and solifluction have been proposed as the prime movers (Kesseli 1941), but the high basal shear stress developed in active rock glaciers seems to require involvement of other processes. Wahrhaftig and Cox (1959) stress that motion is induced by flow of the interstitial ice and probably functions best under permafrost conditions.

Environmental and Engineering Considerations

As mentioned earlier, human expansion into the arctic and subarctic regions is inevitable. This impending population growth has generated a growing concern as to whether our perception of geomorphic systems operating in these areas is sophisticated enough to develop the regions and still preserve their fragile environmental balance. The concerns are most evident in the engineering community that deals with construction problems in arctic regions on a day-to-day basis. The technical problems are real; our ability to cope with them depends on how well we understand the processes functioning within the system and how those processes are ultimately related to the environmental variables.

Most engineering problems are associated with permafrost. According to Brown (1970), any one of four basic engineering approaches may be employed when dealing with the permafrost condition: (1) disregard the permafrost; (2) use an *active* approach in which the permafrost in the near surface is eliminated by keeping the soil continuously thawed or by removing the natural soil and replacing it with permafrost-resistant material; (3) a *passive* approach in which permafrost is preserved by keeping the soil frozen at all times and thereby eliminating the annual thaw; and (4) designing structures to withstand frost action.

Which of the above methods is best depends on the local situation. In discontinuous permafrost, it may be possible to ignore the condition if the parent material is bedrock or a sandy or gravelly soil. Active approaches, such as simply removing the vegetation and its insulating effect, are also feasible on a local scale. In zones of continuous permafrost, the potential problems cannot be disregarded. There the most practical method is to keep the area continuously frozen, primarily by ventilation or insulation.

Table 11.5 Landforms that indicate ground conditions in arctic and near-arctic regions.

Feature and description	Associated ground conditions
Polygonal ground (ice wedges)—Usually indicates the presence of a network of ice wedges—vertical wedge-shaped ice masses that form by the accumulation of snow, hoarfrost, and meltwater in ground cracks that form owing to contraction during the winter. Wedge networks are also common in wet tundra where no surface expression occurs. (Subject to extreme differential settlement when surface disturbed.)	Typically indicates relatively fine-grained unconsolidated segments with permafrost table near the ground surface; also known from coarser sediments and gravels where wedge ice is less extensive.
Stone nets, garlands, and stripes—Frost heaving in granular soils produces net-like concentrations of the coarser rocks present. If the area is gently sloped, the net is distorted into garlands by down-slope movement. If the slope is steep, the coarse rocks lie in stripes that point down hill.	Indicate strong frost action in moderate well-drained granule sediments that vary from silty fine gravel to boulders. Surficial material commonly susceptible to flowage.
Solifluction sheets and lobes—Sheets or lobe-shaped masses of unconsolidated sediment that range from less than a foot to hundreds of feet in width that may cover entire valley walls; found on slopes that vary from steep to less than 3°.	Indicate an unstable mantle or poorly drained, often saturated sediment that is moving downslope largely by seasonal frost heaving. On steeper slopes they often indicate bedrock near the surface, and on gentle slopes, a shallow permafrost table.
Thaw lakes and thaw pits—Surface depressions form when local melting of permafrost decreases the volume of ice-rich sediments. Water accumulates in the depressions and may accelerate melting of the permafrost. Often form impassible bogs.	Usually indicate poorly drained, fine-grained unconsolidated sediments (fine sand to clay) with permafrost table near the surface.
Beaded drainage—Short, often straight minor streams that join pools or small lakes. Streams follow the tops of melted ice wedges, and pools develop where melting of permafrost has been more extensive.	Indicate a permafrost area with silt-rich sediments or peat overlying buried ice wedges.

Table 11.5 Continued.

Feature and description	Associated ground conditions
Pingos—Small ice-cored circular or elliptical hills that occur in tundra and forested parts of the continuous and discontinuous permafrost areas. They often lie at the juncture of south and southeast-facing slopes and valley floors, and in former lakebeds.	Indicate silty sediments derived from the slope or valley, also groundwater with some hydraulic head that is confined between the seasonal frost and permafrost table or is flowing in a thawed zone within the permafrost. Those in former lakebeds indicate saturated fine-grained sediments.

(From Ferrians et al. 1969).

Some heaving and settling will occur in the first several years after construction regardless of the method employed, because the permafrost requires some time to adjust into a new equilibrium condition. Thus, any project should prepare for the initial response in the system with a design that can withstand the stresses generated. Often the periglacial features described earlier give tangible clues about the near-surface ground conditions at any locality (table 11.5), and their recognition is an important first step in a site analysis. Air photos and remote sensing data interpreted by a geomorphologist can provide a valuable contribution in the planning stage.

Building Foundations

The bearing strength (ability to support load without plastic deformation) of a surface varies significantly with the type of material, both when it is frozen and when it thaws (Swinzow 1969). The loss of strength during thaw is most dramatic in clay and silt soils that are not permeable and so retain a large portion of the water released by melting. Bedrock or coarse-grained deposits are reasonably stable during the thaw phase because they drain easily. Because water cannot permeate below the permafrost table, the active layer in fine-grained soils becomes a quagmire that cannot support weight and therefore allows differential settling of overlying buildings, bridge abutments, etc. In addition, the lack of lateral drainage and the high suction potential will accentuate differential heaving during the next freeze. Building foundations, therefore, present enormous problems. In the case of small dwellings it may be adequate simply to live with the frost activity by pursuing an annual maintenance program. Large buildings, however, become unsafe if they are allowed to endure substantial heaving and settling, and preventive measures must be taken. Lobacz and Quinn (1963) were able to show that residual thaw zones persist beneath buildings that have no ventilation at their base, and that when buildings are elevated, seasonal frost will extend from the surface to the permafrost table.

Because of these conditions, most building constructions today in areas of continuous permafrost follow the passive approach and attempt to maintain the permafrost intact. The floor of the building is built on top of pilings driven into the permafrost layer, so that it normally stands 1–2 m above the ground surface, and air can circulate freely in the open space between the floor and the ground. This keeps the ground frozen in the winter because the heat radiating from the building is removed by cold air moving through the open space. Thawing does occur in the summer, but it is minimized because the surface is shaded by the building. Nonetheless, the piles supporting the building are susceptible to some heaving in the active layer; Johnston (1963) suggests they should be driven to a depth at least twice the thickness of the active layer in order to keep the heaving effect at a minimum.

Whenever possible, buildings should be located on a sand and gravel substrate, for reasons explained above. In coastal regions some communities, built prior to any understanding of periglacial processes, have been built on silty, deltaic plains. Further development may be impossible in such a setting; in fact, some small towns may have to be relocated if the problems associated with the permafrost cannot be controlled. Relocation was actually accomplished for one community that originally existed on the deltaic plain of the Mackenzie River. The Canadian government physically moved the community 56 km to a new townsite that was chosen because it had abundant sand and gravel for building sites and roads. Although construction of the new town was costly, its success demonstrated that a reasoned approach and proper regulations can allow development to proceed without harmful degradation of the system (for details see Pritchard 1962; Price 1972a; Cooke and Doornkamp 1974).

Roads and Airfields

The outbreak of World War II made the U.S. and Canadian governments aware that defending Alaska and the Northwest Territories from attack would be a monumental task. In an attempt to rectify this vulnerability, many necessary but hastily conceived projects were completed to facilitate the transportation of men and supplies to the area. Railroads, pipelines, and airfields were constructed, and the Alcan Highway, stretching almost 3000 km, was completed in only eight months, a truly remarkable engineering feat. The highway, however, proved to be a case study in permafrost mechanics, as the construction and maintenance problems were often baffling and disruptive (Richardson 1942, 1943).

We now know that problems associated with roads or airfields vary greatly because they are affected by microenvironmental conditions. In general, the most common ailments are (1) differential heaving and settling that creates a washboard surface (fig. 11.15), (2) sinking of portions of the surface, (3) destruction of bridges by spring floods, (4) burial of the pavement by slumping or other mass movements, and (5) icing. Icing is by far the worst problem. It occurs when water runs onto road surfaces during the winter, quickly freezes over large

areas, and often accretes to a considerable thickness. Groundwater under hydrostatic pressure may continue to flow through most of the winter and, with improper planning, may also run onto the road surface if road cuts open springs and seeps. Even something as innocuous as a river freezing may result in icing of a nearby highway surface. This happens when the river freezes from the surface downward and creates a pressure head on the bottom water. The water, thus mobilized, moves to the channel sides where it rises to the surface and overflows the channel banks onto the highway.

The best road protection requires wise planning and careful pavement design. These can reduce frost action in the active layer and prevent excess water from reaching the road surface. The highway route should avoid seeps and

Figure 11.15. Gravel road near the Umiat Airstrip showing severe differential subsidence caused by thawing of ice-wedge polygons in permafrost. Anaktuvuk district, Northern Alaska region, Alaska, Aug. 1958. (Photo from Ferrians et al. 1969, fig. 25)

springs by diverting around hills. If cuts must be placed in hillslopes, the pavement can be protected from icing by integrating a system of culverts and drainage ditches that will divert free water away from the road.

Pavement design usually begins with a layer of gravel 0.6–1.5 m thick spread directly on the tundra to insulate the underlying surface and so minimize thawing. Some melting must be expected, nonetheless, and the design should allow for reduced strength in the summer and for heaving in the winter (Linell and Johnston 1973; Hennion and Lobacz 1973). Even painting the pavement surface white may help to reduce the thaw of the substrate.

Utilities—Water and Sewage

Some of the most burdensome problems facing communities in permafrost regions involve utilities and services that are taken for granted in the mid-latitudes. Water supply, for example, is complicated by deep winter freezing, and large lakes may be the only sufficient, continuous, and dependable water source. Taliks within the permafrost may provide some water, but their distribution is not regular and they may not contain enough water to support even a small settlement. Although taliks beneath the permafrost can provide large amounts of water, it is very expensive. Drilling through the permafrost is costly, and pipes must be cased to prevent freezing. In addition, some deep subpermafrost water is brackish or highly mineralized (J. R. Williams 1970).

Locating an adequate, continuous water supply solves only half the problem, because the water must still be delivered to the place where it is to be used. People in tiny settlements have traditionally obtained water by carrying it in tanks from a nearby river or lake. In the winter, when the source freezes over, blocks of ice are melted. Larger communities cannot rely on such primitive methods and most have piped-in water. Either the pipes must be insulated or the water must be continuously circulated to prevent freezing. Many areas now utilize an enclosed heated and insulated conduit, called a *utilidor,* for water delivery. The utilidor complex, placed above the ground, is designed to hold water pipes, electric cables, and sewage disposal pipes, but its installation is very expensive. For example, the utilidor system in the town of Inuvik (Northwest Territories), Canada, cost approximately $750 per linear meter, and connections to individual homes about $400 a meter.

Sewage disposal is still another perplexing problem. Some small villages utilize individual buckets as collectors of all kinds of waste, human and otherwise. These are carried to a specified location on a lake or river where they are dumped in expectation of the spring thaw. Essentially, nature cleans the system at winter's end. For sanitation alone, larger communities must be more sophisticated and most now employ the utilidor system. In practice, however, utilidors solve only the immediate necessity of removing wastes from individual doorsteps. The final disposal of wastes is a most difficult problem. Few communities have efficient treatment plants, and most wastes are ultimately dumped as raw sewage into the nearest waterway. The problem is not nearly as acute in discontinuous permafrost areas, where septic tanks and cesspools are common. In continuous permafrost zones, however, alternative techniques must be developed. Sewage lagoons (large dug-out hollows) do allow waste to decompose anaerobically (Brown 1970) and are becoming a common disposal method. One novel idea, tried at a military base, was to use fuel oil instead of water in the sewage system. When the wastes are collected, they are sufficiently mixed with the oil to be injected directly into a furnace where incineration eliminates the problem and simultaneously generates heat and electricity for the base (Alter 1966).

Pipelines

It is perhaps fitting to end our discussion of periglacial geomorphology with a brief description of the one project that focused so much attention on this environment in recent years. The discovery of oil on the north shore of Alaska near Prudhoe Bay was exciting news to an energy-addicted nation. The conflict between energy and environment that ensued created a furor that ran from reasoned arguments to pure emotionalism. Nonetheless, it did force us to analyze in a systematic way the problems associated with construction of a pipeline in a permafrost region.

The Trans-Alaska pipeline was not the first pipeline constructed by the United States in a periglacial environment. The Canol pipeline system was built during World War II across 2575 km of discontinuous permafrost. Its purpose was to pump crude oil along the main pipe to a refinery located along the Alcan highway near Whitehorse, B.C. From there gasoline was pumped to airfields and other military establishments. The Canol project, like the Trans-Alaska pipeline, was steeped in controversy (Richardson 1944), and its use was discontinued in May 1945 after only 13 months of operation, presumably because of the high cost of maintenance and its decreased importance in the war effort.

The Trans-Alaska pipeline is different in most respects from the older Canol project, and little applicable knowledge was derived from the earlier work. The Trans-Alaska line is much farther north, and much of its 1270 km route from Prudhoe Bay to Valdez crosses zones of continuous permafrost. The 1.2 m diameter pipe is larger than the Canol pipe and, fully loaded with oil, weighs over 900 kg per meter (Harwood 1969). The crude oil enters the pipe at a temperature of about 58°C, but friction along the flow path increases this temperature considerably. Cooling devices are placed along the route to keep the temperature at a constant 63°C, but the oil is still hot and so presents the most difficult problem in preserving the permafrost environment intact.

Given the above, the planning had to consider the possibility of major changes in the physical environment as well as a host of engineering problems (Lachenbruch 1970; Kachadoorian and Ferrians 1973). A pipe resting on the surface and carrying hot fluids would certainly cause thawing of the underlying permafrost. Without preventive measures, the thaw would probably expand outward with time as shown in figure 11.16, perhaps at a decreasing rate, but possi-

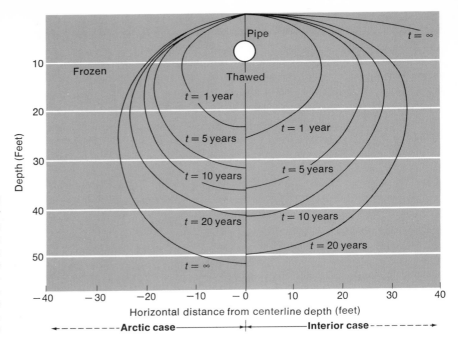

Figure 11.16. Theoretical growth of thawed cylinder around heated pipe placed in silty soil. Pipe is 1.2 m in diameter and kept at a temperature of 80°C. The curves at the left apply to conditions near the Arctic coast, those on the right to the southern limit of permafrost. (After Lachenbruch 1970)

bly never reaching an equilibrium state (Lachenbruch 1970). The soil could also become liquefied and unable to support the heavy pipe, and any solifluction induced by the liquefaction might carry the pipe along with the soil, producing ruptures and oil spills. As the pipe crosses materials of diverse texture and composition, differential heaving and settling are inevitable, and stresses generated in this manner could bend and rupture the pipe. This problem is especially pertinent where the pipeline crosses slopes underlain by ice wedges that melt more quickly and undermine the support rapidly. In addition, the actual construction could have inadvertently altered the delicate environmental balance. For example, destruction of the surface vegetation or diversion of river courses would have disrupted the thermal regime and changed the thickness of the active layer.

Faced with these potential hazards, the designers of the pipeline anticipated the problems based upon the best available knowledge of how the system would respond. In general three approaches are employed in different places: (1) the pipe is buried, (2) the pipe is suspended above the ground, and (3) the pipe is placed along the edge of a road that parallels the route of the pipeline. Burying the pipe is perfectly safe in coarse-grained soils that are well drained and contain little ground ice. In other zones a refrigerant is run along the pipe to keep the ground frozen. Suspending the pipe above the ground is necessary where the soil is particularly susceptible to thawing and also where the route crosses the path of migratory animals such as caribou. The above-ground suspension requires a dense network of pilings to provide the needed strength, and these are inserted to great depth. In addition, some of the supporting piles are refrigerated and others are distributed in a zigzag fashion to absorb the effect of expansion and contraction. Perhaps the most efficient approach is where the pipeline is combined with the necessary access road. The heat of the pipe, however, demands that the roadbed be adequately filled with gravel. The pipe is also insulated and covered with more fill.

Only time can tell us whether the design of the pipeline and the careful planning of the overall project will provide a continuous supply of crude oil with a minimum of hazards to the environment. Initial operation of the pipeline has been encouraging, although some mishaps have occurred. Most problems, however, have been related more to human error or sabotage than to poor design.

Summary

The processes and landforms found in periglacial environments have been briefly examined. In general, periglacial conditions are those existing in any cold, nonglacial setting regardless of latitude, but most regions of intense periglaciation are polar or subpolar. In many cases periglacial zones possess a unique property of permanently frozen ground, called permafrost. Permafrost adds a complicating factor in the activity of periglacial processes, and leads to the formation of diagnostic features such as ice-wedge polygons, thermokarst, and pingos that develop only where permafrost is present.

The driving processes in periglaciation are frost action and mass movements. Frost action includes wedging, heaving, and cracking. Mass movements usually involve frost creep and a form of soil flowage called solifluction. Both processes act in the development of most periglacial features, although certain forms seem to be more common where one or the other process is dominant.

The rigorous climate and the presence of permafrost lead to unusual difficulties in urban development that require engineering techniques not employed in other regions. The expected population expansion in arctic zones demands that we develop a sophisticated understanding of periglacial processes in order to maintain the fragile environmental balance.

Suggested Readings

Additional readings that provide greater detail about the topics discussed are suggested in the following abbreviated bibliography.

Benedict, J. B. 1970. Downslope soil movement in a Colorado alpine region: Rates, processes and climatic significance. *Arc. and Alp. Res.* 2:165-226.

Black, R. F. 1974. Ice-wedge polygons of northern Alaska. In *Glacial geomorphology,* edited by D. R. Coates, pp. 247-75. Proc. 5th Ann. Symposium, S.U.N.Y., Binghamton.

———. 1976. Periglacial features indicative of permafrost: Ice and soil wedges. *Quat. Res.* 6:3-26.

Embleton, C., and King, C.A.M. 1975. *Periglacial geomorphology.* New York: Halsted Press.

Ferrians, O. J.; Kachadoorian, R.; and Greene, G. W. 1969. Permafrost and related engineering problems in Alaska. U.S. Geol. Survey Prof. Paper 678.

Flemal, R. C. 1976. Pingos and pingo scars: Their characteristics, distribution, and utility in reconstructing former permafrost environments. *Quat. Res.* 6:37-53.

Goldthwait, R. P. 1976. Frost sorted patterned ground: A review. *Quat. Res.* 6:27-35.

Péwé, T. L. 1969. *The periglacial environment*. Montreal: McGill-Queen's Univ. Press.

Price, L. W. 1972. The periglacial environment, permafrost, and man. Assoc. Am. Geog., Comm. on College Geog. Resource Paper 14.

Tricart, J. 1969. *Geomorphology of cold environments*. Translated by Edward Watson. New York: St. Martin's Press.

Washburn, A. L. 1973. *Periglacial processes and environments*. London: Edward Arnold Ltd.

In regions of carbonate rocks and evaporites, weathering and erosion produce unique landforms called *karst* or, when widespread, *karst topography*. In contrast to most processes studied before, *karstification* (the processes that develop karst topography) is not easily observed because much of the geomorphic work is accomplished well below the ground surface. In fact, some modern textbooks of geomorphology completely ignore karst or treat it in a few descriptive paragraphs because it is primarily driven by the solution process and can justifiably be considered under the topic of chemical weathering. However, the magnitude of solution in the development of karst is so great, and the unique topography resulting from the process so widespread, that it seems to deserve special treatment.

Several excellent books are available that deal specifically with the topic of karst (Jennings 1971; Herak and Stringfield 1972; Sweeting 1973), and information contained in those sources was used as a framework for this chapter. In general, however, syntheses of literature dealing with karst, and published in English, are rather uncommon for reasons that will become apparent.

Introduction

Karst—Processes and Landforms 12

Definitions and Characteristics

Karst is defined by Jennings (1971) as "terrain with distinctive characteristics of relief and drainage arising primarily from a higher degree of rock solubility in natural water than is found elsewhere." The definition stresses two main points: (1) distinctive landforms and other surface characteristics developed on highly soluble rocks, and (2) a unique type of drainage pattern resulting from the karst processes. As all rocks are soluble to some extent, karst must develop only on those rocks that are particularly susceptible to solution. In such situations, the solution process can create and enlarge cavities within the rocks. This leads to the progressive integration of voids beneath the surface and allows large amounts of water to be funneled into an underground drainage system while simultaneously disrupting the pattern of surface flow. Because hydrology exists in physical "symbiosis" with solution, it becomes a very important aspect of karst phenomena. Solution integrates spaces, allowing pronounced underground circulation of water which, in turn, promotes further solution. A true karst area, therefore, possesses a predominantly vertical and underground drainage with a poorly developed surface network of streams. As our interest is primarily in processes, the basic concepts may be applicable in many areas that have surface drainage and have not developed a strong enough topography to be considered as karst. Importantly, however, the area may still be affected by karst-forming processes.

Karstlands (or karst landforms) are found in almost every region of the world, including arctic and arid zones, but are most likely to occur in temperate or tropical climates. In the United States, karst has developed wherever conditions are favorable, but the major concentrations of features exist in several general areas: (1) the Valley and Ridge province of the Appalachian Mountains (Pennsylvania, Maryland), (2) central Florida, (3) the plateau region of east-central Missouri, (4) a belt extending from south-central Indiana into west-central Kentucky, (5) the Edwards Plateau (Texas).

The term *karst* is a German adaption of the Slavic word *kras* or *krs* and the Italian word *carso* which literally mean "a bleak waterless place" (Monroe 1970) and also connote a bare rock surface. The early and classical description of karst was derived from a high plateau area near the Adriatic Sea between northwest Italy and Yugoslavia. This region is characterized by irregular topography containing many closed depressions and interrupted stream valleys; thus, early observers of the region considered karst as a geomorphic freak with chaotic and disordered topographic expression. Most geologists and geographers thought of karst as a curiosity rather than a topic for serious scientific investigations. It was not until 1893, when Jovan Cvijić (name rhymes with "screech") published his book *Das Karstphanomen,* that karst geomorphology was given true scientific status. Cvijić's work clearly defines karst landforms and demonstrates the predominance of solution in their development, even though the solution effect had been alluded to in earlier studies (Sawkins 1869; Cox 1874). Cvijić viewed karst development as a surface embellishment on soluble rocks; that is, the topography formed by differential solution from the surface downward.

Between 1900 and 1920 two important concepts bearing on karst geomorphology were introduced. Grund (1903) proposed the existence of "karst groundwater" which would rise into rock fractures during a wet season, accomplish solution, and then sink again with the dry season. This concept obviously disagreed with Cvijić's view that solution proceeded from the surface downward, and the discussions that followed led to the realization that hydrology was indeed a significant part of the karst system.

The second major innovation during the early 1900s was the introduction of cyclic analysis of karst landscapes. Both W. M. Davis and A. Penck viewed valleys without rivers as developing initially by normal fluvial action. The widened valley floor was subsequently altered by solution activity that occurred in a distinct stage of development (see Roglić 1972 for detail). The idea of cyclicity was further refined by Grund (1914), who suggested that individual landforms proceeded through stages toward the groundwater level which represented an ultimate corrosional level, i.e., a karst peneplain. Most geomorphologists have now abandoned this approach, stressing instead the equilibrium condition attained between process and form.

Both these early ideas engendered considerable controversy and stimulated much thought about karst and karst processes until the pressures of World War I took scientists in other directions. After the war, karst investigators followed many new routes. Of special note is the climatic approach to karst morphology, conceived and promulgated most forcibly by H. Lehmann, the great German geographer. The climatic control on karst processes is still accepted in some geomorphic circles, but many recent studies have abandoned it in favor of detailed analyses of the solutional process and their relation to hydrology. In the United States, most scientific effort did not begin until after World War I, and has historically been concerned with the origin of caves. The contributions of Davis, Bretz, and others are discussed later in the chapter.

This brief historical sketch highlights an important fact that is often overlooked. The field of karst geomorphology has its roots in central and eastern Europe, and contributions by English-speaking geomorphologists were rather meager during the formative years of the discipline. In one sense, this has complicated efforts by American geomorphologists to do research in the field because so much of the critical basic literature is in uncommon languages. In recent years, however, U.S. geomorphologists have added greatly to our knowledge of the field, and many articles are now appearing in scientific journals, especially the *Journal of Hydrology* and *Water Resources Research*. The historical development of the karst discipline also explains the proliferation of terminology that has arisen for karst features and processes. Although the Yugoslavian terrain inspired the study of karst, not all the names for individual karst features come from there, and almost every country has developed a particular terminology of karst geomorphology. Table 12.1 presents an abbreviated glossary of karst terminology adopted by the U.S. Geological Survey and followed in this discussion. The complete listing of terms can be found in Monroe (1970).

Table 12.1 Abbreviated glossary of common karst terminology

Term	Definition
aggressive water	Water having the ability to dissolve rocks. Especially water containing dissolved CO_2.
blind valley	A valley that ends suddenly where its stream disappears underground.
cave	A natural underground room or series of rooms and passages large enough to be entered by a person.
chamber	The largest order of cavity in a cave or cave system.
closed depression	Any closed topographic basin having no external drainage, regardless of origin or size.
cockpit	1. Any closed depression having steep sides. 2. A star-shaped depression having a conical or slightly concave floor.
cockpit karst	Tropical karst topography containing many closed depressions surrounded by conical hills. Similar to *cone karst*.
cone karst	Tropical karst with star-shaped depressions at base of many steep-sided, cone-shaped hills.
doline	A basin or funnel-shaped hollow in limestone ranging in diameter from a few meters to a kilometer and in depth from a few to several hundred meters. May be distinguished as "solution" or "collapse" if precise origin is known. In U.S., most dolines referred to as *sinks* or *sinkholes*.
exsurgence	Point at which underground stream reaches the surface if the stream has no known surface headwaters.
karren	Channels or furrows caused by solution on massive bare limestone surfaces. Synonym *lapiés*.
karst plain	A plain on which closed depressions, subterranean drainage, and other karst features may be developed. Also called *karst plateau*.
karst topography	Topography dominated by features of solutional origin.
karst valley	1. Elongate solution valley. 2. Valley produced by collapse of a cavern roof.
karst window	Depression revealing a part of a subterranean river flowing across its floor, or an unroofed part of a cave.
karstic	Adjective form of karst.
karstification	Action by water, mainly chemical but also mechanical, that produces features of a karst topography.
mogote	A steep-sided hill of limestone generally surrounded by nearly flat alluviated plains. Generally used for karst residual hills in the tropics. Synonym *pepino*.
polje	A very large closed depression in areas of karst topography, having flat floors and steep perimeter walls.
resurgence	Point at which underground stream reaches the surface. Reemergence of a river that has earlier sunk upstream.
room	A part of a cave system that is wider than a normal passage. Similar to *chamber*.
speleothem	A secondary mineral deposit formed in caves.
swallet, swallow hole	A place where water disappears underground in a limestone region. A swallow hole generally implies water loss in a closed depression or blind valley. A swallet may refer to water loss in a streambed even though there is no depression. Also *ponor, sink, sinkhole, stream sink*.
terra rossa	Reddish-brown soil mantling limestone bedrock; may be residual in some places.
tower karst	Karst topography characterized by isolated limestone hills separated by areas of alluvium. Towers generally steep-sided and forest-covered hills, often with flat tops.
uvala	Large closed depression formed by the coalescence of several dolines; compound doline.

After Monroe 1970.

Karst Rocks—The Resisting Framework

It is tempting to say that karst develops primarily on limestones and leave it there without further elaboration. Although technically correct, such a statement is grossly misleading because some limestones are not potential harbingers of karst topography. Only certain limestones have the unique combination of properties that allow them to succumb to karstification and foster karst topography. We will briefly examine what rock properties are most conducive to karstification and why.

Lithology   As a general rock group, limestones show great variability, but the accepted definition is that a limestone is a rock containing at least 50 percent carbonate minerals, most of which occur in the form of calcite ($CaCO_3$). Although very young limestones may contain some aragonite, the two most common carbonate minerals in limestones are a low-magnesium *calcite*, containing 1–4 percent magnesium, and *dolomite* (Sweeting 1973). If more than 50 percent of the carbonate minerals are calcite, the rock is called limestone; if more than 50 percent of the carbonate minerals are dolomite, the rock is called dolomite (Leighton and Pendexter 1962). The purer the limestone is with respect to $CaCO_3$, the greater will be its tendency to form karst. Corbel (1957), for example, suggests that 60 percent $CaCO_3$ is needed before any karst will form, and about 90 percent is required to expect a fully developed karst region. It should be noted, however, that even pure limestones may not produce a karst terrain because the processes also depend on permeability and rock strength.

Rocks other than limestone can produce karst if ancillary conditions are proper and the material is sufficiently soluble. Dolomites are commonly karstified, but unless they are very pure their porosity and permeability tend to be somewhat low. Evaporites such as gypsum and halite are also prone to karstification. In general, however, the occurrence of karst in dolomites and evaporites is minor compared to the widespread distribution of limestone karst regions.

The texture in limestones is an equally important control on karstification. Because most limestones are marine rocks, they usually contain, in varying proportions, insoluble detrital particles and chemically or biochemically precipitated material. In addition, a significant percentage of cement often binds the particles of different origin together. The amounts of these major constituents are the basis for classification of limestones (Folk 1959; Dunham 1962) and tend to produce textural variations that directly influence the karst processes. Folk (1959) classifies limestones on the basis of the three components mentioned above, which he calls allochem grains, microcrystalline calcite matrix, and sparry calcite cement (fig. 12.1). *Allochem grains* consist of relatively large particles such as mineral clasts of varying size and composition, fossil fragments, oölites, and aggregates of fine grains called pellets. *Microcrystalline calcite matrix* consists of fine-grained particles 1–4 microns in diameter, that usually accumulate as a chemically or biochemically precipitated ooze in shallow, calm seas. *Sparry calcite cement* is composed of calcite crystals that are 10 microns or more in diame-

The Processes and Their Controls

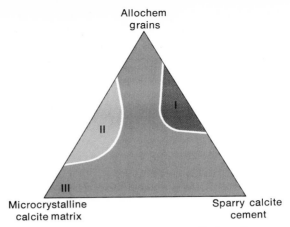

Allochem
grains

II

III

Microcrystalline
calcite matrix

Sparry calcite
cement

Figure 12.1. The classification of limestones. (Adapted from Folk 1959. Used with permission of the American Association of Petroleum Geologists)

I = Allochems with sparry calcite cement; called sparites

II = Allochems with ooze matrix

III = Few or no allochems; mostly fine, ooze material; called micrites

ter and may or may not represent postdepositional precipitation in spaces between the other constituents.

Without going into a discussion of limestone terminology, it must be stressed that certain combinations of these constituents seem to produce rocks with open textures, higher porosities, and greater susceptibility to karstification. Porosity is merely the percentage of void spaces in a rock or

$$ P = \frac{V_v}{V} \times 100 $$

where P is porosity in percent, V_v is the volume of voids, and V is the total volume of the material. Textures associated with rocks having a predominance of microcrystalline matrix are notoriously low in porosity, usually less than 2 percent. Where sparry cement dominates, the porosity usually increases to 5–8 percent. In ideal situations, allochem grains with little matrix and a small amount of sparry cement are the prime constituents, and porosity might increase to as much as 25 percent. Presumably, open textures and higher porosities should promote the solution process and the development of karst because the rock will hold more water. Some evidence exists to support this contention (Sweeting 1973).

Secondary Features   The porosity just discussed, which relates to intergranular void spaces, is known as *primary porosity*. The space associated with primary textures is commonly decreased with time by precipitation of cement, recrystallization, and change in mineralogy. In older limestones, for example, calcite tends to be replaced by dolomite, a process that usually decreases the primary porosity (Powers 1962). In addition, metamorphism may decrease carbonate porosity by inducing an increase in calcite grain size and reducing void size by pressure. Thus, primary porosity may be an ephemeral characteristic of limestones and perhaps is not as important in karstification as secondary porosity and permeability.

Secondary porosity comes from openings in rocks that occur along bedding planes or as fractures, such as joints and fault zones. It is generally agreed that the nature and pattern of these openings may be the single most important factor in karstification. Not only do they allow the rocks to hold more water, but they also promote circulation within the system by increasing permeability. Because permeability (the capacity to transmit water) depends on the continuity of voids, even rocks with high primary porosity may develop little karst if no secondary avenues of flow are present (Tricart 1968).

Porosity, then, has real importance only if the system is also permeable. Zones of weakness increase porosity but, more important, they have a decided influence on the permeability. This explains why permeability rates vary by as much as five orders of magnitude depending on the size and interconnection of fractures and bedding planes.

The character of bedding within soluble rocks influences the ease with which they can be sculptured into karst. Generally thin-bedded limestones are less likely to form karst than are units with massive bedding. This is because insoluble materials concentrated along bedding planes and shale interbeds tend to restrict free circulation of water through the thin-bedded units. Massive units are conducive to large karst features such as caves provided they are strong enough and compositionally pure. For example, massive formations of chalk, a rock often containing 95 percent $CaCO_3$, commonly do not develop karst because the material is too weak to maintain the topographic forms.

Joints are important avenues of water transport and enhanced permeability. Usually joints occur in patterns with one dominant direction and a secondary set of joints intersecting the main set at angles between 70° and 90°. The spacing of the joint sets can be very significant in the genesis of karst. If intersecting planes are too close, the rock may be highly permeable but too weak to allow the full development of karst.

Faults also transmit water effectively, but their precise role in karstification varies according to local conditions (Stringfield and LeGrand 1969). For example, fault zones sometimes have low permeability where voids are occupied by secondary mineralization associated with ore deposits. Such a deterrent to karstification, however, may be partly offset if the ore includes sulfide minerals, because oxidation produces sulfuric acid which makes the permeating fluids more aggressive in the solution process (Pohl and White 1965; Morehouse 1968).

In summary, for full development of karst, the rocks should have the following characteristics (Sweeting 1968).

1. The rocks should be massive, pure limestones that are hard and crystalline. Noncrystalline varieties such as chalk may not have enough strength to develop the classic karst forms.
2. The soluble strata should be thick, preferably a sequence greater than 100 meters.
3. The rocks should have well-defined bedding surfaces and numerous fractures.
4. The available relief (height above base level) should be great enough to permit free circulation of the water in the system.

The Driving Mechanics and Controls

Climate and Vegetation   We know intuitively that karstification requires abundant water that is free to circulate through the karst rocks. The water not only serves as the solvent in the development of karst, but also encourages the growth of vegetation and the soil microbial activity that add extra CO_2 to the system. Regions of low rainfall coupled with high temperature and evaporation rates will, therefore, be less susceptible to karst development. This is not to say that karst features never form in arid or semiarid regions. They do—Carlsbad Caverns in New Mexico and karst of the Nullarbor Plain in Australia attest to that fact. Normally, however, the topography produced is quite subdued, and features such as collapses and deranged surface drainage are not as striking. It is also possible that the karst in these areas formed at an earlier time when the climate was more humid than at present.

In extremely cold arctic or subarctic regions, the full development of karst is hindered by the presence of permafrost and by the low vegetal productivity which retards microbiologic activity (D. I. Smith 1969). Although water is present, it is often frozen. No free circulation exists and the important aspect of biogenic CO_2 (which we will discuss later) is missing. Karst is therefore a rather anomalous phenomenon in these areas.

At the other extreme, a tropical humid climate, with the ideal combination of temperature and precipitation to drive the solution process, should be most conducive to karstification provided enough relief exists to promote downward and circulatory movement of groundwater. Chemical reactions proceed more rapidly at higher temperatures, and lush vegetal cover combined with intense microbial activity impresses the tropical soil water with high partial pressures of carbon dioxide (P_{CO_2}). The subsurface water in such regions is very aggressive in solution, and a great abundance and variety of karst features are found in tropical regions (Jennings and Bik 1962; Monroe 1976).

The large variation in the controlling factors of karst in different climatic regimes has perpetuated the climatic geomorphology thrust initiated in the 1930s (Lehmann 1936). For example, Corbel (1959) analyzes rates of karst denudation in relation to climatic parameters, and others (P. W. Williams 1963; Douglas 1964) have modified the details of Corbel's original approach while still maintaining its dependence on climatic variables. Nonetheless, regions with very similar climates do produce karst areas that display wide variations in the spatial characteristics of the features. Even within a single karst domain, small differences in lithology or fracture patterns can inflict dramatic changes on the form of the resulting features. Consequently, recent work has placed less emphasis on the climatic variables even though their imprint on the karst regime cannot be denied.

The Solution Process   The solution process itself is in reality the critical function in the entire analysis of karst. Regardless of how conducive climate, lithology, fractures, and other variables are to karstification, karst topography would never develop if the solution process were somehow rendered inoperative. Its function or malfunction is fundamental to the topic, and we must attempt to understand its mechanics, at least in its simplest terms.

Laboratory studies tell us that the mineral calcite, like all common minerals, is soluble in pure water. At saturation it is soluble to the extent of about 12–15 ppm depending on the temperature of the water. This solubility is rather startling when compared to natural river waters where concentrations of Ca^{+2} and bicarbonate (HCO_3^-) are much greater (Livingstone 1963), indicating a substantial increase in the solubility. Since we are considering the same substance (calcite), it is obvious that the solvent in the natural system is not pure water. Rainwater, in fact, is not pure because it incorporates a variety of chemical constituents as it passes through the atmosphere. The most important of these is carbon dioxide (CO_2), which is soluble in pure water; some of the dissolved CO_2 reacts rapidly with the water to form a weak acid (H_2CO_3), called carbonic acid

$$CO_2 \text{ (dissolved)} + H_2O \rightleftharpoons H_2CO_3 \qquad (1)$$

This acid, however, is always dissociated into its ionic state, and the above reaction can be expressed more realistically as

$$CO_2 \text{ (dissolved)} + H_2O \rightleftharpoons H^+ + HCO_3^- \qquad (2)$$

The amount of CO_2 actually dissolved in water depends on the partial pressure of the carbon dioxide (P_{CO_2}) in the air standing at the air-water interface, and on the temperature of the water. The air in contact with the water can be in the atmosphere, or in spaces within the soil, or in subterranean cavities such as caves. In any case, the amount of CO_2 dissolved in the water increases as the P_{CO_2} of the air increases and as the temperature of the water decreases. Colder water will dissolve more CO_2 than warm water at any given P_{CO_2} value. In the atmosphere P_{CO_2} is rather small, having values averaging about 0.03 percent of volume (3×10^{-4} bar); similar values are found in most caves (Holland et al. 1964). Anomalously high P_{CO_2} values are found in air contained within soil or the vegetal litter covering it. Values of 1–2 percent of volume are common, and some poorly ventilated tropical soils may contain as much as 20–25 percent (Jennings 1971). The abnormal CO_2 values in soil air, and the concomitant large amounts of dissolved CO_2 in the soil water, stem from microbial action involved in the decomposition of vegetal matter. This "biogenic CO_2" is regarded by most karst experts as being the prime ingredient in the solution process.

Calcite itself is dissociated into an ionic state such that

$$CaCO_{3(calcite)} \rightleftharpoons Ca^{+2} + CO_3^{-2} \qquad (3)$$

However, the CO_3 ion produced quickly reacts with the H^+ formed when CO_2 is dissolved in water (reaction 2), and thus the dissociation of calcite also produces a bicarbonate ion because

$$CO_3^{-2} + H^+ \rightleftharpoons HCO_3^- \qquad (4)$$

It is clear from these reactions that the solution of limestone revolves around the $CaCO_3$–CO_2–H_2O chemical system. This system is extremely complicated, and its mechanics much more sophisticated than this introductory treatment can show. We are not dealing with a single reaction that produces solution of calcite, but a process that involves a series of reversible and mutually interdependent reactions, all proceeding at different rates, and each regulated by different equilibrium constraints. We will therefore explain the process only in the most general terms.

If we combine reaction (2) and (3), a general form of the process can be expressed by the following:

$$CaCO_3 + H_2O + CO_{2(dissolved)} \rightleftharpoons Ca^{+2} + 2HCO_3^- \qquad (5)$$

The bicarbonate ions (HCO_3^-) are derived from two sources shown in equations (2) and (4). In equation (4), the reaction of carbonate ions (from the dissociation of $CaCO_3$) and the H^+ (from the dissolving of CO_2 in water) produces a disequilibrium between the P_{CO2} in the air and the P_{CO2} in the water. This causes more CO_2 to diffuse from the air into the water and allows further solution of the calcite because reaction (5) is driven to the right side of the reversible equilibrium. The diffusion effect, which operates only where air is in contact with the water, proceeds faster if the water movement is turbulent. That is, water moving through the system rapidly and with some turbulence will ultimately dissolve more limestone; whereas stagnant water may become supersaturated with respect to $CaCO_3$ as equilibrium conditions are attained throughout the various reactions in the system (Kaye 1957; Weyl 1958).

To a large extent, the amount of CO_2 dissolved in water regulates the solubility of limestone. Where the amount of CO_2 dissolved in water is high, the fluid will aggressively attack the calcite. Soil water with its high P_{CO2} is therefore the most effective solution agent (Drake and Wigley 1975).

As mentioned, the general scheme of solution presented above is complicated by other factors. For example, each reaction functions at a different rate.

Sweeting (1973) points out that nearly all the possible limestone solution is accomplished within the first minute of contact, but for all the reactions to establish equilibrium throughout the system may take anywhere from 24 to 60 hours. Thus, the rate at which the solvent moves past the surface of the calcite grains is important in determining the total amount of limestone removed in solution.

Bögli (1964a) has described another important mechanism in the solution process which he called *mixing corrosion*. Essentially, when two bodies of water at equilibrium with different CO_2 contents are mixed, the resulting fluid needs less CO_2 to establish equilibrium than the sum of the CO_2 contained in the two original fluids. Some CO_2 is released and becomes available to promote further solution. Thrailkill (1968), however, cautions that such a mechanism really depends on the type of water masses being mixed. Where water above the water table is slowly permeating downward—a fluid called *vadose seepage* (Meinzer 1923)—the mixing effect may be completely defeated for a variety of reasons. However, water rapidly introduced to the subsurface, called *vadose flow*, may trigger considerable mixing corrosion when it meets the liquid standing at the water table (see Thrailkill 1968). In addition, the mixing of vadose seepage and vadose flow may cause pronounced undersaturation of the underground water and promote the solution process (Thrailkill 1972).

Surface Flow

Karst Hydrology and Drainage Characteristics

Karst regions are unique because their drainage networks characteristically are disrupted, and few rivers can traverse such areas in a continuous and unsegmented manner. The reason for this strange fluvial behavior is the facility with which surface flow can be diverted into the underground system. Depending on soil types and joint spacing, overland flow on hillslopes may be drastically reduced by infiltration, and in extreme cases no water will enter the nearby channels. The effect of infiltration is conditioned by the type of vegetation covering the surface. Holmes and Colville (1970) found evaporation from forested slopes in karst areas of southern Australia to be 2.2 times greater than that from grassy slopes, accounting easily for the low recharge of karst groundwater in forested regions as compared with land under pasture (Colville and Holmes 1972).

Rivers also lose water when some of the flow descends into *swallow holes* or *swallets,* shown on the map in figure 12.2. These are nothing more than open cavities on the channel floor that are capable of pirating a portion of a river's discharge or even the entire river (especially during low flow) into the underground system. Thus, a large part of the total flow of rivers in karst regions may follow a subsurface route that may or may not parallel the path of the river valley on the surface. Rivers that are able to cross a karst terrain as continuous surface entities have distinct hydrologic characteristics. For example, flood records for 114 basins of different sizes show that the mean annual floods (recurrence interval of 2.33 years) in carbonate basins of Pennsylvania were considerably lower than those in basins underlain by different rock types (White and Reich 1970). Commonly, however, the peak flow is spread over a longer period as the subsurface water is slowly released to the rivers. It appears likely that the precise hydrologic character of surface rivers in karstic areas depends greatly on the state of development of the underground drainage, especially the degree of interconnection between subsurface passageways (Ede 1975).

Underground Water—The Normal Hydrogeologic Setting

The increased demand for water that accompanied population growth, industrial expansion, and extended irrigation in the United States has brought with it a marked increase in the utilization of groundwater. The volume of water con-

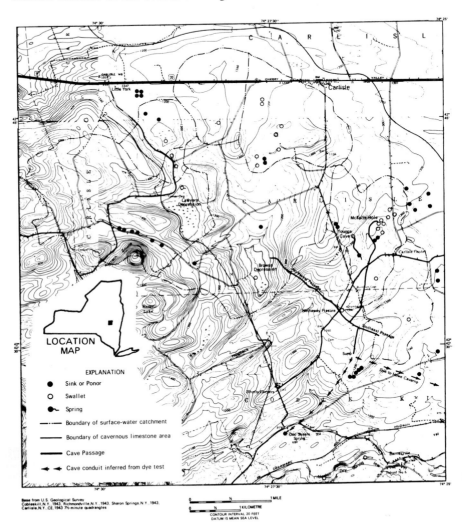

Figure 12.2. The distribution of swallets, sinks, and springs in a karst region of southeast New York. (From Baker 1976, fig. 15)

tained in cracks and pores in the Earth's underground reservoir is enormous, being estimated at almost 8 million km³ in the outer 5 km of the crust (Todd 1970). Unfortunately, this vast resource is not evenly distributed, and some regions are blessed with abundant groundwater while others are seriously deficient. Ironically, many of the most rapidly expanding areas of the United States are in regions with low reserves of surface and groundwater. The lack of available water, coupled with a growing demand, poses a real challenge to geologists and requires that we continue to expand our knowledge concerning the distribution, movement, and utilization of groundwater. Groundwater geology is a separate discipline in itself; however, we will diverge briefly to explore the basics of how normal groundwater systems work and how misuse of the system can have dire results.

The Groundwater Profile   Groundwater in porous and permeable rocks or unconsolidated debris usually has a rather distinct distribution that can be visualized as a vertical zonation known as the *groundwater profile,* shown in figure 12.3. Below the *zone of soil moisture,* water moves downward under the influence of gravity into and through the *zone of aeration.* In this zone, pore spaces are partly occupied by water, called *vadose water,* and partly by air that is physically connected to the atmosphere. At some lower depth all the pore spaces are occupied by water, marking the beginning of a thin, saturated zone, the *capillary fringe,* where water, held in tension, will not drain freely into a well. Beneath the capillary fringe the spaces are also filled with water in a zone known as the *phreatic zone.* The distinction between these two saturated zones is that phreatic water will drain freely into a well because the hydrostatic pressure in the phreatic zone is greater than atmospheric pressure. The *water table* marks the top of the phreatic zone and represents the level at which the hydrostatic pressure is equal to the atmospheric pressure.

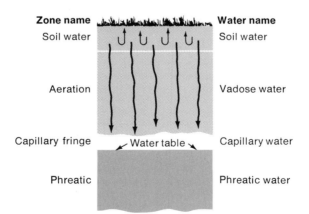

Zone name		Water name
Soil water		Soil water
Aeration		Vadose water
Capillary fringe	Water table	Capillary water
Phreatic		Phreatic water

Figure 12.3. Schematic diagram of zones and water types in the groundwater profile. (Not to scale.)

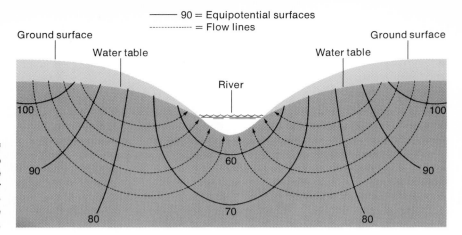

Ground surface   Ground surface

Water table   Water table

River

100   100

90   90

60

70

80   80

Figure 12.4. Movement of groundwater according to distribution of potential in the underground system. Water moves from high to low potential and perpendicular to the equipotential surfaces.

Movement of Groundwater   After descending to the phreatic zone, groundwater does not stagnate but is capable of further movement. Unlike flow in the zone of aeration, however, movement in the phreatic zone is not entirely controlled by gravity. Each unit volume of water contained in the phreatic zone possesses a certain amount of potential energy, called its *potential* or *head*. The amount of potential varies from droplet to droplet, being dependent on each drop's pressure and elevation above some datum. When pressure and elevation are known, the potential for each unit volume of water can be calculated, and water particles having the same potential can be contoured along surfaces known as *equipotential surfaces* (fig. 12.4). Although some diffusion occurs, ground-water particles move along paths that are perpendicular to the equipotential surfaces.

The movement of groundwater is a mechanical process whereby some of the initial potential energy of the water is lost to friction generated as the water moves. It follows that water moving from one point to another must have more potential energy at the beginning of the transport route than at the end. Thus, combining the above, groundwater always moves according to the following rules: (1) it moves from zones of higher potential towards zones of lower potential, and (2) it flows perpendicular to the equipotential surfaces (Hubbert 1940).

The velocity and discharge of groundwater flow is directly proportional to the loss of potential (head) that occurs as water moves from one point to another (fig. 12.5). This was demonstrated by Darcy in his experiments on flow through permeable material and led to his law of groundwater flow expressed as

$$V = K \frac{h_1 - h_2}{L}$$

where V is velocity, h is the head, L is the distance between points 1 and 2, and K is a constant of proportionality representing the permeability (or the hydraulic conductivity) of the medium. In figure 12.5, the velocity of flow can be calculated from the difference in hydrostatic level between two wells ($h_1 - h_2$) if the permeability of the material is known. Discharge can also be calculated by adding the cross-sectional area to the equation so that

$$Q = KIA$$

where Q is discharge, A is the cross-section area of an aquifer ($w \times d$), and I, called the hydraulic gradient, equals $\dfrac{h_1 - h_2}{L}$.

As mentioned above, the top of the phreatic zone is an important hydrologic feature known as the *water table* (fig. 12.3). Because hydrostatic pressure everywhere along the surface of the water table is equal to atmospheric pressure, the potential of water there is completely a function of elevation, and water at that surface will always move from higher to lower elevations. This partially explains why the water table is generally a mirror image of surface topography. Where surface water and groundwater systems are physically connected, the levels of rivers, lakes, swamps, etc., are merely surface extensions of the underground water table (fig. 12.4).

Aquifers, Wells, and Utilization Problems   *Aquifers* are lithologic bodies that store and transmit water in economic amounts. The most common aquifer is called an *unconfined aquifer* because it is open to the atmosphere and its hydrostatic level (the level at which water stands in an open hole) is within the water-bearing unit itself. The hydrostatic level in an unconfined aquifer is the water table. In some other aquifers the water is held in a porous and permeable unit that is not connected vertically to the atmosphere but is overlain and underlain by impermeable layers called *aquitards*. Water contained in these *confined aquifers* (fig. 12.6) will rise above the top of the aquifer when it is penetrated by an open hole. The level to which water will rise is called the *piezometric surface,* and its height above the aquifer itself depends on the difference in potential at the point where precipitation enters the aquifer (recharge zone) and the position of the hole (fig. 12.6). Sometimes the piezometric level is above the elevation of the ground surface, and water will flow freely out of the hole without pumping as *artesian flow*.

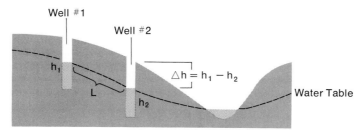

Figure 12.5. Water table and loss in head as water moves from well 1 to well 2 in unconfined aquifer.

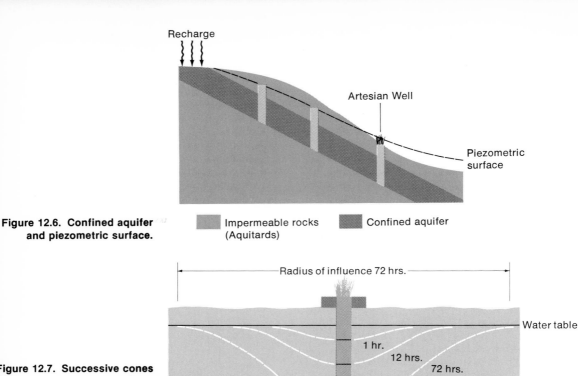

Figure 12.6. Confined aquifer and piezometric surface.

Recharge

Artesian Well

Piezometric surface

Impermeable rocks (Aquitards)   Confined aquifer

Radius of influence 72 hrs.

Water table

1 hr.

12 hrs.

72 hrs.

Figure 12.7. Successive cones of depression caused by drawdown of water table during 72-hour pumping of unconfined aquifer.

The development of an aquifer for water supply requires wells; the larger the demand, the larger and more numerous the wells. As a well is pumped, the hydrostatic level (water table or piezometric surface) surrounding the well is molded into an inverted cone known as a *cone of depression* (fig. 12.7). The cone develops because the release of water (or of pressure in a confined aquifer) is greatest near the well, causing pronounced lowering of the hydrostatic level, called *drawdown,* adjacent to the well. The effect of pumping decreases away from the well, and so the drawdown is less at the perimeter of the cone. In the initial phase of pumping, the rate of drawdown is high, but as pumping continues the rate gradually decreases until the cone attains a nearly constant form. The dimensions of the quasi-equilibrium cone depend on the rate at which the well is pumped and the hydrologic properties of the aquifer.

Utilizing groundwater can result in a number of environmental problems if the system is not carefully studied before development. For example, excessive drawdown can occur locally if wells are placed too close to one another and the radius of influence (maximum diameter of the cone of depression) of adjacent wells overlaps. This produces abnormally high drawdown in the zone of overlapping because the actual lowering of the water level is the sum total of drawdown produced by all the interfering wells. On a regional scale, most problems arise when more water is pumped from the aquifer over a period of years than is returned to the aquifer by natural or artificial recharge, a practice known as *overdraft*. In such a case the aquifer is actually being mined of its water, and on a long-term basis the normal hydrostatic level may be drastically lowered.

The effects of continued overdraft differ, but two types of responses will demonstrate the problems that can result from misuse of the groundwater system. First, in confined aquifers, the drawdown of the piezometric surface reflects the decrease of pressure within the aquifer. Because the pressure is lowered, water from the overlying aquitard seeps downward into the aquifer. As the aquitard drains, the normal load exerted by the weight of the overlying rocks compacts the aquitard and decreases its thickness. Ultimately, the process culminates in measurable subsidence of the ground surface. Some areas (Mexico City, Las Vegas, Central Valley, California, Houston-Galveston area) have experienced 1–5 meters of overdraft subsidence, creating a variety of annoying and hazardous conditions such as cracking of buildings, strain on buried pipelines, and destruction of well casings.

A second major effect of overdraft usually occurs in coastal regions where the aquifer is physically connected to the ocean. There two fluids (ocean water and fresh water) having different densities are separated by a sharp boundary, as diagrammed in figure 12.8. The location of the interface depends on the hydrodynamic balance between fresh water (density = 1.000 g/cm^3) and salt water (density = 1.025 g/cm^3). In general, the depth below sea level to the saltwater boundary is about 40 times the height of the water table above sea level. In such a situation, continued overdraft can cause pollution of the aquifer because minor lowering of the water table necessitates a much greater rise of the saltwater-freshwater interface, a phenomenon known as *saltwater intrusion*. For example, a 2 m drawdown of the water table requires a concomitant 80 m rise of the salt water, and any well extending to a depth greater than the new interface level will be polluted with nonpotable water.

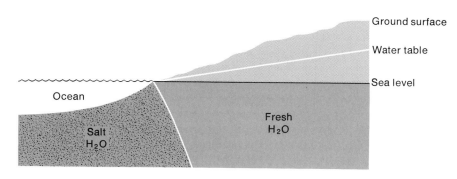

Ground surface

Water table

Sea level

Ocean

Salt
H$_2$O

Fresh
H$_2$O

Figure 12.8. Relationship of salty and fresh groundwater in an aquifer of a coastal region. Lowering of water table by overdraft requires a 40-fold rise in the boundary between the fresh and salt water and increases the possibility of pollution of the aquifer by salt water.

Many cities along the coasts of California, Texas, Florida, New York, and New Jersey have been affected by saltwater intrusion. The phenomenon, however, can function wherever two fluids of different density exist within the same aquifer. For example, the water supply of Las Vegas, Nevada, is in jeopardy from pollution by high-magnesium groundwater located about 24 km south of the city. The impending intrusion is due to the large overdraft from the aquifer beneath Las Vegas.

Karst Aquifers and Groundwater

This brief look at normal hydrogeology allows us to examine the differences between karst aquifers and the groundwater conditions in other porous and permeable materials.

Historically, karst geomorphologists have viewed the hydrology in karstic systems in discordant ways. Most conflicting opinions concerned the type of water movement, the depth to which groundwater would penetrate, and most important, whether or not a water table as we normally conceive it exists in karstic terrains. In the first two decades of the twentieth century, two schools of thought developed concerning the hydrologic system.

In one model, groundwater in karst regions is considered to be similar to the normal groundwater found in unconfined aquifers composed of other rocks or unconsolidated debris. The water presumably exists in a vertical stratification. Water in the zone of aeration moves downward to a pronounced water table by seepage or by flow in various types of fissures and cavities. Below the water table the phreatic zone is completely saturated, although the mechanics of how water moves through this zone are very controversial. The water table itself is subject to significant fluctuations, rising and falling on a seasonal basis.

The second major conceptual model of karstic groundwater completely denies the existence of a water table and with it the distinct vertical zonation of the water. Workers accepting this view believe that the distribution and movement of groundwater are controlled entirely by the spatial characteristics of the network of interconnected passageways within the rocks. Several points support this conclusion: (1) adjacent wells drilled into limestone often have hydrostatic levels that are significantly different in elevation; (2) tunnels reveal dry fissures immediately next to cracks that are filled with water; (3) tracing of water with dyes shows that paths of movement often cross one another (obviously flowing at different levels) and even pass under a surface stream from one side of a valley to the other, a physical impossibility in normal unconfined aquifers; and (4) poljes (see table 12.1) at the same level do not behave in a similar manner, some flooding in winter while others remain dry. Thus, the underground system is thought to be a collection of conduits functioning like three-dimensional rivers. The passages may be totally interconnected, or in some cases they may function like a single confined aquifer having a recharge area, a discernible pathway, and a separate point where the water emerges once again to the surface.

In an attempt to compromise these two concepts, Lehmann (1932) suggested that they represent two end members in a sequential development of the underground system. Initially the groundwater behaves like that in an unconfined aquifer, but as solution progressively widens and links fractures within the limestone, the water system begins to be controlled by structure. Thus, time or stage of development changes the groundwater condition from a young, unconfined type to an old, structurally controlled type. Although recent evidence shows that the state of development controls the pattern of groundwater movement through some karst regions (Ede 1975), and time is an important control of karstification, most geomorphologists now believe that these two basic models can simultaneously coexist in any given karst area.

As a result W. B. White (1969) has classified carbonate aquifers according to their hydrogeologic properties (table 12.2) as *diffuse flow, free flow* or *confined flow* types, each type having subtypes; the main classifications are depicted in figure 12.9. In diffuse flow, cavities are limited in size and numbers, caves are rare, a well-defined water table is present, and flow obeys or nearly obeys Darcy's law. In free-flow aquifers, the water moves through integrated conduits under the influence of gravity, often attaining turbulent flow. The flow is capable of transporting sediment as discharge is enlarged by surface runoff sinking into fractures (swallets). Discharge of the groundwater is usually through large springs that accumulate, at a single outlet, the water flowing through vast areas of underground drainage. Confined-flow aquifers are characterized by water that moves in response to pressure. They may be true confined aquifers as described earlier and may contain water under considerable hydrostatic head; thus, artesian flow conditions are possible in these aquifers.

It is now becoming clear that the different types of aquifers recognized by White may also be distinguishable on the basis of the chemical behavior of their water. Shuster and White (1971), analyzing 14 springs in the central Appalachians, suggest that diffuse-flow aquifers maintain a constant hardness and have spring water nearly saturated with respect to $CaCO_3$. Aquifers that utilize conduits are undersaturated and show considerable variability in hardness. This fits Thrailkill's (1972) suggestion that degree of saturation and P_{CO_2} are functions of discharge. Drake and Harman (1973) also found a relationship between the hydrogeologic classification and the chemical properties of the aquifer water, suggesting that the degree of calcite saturation and the equilibrium P_{CO_2} are sufficient to distinguish the groups.

The Relation between Surface and Groundwater

Assuming that flow through conduits is the unique characteristic of karst hydrology, we should briefly examine how water is exchanged between the surface and the underground system and vice versa, and how the properties of the passageways control the movement.

Table 12.2 Types of carbonate aquifer systems in regions of low to moderate relief.

Flow Type	Hydrological Control	Associated Cave Type
Diffuse flow	Gross lithology. Shaley limestones; crystalline dolomites; high primary porosity.	Caves rare, small, have irregular patterns.
Free flow Perched	Thick, massive soluble rocks Karst system underlain by impervious rocks near or above base level.	Integrated conduit cave systems. Cave streams perched—often have free area surface.
Open	Soluble rocks extend upward to level surface.	Sinkhole inputs; heavy sediment load; show channel morphology caves.
Capped	Aquifer overlain by impervious rock.	Vertical shaft inputs; lateral flow under capping beds; long integrated caves.
Deep	Karst system extends to considerable depth below base level.	Flow is through submerged conduits.
Open	Soluble rocks extend to land surface.	Short tubular abandoned caves likely to be sediment-choked.
Capped	Aquifer overlain by impervious rocks.	Long, integrated conduits under caprock. Active level of system inundated.
Confined Flow	Structural and stratigraphic controls	
Artesian	Impervious beds which force flows below regional base level.	Inclined 3-D network caves.
Sandwich	Thin beds of soluble rock between impervious beds.	Horizontal 2-D network caves.

From White 1969, Table 1, p. 16.

Springs   We have already described briefly how overland flow and river water are diverted to the groundwater system through fractures and swallets. In addition, interconnected spaces in the zone of aeration permit slow percolation of vadose seepage, and this water is added to the total accumulation of groundwater. We have not examined, however, how the moving water is discharged from the subsurface back to the surface. In diffuse aquifers, water will emerge according to the flow established under the constraints of potential. Some question lingers about the depth to which phreatic water will penetrate, and we will discuss this problem when we review the origin of caves. In conduit systems, however, water usually reaches the ground surface in the form of springs, and the characteristics of these interesting geomorphic features commonly reveal the distribution and properties of the aquifer itself.

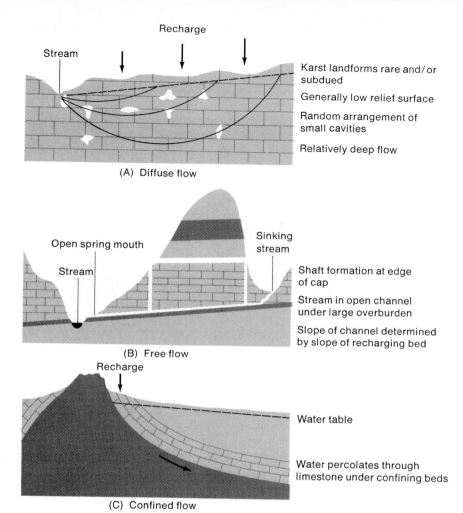

Recharge

Stream

Karst landforms rare and/or
subdued

Generally low relief surface

Random arrangement of
small cavities

Relatively deep flow

(A) Diffuse flow

Open spring mouth

Sinking
stream

Stream

Shaft formation at edge
of cap

Stream in open channel
under large overburden

Slope of channel determined
by slope of recharging bed

(B) Free flow

Recharge

Water table

Water percolates through
limestone under confining beds

(C) Confined flow

**Figure 12.9. Classification of
flow types in karst aquifers.
(See table 12.2.) (From W.B.
White 1969)**

Springs in karst areas are usually larger and more permanent than those in other regions because of the high infiltration associated with karst rocks. Those springs or seeps stemming from flow in diffuse aquifers are called *exsurgences* (fed by seepage), in contrast to those fed by groundwater moving through distinct conduits, which are called *resurgences*. This distinction is often confused since normal diffused flow may encounter cavities during its movement and ultimately emerge mixed with water derived from sinking streams (Thrailkill 1972). In general, however, resurgences are much more inconsistent in their discharge and chemistry. In some humid areas, springs respond quickly to changes in surface flow (White and Schmidt 1966; Baker 1973). Baker (1973), for example, found high, turbid discharges in New York springs during the spring season when snowmelt and rainfall produced high flow in the surface streams. During fall and winter, however, the discharge from springs was less by two orders of magnitude, and the water was clear.

The chemistry of spring waters and streams crossing karst areas can be very complicated because the dissolved load depends on so many auxiliary variables. For example, some investigators find seasonal variations in water chemistry (Groom and Williams 1965; White and Stellmack 1968; Ede 1975), and others find none (Thrailkill 1972). The type of aquifer (Shuster and White 1971; Drake and Harmon 1973), the equilibrium P_{CO_2} and source of CO_2 (Thrailkill 1972; Drake and Wigley 1975), and the temperature (Harmon et al. 1975) may mask any sensible variations in the water chemistry.

The physical controls on springs that rise from conduit systems are themselves quite variable. Some springs represent the emergence of an aquifer stream that flows through a cave system under the influence of gravity. Other springs reach the surface under pressure by rising up as cavities or fractures. In most cases, however, springs are ultimately controlled by large structural features or by stratigraphic relationships. Faults sometimes serve as the locus of large artesian springs, especially if the structural relationships produce a large potential difference between the recharge area and the fault (Burdon and Safadi 1963). In other situations, faults may deflect underground flow until it emerges into a valley that stands well below the level of the recharge zone (fig. 12.10). An excellent example of this is in the area of the Kaibab Plateau in Arizona (Huntoon 1974). The plateau north of the Grand Canyon supports no surface streams, and all precipitation not consumed by evaporation infiltrates into porous surface rocks. Huntoon (1974) observes that water collected in those rocks is transmitted to faults that act as drains, funneling the water downward to carbonate rocks that stand 3,000 feet below the plateau surface. The water is then discharged through springs at the base of the canyon.

Regional stratigraphy and the dip of beds also may control the position of springs. Baker (1973) showed that in New York state, spring placement often results from water moving down dip in well-integrated conduits floored by rocks with low permeability (fig. 12.11).

Figure 12.10. Fault control on the position of springs. Rainwater enters shales exposed at the surface along faults. Water follows fault zones into and through limestones until it emerges as springs. (Arrows show movement paths of water.)

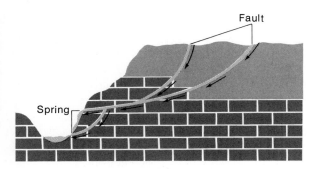

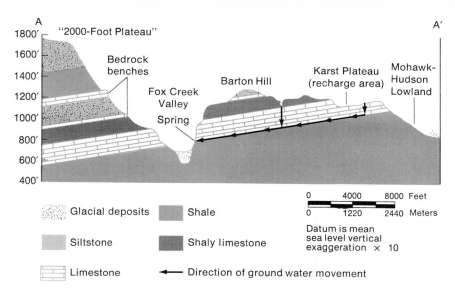

Figure 12.11. Stratigraphic control on the movement of groundwater. Water moves down dip at contact of limestones and impermeable shale. (From Baker 1976, fig. 2)

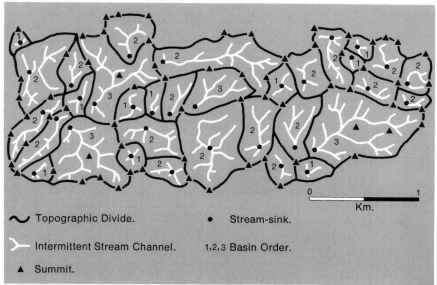

Figure 12.12. Components of karst morphometry and hydrology in New Guinea as ordered by P.W. Williams. (After Williams 1972b. Used with permission of the Geological Society of America)

Morphometry of Karst Drainage   A recent trend in karst geomorphology has been the attempt to describe karst drainage and topography in quantitative terms (P. W. Williams 1966b, 1971, 1972a,b; LaValle 1967, 1968; Baker 1973). Results of these studies show karst drainage to be well organized rather than having the chaotic nature assumed by early investigators. Williams devised a method, shown in figure 12.12, of ordering stream segments following the

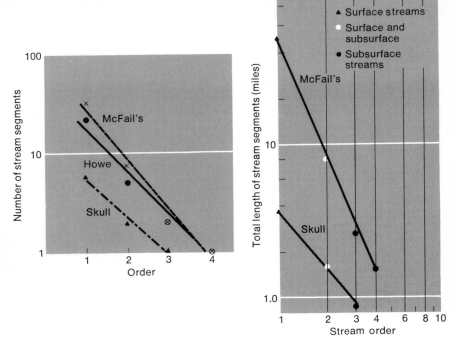

Figure 12.13. Morphometric relationships of the karst hydrology, including subsurface segments, in an area of eastern New York state. (From Baker, V.R., *Water Resources Research* 9, figs. 7 and 8, 1973, copyrighted by American Geophysical Union)

Strahler method such that every swallet would accept the drainage of a particular order. After each sink is ordered, measurements of morphometric parameters are made and plotted against the order hierarchy to demonstrate a statistical relationship (fig. 12.13). Baker (1973) extended the Williams methodology by including in the analysis underground links between the swallets and the springs that were identified by mapping and dye tracing. Stream orders so designated also showed a high correlation to the number of streams and total length of streams.

It is also known now that subsurface karst passageways have a tendency to meander like surface rivers (Deike and White 1969; Baker 1973), although lithologic and structural properties may cause the tendency to be fulfilled less completely than is the case with free meanders of surface streams (Ongley 1968).

Surficial Landforms

The solution process, working on rocks with diverse properties, results in a number of surficial landforms that define a true karst. The features range in size from tiny modifications of exposed limestone outcrops to large depressions and hills that dominate the topography. However, they all manifest the corrosional process. Unless otherwise noted, the features discussed originate in the humid-temperate climatic regime; a special section will consider humid-tropical karst because of the widespread occurrence and anomalous topographic forms generated under tropical conditions.

Closed Depressions

If someone were to ask what kind of landform best typifies a karst terrain, the answer would have to be closed depressions. Although these depressions range in size from tiny holes to those covering wide areas, they all have in common the property of supporting no external surficial drainage.

Dolines   By far the most common karst landform is a closed hollow of small or moderate size called a *doline* (in the United States often called a *sink* or *sinkhole*). Dolines are usually wider than they are deep, having diameters ranging from 10 to 100 m and depths between 2 m and 100 m. In plan they are circular or elliptical, but their cross-profile shape can vary considerably from the normal funnel-like form to shapes resembling a disc, bowl, or cylinder. Occasionally isolated dolines occur, but more commonly they are abundant enough to provide a karst terrain, such as the one in figure 12.14, with a strongly pitted appearance.

Figure 12.14. Ground surface pitted by dolines. Monroe County, Ill. (Photo by C. William Horrell)

For example, Malott (1939) estimates that as many as 300,000 dolines exist in the karst region of southern Indiana. As figure 12.15 illustrates, dolines can justifiably be considered as the fundamental element of karst because when present in large numbers they substitute for the valleys that dominate the normal fluvial environment.

The term "doline" has suffered through enough different connotations to prompt some authors to call for its elimination, but its use is so widespread that its removal is virtually impossible. In fact, Cvijić (1893) used the term in his classic book on karst, and classifications of doline types are firmly established (Cramer 1941). A variety of doline types have been described, including *solutional dolines, collapse dolines, alluvial dolines,* and *solution subsidences* (Cramer 1941). We will focus on the solutional and collapse dolines, which are the predominant forms.

Figure 12.15. Area of southern Indiana showing well-developed karst topography. Note the predominance of dolines and the absence of surface drainage. (Corydon East Quadrangle, Indiana. U.S. Geol. Survey 7 1/2 ' quadrangle. Contour interval = 10 ')

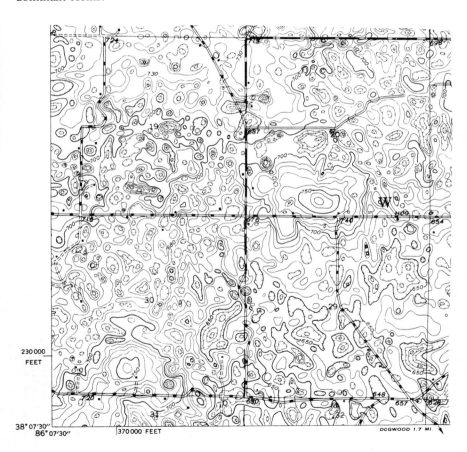

Solutional Dolines   Waters infiltrating into joints and fissures enlarge the cracks by solution and create a closed surface depression called a solutional doline. Many reports have documented the surface-downward origin, and it seems clear that this form represents the paramount doline type. As joints enlarge, especially at the intersection of joint trends, the surface is depressed, and internal drainage fosters continued development of the feature.

Like all geomorphic features, solutional dolines develop best where controlling factors are combined in a particular way. The factors most conducive to the formation of solutional dolines are these:

1. *Slope*. Since ponding or retardation of flow accelerates infiltration, the frequency of solutional dolines is inversely proportional to the surface gradient. Steep slopes promote rapid flow across the surface, and so, valley floors or gently undulating plains are the best places for solvent action to initiate the process. Dolines formed on steeper slopes tend to be asymmetric; however, that phenomenon can also be produced in other ways.
2. *Lithology and structure*. Porous limestones are less susceptible to solutional doline formation than dense limestones that are well jointed. The joints allow selective solution rather than a uniform corrosion over the entire surface. Solutional dolines can form in dolomites, but the features are usually deeper with steep, rocky sidewalls. Structures tend to align and elongate the dolines parallel to the major trends (LaValle 1967; Matschinski 1968; Kemmerly 1976), but the degree of control depends on many variables.
3. *Vegetation and soil cover*. Soil and vegetal cover usually increase solution activity because of the CO_2 factor. Other factors being equal, solutional dolines will develop more rapidly under an organic-rich cover than where surfaces are bare. Trees seem to be especially important in this process (fig. 12.16). In the central United States, Malott (1939) found that most

Figure 12.16. Doline near Dongola, Union County, Ill.

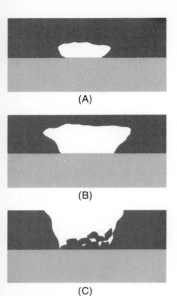

(A)

(B)

(C)

Figure 12.17. Sequential development of a collapse doline.

dolines were formed beneath a residual soil known as *terra rossa* and developed as symmetrical depressions with flat bottoms and gentle side slopes. This is somewhat different from the downward-tapering conical shape expected in solutional dolines, but the accumulation of sediment at the base of the opening often ponds water and allows the bottom to spread laterally. Hall (1976) points out that infiltration through dolines is often controlled by the water demands of sediment accumulated in the feature. In his study, no groundwater recharge was accomplished until loess within the doline was completely saturated.

Collapse Dolines   Collapse dolines differ from solutional dolines in that the depressions are initiated by solution that occurs beneath the surface. Expansion of caverns, caused by corrosion and by the roof material falling under gravity, decreases the support of the overlying rock material. When the subsurface hole widens, the lithologic load is held up at points that are separated by considerable distance; eventually the weight of the overlying rocks exceeds the strength of the subsurface base, and collapse occurs, as diagrammed in figure 12.17. Collapse dolines tend to have greater depth:width ratios than solutional dolines. Their sidewalls are characteristically steep and rocky, and the bottom is filled with fragments of the collapsed debris.

The collapse process is facilitated in humid environments where underground drainage is well established, and especially in the downstream portion of that drainage, where dolines seem to be deeper and more numerous (LaValle 1967). Some evidence suggests that collapsing can be initiated by rapid lowering of the water table. Although geological processes such as river entrenchment can cause this, the association of drawdown with collapse dolines sometimes introduces human action as a geomorphic agent, commonly with catastrophic results. A classic example of human intervention into the natural balance was reported by Foose (1967). Near Johannesburg, South Africa, a mining company lowered the water table by an extensive pumping program in order to have access to deeper ore that existed in the phreatic zone. Several years after the project's completion, large collapse dolines, some 125 m in diameter and 50 m deep, began to form suddenly and with disastrous results. In December 1962, an ore refining plant dropped 30 m into a doline as the surface suddenly collapsed, and 29 men were killed in one terrifying moment.

Other Doline Types   In addition to collapse and solutional dolines, other types have been distinguished, primarily by the nature of material covering the soluble limestones. Where the limestones are overlain by unconsolidated debris, solution along joints in the limestone causes the overlying sediment to sag inward and form shallow depressions at the surface known as *alluvial dolines*. *Solution subsidences* are analogous except that the overlying material is lithified, being usually shales or sandstones. Perhaps the best example of this phenomenon is the subsidence revealed in the Millstone Grit of England. Solution is often concentrated in sags along an unconformable contact between the Millstone Grit and the underlying Carboniferous Limestone, causing subsidence of the grit unit (Thomas 1973).

Uvalas and Poljes   Closed depressions of larger size than dolines are called *uvalas* or *poljes*. Presumably, as dolines enlarge they coalesce into hollows with undulating floors, the irregularity being produced by the differences in size of the integrated dolines. Jennings (1967) reports a single uvala that was constructed from 14 separate dolines of diverse size and shape. Uvalas have no specific size requirements as they range from 5 to 1,000 m in diameter and from 1 to 200 m in depth. Their plan shape can be highly irregular as a result of their unique origin.

Poljes are relatively large closed depressions with flat bottoms and steep sides. They are irregular in plan and usually are elongated along the strike of bedding or some zone of structural weakness. Thus, they can be structurally or lithologically controlled, and some expand by pronounced lateral corrosion when they are temporarily filled with water. Gams (1969) places a rather vague minimum size requirement on poljes, suggesting that they must be at least several square kilometers in area.

Poljes often abut against impermeable and nonsoluble rocks, and rivers flowing through those rocks may extend partly across the polje surface before they sink. In times of high flow the sink may not be able to absorb the discharge, and shallow lakes occupy the polje basin. In dry seasons evaporation may destroy the lakes. As discussed earlier, polje lakes may also form and disappear with changes in the underground hydrology.

Karst Valleys

The second major topographic group in karst topography is karst valleys, which show a variety of characteristics. In general they can be divided into several types with clearly discernible properties.

Allogenic Valleys   *Allogenic valleys* head in impermeable rocks adjacent to the karstic area. As the surface flow originating in the nonkarstic rocks enters the karst region, it forms spectacular gorges with steep, canyonlike walls. For instance, the Tarn gorge in the Grands Causses area of France is 2 km wide and 300 m deep. Such magnificent valleys obviously require considerable discharge to develop, with a combination of solution and fluvial abrasion as the driving mechanics.

The longitudinal extent of allogenic valleys depends on the sustainable discharge. Small streams tend to disappear rapidly into the karstic terrain, and their gorges are quite short. Larger rivers may extend valleys for some distance, and when they sink, may provide enough subsurface flow to produce the meandering caves and natural bridges that are commonly associated with allogenic valleys.

These latter features, however, may have a variety of origins and do not automatically provide evidence of an allogenic valley.

The steep valley sides probably stem from the combination of lack of slope wash in the limestone area (because surface flow is rapidly infiltrated) and lateral undercutting by the allogenic river. Thus, the side slopes are not allowed to flare as the valley widens, but the precise shape is somewhat dependent on the limestone itself, massive units perfecting steeper slopes.

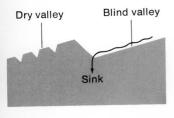

Figure 12.18. Longitudinal profile of a blind valley and dry valley.

Blind Valleys   Most surface rivers traversing a karst surface, no matter where they originate, eventually sink into the underground system. They may disappear into holes within the channel (swallets or swallow holes), or the entire drainage may be inside a doline and so all surface water (confined in a channel or not) sinks into the base of the doline. As figure 12.18 shows, rivers flowing across a karst surface in well-defined valleys tend to lower the valley floor upstream from the sink faster than the reach downstream from the sink because the flow is drastically diminished at the point of infiltration. Eventually a limestone scarp may develop, separating the two channel reaches. This scarp can vary in size from a few meters in small streams to tens of meters in larger rivers (Malott 1939). Usually the cliff increases in height with age. Rivers terminating at the cliff face are said to be occupying *blind valleys*. If some flow is able to overtop the threshold cliff during snowmelt- or rain-induced floods, the valleys are called *semiblind* (although this distinction adds little to the understanding of the formative process). In fact, if the sink cannot accept water fast enough, lakes may temporarily exist in blind or semiblind valleys after a flooding event.

Blind valleys may extend for considerable distances across a karst plain and may be entrenched well below its general level. The amount of incision and the shape of the valley, however, seem to depend on the original hardness (carbonate content) of the water (Gams 1965). Water with a low initial hardness will dissolve more limestone, thereby making the valleys longer and wider. An alluvial fill in the channel aids in the widening process because it effectively retards corrosion of the underlying limestone and facilitates lateral corrosion.

Pocket Valleys   *Pocket valleys* are essentially the opposite of blind valleys in that they begin where groundwater resurges rather than where it sinks. They are normally associated with large springs that resurge on top of an impermeable substrate at the foot of a thick exposure of karst limestone.

Pocket valleys, sometimes called *steepheads,* are usually U-shaped in cross-profile, having steep sidewalls. They characteristically have a steeply inclined headwall that may be recessed by spring sapping which undermines the overlying limestone. Their dimensions vary depending on the spring discharge and the nature of the karst rocks, but some are 8 km long, 1000 m wide and 300–400 m deep (Sweeting 1973).

Dry Valleys *Dry valleys* have all the properties of normal fluvial valleys except, as the name implies, they have no well-defined watercourses or carry only ephemeral flow in response to massive floods upstream. They probably represent the most common form of karst valleys and, as mentioned earlier, are the most favorable locale for doline formation because rainwater tends to pond on the valley surface and sink into the rock fractures. Like all karst valleys, their sides tend to be steep but factors of lithology and age cause some variability in their cross-profile shape.

The origin of dry valleys is complex and varied. In the simplest and most obvious case they represent the downstream reach of a blind valley that absorbs all the surface flow at a particular sink (fig. 12.18). Some dry valleys, however, have a well-integrated drainage pattern with numerous dendritic branches. These probably result from superposition of a fluvial system flowing on impermeable rocks that once covered the limestone strata. Eventually, stripping of the covering rocks and incision into the limestone induce the flow to submerge, and the valleys become dry. Some dry valleys are also genetically related to periglacial conditions where permafrost melting and summer rain provide the flow needed to erode the valleys (Brown 1969). In addition, when climate change or tectonics initiates downcutting of allogenic valleys, tributaries may become dry and stand as hanging valleys above the main valley level. Entrenchment also causes swallow holes to migrate upstream and springs to migrate downstream, thereby extending the intervening zone occupied by dry valleys (Jennings 1971).

Minor Solution Features

In many karstic regions, a myriad of minor solution effects sculpture and etch the surfaces of exposed or thinly covered limestones. Many of the original descriptions of such features came from work in different portions of southern Europe, resulting in a mind-boggling variety of terms and classifications (for example, Bögli 1960). It is generally accepted, however, that the German word *karren* should be employed to encompass all minor solutional forms (table 12.3). Karren forms exist in a staggering diversity of grooves, pits, flutes, and pinnacles, and their superposition or intermingling gives the limestone an irregular, low-relief topography (fig. 12.19). In general, the vertical relief produced on the surface ranges from centimeters to a few meters. The length of linear features depends primarily on the dimensions and slope of the exposed limestone, but they are rarely longer than 25 m.

The type and density of karren depend on a variety of fundamental controlling factors. As Jennings (1971) points out, these controls can be passive in nature, such as lithology, fracturing, and soil or vegetal cover; or they may be active controls such as temperature and precipitation. In addition, it is clear that some karren are not adjusted to the present environmental conditions, indicating that geomorphic history involving climate change, erosion, or human interven-

Figure 12.19. Intense karren development on surface of limestones in Puerto Rico. (Photo from Monroe 1976, fig. 50)

Table 12.3 Classification of karren.

Linear or Curvilinear		Circular or Crescentic	
Solution Flutes	Straight, fine-textured hollows 1–2 cm deep, 2 cm wide; uniform spacing; 10 cm–several meters long; hollows separated by sharp-crested ridges, steep outcrop faces; oriented parallel to slope direction.	Rainpits	Small holes on flat bedrock surface; < 3 cm in diameter, 2 cm deep; formed by dripping action.
Solution Runnels	Definite channels, 40 cm deep, 40–50 cm wide, up to 20 m long; sometimes called *grikes* when they form as solution-widened joints or bedding planes.	Solution Pans	Flat-bottom depressions < 1–50 cm deep, 3 cm–3 m wide. Form under local vegetal cover on flat bedrock.
Solution Ripples	Perpendicular to slope direction; 2–3 cm high, 10–50 cm long; form on steeply dipping faces.	Solution Bevels	Smooth, flat treads and scarps; treads 20 cm–1 m long; scarps 3–5 cm high. Formed by water moving over bedrock faces with moderate slopes.

tion is important. Jennings (1969) describes a variety of karren in an arid zone of Australia that must have been initiated during the pluvial climate associated with the Pleistocene epoch.

Perhaps the most fundamental passive control is whether the responsive limestone surface is completely bare or covered by soil or vegetation. Each situation produces different karren forms, and each utilizes water with different CO_2 contents and, therefore, differing corrosive abilities. The other major passive control is the nature of the limestone itself. Karren forms are usually less well developed on dolomitic limestones or on pure limestones that are porous or contain a framework of insoluble grains (Pluhar and Ford 1970). The lithologic control, however, can be offset by pronounced fracturing and well-defined bedding. Joints and exposed dip slopes that are steep tend to promote elongate karren types. These contrast with flat, unjointed surfaces which are likely to exhibit solution pits or basins.

The active climatic factors are most evident in the effect of variations in precipitation type and intensity. Karren develop readily in mountainous, snowy regions because water permeating through snowbanks has a slightly higher CO_2 content than free water. The character of rainfall seems to be important because regions with high-intensity rains tend to produce a greater abundance of karren than those receiving a more regular, drizzling type of precipitation. Thus, regions that have pronounced seasonality combined with torrential storms are very likely to develop karren, while dry areas or regions with a constant soaking rain, such as Great Britain, are notable for their lack of minor solution features.

Tropical Karst

The normal karst features just described can be compared with the spectacular karst topography developed in humid-tropical climates. Once again, we find that studies of tropical karst carried on in diverse regions of the world have resulted in a confusing array of terms describing the same features. Nonetheless, it is safe to say that every landform recognized in karst terrains of temperate regions is present in tropical karst. What sets tropical karst apart seems to be the fact that the general landscape is dominated by residual hills rather than the closed depressions so characteristic of the temperate karsts. Most of the residual forms occur as steep-sided, cone-shaped hills which, combined with surrounding depressions, constitute a particular type of karst topography known as *cone karst* (Ger. *kegelkarst*). The subforms within the general category of cone karst are numerous and transitional, but two topographic components called *cockpits* and *towers* seem to be the uniquely diagnostic elements of tropical karst; in fact the terms "cockpit karst" and "tower karst" are deeply entrenched in the literature.

Cockpits are similar to temperate-climate dolines except that they are usually irregular or star-shaped depressions that surround the residual hills (fig. 12.20). Cockpit karst was first described and named in Jamaica and was originally ascribed to solution along joints and faults (Lehmann 1936; Sweeting 1958). Later

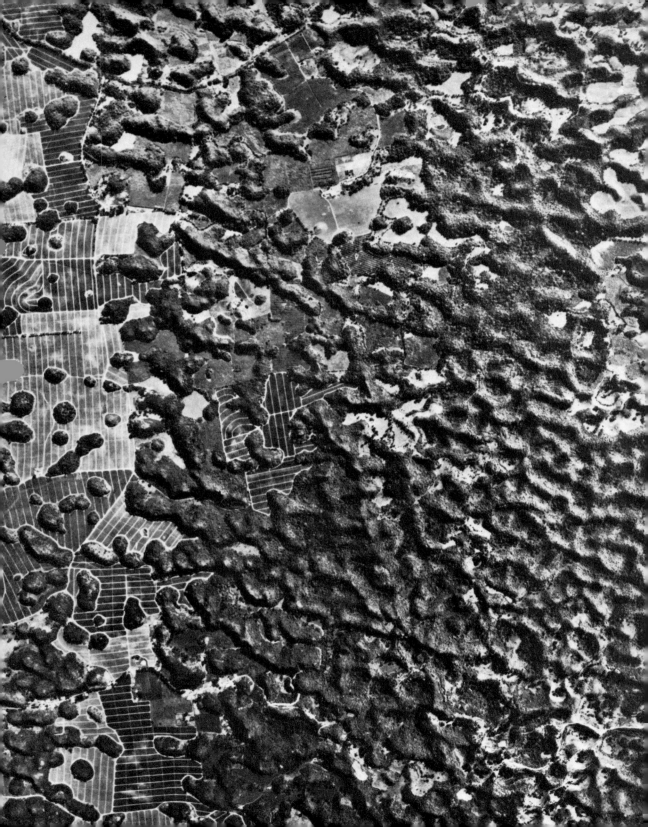

work by Aub (as quoted by Sweeting 1973) and P. W. Williams (1971) suggests that the star shape of cockpits arises from gulley erosion of centripetal streams flowing into the depressions. It seems certain, however, that cockpits, like most normal dolines, develop by solution from the surface downward. Indeed, Monroe (1976) found that circular solutional dolines are the most common closed depression in Puerto Rico, although collapse dolines and uvalas are also present. Where cockpit karst ("cone karst" of Monroe 1976) is prevalent, the residual hills are often joined together to form linear ridges that in turn are cut by gullies into a sawtooth configuration. The alignment of the ridges and the cockpits are apparently so random that joint control seems to be ruled out, although the gullies may be related to zones of structural weakness.

Towers are steep-sided hills and can be vertical or even overhanging. The steep inclination can change where surface erosion piles talus at the slope base (McDonald 1975). According to Jennings (1971), tower karst differs from cockpit karst in the steepness of the residual hills and the presence of swampy, alluvial plains (often similar to poljes) surrounding the towers rather than the depressed cockpits. Not all towers are set in alluvial plains, however; some rise above granites or surfaces underlain by other impermeable rocks. Towers can be of dramatic size, sometimes rising several hundred meters above the surrounding plain (Wilford and Wall 1965). Towers have also been called *pepinos, haystacks* and, commonly, *mogotes,* a term used in Cuba, Vietnam (Silar 1965), and Puerto Rico (Monroe 1976). In fact, Monroe (1976) suggests that the term *mogote* is probably more appropriate for the feature and should be universally accepted.

Mogotes in Puerto Rico rise from blanket sand deposits as hills between 30 and 50 m high (figs. 12.21, 12.22). Some have solution rock shelters etched into the mogote sides, but caves passing entirely through the hill, a common phenomenon in towers of other regions, are rare. Many of the Puerto Rican mogotes are asymmetric in their cross-profiles (fig. 12.23), which Thorp (1934) ascribed to the wind-driven storms causing greater solution and gentler slopes on the windward side. Monroe (1976), however, attributes the asymmetry to case-hardening of the limestones on the windward slope by repeated solution and reprecipitation of $CaCO_3$. The downwind slopes, being less resistant, are oversteepened and even overhanging due to slumping of the less-indurated rocks. Some controversy exists as to what process causes mogotes or towers to rise above the alluvial plain. Theories suggested have been collapse of caves, river erosion, and differential solution of the karst rocks; probably the features are polygenetic (Panoś and Štelcl 1968). In Puerto Rico, studies show clearly that the residual hills are composed of the same limestone that underlies the blanket sands. The one significant factor, however, is that the material holding up the mogotes has been indurated by solution and reprecipitation of $CaCO_3$ during alternating episodes of wetting and drying (Monroe 1969, 1976; Miotke 1973). The original limestone surface beneath the alluvial plain was probably irregular,

Figure 12.20. Aerial view of cockpit karst. (Photo by U.S. Geological Survey)

Figure 12.21. Large mogote (tower) in Puerto Rico. (Photo from Monroe 1976, fig. 23)

and so the sand deposits preferentially collected in minor depressions during their initial accumulation. As soon as plants became fixed on the sands, the limestones were corroded more rapidly and this, combined with the case-hardening of exposed limestone knobs, produced the mogote topography.

Other karst features in tropical environments include a special type of karren called *phytokarst* (Folk et al. 1973) that develops into spongelike pinnacles under the biochemical action of blue-green algae. Another special karst form that seems to be unique to tropical regions are groups of linear trenches called *zanjones*. These features are probably joint controlled and similar to other types of linear karren (Sweeting 1973) except they are considerably larger. Monroe (1976) describes them as being vertical-sided trenches, 100 m or more long, 1–4 m deep, and ranging in width from a few centimeters to 3 m.

The origin of tropical karst is probably related to the higher values of temperature, total rainfall, and rainfall intensity, which promote rapid and prolonged corrosional activity. In addition, rapid plant growth and decay combine with extreme microbial activity to supercharge infiltrating water with CO_2 and, of course, intensify the solution process. This viewpoint, held by most karst experts, has been challenged by Corbel (1959) on the grounds that because less CO_2 can be dissolved in warm water, the saturation equilibrium for CO_2 is lower in tropical climates and less limestone will be dissolved. Nonetheless, it seems certain that the temperature control noted by Corbel is subordinate to the effect of high P_{CO_2} in tropical soils. Corbel also believed that tropical regions were virtually unaffected by Pleistocene climatic changes and so have had considerably more time for karst development than temperate-zone karsts. Such an argument

Direction of wind ⟵

Blanket sand

Limestone indurated by solution and precipitation

Soft chalky limestone

o   Empty solution cavities

  Solution cavities partly filled with sinter

  Solution cavities entirely filled with sinter

  Stalactite

Figure 12.22. Mogotes rising from pineapple fields in Puerto Rico. (Photo from Monroe 1976, fig. 29)

Figure 12.23. Diagram showing characteristics of an asymmetric mogote in Puerto Rico. (From Monroe 1976)

is appealing in cases where karst topography is definitely Tertiary in age, but it cannot explain tropical karst that developed on late Pleistocene limestones or other very young karst regions (Jennings and Bik 1962).

The fact that soluble limestones form ridges in tropical karst adds another dimension to the genetic problem, but Monroe's work in Puerto Rico (1969, 1976) sheds considerable light on the formative mechanics. He suggests that in tropical conditions, limestones often form a thick crust of reprecipitated $CaCO_3$ that essentially armors the less resistant limestone beneath it. Soil on ridges and hills tends to be washed away in torrential rains, leaving a bare limestone surface. The solution of ridges is very slow compared with corrosion beneath plant-rich soils where biogenic CO_2 is plentiful. In addition, dissolved $CaCO_3$ is reprecipitated when water and CO_2 are driven off as rainfall is evaporated (see reversible reaction 5). The combined effect of all these factors is to develop pronounced differential solution; soil-covered, plant-rich limestones are rapidly dissolved into depressions, while bare, recemented limestones resist corrosion and develop into hills.

Morphometry of Karst Landforms and Landscapes

As mentioned earlier, the belief that karst forms are controlled by climate was the prevailing concept in karst geomorphology after World War I. This reliance on climate to explain karst processes and form maintained its dominance until investigators became concerned that too much variability of form existed in the same local region where climatic controls were identical (Jennings and Bik 1962; Verstappen 1964). Since the 1960s, investigators have attempted to explain karst in terms of local geological variations, and studies of spatial morphology of the karst features have appeared. For example, Hack concluded as early as 1960 that doline density varied significantly with rock type in the Shenandoah Valley of Virginia.

Perhaps the most easily demonstrated morphometric relationships are those involving linear karst features and structural controls but, interestingly, most morphometric work has concentrated on the spatial analyses of closed depressions, especially dolines. LaValle (1967, 1968) made detailed studies of the karst depressions in south-central Kentucky in which he related doline orientation, elongation, depth, area, and flank slope to geological and hydrologic variables. In general, he found that the greatest elongation, determined by the ratio of the long axis to the short axis, correlated well with the limestone lithology (especially a high content of insolubles), with proximity to the mouth of the drainage basin, and with a high hydraulic gradient. The most important geologic controls on alignment were structural trends, a conclusion reinforced by Matschinski (1968). In a recent study of Tennessee karst, Kemmerly (1976) found a relationship between doline length and mean width that can be expressed as a simple power function

$$L = 1.656 W^{1.024}$$

where L is the length of the long axis in meters and W is the width in meters. Kemmerly suggests that the dolines here represent a case of dynamic allometry, i.e., a relationship where the two variables change at different rates, length increasing faster than width. This suggests that length is influenced by structure more than width, a conclusion substantiated by the correlation between joint trends and the long-axis orientations.

In a series of papers, Williams (1966b, 1971, 1972a, 1972b) developed a more sophisticated spatial analysis of tropical karst topography. Using aerial photographs or topographic maps, the karst terrain can be separated into divides, summits, channels, and stream sinks. Each closed depression is then given a number representing the highest stream order of the drainage that disappears into the sink (fig. 12.12). The topographic divides surrounding the depressions form a polygonal network, usually with a pentagonal pattern, indicating that the terrain is completely partitioned into separate and adjoining basins. This topography, which Williams calls *polygonal karst,* is dominated by the hills but dynamically controlled by the position of the sinks.

Williams analyzed the pattern of sinks by measuring the average distance from each sink to its closest neighbor and comparing that value to the expected mean distance, which is determined from a density analysis of the sink population. The index ratio $\overline{La}/\overline{Le}$ (where $\overline{La}$ is the mean actual distance and $\overline{Le}$ is the mean expected distance) tells whether the sinks have a random or uniform distribution. In New Guinea the stream-sink dispersion is highly uniform, which Williams (1971, 1972a,b) interprets as the best accommodation that can be made as depressions compete for space when the topography evolves.

The karst basins are not perfectly polygonal, suggesting that some irregularity is introduced by rivers flowing parallel to the general slope of the land surface and by structures causing a preferred direction of erosion. Nonetheless, Williams's conclusion that the polygonal cells are stabilized as the most likely pattern and exist in a steady-state condition is very intriguing. Presumably the distribution, once established, should not change with time unless one or more of the major controlling factors change as the region downwastes.

An emphasis on morphometry does not mean that climate is irrelevant in karst processes. Obviously it is an important factor. Time and geomorphic history are also significant in the geomorphology of karst, as has been demonstrated in both temperate regions (Palmer and Palmer 1975) and tropical zones (H. J. Cooke 1973). It is also known that certain karsts can be strongly influenced by surface erosion if sinks or depressions are plugged with impermeable terra rossa soils (Gerson 1974). What morphometry should demonstrate is that karst landforms, like most surface features, may approach and possibly attain a dynamic equilibrium, the properties of which are determined by local controlling factors. In addition, even the few morphometric analyses that have been accomplished demonstrate that karst terrains are not the chaotic mess assumed by the original workers. In fairness to them, however, it should be noted that the Dinaric Alps of Yugoslavia is a region of very complex structure. Solution in such areas produces a landscape in which order is not readily apparent.

Limestone Caves

Any survey of karst processes and landforms must include a brief discussion of limestone caves. Caves are natural underground cavities which include entrances, passages, and rooms that can be traversed by a human explorer (W. B. White 1976). Technically, caves, being underground features, are not part of karst topography. They may, however, create surface topography by facilitating collapse, and they both influence and reflect the mode of karst hydrology that exists in any particular region. So caves can justly be considered as part of the karst system.

Cave Physiography

As defined, caves have entrances, passages, and rooms, and unless they intersect the surface in the downflow direction, they must terminate in some way. The assemblage of these components in different combinations produces caves with a variety of shapes and patterns.

Entrances and Terminations   Cave entrances can be found in places such as doline bottoms, hillsides, spring mouths, roadcuts, and quarries. Perhaps the most spectacular entrances to cave systems are openings into vertical voids called *shafts* or *chimneys*. Shafts (Pohl 1955) are cylindrical in shape, evidently formed by solution when a film of water moves rapidly down their walls (Brucker et al. 1972). They are most common in flat-lying rock sequences such as those in the Appalachian Plateau, where shafts sometimes are more than 100 m deep and 15 m wide. Shafts characteristically drain at the bottom through a narrow opening that may or may not be connected to a larger cave system. Thus, the term has several connotations. In one sense, large shafts represent an entire cave, especially where the bottom opening is too small to traverse. In other cases, shafts are merely minor vertical entrances into a much larger cave system.

Chimney is a term used by White (1976) to encompass all vertical or nearly vertical openings that do not have the cylindrical shape of a shaft. Commonly chimneys follow steeply dipping bedding planes.

Caves terminate when the passages narrow to the point that a person can no longer follow the opening. These terminations may be due to collapse of overlying rocks, narrowing of the voids into permeable but impassable units, or clay and silt deposits that fill to the ceiling of the cave.

Passages, Rooms, and Patterns   Passages are the main physiographic component of caves. They have various shapes, sizes, and patterns, shown in figure 12.24, that are developed at the same or different times by a single source of water or by a complexly integrated drainage network. In plan view individual passages are either *linear, angulate,* or *sinuous* depending on the structural and hydrologic controls. Highly interconnected segments form a maze pattern which Palmer (1975) has divided into three subtypes known as a *network maze,* an *anastomotic maze,* and a *spongework maze.*

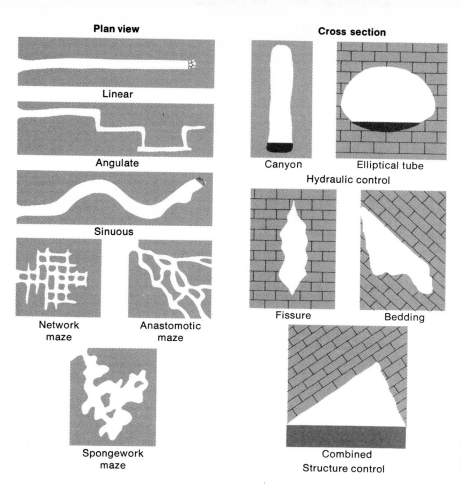

Plan view

Linear

Angulate

Sinuous

Network maze

Anastomotic maze

Spongework maze

Cross section

Canyon   Elliptical tube

Hydraulic control

Fissure   Bedding

Combined
Structure control

Figure 12.24. Plan view and cross-section shapes of cave passages. (Adapted from W.B. White 1976)

In cross section, the shape of a passage represents an accommodation between hydraulics and rock properties such as lithology, bedding, and jointing. Where flow velocities are high and limestones are thick, the passages are usually controlled by hydraulics and exist mainly as narrow, vertical slits called *canyons* or more circular voids up to 30 m in diameter known as *elliptical tubes*. Passages that are controlled by rock variations are much more irregular in cross section (fig. 12.24). Realistically, all variations exist in between the different types, and strict categorization of passage shapes is probably meaningless. *Rooms* are usually nothing more than zones of enlarged passages caused by intersection or enhanced solution.

Almost all passages show a variety of small sculptured markings on their walls, ceilings, and floors. The features are too numerous to describe here (see Bretz 1942 for details). Some are formed by solution and others by abrasion. In addition, many rooms and passages have an abundance of speleothems (chemically precipitated dripstone deposits) that inspire the popular fascination with caves. These features (stalactites, stalagmites, etc.) are composed of a $CaCO_3$ substance called *travertine*. They form when downward-permeating water, saturated with respect to calcite, reaches the cave passage. At this point CO_2 is diffused from the water to the cave atmosphere because the P_{CO_2} in the water is considerably greater (Holland et al. 1964). Some water may also be lost by evaporation. In either case (loss of water or loss of CO_2), the $CaCO_3$ must precipitate as shown in the general reversible reaction (5).

The Origin of Limestone Caves

The origin of limestone caves has been steeped in controversy since the features were first recognized, and disagreement still rages today. The differences of opinion probably stem from several factors. (1) Historically, theories about cave formation have tried to fit all caves into genetic models with little regard for geologic differences. For example, there is no compelling reason that caves in highly deformed rocks of the Alpine region should develop in precisely the same way as those in the plateau areas of southern Indiana. (2) Most cave models are based on hydrological schemes, yet there is remarkable lack of hydrologic data in the arguments. In fact, most theories are founded on physiographic evidence. (3) Very few caves have been examined in great enough detail to substantiate a local origin, let alone an all-encompassing genetic model.

In light of the above difficulties, it is indeed refreshing to examine Thrailkill's 1968 study of caves as scientific phenomena, involving detailed analyses of the chemical and hydrological constraints exerted in cave development.

Theories of Cave Development   All classical theories of cave development are concerned primarily with the position of the solvent water relative to a water table. Three main ideas have evolved over the years, suggesting that caves form (1) above the water table by the corrosive action of vadose water, (2) beneath the water table by deep circulation of phreatic water, or (3) at the water table or in the shallow phreatic zone, often associated with fluctuations of the water table itself.

Vadose Water Theory   The concept that caves develop above the water table by vadose water appeared early in this century and has been complicated by many researchers adopting slightly different variations of the theme (Martel 1921; Malott 1921, 1938; Piper 1932; Gardner 1935). In general, most workers recognized the solution effect of vadose flow, and many suggested that rivers flowing with some velocity under hydrostatic head were able to enlarge caves by abrasion. Some saw importance in the water table level since free-flowing water is essentially on the top of the water table, but others (Malott 1921; Addington 1927) denied its importance.

Deep Phreatic Theory   In 1930 W. M. Davis published his classic "two-cycle theory" of cave development, which depends on deeply circulating phreatic water. Davis argued that most caves are now in a phase of active deposition rather than solution, as evinced by the abundance of dripstone. Thus, cave erosion must have occurred at some earlier time when the water table was higher than the present. This time was, in Davis's logic, during the old age phase of the geographical cycle when relief was low, streams were not deeply entrenched, and the water table was therefore near the surface. According to Davis, solution of the caves took place along curving flow lines in the phreatic zone (fig. 12.25a) which descended deep below the surface and emerged under the major river. Subsequently, the region was rejuvenated by uplift, rivers entrenched into the surface, and the water table was lowered to join with the new level of the rivers. This, of course, exposed the already formed caves and allowed dripstone deposition to begin.

The two-cycle theory was given support by Bretz (1942) who felt that clay filling of the caves also occurred at the end of the first stage, and he suggested that minor solution features on cave ceilings and walls were produced by both phreatic and vadose water. Hubbert (1940) also added credence to Davis's conceptual model by demonstrating that water beneath a water table should move along deeply curving flow lines. Rhoades and Sinacori (1941) suggested a modified but similar model of flow and related it to cave development. Critical questions, however, must be asked about the deep phreatic theory. For example, caves are usually not curved but are mostly horizontal or vertical. More importantly, is deep phreatic water so undersaturated with respect to calcite that it can accomplish significant corrosion? Gardner (1935) finds little evidence for large

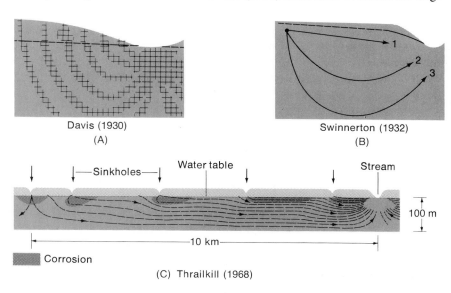

Figure 12.25. Models of groundwater flow in the development of caves. (Used with permission of the Geological Society of America)

cavities in deep drill holes, and most vadose seepage is saturated with respect to $CaCO_3$ before it even reaches the water table.

Shallow Phreatic Theory   The third major model, and its many variations, place the main zone of cave enlargement in the reach where the water table fluctuates with the seasons. Swinnerton (1932) accepted earlier ideas that stressed the importance of the water table. He believed that if water moves along all possible paths beneath the water table, the zone just below the table would be the most undersaturated and would have the greatest discharge (fig. 12.25b). Thus, cave formation would proceed most efficiently in the shallow phreatic zone and should be related to the surface stream level. Other studies seem to support this model (Sweeting 1950; Davies 1960; White 1960; Wolfe 1964) but the most convincing evidence is presented by Thrailkill (1968).

Thrailkill provides a sound theoretical base for the shallow phreatic model by pointing out that where no water crosses the water table, the flow will be nearly horizontal and water just beneath the water table will follow very shallow flow routes, especially if the recharge occurs from a number of discrete point sources (fig. 12.25c). The water at the water table becomes undersaturated when it mixes with vadose water (mixing corrosion), when it becomes cooled (dissolves more CO_2), or when water is introduced by backflooding of nearby rivers. The combination of lateral flow and undersaturation dissolves more limestone during floods, and therefore, the position of cave formation is most likely to be the average level of the flood water table, which he considers to be the upper limit of the shallow phreatic zone. Thrailkill's work explains nicely the properties of Mammoth Cave, Kentucky. More detailed studies of this type are needed before we can reach a complete understanding of cave formation.

Summary

Karst topography develops by extensive solution of limestones and other soluble rock types. Limestones that are most susceptible to karstification are massive, crystalline, high in calcite content, and intensely fractured. The limestones should also exist in a thick stratigraphic sequence that stands well above base level. The karst processes function best in humid temperate and humid tropical climates where abundant water is available and biogenic CO_2 is freely added to the water from a thick vegetal cover.

Karst areas are characterized by a distinct hydrology in which surface flow is partially or totally disrupted when water is diverted into the underground system. Groundwater movement in karst aquifers differs from normal situations in that the flow is often controlled by the orientation of fractures and interconnected solution cavities.

The most common karst landforms are enclosed depressions called dolines (sinks or sinkholes in the U.S.). Other distinct forms are minor solution features, called karren, and a variety of surface valleys that are unique to karst regions. Tropical karst differs from temperate karst in that hills rather than depressions dominate the topography. In most regions, karst landforms and hydrologic parameters seem to possess regular morphometric relationships that probably reveal some type of dynamic equilibrium between process and form. Limestone caves are subsurface manifestations of karstification. The mechanics of cave formation are not completely understood. Most interpretive disagreement has centered on where the solution process is most efficient in relation to the water table. Models proposed suggest that caves form either above the water table, in the deep phreatic zone, or immediately below the water table.

The following references are suggested to provide more detail concerning the topics examined in this chapter.

Suggested Readings

Baker, V. R. 1973. Geomorphology and hydrology of karst drainage basins and cave channel networks in east-central New York. *Water Resour. Res.* 9:695-706.

Jennings, J. N. 1971. *Karst*. Cambridge, Mass.: M.I.T. Press.

Monroe, W. H. 1976. The karst landforms of Puerto Rico. U.S. Geol. Survey Prof. Paper 899.

Sweeting, M. M. 1973. *Karst landforms*. New York: Columbia Univ. Press.

Thrailkill, J. 1968. Chemical and hydrologic factors in the excavation of limestone caves. *Geol. Soc. America Bull.* 79:19-45.

White, W. B. 1976. Geology and biology of Pennsylvania caves. *Pennsylvania Geol. Survey Gen. Rep.* 66:1-71.

Williams, P. W. 1972. Morphometric analysis of polygonal karst in New Guinea. *Geol. Soc. America Bull.* 83:761-96.

In the preceding chapters we have examined a variety of processes that mold the exposed surface of the Earth into diagnostic landforms. However, 70 percent of our planet is covered by ocean water, and the sea possesses immense energy (Inman and Brush 1973). The application of this energy on the adjacent land can produce rapid and enormous changes in the nearshore physical environment. Therefore, the erosional and depositional processes that function at the interface between the ocean and the land represent another major realm of geomorphology.

Coastal zones respond to forces as does any other geomorphic system. Viewed on a short-term basis, parts of the coast might be considered as quasi-equilibrium forms. In that sense, repeated movement of sediment and water constructs a beach profile that reflects some balance between the average daily or seasonal wave forces and the resistance of the landmass to wave action. Considered over a longer time span (graded time), however, beaches or entire coastlines may be imperceptibly changing toward a larger equilibrium form. On the even larger geologic time scale, marine transgressions and regressions may represent dramatic alterations of the position and character of the coast. Our attention here will focus mainly on the dynamics of beach processes over steady and graded time spans.

The study of beaches and coasts is not strictly academic. Most of the largest U.S. cities are near the ocean, and three-fourths of the people in the country live in coastal states (Soucie 1973). The concentration of people in coastal regions and the pressures involved in using this land for recreation and development have placed a genuine strain on the system. Whether we can utilize the environment completely and still prevent damaging changes in its character depends on how well we understand the functional processes. Engineers, geographers, oceanographers, geologists, and other scientists have made significant contributions to our present understanding of the coastal environment. As usual, however, interdisciplinary communication has not been great because each discipline is engrossed in specific problems. Nonetheless, if demands on the coastal system continue to increase, cooperative efforts of all coastal workers will be needed to avert environmental disaster.

Coastal Zones— Processes and Landforms

13

For those interested in coasts or beaches, there exist a monumental wealth of data and many exceptionally good syntheses. The following references treat the topic with a number of different approaches and at many levels of sophistication: D. W. Johnson 1919; Guilcher 1958; Steers 1962; Shepard 1963; Wiegel 1964; Bascom 1964; Ippen 1966; Zenkovich 1967; Manley and Manley 1968; Bird 1969; Muir Wood 1969; Shepard and Wanless 1971; King 1972; Davies 1973; CERC 1973; Komar 1976. Much of the material presented here has been drawn from these excellent sources.

Coastal Zones

A coast is a relatively large physiographic zone that extends for hundreds of kilometers along a shoreline and often several kilometers inland from the shore. Like all large physiographic entities, coasts have not escaped the irrepressible urge of scientists to classify, and many attempts have been made to categorize their features (see King 1972). Coasts are particularly susceptible to classification because their regional properties are clearly documented on available maps and aerial photographs. In some cases, no extensive field study is needed to systematize coastal properties into a viable classification scheme.

Some of the commonly used classifications of coasts combine, in different ways, a few basic ingredients that serve as the fundamental criteria. Most important of these are (1) form of the land-sea contact, i.e., the configuration of the shoreline; (2) stability or relative movement of sea level; and (3) influence of marine processes. Some of the classifications are purely descriptive and may have little application in dynamic geomorphology. Others are genetic and so are allied closely with the processes involved in developing the diagnostic coastal properties. For example, Inman and Nordstrom (1971) classify the morphology of coasts on the basis of modern plate tectonic theory. On the other hand, as Russell (1967a) suggests, perhaps we have not yet collected enough precise data concerning coastal properties to warrant any classification at this time. In addition, all coasts have a past as well as a present, and so time also becomes an important consideration in coastal classification (Bloom 1965). That is, coasts may reflect evolution and contain relict parts that are not in equilibrium with modern, or even Holocene, processes. For instance, marine terraces standing well above the ocean (Bradley 1957; Orme 1974) are not related to modern wave attack, but do indicate a coast that has undergone movement relative to sea level.

Because of all these considerations, no specific classification will be advocated here, and no attempt will be made to analyze entire coastal regions. Instead, we will examine only parts of the entire coast that are being actively affected by modern processes. In that sense we will talk about "erosional" or "depositional" coasts, but these terms are applied only on a local basis and have no regional connotation.

Beaches

The difficulties of associating entire coasts with formative processes tend to direct our attention to smaller components of the coast where processes are more easily understood. The feature most amenable to process analysis is the *beach,* which is most simply defined as the relatively narrow portion of a coast that is directly affected by wave action. It usually terminates inland at a sea cliff, a dune field, or at the boundaries of permanent vegetation. Oceanward, under the constraints of our definition, part of the beach is continuously submerged because it lies beneath low-tide sea level. It is still part of the beach, however, because the bottom is subjected to wave action. In that sense the term *beach* is synonymous with the *littoral zone,* an expression used commonly in geological work.

The beach environment is ideally suited for our approach because it is the place where forces generated in the ocean are exerted against the resisting geologic framework. Beach systems can be considered in two different ways: (1) as large parts of the coastal physiography, or (2) as small open systems in which water and sand are moved on and off the beach surface. As major coastal elements, beaches exhibit rather large geomorphic features that indicate whether the environment is dominated by erosion or deposition. As small open systems, beaches have properties that are the result of breaking waves transporting sediment perpendicular to the shoreline, or of currents that cause sand to drift parallel to the shore.

Waves

The paramount driving forces in shoreline processes are waves, which expend the energy they obtained in the ocean against the margins of the land. A thorough analysis of wave theory is steeped in complex physics and mathematics. Here we can introduce only a simplified version of the basic concepts needed to understand the geomorphic role of waves, recognizing that such a discussion is a broad generalization of a very complicated phenomenon.

Wave Generation   Most waves that do geomorphic work are generated by strong winds blowing across large portions of the open ocean. Precisely how energy is transferred from the wind into ocean waves is not completely understood (discussed in Komar 1976). Nonetheless, empirical studies indicate that wave properties reflecting this energy exchange (fig. 13.1) depend primarily on the wind velocity, the wind duration, and the *fetch* (the distance over which the wind blows). The fetch is particularly important in determining the height of waves and their *period,* a parameter that is merely the time interval between two successive wave crests passing a fixed point. Extremely high waves with long periods can be generated only when all three controlling factors are at a maximum.

The Driving Forces

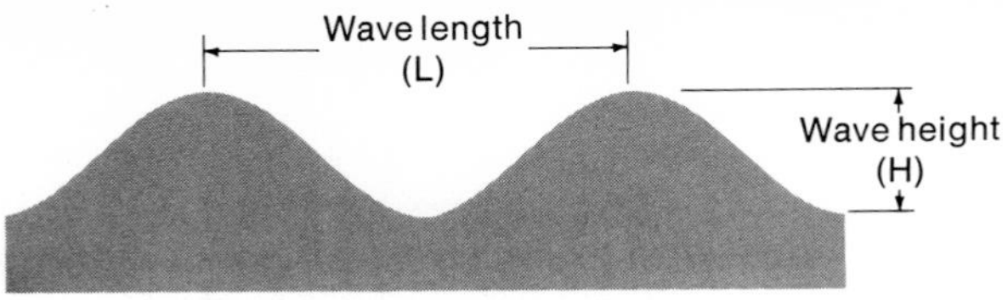

Figure 13.1. Fundamental
parameters of waves.

In the area where they are generated, waves are irregular, and individual crests are discernible for only short distances before they disappear in a maze of interfering waves. This chaotic state occurs mainly because small and large waves generated during the same storm are out of phase. Waves passing through the system tend to interfere with one another, adding to the height of some waves and subtracting from others. When waves are enhanced, the heights developed can be truly awesome. Storm waves are commonly more than 20 m high under severe winds. The greatest documented wave height, 34 m (112 ft), was measured in a February 1933 storm generated in the open water in the deep part of the South Pacific.

As waves move away from their source area, they begin to separate from one another according to their various periods. This process, called *wave dispersion,* causes regularly spaced successions of waves with rounded crests to migrate from the source zone. Emerging waves typically have a low ratio of wave height to wave length (a parameter known as the *wave steepness*) and appear as the long, low waves commonly referred to as *swell*. In swell, water particles assume the circular orbital paths that characterize deep-water *waves of oscillation* (fig. 13.2). Although individual water particles in oscillatory waves have little forward motion, the waves themselves advance with a typical trochoid form. That is, the wave form is moving forward, not the ocean water.

The dispersive process, shown diagrammatically in figure 13.3, works because wave length and velocity are both a function of the period such that

$$L = \frac{gT^2}{2\pi}$$

and

$$V = \frac{gT}{2\pi}$$

where g and π are the well-known constants, L is the wave length in meters, V is velocity in m/sec, and T is the period in seconds. Because of these relationships, long-period waves with greater length and velocity will separate from short-period waves. For example, a wave with a period of 6 seconds will have a length of 56 m (184 ft) and a velocity of 9 m/sec (30.7 ft/sec). Waves generated in the same storm that have a period of 14 seconds have a length of 303 m (1000 ft) and a velocity of about 22 m/sec (71.5 ft/sec). It is apparent that waves disperse from the generation zone simply because longer waves outrun the shorter ones during any given interval of time.

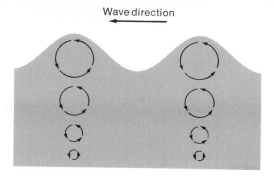

Wave direction

Figure 13.2. The orbital motion of water particles in a wave of oscillation.

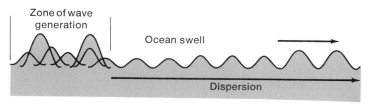

Zone of wave generation

Ocean swell

Dispersion

Figure 13.3. Dispersion of waves from an area of wave generation. Waves having different periods separate from one another to create ocean swell.

Waves with identical periods travel away from their source as distinct groups. It is important to note that storm-generated waves are not the same as waves produced by a point-source impulse such as a pebble tossed into a pond or a tsunami created by rock displacement beneath the ocean floor. In such cases, waves radiate in all directions from the point source. In contrast, ocean swell follows a pathway that encompasses a substantial area of the ocean surface but also has finite lateral boundaries; the direction of the well-defined corridor is determined by the direction of the generating wind. Although the wave path does spread somewhat, it is quite possible for swell to strike a few score kilometers along a coast while leaving nearby reaches virtually unaffected by the storm.

Studies show that swell can travel great distances across the open ocean without losing much of the original energy (Snodgrass et al. 1966). Most energy loss, especially in the shorter period waves, occurs near the generation zone. Once the swell condition is attained, the long-period waves experience little additional dissipation of energy and may traverse an entire ocean basin.

Waves in Shallow Water   As swell approaches a landmass, the waves begin to "feel" the effect of the ocean bottom. When the water depth is approximately half the wave length, the oscillatory waves begin a transformation into steplike forms called *waves of translation* (fig. 13.4). Translation waves develop when the orbital path of water particles is intercepted by the ocean floor. The orbits begin to flatten noticeably and their axes of rotation rise to higher levels. Eventually the orbital path is destroyed, producing the different wave type. Waves of translation are different from waves of oscillation in that the water particles have a distinct forward motion without the corresponding backward movement that

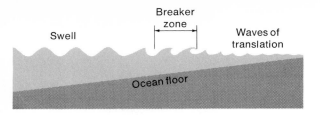

Figure 13.4. The transformation of oscillation waves into waves of translation as swell approaches the nearshore environment.

characterizes oscillation. Once in the form of translation, the velocity and length are determined as

$$V = \sqrt{gh}$$

and

$$L = T \sqrt{gh}$$

where h is the depth of the water.

The change in wave types is accompanied by the phenomenon of breaking waves. Deep-water oscillatory waves remain stable only if the wave steepness (H/L) is lower than 1/7 (.14). As waves begin to "feel bottom," their heights increase as the rounded crests become more peaked, and their velocities and lengths decrease as the wave crests bunch up; only the wave periods remain constant (fig. 13.4). The combined effect of an increasing height (H) and decreasing length (L) causes oversteepening of the wave; the critical H/L value is attained and the wave breaks. The distance over which breaking occurs depends primarily on the bottom slope, i.e., how rapidly the water depth decreases toward the shoreline.

Three common types of breakers, called *spilling, plunging,* and *surging,* have been observed and are depicted in figure 13.5 (Wiegel 1964). In spilling breakers, the top of the wave crest becomes unstable and flows down the wave front as an irregular foam. In the plunging type, the wave crest curls over the front face and falls with a splashing action into the base of the wave. In surging, the wave crest remains essentially unbroken but the base of the wave front advances up the beach. A fourth type of breaker, called *collapsing,* has been identified by Galvin (1968); its characteristics are intermediate between those of the plunging and the surging types (fig. 13.5). Actually, breaking waves probably occur in a complete spectrum of types that depend on the bottom slope, $H/L,$ and the period. Generally, however, spilling occurs when steep waves encroach on a beach with a low slope, whereas surging develops with low wave steepness and high-gradient beaches (fig. 13.5). Other types develop when steepness and bottom gradient are between the end-member conditions represented by surging and spilling.

Breakers mark the oceanward limit of a zone, called the *surf,* in which the original energy given to the waves receives its final transformation. High velocities and substantial impacts occur under breakers, and thus the creation of

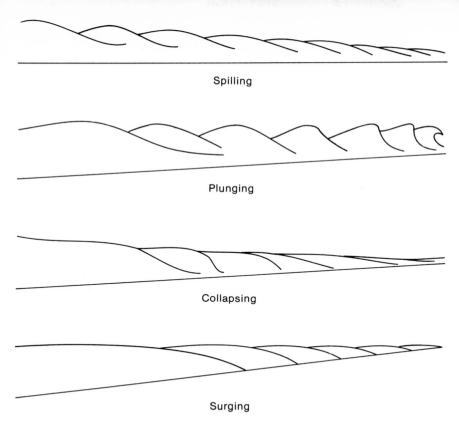

Spilling

Plunging

Collapsing

Surging

Figure 13.5. Types of breaking waves. (Adapted from Galvin 1968, p. 3652, © American Geophysical Union. Used with permission of the *Journal of Geophysical Research*)

translation waves provides the water with kinetic energy that is capable of doing geomorphic work. Once formed, the waves of translation (and the water they contain) move forward to their inevitable collision with the landmass. The surf ends when the wave form is lost as it impinges on the beach face. From there, water carried by its own momentum continues to slide up the beach as *swash* until, at its highest encroachment, the force gathered in the open ocean is finally and totally dissipated. The *swash zone* is alternately covered as the water rushes up the beach face (swash) and exposed as the water moves back down the beach (backwash) under the influence of gravity.

The position of the surf zone may change frequently because breaking waves not only reflect the bottom topography, but also use their energy to realign it. That is, swell approaching a shoreline will break according to the existing shoaling configuration, but at the same time, the breakers may shift the bottom sediment so as to produce a new sea-floor topography for the next train of storm-generated waves. In addition, the surf zone changes in response to vertical displacement of the ocean level due to tides, storms, and so forth.

In some cases two sets of swell generated in different areas arrive at a shore simultaneously. This produces a systematic variation in the heights of waves striking the shore, a phenomenon known as *surf beat* (Munk 1949; Tucker 1950). In surf beat, successive waves gradually increase in height until they reach a maximum, then systematically decrease in height to a minimum value. Thus a pronounced periodicity develops whereby one large wave will appear with predictable regularity. In many cases the dynamics are such that every sixth to eighth wave will be at the maximum height, but the precise beat depends on the wave periods of the two swell systems and their resulting harmonics. The variation of breaker heights produced by surf beat affects beach processes because it changes the prevailing water level and notably alters the velocity of nearshore currents.

In addition to these changes in wave mechanics, ocean swell entering shallow water also changes direction. Because deflection of the wave crests is a function of the water depth, the waves adjust to the contours of the bottom topography and so bend according to the configuration of the shoreline. The process, known as *refraction,* works because some sections of a wave crest approaching a coast obliquely are moving in shallow water and at lower velocities than other portions of the same wave, which are traveling in deeper water (Silvester 1966). As shown in figure 13.6, the waves converge on a landmass that juts into the ocean. When convergence occurs, the energy per unit length of wave is increased, causing higher waves and increased energy to impinge on the headland region. Waves can also diverge over embayments or submarine canyons, resulting in lower waves and a concomitant spreading of the energy.

Tsunamis, Storm Surges, and Seiches   Like the waves produced in the open ocean by winds, other waves generated in different ways may have geomorphic significance. *Tsunamis* are waves formed by sudden impulses be-

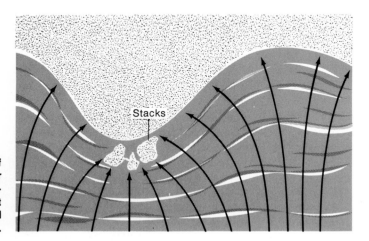

Stacks

Figure 13.6. Refraction of waves along an irregular coastline. Wave energy converges on landmasses that project oceanward and diverges in recessed areas.

neath the ocean that cause trains of waves to radiate in all directions from the point source. Tsunamis are usually initiated by earthquakes of more than 6.5 magnitude on the Richter scale with foci located less than 50 km beneath the ocean floor (Van Dorn 1966). Submarine landslides, volcanic eruptions, and slumping have also been cited as causes. In any case, sudden movement displaces the overlying water column, causing it to oscillate up and down as the water tries to reestablish mean sea level.

The waves leaving the source zone of a tsunami have distinct characteristics. They are extremely long (as much as 240 km), their period may be as much as 1000 sec, and in open water they may be only a meter high. Tsunamis obey the same laws that control shallow-water waves, and therefore their velocity is proportional to the water depth. In 10,000 feet (3000 m) of water, the wave velocity will be

$$
\begin{aligned}
V &= \sqrt{gh} \\
&= \sqrt{32 \cdot 10{,}000} \\
&= 566 \text{ ft/sec or } 386 \text{ mph (approximately 618 km/hr)}.
\end{aligned}
$$

In the open ocean, such waves pass quickly and with little notice. As waves approach the shore, however, they seem to trigger a harmonic oscillation that does not follow the bottom configuration. In a typical event, a moderate rise or recession of sea level is followed by three to five major wave fronts that are tens of meters high and capable of great destruction. For example, N. H. Heck (as quoted by Bascom 1964) describes a remarkable case in which a U.S. warship anchored in a Peruvian port in 1868 was picked up by a tsunami wave, carried over the top of the small port city (Iquique), and finally dropped 400 meters inland. After the major waves pass, the nearshore system gradually returns to normal (Van Dorn 1965, 1966). Geomorphic effects of tsunamis are dramatic, but probably short-lived; tsunamis occur rarely and normal waves rework the coast according to more prevalent controls.

Storm surges are associated with large storms such as hurricanes or typhoons. Beneath the low-pressure centers of such storms there tend to be compensating upward bulges of the ocean levels. When a storm enters shallow water, the bulge may couple with high tides to form large waves that inundate the coastal lowlands. Surges of this type can cause vast property damage and often a great loss of life; for example, a 1900 storm surge in Galveston, Texas, killed 6,000 people.

A *seiche* is still another wave type that is not directly related to a prevailing open ocean wind. Seiches are free oscillations of water in enclosed or semienclosed basins. Although originally observed in lakes, the seiche phenomenon occurs also in harbors, where it is often called surging, and along open coasts with a broad, shallow continental shelf. A seiche is recognized as a repeated rise and fall of the water level; the oscillatory motion begins when some force displaces the water from its equilibrium position. The driving impulse may be heavy rainfall, flood discharge from nearby rivers, long-period waves such as tsunamis or surf beat, or rapid pressure fluctuations associated with storm surges (B. W. Wilson 1966). In any case, when the initiating force passes, the oscillations gradually decrease until the equilibrium level is once again attained.

Tides and Currents

Although waves are the dominant force influencing the coastal environment, they are not the only significant water motion. Tides and currents each constitute a type of movement that can modify coastal properties. Although our treatment of these forces will be necessarily brief, a complete understanding of nearshore geomorphic mechanics requires detailed consideration of these factors.

Tides   As any dedicated beachcomber knows, tides usually occur as a twice-daily rise and fall of sea level. Away from coastal areas, this movement is of little concern to most laymen, but the tidal effect is of consequence to the coastal geomorphologist for several reasons. First, because of the continuous change in water level, the position of wave attack migrates through a notable vertical range and a corresponding lateral shift, increasing the size of the beach and complicating our understanding of beach processes. Second, tides initiate currents that flow into and out of constricted reaches of the shoreline such as bays or lagoons. Many times this ebb and flow can keep drifting beach debris from closing the entrance to the embayment, and in fact, some tidal currents are capable of eroding coastal rocks. This is especially true in narrow bays such as the Bay of Fundy in eastern Canada, where the inland constriction produces a maximum range of 15.6 m, the highest value on Earth. In some constricted estuaries tides move inland with a pronounced wave front called a *tidal bore*. Often the bores actually break, providing observers with a spectacular show. For example, the tidal bore on the Amazon River looks like an 8 m high waterfall advancing upstream for 480 km at a rate of 12 knots.

The tides, of course, are driven by the gravitational effect exerted on the Earth by the sun and the moon (for a nonmathematical discussion see Bascom 1964; for an understandable mathematical discussion, Komar 1976). Here we need only recognize that in most coastal regions the lunar influence results in two high tides and two low tides daily. The gravitational attraction of the sun complements or detracts from that of the moon. As figure 13.7 illustrates, every two weeks the moon and sun are aligned, causing a higher tide than normal, called the *spring tide*. In the intervening period the moon and sun reach positions that are 90° apart. The solar pull detracts from the lunar effect, and the tide is lower than normal, the so-called *neap tide*. Spring tides are about 20 percent greater than the average tidal range and neap tides about 20 percent less.

Not all places on Earth experience the tidal motion that pure astronomical and gravitational theory predicts. Some areas, for example, have only one high and low tide, which occur at the "wrong" times. The tide at any locality also varies in magnitude because of other factors, such as perturbation of the lunar orbit, tilting of the Earth's axis, and ocean bottom topography. Equilibrium theory, therefore, cannot account for all tidal variations.

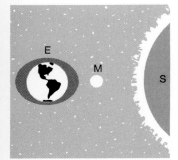

Spring tide

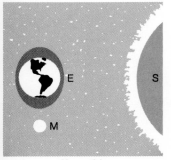

Neap tide

Figure 13.7. Tidal distribution at various stages of the lunar cycle (S = sun; E = earth; M = moon). During spring tide the moon and sun are aligned and cause larger tides. During neap tide the moon and sun are not aligned, and tides are lower than average. (Not to scale.)

Nearshore Currents   Another type of water motion complicates the near-shore system and our understanding of the mechanics in the surf and swash zones. Other than wave action itself, two wave-induced currents control water movement in the beach zone: (1) a cell circulation consisting of *rip currents* and their associated longshore currents; and (2) longshore currents that usually are generated by waves striking at an angle to the prevailing direction of the shoreline, but may also be caused by tides or storms.

Rip Currents and Cell Circulation   Rip currents are narrow zones of strong flow that move seaward through the surf (fig. 13.8). They return to the offshore zone water that was moved toward the beach by waves of translation. The rips are caused by small currents that move on the beach face parallel to the shoreline; these originate about halfway between two adjacent rips. The velocity of the feeder current reaches a maximum at the entrance to the rip zone. Velocity in the rip current itself can be significant, with known speeds reaching approximately 2 m/sec (Sonu 1972).

The development of the cell formed by a longshore current and a rip is related to variations in the rise of mean water level above the level normally attained under still water, a phenomenon called *wave set-up*. Circulation begins because the height of incoming waves varies in a longshore direction, and higher waves create greater wave set-up. Longshore currents therefore flow from zones of the highest breakers and return oceanward at the position of the lowest breakers. Waves of similar height also initiate cells, but the process probably depends on the type of breakers involved (Sonu 1972).

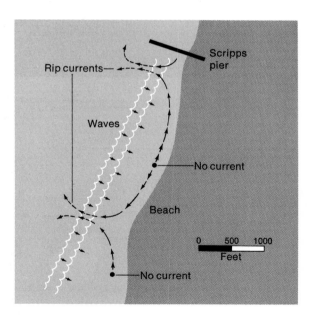

Figure 13.8. Nearshore circulation pattern at Scripps Beach, La Jolla, Cal., showing rip currents and longshore currents that feed the rips. (Adapted from Shepard and Inman 1950, p. 199, © American Geophysical Union. Used with permission of the American Geophysical Union)

Shepard and his co-workers (1941) realized long ago that the position of rip currents is controlled by the bottom topography in the surf zone. We can expect, therefore, that areas of wave convergence, where waves characteristically are higher, are the starting points of the longshore component in the cell. Rips occur away from these zones where breaker heights are smaller. Although this analysis has been shown to be correct, rip currents also exist on long, straight beaches with smooth bottoms. Therefore, bottom topography is not the only cause of cell circulation. For example, it is now known that ocean swell by itself produces secondary waves in the surf zone (Bowen and Inman 1969; Huntley and Bowen 1973). These waves, called *edge waves,* have crests normal to the shoreline and wave lengths parallel to the shore; they oscillate with an up-and-down motion. Some authors have suggested that edge waves may be the triggering devices of rip currents (see Komar 1976), but this suggestion is by no means universally accepted (Hino 1975). Regardless of their precise origin, rip currents tend to perpetuate themselves because they can modify the bottom topography so that it enhances the circulation pattern.

Longshore Currents from Oblique Waves   Longshore currents are also generated by waves that strike the beach obliquely. These currents are extremely important in beach mechanics because they tend to move sediment parallel to the shoreline for considerable distances and thereby present coastal engineers with innumerable problems. A number of attempts have been made to derive equations that relate wave properties to the velocity of the longshore current (for reviews, see Galvin 1967; Komar 1976). In general, most equations fail in a predictive sense because they either (1) do not distinguish the longshore current associated with cell circulation from that produced by obliquely striking waves, or (2) employ invalid criteria as the fundamental theoretical base.

One approach, based on momentum analysis, has produced good results and probably represents the best hope for a usable predictive model. Momentum, unlike energy, is preserved as waves break and is separated into two components, one directed toward the shoreline and one directed parallel to it. Thus, the flux in momentum directed along the shoreline should be proportionately related to the velocity of the longshore current. This concept has been developed thoroughly by many authors, but most completely by Bowen (1969), Longuet-Higgins (1970), Komar and Inman (1970), and Komar (1975). Komar (1976) suggests that the best estimate of longshore currents at the mid-surf position follows the equation

$$\overline{V}_\ell = 2.7 u_m \sin \alpha \cos \alpha$$

where u_m is the maximum orbital velocity at the breaker zone, $\overline{V}_\ell$ is mean velocity in cm/sec, and α is the breaker angle, i.e., the angle of incidence with a line parallel to the shoreline (details of the derivation have been omitted).

It should be pointed out that tides and winds blowing in the longshore direction may complicate the system. At low tide, for example, water can be trapped in troughs that parallel the shoreline. Continued spilling of waves into the troughs drives a circulation pattern in which water moves alongshore confined within the troughs until it reaches a rip channel cut through an offshore bar. There the water turns seaward as part of the rip current. In these cases the longshore current may have a velocity that obeys a mass continuity law rather than a momentum law (see Inman and Bagnold 1963; Bruun 1963; Galvin and Eagleson 1965).

Beaches as Dynamic Equilibrium Forms

Beaches represent dynamic systems where loose granular debris is moving steadily under the attack of waves and currents. If water motion could be held constant, this debris would be molded into a characteristic profile that would reflect an equilibrium between the driving forces (waves and currents) and the properties of the beach sediment. We can thus visualize an equilibrium beach profile for any set of water and sediment conditions. Waves and currents, however, do not remain constant but change their properties on a daily (see Zeigler and Tuttle 1961) or seasonal basis, requiring some response in the process of sediment transportation and ultimately in the beach profile. Although we can think of the equilibrium profile as the ideal case, under the dynamics of the natural setting its properties are constantly changing along with the driving forces.

The Beach Profile

The beach profile, shown in figure 13.9, consists of a number of component parts, each of which develops its own diagnostic characteristics. In its entirety the profile represents a topographic form that induces waves to dissipate energy by breaking. Thus, the larger the beach, the more protection it provides against coastal erosion, because energy will be lost before water can reach the vulnerable base of the sea cliff.

Berm   The *berm* is a nearly horizontal surface on the backshore portion of the beach. Some beaches have more than one berm; others have none, especially if the sediment is coarser than sand. Landward the highest berm terminates at the base of the sea cliff, and oceanward it joins the *beach face* (fig. 13.9). Because the berm is formed by deposition of sediment during backwash, its elevation is

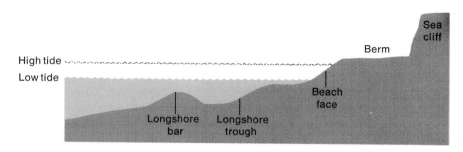

Figure 13.9. Diagram of idealized beach profile.

determined by how high the swash runs up the beach and by the grain size of the sediment (Bagnold 1940). As swash moves up the beach, it loses velocity because of friction and the loss of water that permeates into the beach debris. Continued upward growth of the berm surface would require ever higher waves and swash run-up. In this sense berms are analogous to vertically accreted floodplains, and their rate of upward growth must decrease with time, for only infrequent storm waves can add sediment to the surface. As Bascom (1964) points out, however, storm waves also erode the front edge of the berm, thereby reducing its horizontal length while simultaneously building the remaining surface to a higher level.

Beach Face   The beach face is the sloping section of the beach profile immediately seaward of the berm. The slope of the beach face is controlled by many factors and so may range from nearly horizontal to gradients that approach the angle of repose for unconsolidated sediment. Sediment moves up the beach face with the swash and down the beach face during the less powerful backwash. The beach face therefore represents a surface striving to attain some balance between onshore and offshore sediment transport. Its slope adjusts to provide the balance.

Beach face slopes are directly proportional to the size of the particles being moved in the swash zone. Large particle sizes tend to maintain steep slopes and vice versa (Bascom 1951; Wiegel 1964; McLean and Kirk 1969; DuBois 1972). We would expect such a relationship because slope provides the backwash velocity needed to transport sediment of any given size, but the phenomenon is much more complicated. Actually the slope depends primarily on the amount of water lost by percolation into the beach face material. Because coarse sediment is more permeable than fine sediment, more water is lost to the larger material as swash traverses the beach face. The backwash under these conditions has a lower discharge and competence; the higher slope that develops represents the compensation needed to transport coarse debris with less water. Corroboration of this process is found on beaches having mixed particle sizes and poor sorting, factors that tend to lower the permeability. When two beaches have the same mean grain size, the beach with the poorest sorting will usually have a lower slope (McLean and Kirk 1969). Because both beaches have the same mean size, the variation in slope must occur because less water infiltrates the poorly sorted debris, and so the backwash has greater flow to transport sediment on a lower gradient.

Slope angles also show an inverse relationship with the wave energy as it is revealed by H/L (Rector 1954; W. Harrison 1969) or the wave height alone (see Komar 1976). In either case, higher waves or greater wave steepness will promote lower beach face slopes, presumably because more backwash ensues. Other factors such as the tidal range and the level of the water table (Duncan 1964; W. Harrison 1969) affect the beach slope, but their influence is less pronounced and poorly understood.

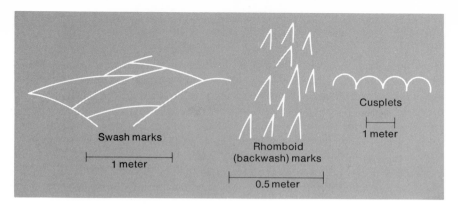

Figure 13.10. Minor features found on the beach face.

The beach face is also the locale of minor beach features formed under the swash-backwash cycle (fig. 13.10). Most pronounced of these are *swash marks, backwash rills,* and *cusplets*. Swash marks are lines of sand particles pushed up the beach face by the leading edge of the swash and deposited as the forward motion ceases. Diamond-shaped rill channels develop during backwash for reasons that are still not clear. Cusplets are small (<1 m) crescentic indentations on the beach face caused by swash erosion and deposition. They are similar in origin to a variety of cuspate features that will be discussed later in the chapter.

Longshore Bars and Troughs   Sediment transported from the beach face to the offshore zone normally deposits as a submerged *longshore bar*. An associated trough usually develops between the bar and the beach face. Longshore bars may be absent from an equilibrium profile, especially if the beach is steep. On the other hand, where the beach gradient is low, several bars may be present. The creation of multiple bar systems may be accomplished in a variety of ways: (1) each bar may reflect the breaker position for waves of different sizes; (2) the bars may relate to shifting breaker positions associated with high and low tide levels; or (3) oscillation waves that break far offshore on low-gradient beaches may re-form over the trough and break again closer to shore, each episode of breaking creating an associated bar.

These hypotheses concerning bar formation are all based on the assumption that breaking waves are intimately involved in the formative mechanics. A number of early laboratory and field studies support that contention (Evans 1940; Keulegan 1948; King and Williams 1949; Shepard 1950), and the relationship is now generally accepted as being real. Usually breakers establish the size, position, and depth of the bars and troughs. Larger breakers normally produce features in deeper water. These generalizations, however, must be conditioned by evidence showing that bar positions may migrate shoreward or seaward according to variations in wave height, steepness, breaker type, or even by a complex process of wave reflection off the beach face (see Lau and Travis 1973; Carter et al. 1973).

Some bars are continuous parallel ridges that extend unbroken for tens of kilometers. Bars do not usually display such regularity, however, because shifting wave directions realign parts of the bar system and break ridges into segments with a chaotic distribution. Violent storms also produce drastic changes in bar directions and patterns.

The maximum depth of bar formation depends on the depth to which wave action is able to agitate the bottom sediment. Under average wave conditions, this depth is approximately 10 m below low tide, but large storm waves are known to move bottom sediment in water as deep as 25 m. The deepest bars, molded in the fury of violent storms, maintain their position and shape for long intervals during which normal waves do not influence the bottom. In considering bars as products of wave action and as controlling factors in the breaking process, the shallow shoreward bars and troughs are therefore decidedly more important in nearshore wave dynamics.

Complications of the Beach Profile

The equilibrium profile described earlier is subject to various complicating factors such as tidal changes, storm waves, and longshore movement of sand. Any of these factors will produce changes in the beach profile by depositing new features that are added to the profile or by eliminating, during erosion, dominant components of the profile. In addition, the position of the major features of the beach profile may be changed according to the dynamics of the modifying factors.

Variation in sea level with the tides may produce significant changes in the beach profile. These changes occur on a daily basis (Strahler 1966; Schwartz 1967), and they may be related to water table fluctuations during rising and falling tide stages (Duncan 1964; W. Harrison 1969). Tidal effects may also occur on a longer time scale. Inman and Filloux (1960), for example, noted that in the northern Gulf of California, a region with a large biweekly tidal range, the profile is characterized by a steep, coarse-grained beach face that presumably relates to the high-tide level. Seaward from the beach face is a wide, low-tide terrace composed of poorly sorted fine-grained debris. At the seaward edge of the terrace is another beach face which corresponds to the low-tide wave conditions. Low-tide terraces are sometimes scalloped into a system of low mounds and depressions called *ridges* and *runnels* that seem to form where there is a large tidal range with low wave energy and available sand (King and Williams 1949). As similar features have been described in lakes where tides are negligible (Davis et al. 1972), some confusion exists as to the precise meaning of a ridge and runnel system (see Komar 1976).

Despite these other factors, the most pronounced changes in the ideal profile are brought about by the dynamics associated with storm waves. Where storms are common, the large and vigorous waves tend to destroy or drastically limit the extent of the berm. The eroded material is simultaneously shifted to an offshore position where it collects in a series of longshore bars. In contrast, where storm waves are unusual, the movement of beach sand is shoreward. This tends to destroy the offshore bars and transport the sediment to the upper part of the beach where it rebuilds a broad, flat berm. Because the frequency of storms is a seasonal phenomenon, we can expect the beach profile to change dramatically from one part of the year to another. Many authors, therefore, refer to a *summer profile*, characterized by the absence of bars and a wide berm, and a *winter profile*, with no berm and a series of longshore bars; an example is shown in figure 13.11. It is probably incorrect to think of the summer or winter profiles as occurring specifically during those seasons. What we are really talking about is a profile formed by storm waves versus a profile generated by swell. Many field studies document the offshore movement of sands during storm wave conditions and the landward transport under swell (Shepard 1950; Bascom 1964; Strahler 1966; Gorsline 1966), and the fact of the process cannot be questioned.

The mechanics of the changes in storm and swell profiles are seemingly related to the wave steepness, and several studies have determined critical steepness values at which one profile changes into the other. Unfortunately, these threshold values vary from study to study, probably because the critical H/L is partly dependent on the particle size of the beach debris (Rector 1954). The determination of the threshold H/L is not an esoteric study. The type of beach profile has a decided effect on the amount of sea cliff erosion that will occur. Coastal property invariably is in greater jeopardy when a storm profile is dominant. This is simply because, in the absence of a berm, wave energy and momentum are not dissipated as waves move onto the beach. The ocean can attack the sea cliff with greater success during periods of a continuous storm profile.

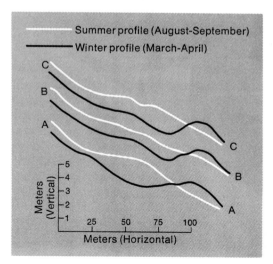

Figure 13.11. Generalized summer and winter profiles along Scripps Pier, La Jolla, Cal. (Adapted from Shepard 1950)

Plan-View Configuration

In addition to the quasi-equilibrium profile that is oriented perpendicular to the shoreline, many coastal geomorphologists now accept Tanner's (1958) premise that beaches develop a shoreline configuration revealing another type of balance between water energy and sediment supply. This plan-view shape is best established where no long-term unidirectional movement of sediment occurs parallel to the shoreline. Like a graded river, the shoreline equilibrium is developed ''over a period of years'' and is adjusted to the prevailing wave characteristics.

It would seem that the most logical environment to preserve an equilibrium configuration would be protected bays where there is no dominant longshore current. In such a locale, waves might produce minor longshore transport of sediment where the wave crests are not completely refracted. The sediment will continue to drift until the shoreline is reoriented parallel to the attacking wave crests at every segment. Offshore topography thus determines wave refraction, and the waves in turn establish the equilibrium shoreline configuration. The complications of even this simple model are staggering. For example, any addi-

Figure 13.12. Sand barrier off the coast of northern Nantucket Island, Mass., separated from the mainland by a lagoon. Large cuspate features have been formed on the lagoonal side of the barrier. (From the University of Illinois Catalog of Aerial Photography and the U.S. Geological Survey)

tional sources of sand to the beach will prevent complete refraction because some longshore transport away from the source will be necessary. In addition, bottom topography is so variable and so susceptible to change that it seems too much to expect that we can precisely describe the form of an equilibrium shoreline.

In light of the above, it is indeed remarkable that many shorelines, open to all types of waves and currents, contain features that are similar in shape and spaced with a regularity that can hardly be attributed to coincidence. Usually crescentic, the features form as periodic seaward projections of the shoreline itself or, on straight shorelines, as narrow pointed accumulations of sediment piled perpendicular to the shore. In either case the seaward projections are separated by a curved embayment (fig. 13.12). A complete hierarchy of these features seems to exist (Dolan and Ferm 1968; Dolan et al. 1974), ranging from minor forms with a wavelength of less than a meter to major cuspate-like indentations of the coastline with spacing measured in hundreds of kilometers. Table 13.1 shows this hierarchy.

Beach Cusps   The most common crescentic forms are *beach cusps.* Cusps develop at the upper part of the beach face and along the outer fringe of the berm. They are usually spaced less than 30 m apart and can form in beach sediment of any size, including boulders and cobbles (Russell and McIntire 1965). Some sorting is produced in the formative mechanics because the cusp projections, called *horns,* are usually more coarse-grained than the intervening embayments.

Beach cusps seem to form most readily where wave crests strike parallel to the shoreline. This perhaps explains why the features tend to remain fixed in their

Table 13.1 Hierarchy of crescentic landforms found on coasts.

Form Characteristic	Cusplet	Cusp	Sand Waves	Secondary Capes	Primary Capes
Spacing	0 to 3 m	3 to 30 m	100 to 3000 m	1 to 100 km	200 km
Material	Fine sand-gravel	Sand-boulders	Sand	Sand	Sand-gravel
Topographic Association	Step	Berm, beach face	Beach berm-offshore bar system	Coastal plains; shores with sufficient sediment	Coastal plain deltas
Rhythmicity	Yes	Yes	Yes	Often	Not always
Motion	Fixed	Normal to beach	Downdrift	Probably downdrift	Slow downdrift
Temporal	Minutes to hours	Hours to days	Weeks to years	Decades	Centuries
Suggested Processes	Swash action on beach face. Groove erosion.	Berm deposition and erosion.	Wave action, nearshore circulation cells, back eddies of longshore transport currents.	Kinematic nature of sediment transport, circulation cells.	Wave action, confluence of coastal currents, back-set eddies, and shoals.

From Dolan et al. 1974. Used with permission of *Zeitschrift für Geomorphologie,* published by Gebrüder Borntraeger, Stuttgart.

positions, although laboratory studies indicate that some longshore migration of the forms is possible if the drift is not excessive (Krumbein 1944). There seems to be some agreement that spacing of cusps is related to the wave height as well as wave direction; the higher the waves, the greater the spacing interval. Even this, however, cannot be pronounced as an inviolate rule; A. T. Williams (1973) found no correlation between wave height and spacing. The spacing of the cusps he studied in a Hong Kong bay were most closely related to the swash distance, i.e., the length between the breaker zone and the highest encroachment of the swash.

Because every study of beach cusps seems to reveal some contradiction of earlier studies, it should come as no surprise that the origin of this feature has been controversial since its earliest description. The simplest explanation of the feature, proposed by D. W. Johnson (1910, 1919) and modified by Kuenen (1948), is based on a process that causes irregular erosion of the beach face. Swash erosion of the beach face initiates cusps, and backwash transports the sediment away. Eroded materials are carried seaward until they deposit as deltaic projections that stand opposite the excavated hollows. Continued swash action progressively transforms the initial hollows into larger embayments until the water crossing the depressions reaches a critical depth. Swash is then refracted in such a way that coarse sediment is deposited on the cusp horns and finer sediment is transported farther offshore. Bays and horns grow until the central area attains a limiting depth, swash action is retarded, and the feature assumes an equilibrium condition.

Since Kuenen's study, this model has been modified, using a different pattern of water and sediment circulation (Russell and McIntire 1965), and rejected (Komar 1976) mainly because the theory does not explain the absence of cusps where conditions are favorable for their development. New theories concerning the origin of cusps have followed. They include these: (1) cusps grow during seasonal changes in wave energy (Russell and McIntire 1965); (2) cusps form by hydrodynamic modification of the breaking waves or the swash (Cloud 1966; Gorycki 1973); (3) cusps are generated by the presence of rip currents (Shepard 1963; Komar 1971b; Hino 1975); (4) cusps develop by breaching of dune ridges or berms (Smith and Dolan 1960); (5) cusps are a function of standing edge waves (Bowen and Inman 1969; Bowen 1973; Komar 1973; Guza and Inman 1975); and (6) cusps arise through the intersection of wave trains having the same period (Dalrymple and Lanan 1976).

The variety of suggested origins indicates that one all-encompassing model for beach cusp development is nowhere in sight, and many more detailed studies are needed.

Rhythmic Topography and Capes   Larger features in the hierarchy of crescentic forms have been called *rhythmic topography* by Komar (1976). They are of two main types: (1) crescentic bars, and (2) rhythmic variations that are controlled by rip currents associated with cell circulation. The latter forms have been referred to variously as *sand waves, giant cusps,* or *shoreline rhythms*. The features occur in widely divergent environments including large lakes (Evans 1938; Krumbein and Oshiek 1950), enclosed seas (King and Williams 1949), and open ocean coastlines in many parts of the world.

Rhythmic topography differs from beach cusps in several important ways. (1) Rhythmic topographic features commonly migrate parallel to the shoreline. The rate of the movement varies but can be a kilometer or more a year. (2) Much of rhythmic topography is submerged. The emergent forms are usually observed as giant cusps or sand waves that have considerably greater spacing than beach cusps. The wavelength of these features ranges from 100 to 3000 m, with horns extending oceanward as much as 25 m. The submerged rhythmic topography is usually in the form of crescent-shaped sand bars or longshore bars that are segmented by rip current channels. The crescentic bars are concave toward the shoreline. They commonly stand opposite the horns of large cusps exposed on the beach face itself, but they also exist off straight beaches, especially in protected bays with a small tidal range (Shepard 1952, 1963; King and Williams 1949; Bowen and Inman 1971). (3) Rhythmic topography seems to be more dependent on the bottom configuration in the offshore surf zone than normal beach cusps.

Uncertainty about the origin of rhythmic topography equals the mystery surrounding beach cusps. There is little doubt that rhythmic topography is a function of wave action, cell circulation with associated rip currents, and longshore transport of sediment driven by obliquely striking waves. Precisely how these combine to construct the features and the details of the processes is simply not clear.

In addition to the cuspate features already examined, many coasts of the world display extremely large shoreline crenulations called *capes*. The map in figure 13.13 shows that they are prominently developed on the southeast coast of the United States, where their spacing is roughly one to two orders of magnitude greater than rhythmic topographic features. Capes may be partially relict and therefore not necessarily adjusted to modern conditions. For example, some of the capes shown on the map are fringed by barrier islands that are Holocene in age. Some also coincide with the position of major rivers (W. A. White 1966; Hoyt and Henry 1971), which may indicate that the seaward projections began as ancient deltas.

Many workers feel that these large-scale rhythms are related to rotational cells that are set up as eddy currents along the western edge of the Gulf Stream. Such eddies have not been proven to exist, however, and the direct cause of the capes is unknown.

Major Beach and Coastal Landforms

Having looked at beach processes as quasi-equilibrium phenomena, we can now examine major components of coastal physiography. In this sense beach forms are not necessarily in equilibrium with marine processes, although it is

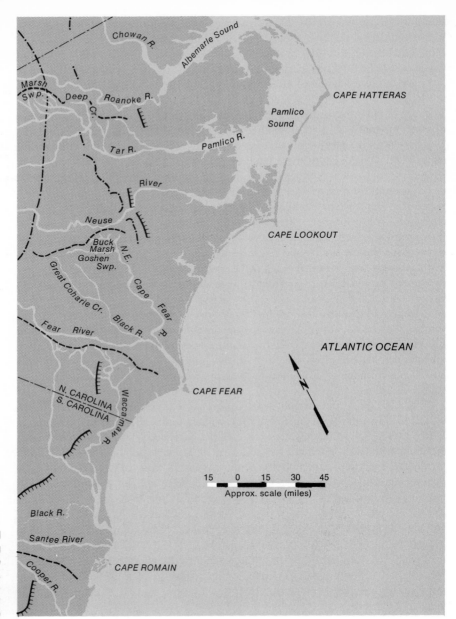

Figure 13.13. Large capes and crescentic recessions along the south Atlantic coast of the United States. (From W.A. White 1966. Used with permission of the Geological Society of America)

probable that they are striving toward that end. Usually the landforms indicate that either erosion or deposition is dominating the physical setting.

It is tempting to envision an evolutionary progression of landforms when considering them on this scale. For example, imagine a coast emerging by uplift or by a eustatic drop in sea level. In the early developmental stage, the coast would have a rugged, irregular configuration controlled by the coastal geology, and beaches would be nonexistent because little, if any, sediment would be available to mold into a profile. Wave energy during this phase would all be used in eroding the shoreline rocks, instead of being consumed when the waves traverse a wide beach zone. With advancing time, however, the coastal rocks are gradually smashed into particles, and the availability of sediment, coupled with recession of the shoreline under erosive attack, allows waves and currents to produce a beach. Geology becomes less dominant in the system as it succumbs to the inexorable marine forces. Finally, when beaches reach their ideal width and profile development, the shoreline is dominated by depositional features as longshore currents spread debris across embayments and generally straighten the shoreline.

An evolutionary model such as this assumes that time is the essential ingredient and that all rocks will eventually surrender in the same way to the ocean forces. Several observations argue against making this assumption. First, sea level is probably too ephemeral to remain at one level long enough for the complete cycle to run its course. Any tectonic or eustatic change resets the cyclic clock, making an orderly progression of landforms hypothetical at best. Second, some rocks are extremely resistant to marine erosion, and enormous time spans are needed to produce the sequence of events. For example, the crystalline rocks along the Cornish coast of Great Britain have probably not receded at all during the last 10,000 years (Bascom 1964). Third, the assumption requires that the ocean impinge on every landmass with similar force and with the same combination of waves, currents, tides, etc. This simply is not the case.

Faced with these contradictory choices about coastal equilibrium, my feeling is that if long-term equilibrium is possible, it comes in all sizes and shapes. We probably should not think in terms of one coastal equilibrium form, but rather of a complete spectrum of forms, each of which reveals a special accommodation between local ocean forces and the geologic framework. Depending on the variables, such an equilibrium might be established quickly or might take more time than we can expect controlling variables to remain constant. That is, some coasts may never reach an ultimate equilibrium. They will continue to change in response to changing conditions over both short and long time spans. Coasts may always be approaching an ultimate equilibrium but never sensibly reaching it because the ingredients of the system change too rapidly for the final consummation to occur. Because of this, we will examine coastal and beach landforms simply as being erosional or depositional, with no connotation of their being either in an evolutionary stage or in equilibrium with modern processes.

Erosional Landforms

Along the shorelines of oceans or large lakes where beaches are poorly developed, wave action directly attacks the rocks or unconsolidated material of the shore. Such action rapidly forms a distinct sea cliff (or lake cliff) and, with time, causes the cliff face to retreat. The erosion is accomplished by a group of geomorphic processes, listed in table 13.2, that function in a complex of interactions to produce a variety of landforms. The final effect sometimes is controlled by ancillary processes such as burrowing action of marine organisms (see Ahr and Stanton 1973), frost action, and mass movement. Cliff retreat is not uniform because rocks vary in resistance depending on lithology, the amount and spacing of fractures, and the orientation of strata in relation to the direction of the attacking waves.

As might be expected, the effectiveness of each process varies with the properties of the shore material and with the particular dynamics of the local ocean or lake system. *Corrosion* affects those rocks that are most susceptible to solution. At normal temperatures seawater is saturated with respect to calcium carbonate and does not dissolve limestone and $CaCO_3$ directly. Solution of these materials, however, may be aided by rainwater or by organisms that create local acidic conditions (Emery 1946). As chunks of coastal or lakeshore rocks are released, *attrition* decreases the size of the particles and allows subsequent waves to drive the sediment into the cliff face. *Corrasion* then assumes a dominant role in the erosive mechanics. *Hydraulic action* is especially important where the rocks are highly fractured. Not only does the force of the wave exert pressure on the cliff face, but the advancing waves may compress air in the rock cavities, producing a pneumatic effect in the cracks. As the wave recedes, external pressure is instantaneously released, and the compressed air within the rocks exerts an outward stress that may disaggregate the outer zones of the cliff face.

As the cliff retreats, it leaves behind a beveled surface, called the *wave-cut platform,* that stands slightly below water level at high tide (fig. 13.14). Although many processes are involved in its creation (see Wentworth 1938), corrasion at the base of the sea cliff is probably responsible for most of the planating action. The platform is not flat but slopes gently oceanward with a declivity that

Table 13.2 Major erosional processes functioning along coasts.

Process	Description
Corrosion	Solution of coastal rocks by chemical action of seawater.
Attrition	Diminution of rock particles as water rolls, bounces, or slides them on a beach or wave-cut platform.
Corrasion	Physical erosion of bedrock caused by the grinding action of rock fragments that are carried in the ocean waves and currents.
Hydraulic Action	Erosion caused by the force of the water itself. Includes wave shock pressure and pneumatic quarrying by air trapped in cracks of the headland rocks.

Figure 13.14. Wave-cut platform near MacLeod Harbor, Alaska. Platform has been raised above wave level by recent tectonism. (Photo by M.J. Kirkby. Published as fig. 16 in Kirkby and Kirkby 1969)

ranges between .02 and .01. Therefore, under a stationary sea level the maximum width of the wave-cut platform depends largely on the depth at which wave abrasion is still a viable process. Bradley (1958) suggests that wave-cut platforms can be eroded in water no deeper than 10 m; under the common slope range, platforms wider than 500 m probably form only if sea level is continuously rising.

The expansion of the wave-cut platform depends on how easily the sea cliff can be eroded. Table 13.3 shows rates of sea cliff retreat that have been determined at various localities under conditions of different cliff materials and ocean forces. Lithology and cohesiveness of the coastal material clearly are of paramount importance in determining the rate of cliff retreat. Unconsolidated debris is eroded most rapidly and, in some cases, almost catastrophically. For example, during a furious storm along the coast near Suffolk, England, a cliff composed of glacial sand was cut back 12.2 to 27.4 m in one night (Williams 1953, as cited in King 1972). Rocks that are nonresistant to wave erosion are usually friable sandstones and shales. Rocks that resist sea cliff retreat are massive igneous rocks, high-rank metamorphic rocks, and certain massive carbonates.

The rate of sea cliff retreat is also affected by other factors such as properties of the ocean, fracturing of the rocks, and orientation of the rock strata. Corrasion, for example, depends to a great extent on wave steepness and the prevailing wave direction. Hydraulic properties change, however, if the

Table 13.3 Representative rates of sea cliff retreat.

Location	Material	Rate (m/100 yrs)	Reference
New England	crystalline rock	0	1
U.S.S.R.	volcanics	0	2
England (Cornish coast)	crystalline rock	0	3
Northern France	chalk	25	3
England (Yorkshire)	sedimentary rocks	9	3
England (Yorkshire)	glacial drift	28	3
Louisiana (coastal islands)	sands and clays	800-3800	4
Southern California	alluvium	30	5
U.S.S.R.	clay	1200	2
New Jersey	sand, clay and gravel	180	6
Cape Cod, Mass.	glacial drift	30	7

1) Johnson 1925
2) in Zenkovich 1967
3) in King 1972
4) Peyronnin 1962
5) Shepard and Grant 1947
6) Rankin 1952
7) Zeigler et al. 1959

shoreline configuration becomes irregular because of refraction. Such irregularity might result from differential erosion of diverse lithologies, but it also can be controlled by the orientation of strata. Where beds strike perpendicular to the shoreline, the coast is etched into irregularity because resistant beds jut oceanward as rock headlands, while nonresistant units are eroded more rapidly into embayments. Where beds strike parallel to the shore, the coast may be straighter as waves abut against resistant units. If the strata dip oceanward, however, rockslides may become a significant erosive factor (Merriam 1960), especially if the alternating units have different permeabilities (Bird 1969).

As the shoreline retreats and irregularities appear, a group of landforms develop that are characteristic of coastal erosion. *Stacks* are isolated parts of the headland formed when narrow oceanward extensions of the coastal rocks are cut into isolated remnants by wave attack (fig. 13.15). The process is accentuated when waves are refracted around the headland reach. Sometimes less resistant rock zones are exposed to local corrasion or scouring, and notches or *sea caves* are formed. As sea caves grow, they may extend completely through the headland to produce a feature known as a *sea arch* (fig. 13.16). Any or all of these features indicate a local erosional environment. As the headland rocks undergo attrition, however, more sediment covers the wave-cut platform. Beach profiles develop, and wave energy dissipates farther offshore as waves break over the longshore bars. This sequence may not occur in every situation. For example, in large lakes such as Lake Erie or Lake Michigan, energy loss over longshore bars is of little or no consequence.

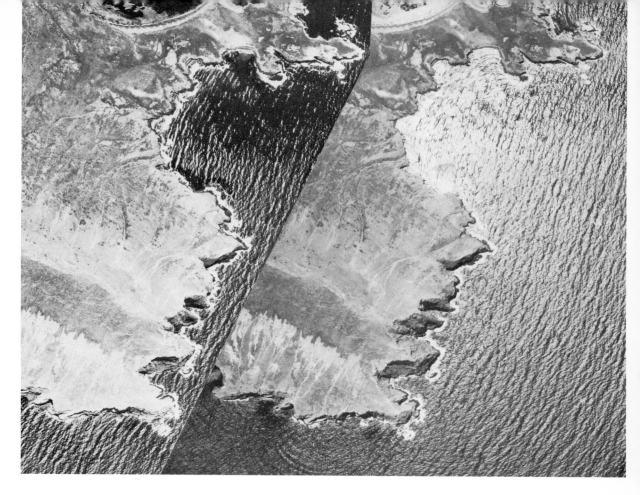

Figure 13.15. Santa Cruz Island off the coast of California near Santa Barbara. Small coves are caused by unequal resistance to wave erosion. Oceanward projections of land have small remnants, or stacks, that have been isolated by erosion. (Photo by U.S.D.A. and University of Illinois Catalog of Aerial Photography)

Figure 13.16. Sea arch or window along coast of Puerto Rico at low tide. Hole is caused by combination of wave action and solution of limestone bedrock. (Photo from Monroe 1976, fig. 47)

The erosion of the coast is a major concern to engineers who are called on to protect property from the ravages of wave attack. Two construction forms commonly are used to protect against sea or lake cliff retreat. The first, a *breakwater,* is a linear barricade of concrete and rocks placed some distance offshore in a direction parallel with or at a slight angle to the shoreline. Its purpose is to absorb the energy of waves in the surf zone, thereby providing a zone of quiet water between the breakwater and the shore. The second, a *seawall,* differs from the breakwater in two main ways. First, it is a concrete or rocky sheath placed directly on top of the cliff rather than offshore. Second, not only does it absorb wave energy, but it also is designed to reflect waves back into the path of the next wave coming toward the shoreline. Either type of construction requires careful design. The choice of which preventive method to use is predicated on the character of normal and storm waves and on the local geology of the sea cliff.

Depositional Landforms

In coastal areas where the supply of sand is abundant and local ocean forces are capable of transporting sediment, a different assemblage of landforms dominates the coastal scene. The shape of features that originate under these conditions is determined primarily by depositional events, but erosional processes are very much involved in the entrainment of the sand before its final deposition. Large depositional beaches usually occur in the form of *spits, baymouth bars,* or *barrier islands.* The first two are definitely related to longshore transport of sediment (also called *littoral drift*), with the site of deposition being in tranquil waters of bays, estuaries, or the open ocean. Barrier islands are large elongate bars that parallel the shoreline but are not physically connected to it. Other depositional features may include some of the cuspate forms discussed earlier (especially cuspate forelands associated with sand waves), but most of these differ in several ways from the large depositional beaches. First, the cuspate forms show a regularity that is not evident in the large beaches and probably relates to different hydrodynamics. It is known that some crescentic forms (rhythmic topography) will migrate in the direction of the longshore transport. However, very strong longshore drift will at first drastically skew the sand waves and giant cusps and then, with increased transport rates, eliminate the form completely. Second, cuspate features may be superimposed on the larger beach types, indicating that in some cases they are secondary rather than primary features.

The location of major beach forms as significant components of a coastline depends, to a large degree, on the availability of sand. The primary source of such debris is the enormous detrital load delivered to the coastal domain by major rivers. For example, the Mississippi River provides more than 3 billion tons of solid material to the Gulf Coast region annually. Sea cliff erosion adds a significant amount of sediment, but normally it is less than 10 percent of the total debris available for accumulation in beaches (Komar 1976). Small additions are made to the detrital load by the wind or by slow onshore transport of sand eroded from unconsolidated shelf regions. In some regions biogenic debris is a significant part of the sand mass. This is especially true in the tropics, where reefs may fringe the coastline and broken shell or coral fragments may be washed onto the beach face and berm.

Most large beaches are created by the drifting of a large detrital load in the longshore direction. Although longshore currents generated by winds or other factors can occasionally move considerable sand, the process of littoral drift is primarily a function of breaking waves striking the beach in an oblique direction. This phenomenon takes place because waves are seldom completely refracted and therefore usually make their final approach at some angle to the shoreline.

The rate of longshore drifting of sand can be significant when considered on a time scale of human life, and the process is often shown in the accumulation of beach debris against man-made structures such as groins (fig. 13.17). The net littoral drift is the sum total of movements initiated by waves arriving at the shoreline from various directions. Waves may carry debris in one direction for a short period and then, as conditions change, return the material to its original position. Such drift reversals may be random or they may be decidedly seasonal. In California, for example, sand usually drifts southward in the winter and northward in the summer. Net littoral drift can be determined only by long-term budget studies that can detail which direction is the prevailing one. Nonetheless, the process of longshore transport perhaps causes more problems for coastal engineers than any other shoreline phenomenon. It places sand where we do not

Figure 13.17. Southern coast of Cape Cod, Mass., showing a series of groins constructed to protect beach from erosion. Groins trap sand that is moving by littoral drift from right to left on photo. Mouth of river is protected by jetties. (Photo by U.S. Coast and Geodetic Survey and University of Illinois Catalog of Aerial Photography)

want it (across harbors and bays) and removes sand from locations where we would like to keep it (resort beaches). Table 13.4 shows net littoral drift along a number of coastlines. The volume of debris involved in the transport demonstrates the tremendous problems engineers face.

The Process of Littoral Drift   Before we examine the forms of major beaches, it will be helpful briefly to discuss the littoral drift process itself. Because engineers and other scientists have faced the problem of longshore transport for many years, they have attempted to construct predictive models to estimate its amount and rate. In most cases, their predictions have relied on empirical correlations between the rate of sand movement and some estimate of the wave power expended in the longshore direction. Figure 13.18, for example, shows the relationship between the wave power and the transport rate on two beaches in California (Komar and Inman 1970). In the diagram, the line fit to the observed data points is defined by the equation

$$I_\ell = KP_\ell$$

where K is a dimensionless constant, I_ℓ is the immersed weight transport rate in dynes/sec, and P_ℓ is the longshore component of wave power in ergs/sec/cm. P_ℓ at the breaker line is estimated as

$$P_\ell = (EV_n) \sin\alpha \cos\alpha$$

where E is the energy density, i.e., the energy content of a wave/unit area. E is a function of the density of the water, the acceleration of gravity, and the wave height. V_n is the wave phase velocity, and α is the angle the breaker makes with the shoreline (for details see Komar and Inman 1970 or Komar 1976).

In many cases such as that shown in figure 13.18, empirical studies show a good correlation between wave variables and transport rates, and parameters used as indicators of the longshore components of wave power or energy are compatible with wave theory (Komar and Inman 1970; Komar 1971a). A complete understanding of the process, however, is complicated by the fact that littoral drift occurs in two different modes of action. When wave steepness is high ($H/L > 0.03$; Bascom 1964) most transport occurs beneath the breaker zone. But when the steepness is low, the drift functions along the beach face in a process known as *swash transport* (beach drift). In this case, waves breaking obliquely to the shoreline push sand grains up the beach face perpendicular to the direction of the wave crests. The backwash, however, affected only by gravity, pulls the water and grains down the beach face in a direction perpendicular to the shoreline. Thus, mobile grains in swash transport move in a spasmodic, zig-zag fashion (fig. 13.19). Besides wave steepness, the angle of wave incidence (α) also is important in the drift rate, maximizing the rate at a 30° angle.

The littoral drift process is further complicated because sediment can be moved in suspension or as bedload. A recent detailed study (Brenninkmeyer 1975) shows that suspended load is important only in a narrow offshore zone.

Table 13.4 Representative rates of littoral drift along coasts of the United States.

Location	Predominant Direction of Drift	Rate of Drift (cu. yds per Year)	Method of Measure of Rate of Drift	Years of Record
Atlantic Coast				
Suffolk Co., N.Y.	W	300,000	Accretion	1946–1955
Sandy Hook, N.J.	N	493,000	Accretion	1885–1933
Sandy Hook, N.J.	N	436,000	Accretion	1933–1951
Asbury Park, N.J.	N	200,000	Accretion	1922–1925
Shark River, N.J.	N	300,000	Accretion	1947–1953
Manasquan, N.J.	N	360,000	Accretion	1930–1931
Barneget Inlet, N.J.	S	250,000	Accretion	1939–1941
Absecon Inlet, N.J.	S	400,000	Erosion	1935–1946
Ocean City, N.J.	S	400,000	Erosion	1935–1946
Cold Spring Inlet, N.J.	S	200,000	Accretion	—
Ocean City, Md.	S	150,000	Accretion	1934–1936
Atlantic Beach, N.C.	E	29,500	Accretion	1850–1908
Hillsboro Inlet, Fla.	S	75,000	Accretion	—
Palm Beach, Fla.	S	150,000 to 225,000	Accretion	1925–1939
Gulf of Mexico				
Pinellas Co., Fla.	S	50,000	Accretion	1922–1950
Perdido Pass, Ala.	W	200,000	Accretion	1934–1953
Galveston, Texas	E	437,500	Accretion	1919–1934
Pacific Coast				
Santa Barbara, Calif.	E	280,000	Accretion	1932–1951
Oxnard Plainshore, Calif.	S	1,000,000	Accretion	1938–1948
Port Hueneme, Calif.	S	500,000	Accretion	1938–1948
Santa Monica, Calif.	S	270,000	Accretion	1936–1940
El Segundo, Calif.	S	162,000	Accretion	1936–1940
Redondo Beach, Calif.	S	30,000	Accretion	—
Anaheim Bay, Calif.	E	150,000	Erosion	1937–1948
Camp Pendleton, Calif.	S	100,000	Accretion	1950–1952
Great Lakes				
Milwaukee Co., Wis.	S	8,000	Accretion	1894–1912
Racine Co., Wis.	S	40,000	Accretion	1912–1949
Kenosha, Wis.	S	15,000	Accretion	1872–1909
Ill. State Line to Waukegan	S	90,000	Accretion	—
Waukegan to Evanston, Ill.	S	57,000	Accretion	—
South of Evanston, Ill.	S	40,000	Accretion	—

From J. W. Johnson 1956. Used with permission of the American Association of Petroleum Geologists.

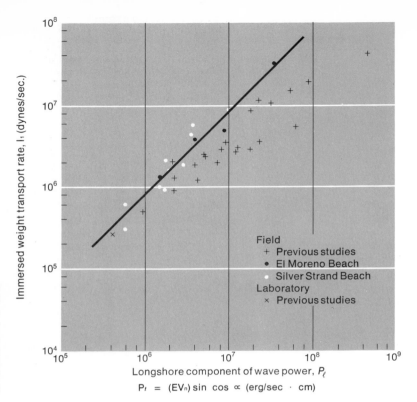

Figure 13.18. Relationship between longshore transport rate and the longshore component of wave energy. (After Komar and Inman 1970, p. 5922, © American Geophysical Union. Used with permission of the *Journal of Geophysical Research*)

$$P_l = (EV_n) \sin \cos \propto (\text{erg/sec} \cdot \text{cm})$$

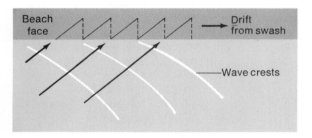

Figure 13.19. Drift of sediment along a beach face caused by waves striking at an angle to the beach.

This analysis seems to contrast with earlier studies which suggested that suspended load was the dominant mode of transport (Fairchild 1973; Thornton 1973). Perhaps some of the difficulty in identifying load types is due to the fact that fine sediment is commonly winnowed from beach debris and carried away from the nearshore environment by rip currents and normal offshore diffusion. Although much sediment seems to be in suspension in the shoreline environment, evidence is beginning to show that the littoral drift process is confined primarily to the coarser bedload debris.

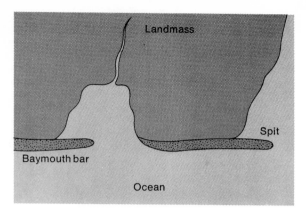

Figure 13.20. Types of depositional features caused by littoral drift of beach sediment.

Briefly, the amount of littoral drift is shown empirically to depend on the longshore component of wave power and the angle at which waves strike the shoreline. Parameters used to estimate wave power, and empirically derived equations relating those parameters to transport rates, are useful. Wave mechanics, a specialized area of physics, provides a sound theoretical basis for these parameters.

Spits and Baymouth Bars   Spits and baymouth bars develop when littoral drift plays a predominant role in the system, provided the drifting sediment enters a zone of slack water where deposition can occur (fig. 13.20). Possibly the erosional and depositional framework in the littoral system can be considered in terms of a littoral power gradient similar to that proposed by May and Tanner (1973). In their study, a wave power model utilizing E, P_l, and dq/dx (q being the quantity of sand transported and x the distance along the beach) as the basic parameters was constructed to predict zones of different littoral transport along a beach. Where dq/dx is greatest, beach erosion is also greatest because a large amount of sediment is being placed in transit. Where dq/dx is negative, deposition is occurring.

Spits and baymouth bars are essentially the same feature. They differ only in that spits extend into the open ocean (fig. 13.21) while baymouth bars cross the gap between two headland reaches. Baymouth bars are also more likely to initiate lagoonal areas shoreward of the bar deposit. These lagoons may gradually change into tidal marshes or swamps as the embayments fill in with fluvially derived sediment.

Although spits and baymouth bars have been attributed to a variety of ocean processes (D. W. Johnson 1919; Lewis 1932), their origin is clearly and most prominently related to littoral drift. The elongate extensions into open water represent continuations of the beach that rests against the coast. They expand continuously in the direction of littoral drift unless other water motions interfere with the growth process. For example, it is not uncommon for wave refraction

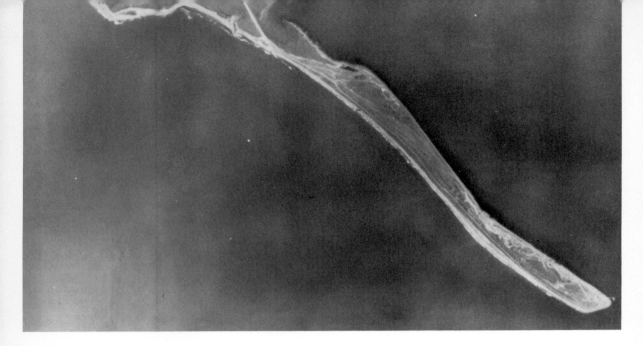

Figure 13.21. Photo of Homer
Spit, Cook Inlet region,
Alaska. (From Waller, R.M.,
U.S. Geol. Survey Prof. Paper
542-D, 1966, fig. 7)

around the free end (Evans 1942) or wave trains approaching from different directions (King and McCullagh 1971) to reorient the terminal end of the spit. In this case the feature may be called a *recurved spit* or a *hook*. Spits or baymouth bars may also widen by progradation associated with the construction of a series of sandy or pebbly ridges on the oceanward side of the feature (fig. 13.22). These ridges or growth lines, called *beach ridges,* will be discussed in the next section. Finally, spits may link the main coastal region to an offshore island, producing the feature known as a *tombolo*. Tombolos, however, also form on normal coastal beaches when wave refraction around the island shapes the beach into a cuspate foreland that eventually extends to the island itself. This process even functions around breakwaters (fig. 13.23) which have been constructed offshore to prevent beach erosion (Inman and Frautschy 1966).

Two engineering methods are commonly used to provide protection against the effect of littoral drift. Where beach erosion is the chief concern, a structure called a *groin* is utilized to prevent sand from drifting away from the beach. A groin is nothing more than a damlike structure, composed of any building material, that is constructed perpendicular to the shoreline. It is usually only a meter or two high and extends oceanward for about 100 meters. A groin is intended to trap drifting sand, and its presence tends to widen a beach on the updrift side of the structure. The beach on the downdrift side of the structure, however, becomes undernourished because its supply of sand has been trapped by the groin. In response, erosion begins and the downdrift beach is narrowed. The solution, of course, is to build another groin to prevent the erosion initiated by the first

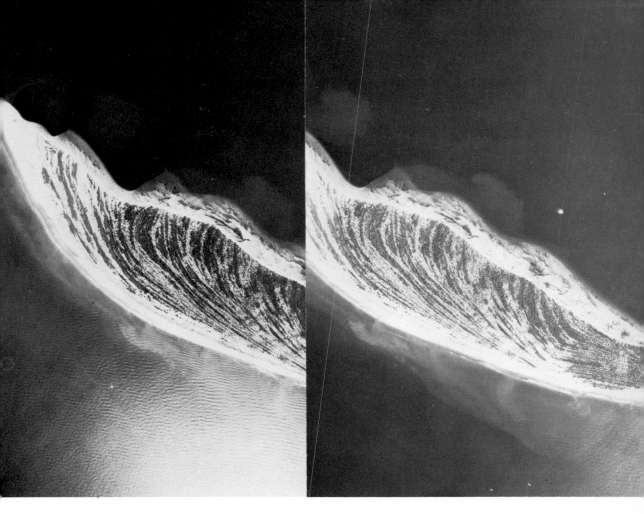

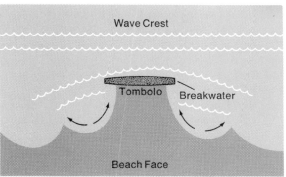

Wave Crest

Tombolo

Breakwater

Beach Face

Figure 13.22. Recurved spit on west coast of Florida. Series of beach ridges show accretionary pattern of spit. (Photo by U.S. Geological Survey and University of Illinois Catalog of Aerial Photography)

Figure 13.23. Connection of breakwater to mainland by the development of a tombolo. (After Inman and Frautschy 1966)

groin, and so it goes until entire shorelines are cluttered with groins (as shown in fig. 13.17). Properly engineered groins can offer some protection against beach erosion, but their overall effect is local and often temporary. More recent attempts at controlling erosion have utilized a ''beach nourishment'' program in which sand is simply trucked onto the beach to replenish that lost by erosion. Although this solution may sound ignorant, it is sometimes cheaper and more efficient to let nature have its way.

At the other end of the littoral conveyor, the problem becomes one of preventing deposition. The growth of baymouth bars has the potential to completely seal off valuable bays or harbors. In such situations, it becomes important to maintain a channel into the harbor that is navigable by large boats. Where there are no tidal inlets, engineers can create a channel by constructing a pair of parallel concrete and rock structures, called *jetties,* in the opening of the bay. These structures are designed so that tidal currents will scour the bay floor, preventing its closure by drifting sand (fig. 13.24). The spacing of jetties requires real engineering talent. If the jetty pairs are too widely spaced, tidal currents will not have the velocity needed to prevent deposition. If they are too close together, the velocities generated may over-erode and undermine the base of the jetties, causing their destruction.

Barrier Islands   Barrier islands are elongate bodies of sand that are not attached to the mainland but are usually separated from it by a lagoon or bay (fig. 13.25). This feature has been called by a variety of names, including barrier beach, offshore bar, and barrier bar. By any name, the feature being described is large, ranging in width from a few meters to a kilometer and in length from hundreds of meters to tens of kilometers. The islands are commonly large enough to support major cities—Atlantic City, Miami Beach, and Galveston are examples. The barrier system may consist of a chain of islands segmented by tidal inlets to the lagoonal area. Tidal inlets presumably develop when the barrier island is breached during strong storm conditions (Pierce 1970). The islands are often surmounted by a pronounced dune system that can raise their elevation considerably above sea level. For example, in a coastal study in southeast Africa, Orme (1973) reports Holocene dunes climbing over older barrier deposits to elevations greater than 100 m.

Barrier islands occur on 13 percent of the world's coasts (King 1972), but they appear to be concentrated in areas where the tidal ranges, wave energies, and offshore gradients are low. Along the boundaries of the United States, they are most prominently displayed along the Gulf Coast and the coast between the middle Atlantic states and Florida.

The origin of barrier islands has always been steeped in controversy, mainly because most of the original interpretations were based on morphology alone. More recent theories rely on subsurface exploration and paleoenvironmental interpretation of facies both shoreward and seaward of the barrier. Thus, there are a number of genetic models that view barrier islands as forming under a wide

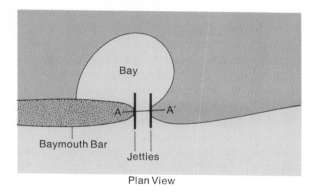

Plan View

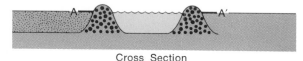

Cross Section

Figure 13.24. The distribution of jetties used to keep a bay open. (Cross section not to same scale as plan view).

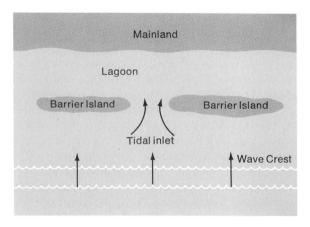

Figure 13.25. Schematic plan view of a barrier island. Lagoon separates island from mainland. Tidal inlet divides island into several components and provides marine water to the lagoon.

variety of sea-level conditions (static, rising, falling). Our modern knowledge of sea-level fluctuations places some limitations on the possible dynamic changes during the late Quaternary. However, if fluctuating sea level is intimately involved in their origin, individual barriers may have an enormous history that represents changes in sea level beginning in the late Pleistocene. In any case, time and stability of conditions cloud the evidence surrounding their origin, and the possibility remains that they may be relict features whose origin is completely unrelated to the modern environment in which they exist.

It seems safe to say that barriers are initiated by wave action working on sand derived from offshore or alongshore. Some clearly begin as spits and are elongated with the littoral drift and subsequently isolated by tidal inlet breaching.

For example, a littoral drift origin for Sapelo Island, Georgia, was convincingly demonstrated by Hoyt and Henry (1967). Other barrier islands, however, are obviously not lateral extensions of the coastline and cannot be attributed to longshore processes. It is these offshore barriers that have presented the greatest interpretive difficulties. The early interpretations of these features seemed to follow D. W. Johnson's (1919) idea that erosion and deposition of bottom sediment in the breaker zone develop a submarine ridge between the breakers and the shoreline. Opposition to this idea was not based so much on the mechanics of the offshore bar development as on the difficulty of raising the ridge above sea level. This led some investigators to believe that barrier islands form on emergent shorelines; that is, the submerged bar is gradually raised above sea level as the entire coast slowly emerges. When data from drilling became available, however, a different concept was formulated. In this hypothesis, barrier islands have a long growth history, beginning as depositional nuclei during the low sea level of the late Pleistocene and growing continuously as sea level rose during the last 15,000–20,000 years. As the coasts became submerged in the postglacial transgression, the barrier islands assumed their present configuration (Shepard, 1960, 1963; Zenkovich 1967; Hoyt 1967; Price 1968; Otvos 1970; J. J. Fisher 1973).

Although a large number of workers seem to agree with the submergence model, the controversy has not completely abated. Differences of opinion exist as to where the initial ridge forms in the coastal setting, and some authors still argue that submergence is not a necessary ingredient in the formative process (C. W. Cooke 1968; Schwartz 1970). Hoyt (1967) suggested the provocative idea that barrier islands do not begin as submerged offshore bars, but are initiated as sand dunes or pebbly beach ridges molded on the upper, landward portions of the beach. As sea level rises, the area behind the dunes or ridges is drowned; the ridges, stabilized by vegetation, become barrier islands and the submerged landward areas become lagoons. The dimensions of the islands in Hoyt's model depend on the amount of progradation and upward growth of the beach ridges and dunes. As mentioned earlier, dunes surmounted on a barrier island can rise to considerable height (Orme 1973). Beach ridges, however, are usually lower in relief, seldom exceeding 6 m. They exist as long, parallel ridges and normally occur in a series, each successive ridge spaced 25 to 500 m from its neighbor (shown in fig. 13.22). An individual beach ridge develops from a single storm, and so the spacing and number of ridges in a series are probably determined by the recurrence interval between large storms.

The model devised by Hoyt does require submergence in the genesis of most barrier islands, but the position of the incipient development is on the exposed portion of the beach rather than in the offshore zone. The model has been challenged (C. W. Cooke 1968; J. J. Fisher 1968) for various reasons. In response to these challenges, Hoyt (1968) defended his concept by pointing out that landward sides of barrier lagoons show no evidence of open-water sedimentation, which suggests that the area was never worked by normal beach processes and that the barrier island could not have developed initially as an offshore submerged bar.

Obviously the origin of barrier islands is a question that has not been resolved; perhaps the feature can form in different ways. In addition, the relict nature of many barrier islands makes the relationship between their geomorphic form and the processes operating in the modern system meaningless; present conditions were not responsible for developing the feature we now observe.

Reefs   A special type of depositional landform, called a *reef,* is common in coastal regions of the tropics (fig. 13.26). Like the barrier islands discussed above, the origin of some reef types depends on relative movements of the ocean and land. Parts of some reefs therefore are relict, and their geomorphic form is not necessarily related to modern conditions.

Reefs are composed of a hard, resistant framework of interwoven skeletons of marine organisms, both living and dead. Although they are commonly referred to as coral reefs, corals may be only part of the living assemblage, with a variety of other marine forms being present within the reef structure. The colonial organisms that build reef structures live under strict environmental controls: (1) they cannot exist at depths greater than 60 m (200 ft) because they require sunlight to function; (2) being immobile and attached to the bottom, they thrive best in water agitated by wave action because the turbulence supplies them with food and oxygen; (3) they require the salinity of normal marine water; (4) they prefer water that has a low sediment concentration; and (5) they can live only where the water temperature is greater than 18°C (64°F) and are best adapted to

Figure 13.26. Reef along Cocos Island in the Mariana Islands near Guam. Part of lagoon is shown on left of photo. Note wave refraction around reef projection. (Photo by K.O. Emery. U.S. Geol. Survey Prof. Paper 403-B, 1962, fig. 13)

water between 25°C and 30°C (77°–86°F). Thus, reefs as coastal geomorphic landforms are geographically restricted to those regions that provide the environment needed for the growth of reef organisms.

As geomorphic features, reefs can vary in width from a few meters to tens of kilometers. They are usually classified, as in figure 13.27, on the basis of their position relative to a nearby landmass. When they are physically attached to the landmass they are called *fringing reefs*. When they are separated from the land by a lagoon they are called *barrier reefs* (for example, the Great Barrier Reef off the northeast coast of Australia). In some cases reefs exist in a ringlike pattern that encircles a lagoon, but no landmass stands above sea level in the interior portion of the enclosing reef. These reefs are called *atolls*. The origin of atolls has been steeped in interpretive controversy, but it seems clear that the immense thickness of the reef structures (in some cases > 1200 m) requires a gradual sinking of the original island and reef complex. Although the organisms can only live in very shallow water, slow sinking of the system would allow the reef colony to grow upward as the island and older reef structures are progressively inundated.

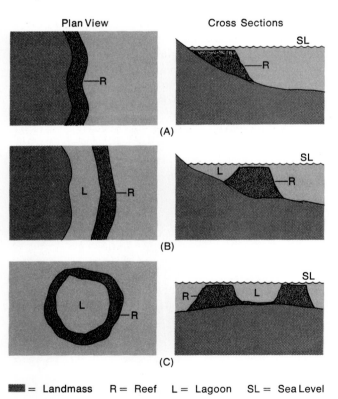

Figure 13.27. Classification of reefs. (A) Fringing reef. (B) Barrier reef. (C) Atoll.

Plan View   Cross Sections

(A)

(B)

(C)

■■■ = Landmass   R = Reef   L = Lagoon   SL = Sea Level

Summary

Coastal zones reflect the balance between the driving forces of ocean waters and the resistance offered by the rocks that form the shoreline. Large lakes may experience the same phenomena as ocean coasts. Beaches are the components of coasts that respond most obviously to the dynamics of the system. Energy possessed by waves, currents, and tides is expended on the beach surface. The response to this activity is the creation of a beach profile that is adjusted to the mean values of the wave properties. Waves striking the beach zone obliquely also cause longshore (littoral) drift of sediment.

Coasts and beaches display a wide variety of geomorphic forms that range in size from minor modifications of the beach face to features that encompass kilometers of the shoreline. Commonly the shoreline is indented with a hierarchy of crescentic or rhythmic forms that are related in some way to cellular circulation of the ocean water. Coasts may be typified as erosional or depositional depending on the kinds of large-scale features that dominate the coastal scene. In some cases (barrier islands, reefs) the features may not be adjusted to modern geomorphic conditions. These relict forms are difficult to relate to shoreline processes as we see them today, and so their origins are often obscure.

Suggested Readings

The following suggested readings are an abbreviated sample of the literature concerning the subjects treated in this chapter. Each reference cited has a more extensive bibliography of topics that may be of particular interest to the reader.

Bascom, W. 1964. *Waves and beaches*. Garden City, N.Y.: Doubleday.

Dolan, R.; Vincent, L.; and Hayden, B. 1974. Crescentic coastal landforms. *Zeit. f. Geomorph.* 18:1-12.

Hoyt, J. H. 1967. Barrier island formation. *Geol. Soc. America Bull.* 78:1125-36.

King, C.A.M. 1972. *Beaches and coasts*. New York: St. Martin's Press.

Komar, P. D. 1976. *Beach processes and sedimentation*. Englewood Cliffs, N.J.: Prentice-Hall.

Komar, P. D., and Inman, D. L. 1970. Longshore sand transport on beaches. *Jour. Geophys. Research* 75, 30:5914-27.

We have now examined a wide variety of geomorphic topics, focusing mainly on processes—how they are affected by driving and resisting elements and how they result in the landforms that characterize the Earth's surface. I believe it is now fair for you to ask why you have been subjected to this abundance of facts, and what significance they have to you as a scientist and as a human being. Before you make such an analysis, allow me to conclude this book with a brief review of two specific cases in which humans have intervened in the geomorphic systematics. The two cases are located within 300 km of one another, and human action began at nearly the same time. It is interesting to note how, when, and where nature has responded to counteract the imbalance created when human actions stressed the geomorphic systems beyond the threshold limits.

Epilogue

Hydraulic Mining in California

During the early days of gold mining in California, the west flank of the Sierra Nevada was cluttered with small prospecting sites where individuals tried their luck in the gold-bearing sand and gravel deposits. The lure of great wealth, however, changed the modus operandi from a single man digging with pick and shovel to intensive hydraulic mining in which large water jets were blasted against the deposits. In this way, almost 1.2 billion cubic meters (1,555,000,000 yd^3) of debris was overturned in a period of about 50 years, a volume of sediment eight times greater than that moved in construction of the Panama Canal.

Much of the silt, sand, and gravel moved as a giant wave of sediment, progressively making its way from the mountain streams into the major rivers crossing the Great Valley of California and finally reaching the tidal waters of the San Francisco bay system. The quantity of debris reaching the Yuba and Sacramento basins was so great that in 1884 farmers in those fertile areas sought and obtained an injunction to restrain mining companies from placing any more mine tailings into the watersheds. A period of 20 years of inactivity passed before the California Miners Association appealed directly to the President of the United States for a detailed investigation and recommendations for procedures that could allow mining to resume while simultaneously protecting the riparian landowners. The request was forwarded to the U.S. Geological Survey, and the assignment given to G. K. Gilbert, the right man to undertake such a monumental task. Gilbert meticulously documented what had happened in the area as the natural system responded to the vast hydraulic mining operation and also to its cessation. The results of his study, published in 1917 (Gilbert 1917), represent a classic piece of geomorphic insight, and the following account is based on that report.

As mine tailings were released from the mountain deposits, the coarse fraction gradually invaded the channels of the major downstream rivers as a sand and gravel bedload. The rivers, unable to transport such an overwhelming load, adjusted by drastically raising their channel bottoms as the material was deposited.

As each segment filled, the gradient increased so that the river acquired the capacity needed to transport the sediment farther down the valley. The rise in channel level stopped at different times in each segment of each river, depending on the distance from the source and the amount of load. For example, the channel of the Yuba River at Marysville (Cal.) rose about 6 m (19.1 ft) between 1849 and 1905, when it reached its highest level. The Sacramento River at Sacramento (Cal.) elevated about 3 m (10.8 ft), attaining its highest level in 1897. It is interesting to note that both rivers continued to fill even after mining ceased in 1884. This happened because the upstream reaches, no longer receiving great volumes of sediment, had excess energy on their steepened gradients and therefore entrenched their channels. The sediment from the entrenching process was transported to the lower river segments. Thus, part of the fluvial system was filling while, at the same time, other parts were entrenching.

The Yuba River received the largest proportion of the mined debris and so had the greatest sediment problem. This prompted California officials to construct a concrete barrier to trap the bedload upstream from the town of Marysville. In 1904 the barrier was 2 m high (6 ft) but filled with sediment in the following year. Another 2.5 m (8 ft) section was added in 1905. By 1907 this also filled. This folly might have continued if the river had not destroyed the barrier in a major 1907 flood. The destruction occurred when water flowing over the barrier, having excess energy because it carried no load, entrenched below the level of the protecting riprap and presumably undermined the structure.

The raising of the channel floors was accompanied by larger and more frequent flooding of the adjacent farmland. Frequent flows of modest magnitude that previously had been contained within the banks began to inundate large tracts of riparian land. Large floods were considerably higher than before and transported significant volumes of sand and gravel onto the adjacent property. It was the flooding that prompted the farmers to seek a restraining order and led to widespread and expensive levee construction to control it. The downstream areas that were not protected suffered greatly, however, because the effect of any levee system is to increase the stage of any flow in the downstream direction. This happens because the natural storage area of flood waters, the floodplain, is prevented from exerting its role in the hydrologic system. Fortunately, the levee construction and the cessation of mining had complementary influences on the transporting ability of the rivers. Excavation of the fills began in most downstream portions of the rivers around the turn of the century.

We must still ask, however, where the suspended load went during this whole sequence of events, and ultimately where the excavated debris went when the rivers began to retrench. It was here that Gilbert showed his true genius. A lesser man would have been limited by the charge given to him, i.e., the clash between hydraulic mining and flooding of the lowland farms. Gilbert, however, understood that all geomorphic systems are interrelated, and therefore he took it upon himself to examine the effect of mining on the San Francisco bay system.

In short, he was able to show that the bays were shoaling and that tidal marshes were expanding where they were not protected by levees. These, combined with the influence of the levee system, were notably affecting the volume of water exchange during tidal fluctuations. The decrease in the tidal volume was in turn adversely influencing the size and position of the Golden Gate bar lying immediately off the entrance to San Francisco Bay. Gilbert thus concluded that while resumption of massive hydraulic mining might be manageable with regard to the agricultural interests, ultimately it could destroy the shipping and trade industries associated with San Francisco harbor.

Pyramid Lake, Nevada

At approximately the same time that the effects of hydraulic mining were reaching their peak in the California basins, another human effort was being completed on the east side of the Sierra Nevada. The Truckee River rises from Lake Tahoe high in the mountains, flows through Reno (Nev.) in the piedmont region, and ends in a naturally enclosed desert lake, called Pyramid Lake, located about 45 km northeast of Reno. The lake, having no outlet, has an elevation that represents the equilibrium between inflow from the Truckee River and water lost by evaporation. Only 10 cm (4 in) of precipitation falls in the area of the lake each year, so its hydrologic influence is negligible.

In 1905 the Bureau of Reclamation constructed a small dam across the Truckee about 20 km upstream from Pyramid Lake, the purpose of which was to divert water into the adjacent drainage basin where it was used for irrigation. Thus, since 1905 half the flow from the river has been lost to Pyramid Lake. As might be expected, the lake level, adjusting to the hydrologic imbalance, has declined about 25 m since the construction of the dam. This in itself is no tragedy, but the chain of events that accompanied the lake-level decline was spectacular and, at that time, completely unpredictable.

As the lake dropped, the Truckee River adjusted to its new base level by simultaneously entrenching and prograding a delta into the lake. Where the river crosses the deltaic plain it changed from a deep, single-channeled pattern to a braided system with wide, shallow channels. Pyramid Lake until that time was world-famous for its population of cutthroat trout, which were abundant enough to support a fishing industry for the local Paiute Indian tribe. The largest cutthroat ever recorded (45 lbs) was taken from the lake. The cutthroat, however, is related to the salmon and, like its relative, must migrate upstream to spawn. The change in channel configuration prevented the trout from their annual migration because the fish foundered on sand bars in the shallow water. In short, the trout are no longer natural inhabitants of the lake. Those that are present have been raised elsewhere and artificially placed into Pyramid Lake. The point here is that the river and the lake are not contaminated with toxic pollutants; the native fish have disappeared because the *physical* environment changed, not because the biochemical environment deteriorated.

Another interesting adjustment relates to the sediment formerly located beneath the lake and now exposed. This unconsolidated debris is not stabilized because precipitation is too low to support a vegetal cover. On the north side of the lake, the prevailing winds are transporting sand from the lake margin as climbing dunes. These have now migrated to the edge of a paved highway, leaving the state of Nevada with the potential dilemma of whether to plow in order to keep the road open or to take preventive measures to stop further movement of the dunes. Stabilizing the dunes with vegetation would require irrigation, and the only available water is in the lake itself. To use it for irrigation, therefore, would simply accelerate the decline of the lake level and expose even more sediment. Perhaps the ultimate solution will be construction of fences to stop the dune migration.

Finally, another ominous result of the dam looms in the near future. Pyramid Lake supports the largest population of white pelicans in North America (Wheeler 1968). The birds nest on an island, called Anaho Island, that stands within the lake. In fact, so many pelicans reside there that in 1913 President Woodrow Wilson declared the island a national wildlife refuge. Now the decrease in lake level has exposed the bottom to the point that the mainland and the island are separated by only a narrow strip of water. The depth of the water in this zone is only about 10 meters, and at the present rate of lake decline the island will be physically connected to the mainland by a land bridge in about 30 years. When that occurs it will be difficult, if not impossible, to prevent predators living in the nearby hills from invading the nesting grounds and destroying the pelican population.

What It All Means

Comparisons and contrasts of these two cases are both enlightening and instructive, and they lead us to the following conclusions: (1) The impulse that upsets the balance in a natural geomorphic system can be extremely large (hydraulic mining) or relatively small (the Truckee River dam). (2) In both cases, environmental damage was least at the locale of the provoking factor. The major geomorphic responses occurred at considerable distance from the original impetus. (3) Both cases demonstrated that geomorphic systems are interdependent. Physical alteration began in the fluvial regime, but it was ultimately transferred to the dynamics associated with lake, ocean, or wind systems. (4) Cause and effect within or between geomorphic systems is analogous to the domino principle. An effect within one system becomes the cause of changes in the others. (5) The time of adjustment varies but can encompass more than the normal span of a human life. Today's activities may not be recognized as a "problem" until your children, grandchildren, or even great-grandchildren reach maturity. (6) Physical changes produce destruction of the biological system even where the biochemistry has not been drastically altered. (7) We commonly spend more time, effort, and money trying to combat nature's response to our original action that we did to initiate the action. Often the costs are greater than the predicted benefits.

The cases described above are not unique. Most of you probably are aware of similar events that have occurred within your region. So, it is now time for you to consider the question posed earlier. I suggest that it is easier to analyze in the context of the human ability to be a catalyst in geomorphic systems. Usually in ignorance, but often with incredible arrogance, we somehow believe we can bend nature's balance to suit our needs. Even worse, we convince ourselves that this can be done without paying a price. Obviously, we cannot revert to a caveman existence, but we should admit to the inherent dangers that exist when we tamper with processes that we do not understand. The trick, then, is to do what we can in such a way that nature will extract from us the minimum cost in return. If we as scientists refuse to provide the guidelines as to how this can be done, then who will?

Bibliography

Abrahams, A. D. 1972. Environmental constraints on the substitution of space for time in the study of natural channel networks. *Geol. Soc. America Bull.* 83:1523-30.

―――. 1975. Topologically random channel networks in the presence of environmental controls. *Geol. Soc. America Bull.* 86:1459-62.

Ackers, P., and Charlton, F. G. 1971. The slope and resistance of small meandering channels. *Inst. Civil Engrs. Proc.,* Supp. 15, 1970, paper 73625.

Addington, A. R. 1927. Porter's Cave and recent drainage adjustments in its vicinity. *Indiana Acad. Sci. Proc.* 36:107-16.

Ahlmann, H. W. 1948. *Glaciological research on the North Atlantic.* Royal Geog. Soc. Res. Ser. 1.

Ahnert, F. 1970. Functional relationships between denudation, relief, and uplift in large mid-latitude drainage basins. *Am. Jour. Sci.* 268:243-263.

Ahr, W. M., and Stanton, R. J. 1973. The sedimentologic and paleoecologic significance of Lithotyra, a rock-boring barnacle. *Jour. Sed. Petrology* 43:20-23.

Allen, J. R. L. 1969. Erosional current marks on weakly cohesive mud beds. *Jour. Sed. Petrology* 39:607-23.

Alter, A. J. 1966. Sanitary engineering in Alaska. In *Proc. Permafrost Internat. Conf.* (Lafayette, Ind., 1963). Natl. Acad. Sci. Natl. Research Council Pub. 1287, pp. 407-8.

American Geological Institute. 1972. *Glossary of geological terms.*

Anderson, D. M. 1968. Undercooling, freezing, point depression, and ice nucleation of soil water. *Israel J. Chem.* 6:349-55.

―――. 1970. Phase boundary water in frozen soils. U.S. Army Corps Engrs., Cold Regions Res. and Eng. Lab. Research Rept. 274.

Anderson, J. G. 1906. Solifluction: A component of subaerial denudation. *Jour. Geology* 14:91-112.

Anderson, R. C. 1957. Pebble and sand lithology of the major Wisconsin glacial lobes of the Central Lowland. *Geol. Soc. America Bull.* 68:1415-50.

Anderton, P. W. 1974. Ice fabrics and petrography. Meserve Glacier. Antarctica. *Jour. Glaciol.* 13:285-306.

Andrews, J. T. 1963. Cross-valley moraines of the Rimrock and Isotoq river valleys, Baffin Island. A descriptive analysis. *Geogr. Bull.* 19:49-77.

Andrews, J. T., and King, C. A. M. 1968. Comparative till fabrics and till fabric variability in a till sheet and a drumlin; a small-scale study. *Yorkshire Geol. Soc. Proc.* 36:435-61.

Andrews, J. T., and Smithson, B. B. 1966. Till fabrics of the cross-valley moraines of north-central Baffin Island. *Geol. Soc. America Bull.* 77:271-90.

Arkley, R. 1963. Calculation of carbonate and water movement in soil from climatic data. *Soil Sci.* 96:239-48.

A.S.T.M. 1971. *Book of American Society for Testing and Materials Standards,* part 2, pp. 225-32.

Atkinson, H., and Wright, J. 1957. Chelation and the vertical movement of soil constituents. *Soil Sci.* 84:1-11.

Atterberg, A. 1911. Die Plastizitat der Tone. *Intern. Mitt. Boden* 1:4-37.

Baas-Becking, L.; Kaplan, I.; and Moore, D. 1960. Limits of the natural environment in terms of pH and oxidation-reduction potentials. *Jour. Geology* 68:243-84.

Bagnold, R. A. 1940. Beach formation by waves—some model experiments in a wave tank. *J. Inst. Civil Eng.* 15:27-52.

―――. 1941. *The physics of blown sand and desert dunes.* London: Methuen and Co.

———. 1953. The surface movement of blown sand in relation to meteorology. In *Desert Research,* UNESCO, pp. 89-93. Res. Council of Israel Spec. Pub. 2.

———. 1960. Some aspects of river meanders. U.S. Geol. Survey Prof. Paper 282-E.

———. 1967. Deposition in the process of hydraulic transport. *Sedimentology* 10:45-56.

Bailey, J. F., and Patterson, J. L. 1975. Hurricane Agnes rainfall and floods, June-July, 1972. U.S. Geol. Survey Prof. Paper 924.

Bajourunas, L., and Duane, D. B. 1967. Shifting offshore bars and harbor shoaling. *Jour. Geophys. Research* 72:6195-6205.

Baker, V. R. 1973a. Geomorphology and hydrology of karst drainage basins and cave channel networks in east-central New York. *Water Resour. Res.* 9:695-706.

———. 1973b. *Paleohydrology and sedimentology of Lake Missoula flooding in eastern Washington.* Geol. Soc. America Spec. Paper 144.

———. 1974. Paleohydraulic interpretation of Quaternary alluvium near Golden, Colorado. *Quat. Res.* 4:94-112.

———. 1975. Urban geology of Boulder, Colorado: A progress report. *Environmental Geol.* 1:75-88.

———. 1976. Hydrogeology of a cavernous limestone terrane and the hydrochemical mechanisms of its formation, Mohawk River basin, New York. *Empire State Geogram* 12; no. 2:2-65.

Baker, V. R., and Ritter, D. F. 1975. Competence of rivers to transport coarse bedload material. *Geol. Soc. America Bull.* 86:975-78.

Bakker, J. P. 1965. A forgotten factor in the interpretation of glacial stairways. *Zeit. f. Geomorph.* 9:18-34.

Baldwin, M.; Kellogg, C.; and Thorp, J. 1938. Soil classification. In *Soils and Men,* U.S. Dept. Agri. Yearbook, pp. 979-1001.

Ballard, T. M. 1973. Soil physical properties in a sorted stripe field. *Arc. Alp. Res.* 5:127-31.

Bandy, O., and Marincovich, L. 1973. Rates of late Cenozoic uplift, Baldwin Hills, Los Angeles, California. *Science* 181:653-55.

Barnes H. A. 1968. Roughness characteristics of natural channels. U.S. Geol. Survey Water Supply Paper 1849.

Barnes, P., and Tabor, D. 1966. Plastic flow and pressure melting in the deformation of ice. *Nature* 210:878-82.

Barnett, D. M., and Holdsworth, G. 1974. Origin, morphology, and chronology of sublacustrine moraines, Generator Lake, Baffin Island, Northwest Territories, Canada. *Can. J. Earth Sci.* 11:380-408.

Barsch, D. 1971. Rock glaciers and ice-cored moraines. *Geogr. Annlr.* 53A:203-06.

Barsch, D., and Updike R. G. 1971. Periglaziale Formung am Kendrick Peak in Nord-Arizona während der letzten kaltzeit. *Geog. Helvetica* 26:99-114.

Barshad, I. 1964. Chemistry of soil development. In *Chemistry of the soil,* edited by F. Bear, pp. 1-70. New York: Reinhold Book Corp.

Bascom, W. N. 1951. The relationship between sand size and beach face slope. *Am. Geophys. Union Trans.* 32:866-74.

———. 1964. *Waves and beaches.* Garden City, N.Y.: Doubleday.

Bates, C. C. 1953. Rational theory of delta formation. *Am. Assoc. Petroleum Geologists Bull.* 37:2119-62.

Battle, W. R. B. 1960. Temperature observation in bergschrunds and their relationship to frost shattering. In *Norwegian cirque glaciers,* edited by W. V. Lewis, pp. 83-96. Royal Geog. Soc. Res. Ser. 4.

Baulig, H. 1957. Peneplains and pediplains. *Geol. Soc. America Bull.* 68:913-30.

Beaty, C. B. 1963. Origin of alluvial fans, White Mountains, California and Nevada. *Ann. Assoc. Am. Geog.* 53:516-35.

———. 1970. Age and estimated rate of accumulation of an alluvial fan, White Mountains, California, U.S.A. *Am. Jour. Sci.* 268:50-77.

———. 1975a. Sublimation or melting observations from the White Mountains, California and Nevada, U.S.A. *Jour. Glaciol.* 14:275-86.

———. 1975b. Coulee alignment and the wind in southern Alberta, Canada. *Geol. Soc. America Bull.* 86:119-28.

Beaumont, P. 1972. Alluvial fans along the foothills of the Elburz Mountain, Iran. *Paleogeogr. Paleoclimatol. Paleoecol.* 12:251-73.

Behrendt, J. C. 1965. Densification of snow on the ice sheet of Ellsworth Land and South Antarctic Peninsula. *Jour. Glaciol.* 5:451-60.

Belly, P. V. 1964. Sand movement by wind. U.S. Army Corps Engrs., Coastal Eng. Res. Center Tech. Memo 1.

Beloussov, V. V. 1970. Against the hypothesis of ocean floor spreading. *Tectonophysics* 9:489-511.

———. 1972. Basic trends in the evolution of continents. *Tectonophysics* 13:95-118.

Benedict, J. B. 1970. Downslope soil movement in a Colorado alpine region. Rates, processes, and climatic significance. *Arc. Alp. Res.* 2:165-226.

————. 1976. Frost creep and gelifluction features. A review. *Quat. Res.* 6:55-77.

Benioff, H. 1955. Seismic evidence for crustal structure and tectonic activity. In *Crust of the Earth,* edited by A. Poldervaart. Geol. Soc. America Spec. Paper 62.

Benson, M. A. 1962. Factors influencing the occurrence of floods in a humid region of diverse terrain. U.S. Geol. Survey Water Supply Paper 1580-B.

Beskow, G. 1947. Soil freezing and frost heaving with special application to roads and railroads. Evanston, Ill.: Northwestern University Tech. Inst.

Bhattacharya, N. 1963. Weathering of glacial tills in Indiana. II heavy minerals. *Jour. Sed. Petrology* 33:789-94.

Bird, E. C. F. 1969. *Coasts.* Cambridge Mass.: M.I.T. Press.

Bird, J. M., and Dewey, J. F. 1970. Lithosphere plate-continental margin tectonics and the evolution of the Appalachian orogen. *Geol. Soc. America Bull.* 81:1031-60.

Birkeland, P. 1968. Mean velocities and boulder transport during Tahoe-age floods of the Truckee River, California-Nevada. *Geol. Soc. America Bull.* 79:137-41.

————. 1969. Quaternary paleoclimatic implications of soil clay mineral distribution in a Sierra Nevada-Great Basin transect. *Jour. Geology* 77:289-302.

————. 1974. *Pedology, weathering, and geomorphological research.* London: Oxford Univ. Press.

Birkeland, P., and Janda, R. 1971. Clay mineralogy of soils developed from Quaternary deposits of the eastern Sierra Nevada, California. *Geol. Soc. America Bull.* 82:2495-2514.

Birkeland, P., and Shroba, R. 1974. The status of the concept of Quaternary soil-forming intervals in the western United States. In *Quaternary environments: Proceedings of a symposium,* edited by W. Mahaney, pp. 241-76. Geog. Monographs, York Univ., Toronto.

Bishop, B. C. 1957. Shear moraines in the Thule areas, northwest Greenland. U.S. Army Corps Engrs., Snow, Ice, Permafrost Res. Est. Rept. 17.

Black, R. F. 1954. Permafrost, a review. *Geol. Soc. America Bull.* 65:839-55.

————. 1963. Les coins de glace et le gel permanent dans le Nord de l'Alaska. *Annales géog.* 72:257-71.

————. 1969. Climatically significant fossil periglacial phenomena in north-central United States. *Biuletyn Perygl.* 20:225-38.

————. 1974. Ice-wedge polygons of northern Alaska. In *Glacial geomorphology,* edited by D. R. Coates, pp. 247-75. S.U.N.Y., Binghamton, Pubs. in Geomorphology, 5th Ann. Symposium.

————. 1976. Periglacial features indicative of permafrost: Ice and soil wedges. *Quat. Res.* 6:3-26.

Black, R. F., and Hamilton, T. D. 1971. Mass-movement studies near Madison, Wisconsin. In *Quantitative geomorphology,* edited by M. Morisawa, pp. 121-79. S.U.N.Y., Binghamton, Pubs. in Geomorphology, Proc. 2nd Symposium.

Blackwelder, E. 1927. Fire as an agent in rock weathering. *Jour. Geology* 35: 135-40.

————. 1931. The lowering of playas by deflation. *Am. Jour. Sci.* 21:140-44.

————. 1934. Yardangs. *Geol. Soc. America Bull.* 45:159-66.

Blagborough, J. W., and Farkas, S. E. 1968. Rock glaciers in the San Mateo Mountains, south-central New Mexico. *Am. Jour. Sci.* 266:812-23.

Blatt, H., and Jones, R. 1975. Proportions of exposed igneous, metamorphic, and sedimentary rocks. *Geol. Soc. America Bull.* 86:1085-88.

Blissenbach, E. 1954. Geology of alluvial fans in semiarid regions. *Geol. Soc. America Bull.* 65:175-90.

Bloom, A. L. 1965. The explanatory description of coasts. *Zeit. f. Geomorph.* 9(4):422-36.

Bluck, B. J. 1964. Sedimentation on an alluvial fan in southern Nevada. *Jour. Sed. Petrology* 34:395-400.

Bögli, A. 1960. Kalkösung and Karrenbildung. *Zeit. f. Geomorph.* Suppl. 2:4-21.

————. 1964a. Mischungskorrosion— ein Beitrag zur Verkarstungsproblem. *Erdkunde* 18:83-92.

————. 1964b. Le Schichttreppenkarst. Revue Belge Géog. 88:63-82.

Born, S. M. 1972. *Late Quaternary history, deltaic sedimentation, and mudlump formation at Pyramid Lake, Nevada.* Univ. Nevada, Reno, Desert Research Inst.

Born, S. M., and Ritter, D. F. 1970. Modern terrace development near Pyramid Lake, Nevada, and its geologic implications. *Geol. Soc. America Bull.* 81:1233-42.

Bottomley, J. T. 1872. Melting and regelation of ice. *Nature* 5:185.

Boulton, G. S. 1967. The development of a complex supraglacial moraine at the margin of Sorbreen, Ny Friesland, Vestspitsbergen. *Jour. Glaciol.* 6: 717-36.

————. 1968. Flow tills and related deposits on some Vestspitsbergen glaciers. *Jour. Glaciol.* 7:391-412.

————. 1970a. On the origin and transport of englacial debris in Svalbard glaciers. *Jour. Glaciol.* 9:213-29.

————. 1970b. On the deposition of subglacial and melt-out tills at the margins of certain Svalbard glaciers. *Jour. Glaciol.* 9:231-46.

————. 1971. Till genesis and fabric in Svalbard, Spitsbergen. In *Till: A symposium,* edited by R. P. Goldthwait, pp. 41-72. Columbus: Ohio State Univ. Press.

————. 1972a. The role of thermal regime in glacial sedimentation—a general theory. In *Polar geomorphology,* edited by R. J. Price and D. E. Sugden, pp. 1-19. Inst. Brit. Geog. Spec. Paper 4.

————. 1972b. Modern Arctic glaciers as depositional models for former ice sheets. *Jour. Geol. Soc. Lond.* 128:361-93.

————. 1974. Processes and patterns of glacial erosion. In *Glacial geomorphology,* edited by D. R. Coates, pp. 41-87. S.U.N.Y., Binghamton, Pubs. in Geomorphology, 5th Ann. Symposium.

Bowen, A. J. 1969. The generation of longshore currents on a plane beach. *J. Marine Res.* 37:206-15.

————. 1973. Edge waves and the littoral environment. *Proc. 13th Conf. on Coast. Eng.,* 1313-20.

Bowen, A. J., and Inman, D. L. 1969. Rip currents. 2. Laboratory and field observations. *Jour. Geophys. Research* 74:5479-90.

————. 1971. Edge waves and crescentic bars. *Jour. Geophys. Research* 76:8662-71.

Bowen, N. 1928. *The evolution of the igneous rocks.* Princeton, N.J.: Princeton Univ. Press.

Bowler, J. M. 1973. Clay dunes: Their occurrence, formation, and environmental significance. *Earth Sci. Rev.* 9:315-38.

Bradley, W. C. 1957. Origin of marine-terrace deposits in the Santa Cruz area, California. *Geol. Soc. America Bull.* 68: 421-44.

————. 1958. Submarine abrasion and wave-cut platforms. *Geol. Soc. America Bull.* 69:967-74.

————. 1970. Effect of weathering on abrasion of granitic gravel, Colorado River (Texas). *Geol. Soc. America Bull.* 81:61-80.

Brenninkmeyer, B. M. 1975. Frequency of sand movement in the surf zone. *Proc. 14th Conf. on Coast. Eng.,* 812-27.

Bretschneider, C. L. 1952. The generation and decay of wind waves in deep water. *Am. Geophys. Union Trans.* 33:387-89.

————. 1966a. Storm surge. In *Encyclopedia of oceanography,* edited by R. W. Fairbridge, pp. 856-60. New York: Reinhold Book Corp.

————. 1966b. Wave refraction, diffraction, and reflection. In *Estuary and coastline hydrodynamics,* edited by A. T. Ippen, pp. 257-79. New York: McGraw-Hill.

Bretz, J. H. 1942. Vadose and phreatic features of limestone caverns. *Jour. Geology* 50:675-811.

Brice, J. C. 1964. Channel patterns and terraces of the Loup River in Nebraska. *U.S Geol. Survey Prof. Paper 422-D.*

————. 1974. Evolution of meander loops. *Geol. Soc. America Bull.* 85:581-86.

Brindley, G.; Bailey, S.; Faust, G.; Forman, S.; and Rich, C. 1968. Report of the Nomenclature Committee (1966-67) of the Clay Minerals Society. *Clays and Clay Minerals* 16:322-24.

Bristol, H., and Howard, R. 1974. Sub-Pennsylvanian valleys in the Chesterian surface of the Illinois Basin and related Chesterian slump blocks. In *Carboniferous of the southeastern U.S.,* edited by G. Briggs, pp. 315-36. Geol. Soc. America Spec. Paper 148.

Brookfield, M. 1970. Dune trend and wind regime in central Australia. *Zeit. f. Geomorph.,* Suppl. 10:121-58.

Brown, E. H. 1969. Jointing, aspect, and orientation of scarp-face dry valleys, near Ivinghoe, Buckinghamshire. *Inst. Brit. Geog. Trans.* 48:61-74.

Brown, R. J. E. 1969. Permafrost in Canada. Canada Geol. Survey Map 1246A.

————. 1970. *Permafrost in Canada.* Toronto: Univ. Toronto Press.

Browning, J. M. 1973. Catastrophic rock slides, Mount Huascaran, north-central Peru, May 31, 1970. *Am. Assoc. Petroleum Geologists Bull.* 57: 1335-41.

Brucker, R. W.; Hess, J. W.; and White, W. B. 1972. Role of vertical shafts in the movement of ground water in carbonate aquifers. *Ground Water* 10(6):5.

Brune, G. 1948. Rates of sediment production in midwestern United States. Soil Conserv. Serv. Tech. Pub. 65.

Brush, L. M. 1961. Drainage basins, channels, and flow characteristics of selected streams in central Pennsylvania. U.S. Geol. Survey Prof. Paper 282-F.

Brush, L. M., and Wolman, M. G. 1960. Knickpoint behavior in non-cohesive material: a laboratory study. *Geol. Soc. America Bull.* 71:59-74.

Bruun, P. 1963. Longshore currents and longshore troughs. *Jour. Geophys. Research* 68:1065-78.

Bryan, K. 1922. Erosion and sedimentation in the Papago County, Arizona with a sketch of the geology. *U.S Geol. Survey Bull.* 730-B:19-90.

————. 1936. The formation of pediments. *16th Internat. Geol. Cong. Rept.,* v. 2, pp. 765-75.

————. 1946. Cryopedology—The study of frozen ground and intensive frost-action with suggestions on nomenclature. *Am. Jour. Sci.* 244:622-42.

Bryan, K., and Albritton, C. 1943. Soil phenomena as evidence of climatic changes. *Am. Jour. Sci.* 241:469-90.

Bryan, R. B. 1968. The development, use and efficiency of indices of soil erodibility. *Geoderma* 2:5-26.

Budd, W. F. 1975. A first simple model for periodically self-surging glaciers. *Jour. Glaciol.* 14:3-21.

Buddington, A. F. 1959. Granite emplacement with special reference to North America. *Geol. Soc. America Bull.* 70:671-747.

Budel, J. 1968. Geomorphology—principles. In *Encyclopedia of geomorphology,* edited by R. W. Fairbridge, pp. 416-22. New York: Reinhold Book Corp.

Bull, W. B. 1963. Alluvial-fan deposits in western Fresno County, California. *Jour. Geology* 71:243-51.

————. 1964a. Alluvial fans and near-surface subsidence in western Fresno County, California. U.S. Geol. Survey Prof. Paper 437-A.

————. 1964b. Geomorphology of segmented alluvial fans in western Fresno County, California. U.S. Geol. Survey Prof. Paper 552-F.

————. 1968. Alluvial fans. *Jour. Geol. Educ.* 16:101-06.

————. 1975a. Allometric change of landforms. *Geol. Soc. America Bull.* 86:1489-98.

————. 1975b. Landforms that do not tend toward a steady state. In *Theories of landform development;* edited by W. N. Melhorn and R. C. Flemal, pp. 111-28. S.U.N.Y., Binghamton, Pubs. in Geomorphology, 6th Ann. Mtg.

Bullard, F. 1962. *Volcanoes.* Austin: Univ. Texas Press.

Burdon, D. J., and Safadi, C. 1963. Ras-el-ain: the great karst spring of Mesopotamia. *J. Hydrol.* 1:58-95.

Butzer, K. W. 1964. *Environment and archeology.* Chicago: Aldine Publishing.

Butzer, K. W.; Helgren, D. M.; Fock, G. J.; and Stuckenrath, R. 1973. Alluvial terraces of the Lower Vaal River, South Africa: A reappraisal and reinvestigation. *Jour. Geology* 81: 341-62.

Cailleaux, A., and Tricart, J. 1956. Le problème de la classification des faits géomorphologiques. *Annales Géog.* 65:162-86.

Caine, N. 1968. *The block fields of northeastern Tasmania.* Australian Natl. Univ., Dept. Geog. Publ. G/6.

Calkin, P. E., and Rutford, R. H. 1974. The sand dunes of Victoria Valley, Antarctica. *Geogr. Rev.* 64:189-216.

Campbell, W. J., and Rasmussen, L. A. 1969. Three-dimensional surges and recoveries in a numerical glacier model. *Can. J. Earth Sci.* 6:979-86.

Carlston, C. W. 1963. Drainage density and streamflow. U.S. Geol. Survey Prof. Paper 422-C.

————. 1965. The relation of free meander geometry to stream discharge and its geomorphic implications. *Am. Jour. Sci.* 263:864-85.

————. 1969. Downstream variations in the hydraulic geometry of streams: Special emphasis on mean velocity. *Am. Jour. Sci.* 267:499-509.

Carmichael, I.; Turner, F.; and Verhoogen, J. 1974. *Igneous petrology.* New York: McGraw-Hill.

Carroll, D. 1958. Role of clay minerals in the transportation of iron. *Geochim. et Cosmochim. Acta* 14:1-27.

————. 1970. *Rock weathering.* New York: Plenum Press.

Carson, M. A. 1969. Models of hillslope development under mass failure. *Geographical Analysis* 1:76-100.

Carson, M. A., and Kirkby, M. 1972. *Hillslope form and process.* London: Cambridge Univ. Press.

Carter, T. G.; Liu, P. L.; and Mei, C. C. 1973. Mass transport by waves and offshore sand bedforms. Am. Soc. Civil Engineers, Coastal Eng. Div., *Jour. Waterways and Harbors,* WW2:165-84.

Casagrande, A. 1932. Research on the Atterberg limits of soils. *Public Roads* 13:121-30.

————. 1948. Classification and identification of soils. *Am. Soc. Civil Engineers Trans.* 113:901-91.

CERC (Coastal Engineering Research Center). 1973. *Shore protection manual.* 3 vols. Washington, D.C.: U.S. Army Corps Engrs.

Chadbourne, B. D.; Cole, R. M.; Tootill, S.; and Walford, M. E. R. 1975. The movement of melting ice over rough surfaces. *Jour. Glaciol.* 14: 287-92.

Chamberlain, R. T. 1928. Instrumental work on the nature of glacial motion. *Jour. Geology* 36:1-30.

Chambers, M. J. G. 1970. Investigations of patterned ground at Signy Island, South Orkney Islands, IV. Longterm experiments. *British Antarctic Surv. Bull.* 23:93-100.

Charlesworth, J. K. 1957. *The Quaternary Era, with special reference to its glaciation.* London: Edward Arnold Ltd.

Chepil, W. S. 1945. Dynamics of wind erosion, II. Initiation of soil movement. *Soil Sci.* 60:397-411.

————. 1959. Equilibrium of soil grains at the threshold of movement by wind. *Soil Sci. Soc. Am. Proc.* 23:422-28.

Chepil, W. S., and Woodruff, N. P. 1963. The physics of wind erosion and its control. *Advances in Agron.* 15:211-302.

Cherkauer, D. S. 1973. Minimization of power expenditure in a riffle—pool alluvial channel. *Water Resour. Res.* 9:1613-28.

Chorley, R. J. 1957a. Illustrating the laws of morphometry. *Geol. Mag.* 94:140-50.

———. 1957b. Climate and morphometry. *Jour. Geology* 65:628-38.

———. 1959a. The geomorphic significance of some Oxford soils. *Am. Jour. Sci.* 257:503-15.

———. 1959b. The shape of drumlins. *Jour. Glaciol.* 3:339-44.

———. 1962. Geomorphology and the general systems theory. U.S. Geol. Survey Prof. Paper 500-B.

———. 1969a. *Introduction to fluvial processes.* London: Methuen and Co.

———, ed. 1969b. *Water, earth, and man.* London: Methuen and Co.

Chorley, R. J.; Beckinsale, R. P.; and Dunn, A. J. 1973. *The history of the study of landforms.* Vol 2. The life and work of William Morris Davis. London: Methuen and Co.

Chorley, R. J.; Dunn, A. J.; and Beckinsale, R. P. 1964. *The history of the study of landforms.* Vol. 1. London: Methuen and Co.

Chorley, R. J., and Morley, L. S. D. 1959. A simplified approximation for the hypsometric integral. *Jour. Geology* 67:566-71.

Chose, B.; Pandy, S.; and Lal, G. 1967. Quantitative geomorphology of the drainage basins in the central Lumi basin in western Rajasthan. *Zeit. f. Geomorph.* 11:146-60.

Church, M. 1972. Baffin Island sandurs: A study of arctic fluvial processes. *Canada Geol. Survey Bull.* 216.

Clague, J. J. 1975. Glacier-flow pattern and the origin of Late Wisconsinan till in the southern Rocky Mountain Trench, British Columbia. *Geol. Soc. America Bull.* 86:721-31.

Clapperton, C. M. 1971. The location and origin of glacial meltwater phenomena in the eastern Cheviot Hills. *York Geol. Soc. Proc.* 38:361-80.

Clark, S. P., and Jager, E. 1969. Denudation rate in the Alps from geochronologic and heat flow data. *Am. Jour. Sci.* 267:1143-60.

Clayton, K. M. 1965. Glacial erosion in the Finger Lake region (New York State, U.S.A.). *Zeit. f. Geomorph.* 9:50-62.

Clayton, L. 1964. Karst topography on stagnant glaciers. *Jour. Glaciol.* 5:107-12.

———. 1967. Stagnant-glacial features of the Missouri Coteau in North Dakota. North Dakota Geol. Survey Misc. Series 30:25-46.

Cleaves, E.; Fisher, D.; and Bricker, O. 1974. Chemical weathering of serpentinite in the eastern piedmont of Maryland. *Geol. Soc. America Bull.* 85:437-44.

Cloud, P. E. 1966. Beach cusps: Response to Plateau's rule? *Science* 154:890-91.

Colbeck, S. C., and Evans, R. J. 1973. A flow law for temperate glacier ice. *Jour. Glaciol.* 12:71-86.

Colby, B. 1963. Fluvial sediments: A summary of source, transportation, deposition, and measurement of sediment discharge. *U.S. Geol. Survey Bull.* 1181-A:21.

———. 1964. Scour and fill in sand bed streams. U.S. Geol. Survey Prof. Paper 462-D.

Coleman, J. M. 1968. Deltaic evolution. In *Encyclopedia of geomorphology,* edited by R. W. Fairbridge, pp. 255-60. New York: Reinhold Book Corp.

Coleman, J. M., and Gagliano, S. M. 1964. Cyclic sedimentation in the Mississippi River deltaic plain. *Gulf Coast Assoc. Geol. Soc. Trans.* 14:67-80.

Colville, J. S., and Holmes, J. W. 1972. Water table fluctuation under forest and pasture in the karstic region of south Australia. *J. Hydrol.* 17:61-80.

Condie, K. C. 1973. Archean magnetism and crustal thickening. *Geol. Soc. America Bull.* 84:2981-92.

Connell, W., and Patrick, W. 1968. Sulfate reduction in soil. Effects of redox potential and pH. *Science* 159:86-87.

Cook, J. H. 1946a. Ice contacts and the melting of ice below a water level. *Am. Jour. Sci.* 244:502-12.

———. 1946b. Kame complexes and perforation deposits. *Am. Jour. Sci.* 244:573-83.

Cooke, C. W. 1968. Barrier island formation: Discussion. *Geol. Soc. America Bull.* 79:945.

Cooke, H. J. 1973. Tropical karst in northeast Tanzania. *Zeit. f. Geomorph.* 17:443-59.

Cooke, R. U. 1970. Morphometric analysis of pediments and associated landforms in the western Mojave Desert, California. *Am. Jour. Sci.* 269:26-38.

Cooke, R. U., and Doornkamp, J. 1974. *Geomorphology in environmental management.* London: Clarendon Press.

Cooke, R. U., and Reeves, R. W. 1972. Relations between debris size and the slope of mountain fronts and pediments in the Mojave Desert, California. *Zeit. f. Geomorph.* 16:76-82.

Cooke, R. U., and Warren, A. 1973. *Geomorphology in deserts.* London: Batsford Ltd.

Cooley, R.; Fiero, G.; Lattman, L.; and Mindling, A. 1973. Influence of surface and near-surface caliche distribution on infiltration characteristics and flooding, Las Vegas area, Nevada. Univ. Nevada, Reno, Desert Research Inst. Proj. Rept. 21.

Cooper, W. S. 1958. Coastal sand dunes of Oregon and Washington. *Geol. Soc. America Mem.* 72.

Corbel, J. 1957. Karsts hauts-Alpins. *Rev. Géogr. Lyon* 32:135-58.

———. 1959a. Les karsts du Yucatan et de la Florida. *Bull. Assoc. Géogr. Fr.* 282-3:2-14.

———. 1959b. Vitesse de l'erosion. *Zeit f. Geomorph.* 3:1-28.

———. 1964. L'erosion terrestre, étude quantitative. *Annales Géog.* 73:385-412.

Corte, A. E. 1966. Experiments on sorting processes and the origin of patterned ground. *Proc. Permafrost Conf.* (Lafayette, Ind., 1963), Natl. Acad. Sci.—Natl. Research Council Pub. 1287, pp. 130-35.

———. 1969. Geocryology and engineering. In *Reviews in engineering geology 2*, edited by D. Varnes and G. Kiersch, pp. 119-85. Boulder, Colo.: Geol. Soc. America.

Costa, J. E. 1974. Stratigraphic, morphologic, and pedologic evidence of large floods in humid environments. *Geology* 2:301-03.

Cox, E. T. 1874. *Fifth annual report of the Geological Survey of Indiana*, pp. 280-305. Indianapolis: Indiana Geol. Survey.

Cramer, H. 1941. Die Systematik der Karstdolinen. *Neues Jb. Miner. Geol. Paläont.* 85:293-382.

Crary, A. P. 1966. Mechanism for fiord formation indicated by studies of an ice-covered inlet. *Geol. Soc. America Bull.* 77:911-30.

Crittenden, M. 1963. New data on the isostatic deformation of Lake Bonneville. U.S. Geol. Survey Prof. Paper 454-E.

Crosby, W. O. 1902. Origin of eskers. *Am. Geologist* 30:1-38.

Crozier, M. J. 1973. Techniques for the morphometric analysis of landslips. *Zeit. f. Geomorph.* 17:78-101.

Cvijić, J. 1893. Das Karstphanomen. *Geogr. Abh.* 5:217-329.

Czudek, T., and Demek, J. 1970. Thermokarst in Siberia and its influence on the development of lowland relief. *Quat. Res.* 1:103-20.

Dahl, R. 1965. Plastically sculptured detail forms on rock surfaces in northern Nordland. *Geogr. Annlr.* 47:83-140.

Dalrymple, T. 1960. Flood frequency analysis. *Manual of hydrology*, part 3, Flood flow techniques. *U.S. Geol. Survey Water Supply Paper* 1543-A.

Dalrymple, G.; Silver, E.; and Jackson, E. 1973. Origin of the Hawaiian Islands. *Am. Scientist* 61:294-308.

Dalrymple, R. A., and Lanan, G. A. 1976. Beach cusps formed by intersecting waves. *Geol. Soc. America Bull.* 87:57-60.

Daly, R. 1933. *Igneous rocks and the depths of the Earth*. New York: McGraw-Hill.

Davies, J. L. 1973. *Geographical variation in coastal development*. New York: Hafner.

Davies, W. E. 1960. Origin of caves in folded limestone. *Natl. Speleol. Soc. Bull.* 22:5-18.

Davis, R. A.; Fox, W. T.; Hayes, M. O.; and Boothroyd, J. C. 1972. Comparison of ridge and runnel systems in tidal and non-tidal environments. *Jour. Sed. Petrology* 42:413-21.

Davis, W. M. 1899. The peneplain. *Am. Geologist* 23:207-39.

———. 1902. River terraces in New England. *Mus. Comp. Zoology Bull.* 38:77-111.

———. 1905. Complications of the geographical cycle. *Report of the 8th Geographical Congress 1904*:150-63.

———. 1930. Origin of limestone caverns. *Geol. Soc. America Bull.* 41:475-628.

Deike, G. H., and White, W. B. 1969. Sinuosity in limestone solution conduits. *Am. Jour. Sci.* 267:230-41.

Demorest, M. 1938. Ice flowage as revealed by glacial striae. *Jour. Geology* 46:700-725.

———. 1942. Glacier regimens and ice movement within glaciers. *Am. Jour. Sci.* 240:31-66.

Denny, C. S. 1956. Surficial geology and geomorphology of Potter County, Pennsylvania. U.S. Geol. Survey Prof. Paper 288.

———. 1965. Alluvial fans in the Death Valley region, California and Nevada. U.S. Geol. Survey Prof. Paper 466.

———. 1967. Fans and pediments. *Am. Jour. Sci.* 265:81-105.

de Swardt, A. M. J. 1964. Lateritisation and landscape development in parts of equatorial Africa. *Zeit. f. Geomorph.* 8:313-33.

Dewey, J. F., and Bird, J. M. 1970. Mountain belts and the new global tectonics. *Jour. Geophys. Research* 75:2625-47.

Dietz, R. S. 1961. Continent and ocean basin evolution by spreading of the sea floor. *Nature* 190:854-57.

Dietz, R. S., and Fairbridge, R. W. 1968. Wave base. In *Encyclopedia of geomorphology*, edited by R. W. Fairbridge, pp. 1224-28. New York: Reinhold Book Corp.

Dietz, R. S., and Holden, J. C. 1970. Reconstruction of Pangaea—Breakup and dispersion of continents, Permian to present. *Jour. Geophys. Research:* 75: 4939-56.

Dionne, J. C. 1974. Polished and striated mud surfaces in the St. Lawrence tidal flats, Quebec. *Can. J. Earth Sci.* 11:860-66.

Dolan, R., and Ferm, J. C. 1968. Crescentic landforms along the mid-Atlantic coast. *Science* 159:627-29.

Dolan, R.; Vincent, L.; and Hayden, B. 1974. Crescentic coastal landforms. *Zeit. f. Geomorph.* 18:1-12.

Doornkamp, J. C., and King, C. A. M. 1971. *Numerical analysis in geomorphology: An Introduction.* London: Edward Arnold Ltd.

Douglas, I. 1964. Intensity and periodicity in denudation process with special reference to the removal of material in solution by rivers. *Zeit. f. Geomorph.* 8:453-73.

————. 1967. Man, vegetation and sediment yields of rivers. *Nature* 215:925-28.

Drake, C., and Girdler, R. 1964. A geophysical study of the Red Sea. *Geophys. Jour.* 8:473-95.

Drake, J. J., and Harmon, R. S. 1973. Hydrochemical environments of carbonate terrains. *Water Resour. Res.* 9:949-57.

Drake, J. J., and Wigley, T. M. L. 1975. The effect of climate on the chemistry of carbonate groundwater. *Water Resour. Res.* 11:958-62.

Drake, L. D. 1968. *Till studies in New Hampshire.* Ph.D. dissertation, Ohio State University.

————. 1971. Evidence for ablation and basal till in east-central New Hampshire. In *Till: A symposium,* edited by R. P. Goldthwait. Columbus: Ohio State Univ. Press.

————. 1974. Till fabric control by clast shape. *Geol. Soc. America Bull.* 85:247-50.

Drake, L. D., and Shreve, R. L. 1973. Pressure melting and regelation of ice by round wires. *Proc. Royal Soc. London,* ser. A:332:51-83.

Dreimanis, A. 1953. Studies of friction cracks along shores of Cirrus Lake and Kasakokwog Lake, Ontario. *Am. Jour. Sci.* 251:769-83.

Dreimanis, A., and Vagners, U. J. 1971. Bimodal distribution of rock and mineral fragments in basal till. In *Till: A symposium,* edited by R. P. Goldthwait. Columbus: Ohio State Univ. Press.

Dryden, L., and Dryden, C. 1946. Comparative rates of weathering of some heavy minerals. *Jour. Sed. Petrology* 16:91-96.

DuBois, R. N. 1972. Inverse relation between foreshore slope and mean grain size as a function of the heavy mineral content. *Geol. Soc. America Bull.* 83:871-76.

Duncan, J. R. 1964. The effects of water table and tidal cycle on swash-backwash sediment distribution and beach profile development. *Marine Geol.* 2:186-97.

Dunham, R. J. 1962. Classification of carbonate rocks according to depositional texture. In W. E. Ham, editor, *Classification of carbonate rocks. Am. Assoc. Petroleum Geologists Mem.* 1:108-21.

Dury, G. H. 1964. Principles of underfit streams. U.S. Geol. Survey Prof. Paper 452-A.

————. 1965. Theoretical implications of underfit streams. U.S. Geol. Survey Prof. Paper 452-C.

————. 1966a. Duricrusted residuals on the Barrier and Cobar pediplains of New South Wales. *Jour. Geol. Soc. Australia* 13:299-307.

————. 1966b. Pediment slope and particle size at Middle Pinnacle, near Broken Hill, New South Wales. *Austr. Geog. Studies* 4:1-17.

————. 1969. Relation of morphometry to runoff frequency. In *Introduction to fluvial processes,* edited by R. J. Chorley, pp. 177-88. London: Methuen and Co.

Dylik, J. 1964. The essentials of the meaning of the term "Periglacial." *Soc. Sci. et Lettres Łódz Bull.* 15: 2:1-19.

Eakin, H. M. 1916. The Yukon-Koyukuk region, Alaska. *U.S. Geol. Survey Bull.* 631.

Eardley, A. G. 1967. Rates of denudation as measured by bristlecone pines, Cedar Breaks, Utah. Utah Geol. and Mineral Survey Spec. Studies 21.

Eaton, J., and Murata, K. 1960. How volcanoes grow. *Science* 132:925-38.

Ede, D. P. 1975. Limestone drainage systems. *J. Hydrol.* 27:297-318.

Eggler, D. H.; Larson, E. E.; and Bradley, W. C. 1969. Granites, grusses, and the Sherman erosion surface, southern Laramie Range, Colorado-Wyoming. *Am. Jour. Sci.* 267:510-22.

Einstein, H. A. 1950. The bedload function for sediment transportation in open channel flows. U.S. Dept. Agri. Tech. Bull. 1026.

Ellison, W. D. 1947. Soil erosion studies, part IV. *Agri. Engineering* 28:349-353.

Elliston, G. R. 1963. Catastrophic glacier advances. *Int. Assoc. Sci. Hydrol. Bull.* 8:65-66.

Elson, J. A. 1968. Washboard moraines and other minor moraine types. In *Encyclopedia of geomorphology,* edited by R. W. Fairbridge, pp. 1213-19. New York: Reinhold Book Corp.

Embleton, C., and King, C. A. M. 1968. *Glacial and periglacial geomorphology.* Edinburgh: Edward Arnold Ltd.

————. 1975a. *Glacial geomorphology.* New York: Halsted Press.

———. 1975b. *Periglacial geomorphology.* New York: Halsted Press.

Emery, K. O. 1946. Marine solution basins. *Jour. Geology* 54:209-28.

Engel, A. 1963. Geologic evolution of North America. *Science* 140: 143-50.

Engel, A.; Itson, S.; Engel, C.; Stickney, D.; and Cray, E. 1974. Crustal evolution and global tectonics: A petrogenic view. *Geol. Soc. America Bull.* 85:843-58.

Enzmann, R. 1968. Geomorphology—expanded theory. In *Encyclopedia of geomorphology,* edited by R. W. Fairbridge, pp. 404-09. New York: Reinhold Book Corp.

Ericksen, G. E.; Pflacker, G.; and Fernandez, J. V. 1970. Preliminary report on the geological events associated with the May 31, 1970, Peru earthquake. U.S. Geol. Survey Circ. 639.

Ericson, D. B., and Wollin, G. 1964. *The deep and the past.* New York: Alfred A. Knopf.

Evans, O. F. 1938. Classification and origin of beach cusps. *Jour. Geology* 46:615-27.

———. 1940. The low and ball of the east shore of Lake Michigan. *Jour. Geology* 48:476-511.

———. 1942. The origin of spits, bars, and related structures. *Jour. Geology* 50:846-63.

Evenson, E. B. 1971. The relationship of macro- and microfabrics of till and the genesis of glacial landforms in Jefferson County, Wisconsin. In *Till: A symposium,* edited by R. P. Goldthwait. Columbus: Ohio State Univ. Press.

Everett, D. H. 1961. The thermodynamics of frost damage to porous solids. *Trans. Faraday Soc.* 57: 1541-51.

Everitt, B. L. 1968. Use of cottonwood in an investigation of recent history of a flood plain. *Am. Jour. Sci.* 266:417-39.

Eynon, G., and Walker, R. G. 1974. Facies relationships in Pleistocene outwash gravels, southern Ontario: A model for bar growth in braided rivers. *Sedimentology* 21:43-70.

Fahey, B. D. 1973. An analysis of diurnal freeze-thaw and frost heave cycles in the Indian Peaks region of the Colorado Front Range. *Arc. Alp. Res.* 5:269-81.

———. 1975. Nonsorted circle development in a Colorado alpine location. *Geogr. Annlr.* 57A:153-164.

Fahnestock, R. K. 1961. Competence of a glacial stream. U.S. Geol. Survey Prof. Paper 424-B:211-13.

———. 1963. Morphology and hydrology of a glacial stream—White River, Mount Rainier Washington. U.S. Geol. Survey Prof. Paper 422-A.

———. 1969. Morphology of the Slims River. In *Icefield Ranges Research Project, Scientific Results #1,* edited by V. C. Bushell and R. H. Ragle, pp. 161-72. Am. Geog. Soc. and Arctic Inst. N. America.

Fairbridge, R. W. 1966. Trenches and related deep sea troughs. In *Encyclopedia of oceanography,* edited by R. W. Fairbridge, pp. 929-39. New York: Reinhold Book Corp.

———. 1968. Land mass and major landform classification. In *Encyclopedia of geomorphology,* edited by R. W. Fairbridge, pp. 618-26. New York: Reinhold Book Corp.

Fairbridge, R. W., and Newman, W. 1968. Postglacial crustal subsidence of the New York area. *Zeit. f. Geomorph.* 12:296-317.

Fairchild, J. C. 1973. Longshore transport of suspended sediment. *Proc. 13th Conf. on Coast. Eng.,* 1069-88.

Fehrenbacher, J. B.; Ray, B. W.; and Alexander, J. D. 1968. Illinois soils and factors in their development. In *The Quaternary of Illinois,* edited by R. E. Bergstrom, pp. 165-75. Urbana: Univ. Illinois Coll. of Agri. Spec. Pub. 14.

Ferrians, O. J. 1965. Permafrost map of Alaska. U.S. Geol. Survey Misc. Geol. Inv. Map I-445.

Ferrians, O. J.; Kachadoorian, R.; and Greene, G. W. 1969. Permafrost and related engineering problems in Alaska. U.S. Geol. Survey Prof. Paper 678.

Feth, J.; Robertson, C.; and Polzer, W. 1964. Sources of mineral constituents in water from granitic rocks, Sierra Nevada, California and Nevada. U.S. Geol. Survey Water Supply Paper 1535-I.

Fieldes, M., and Swindale, L. 1954. Chemical weathering of silicates in soil formation. *Jour. Sci. Tech. New Zealand* 56:140-54.

Finkel, H. J. 1959. The barchans of southern Peru. *Jour. Geology* 67: 614-47.

Fisher, J. E. 1963. Two tunnels in cold ice at 4000 m on the Breithorn. *Jour. Glaciol.* 4:513-20.

Fisher, J. J. 1968. Barrier island formation. Discussion. *Geol. Soc. America Bull.* 79:1421-26.

———. 1973. Bathymetric projected profiles and the origin of barrier islands—Johnson's shoreline of emergence, revisited. In *Coastal geomorphology,* edited by D. R. Coates, pp. 161-79. S.U.N.Y., Binghamton, 3rd Ann. Geomorph. Symposium.

Fisher, R. L. 1975. Pacific-type continental margins. In *The geology of continental margins,* edited by C. Burk and C. Drake. New York: Springer-Verlag.

Fisk, H. N. 1944. *Geological investigation of the alluvial valley of the lower*

Mississippi River. Vicksburg, Miss.: Mississippi River Comm.

————. 1947. Fine-grained alluvial deposits and their effects on Mississippi River activity. 2 vols. U.S. Army Corps Engrs., U.S. Waterways Exp. Sta.

————. 1951. Loess and Quaternary geology of the lower Mississippi Valley. *Jour. Geology* 59:333-56.

Fiske, R. S.; Hopson, C. A.; and Waters, A. C. 1963. Geology of Mount Rainier National Park. U.S. Geol. Survey Prof. Paper 444.

Fitze, P. 1971. Messungen von Bodenbewegungen auf West-Spitzbergen. *Geog. Helvetica* 26:148-52.

Flemal, R. C. 1971. The attack on the Davisian system of geomorphology: A synopsis. *Jour. Geol. Educ.* 19:3-13.

————. 1976. Pingos and pingo scars: Their characteristics, distribution, and utility in reconstructing former permafrost environments. *Quat. Res.* 6:37-53.

Flemal, R. C.; Hinkley, K. C.; and Hesler, J. L. 1976. DeKalb mounds: A possible Pleistocene (Woodfordian) pingo field in north-central Illinois. *Geol. Soc. America Mem.* 136:229-50.

Flint, R. F. 1955a. Rates of advance and retreat of the margin of the late-Wisconsin ice sheet. *Am. Jour. Sci.* 253:249-55.

————. 1955b. Pleistocene geology of eastern South Dakota. U.S. Geol. Survey Prof. Paper 262.

————. 1963. Altitude, lithology, and the Fall Zone in Connecticut. *Jour. Geology* 71:683-97.

————. 1971. *Glacial and Quaternary geology*. New York: John Wiley and Sons.

Flint, R. F., and Bond, G. 1968. Pleistocene sand ridges and pans in western Rhodesia. *Geol. Soc. America Bull.* 79:299-314.

Flint, R. F., and Fidalgo, F. 1964. Glacial geology of the east flank of the Argentine Andes between latitude 39°10′S and latitude 41°20′S. *Geol. Soc. America Bull.* 75:335-52.

Fohn, P. M. B. 1973. Short-term snow melt and ablation derived from heat- and mass-balance measurements. *Jour. Glaciol.* 12:275-89.

Folk, R. L. 1959. Practical petrographic classification of limestones. *Am. Assoc. Petroleum Geologists Bull.* 43:1-38.

————. 1971a. Longitudinal dunes of the northwestern edge of the Simpson Desert, Northern Territory, Australia, 1. Geomorphology and grain size relationships. *Sedimentology* 16:5-54.

————. 1971b. Genesis of longitudinal and oghurd dunes elucidated by rolling upon grease. *Geol. Soc. America Bull.* 82:3461-68.

Folk, R. L.; Roberts, H. H.; and Moore, C. H. 1973. Black phytokarst from Hell, Cayman Islands, British West Indies. *Geol. Soc. America Bull.* 84:2351-60.

Foose, R. M. 1967. Sinkhole formation by groundwater withdrawal: Far West Rand, South Africa. *Science* 157:3792:1045-48.

Forbes, J. D. 1843. *Travels through the Alps of Savoy*. Edinburgh.

Fournier, M. F. 1960. *Climat et érosion*. Paris: Presses Univ. France.

Francis, J. R. D. 1973. Experiments on the motion of solitary grains along the bed of a water stream. *Proc. Royal Soc. London*, ser. A:332:443-71.

Frazee, C. J.; Fehrenbacher, J. B.; and Krumbein, W. C. 1970. Loess distribution from a source. *Soil Sci. Soc. Am. Proc.* 34:296-301.

Free, G. R. 1960. Erosion characteristics of rainfall. *Agri. Engineering* 41:447-449, 455.

French, H. M. 1974. Mass-wasting at Sachs Harbour, Barks Island, N. W. T., Canada. *Arc. Alp. Res.* 6:77-78.

Friedkin, J. F. 1945. A laboratory study of the meandering of alluvial rivers. U.S. Army Corps Engrs., U.S. Waterways Eng. Exp. Sta.

Friese-Greene, T. W., and Pert, G. J. 1965. Velocity fluctuations of Berksackerbrae, east Greenland. *Jour. Glaciol.* 5:739-47.

Frye, J. C. 1961. Fluvial deposition and the glacial cycle. *Jour. Geology* 69:600-603.

Frye, J. C., and Leonard, A. R. 1954. Some problems of alluvial terrace mapping. *Am. Jour. Sci.* 252:242-51.

Fuller, M. L. 1914. The geology of Long Island, New York. U.S. Geol. Survey Prof. Paper 82.

Galvin, C. J. 1967. Longshore current velocity: A review of theory and data. *Revs. in Geophys.* 5:3:287-304.

————. 1968. Breaker type classification on three laboratory beaches. *Jour. Geophys. Research* 73:3651-59.

Galvin, C. J., and Eagleson, P. S. 1965. Experimental study of longshore currents on a plane beach. U.S. Army Corps Engrs., Coastal Eng. Res. Center Tech. Memo 10.

Gams, I. 1965. Types of accelerated karst corrosion. In *Problems of the speleological research,* edited by O. Štelcl, pp. 133-39. Prague.

————. 1969. Some morphological characteristics of the Dinaric karst. *Geogr. Jour.* 135:563-72.

Gardner, J. H. 1935. Origin and development of limestone caverns. *Geol. Soc. America Bull.* 46:1255-74.

Garland, G. D. 1965. *The Earth's shape and gravity*. London: Pergamon Press.

Garner, H. F. 1967. Rivers in the making. *Sci. Amer.* 216, no. 4:84-94.

———. 1974. *The origin of landscapes.* London: Oxford Univ. Press.

Garwood, E. J. 1899. Additional notes on the glacial phenomena of Spitzbergen. *Quart. Jour. Geol. Soc. Lond.* 55:681-91.

Gentilli, J. 1968. Exfoliation. In *Encyclopedia of geomorphology,* edited by R. W. Fairbridge, pp. 336-39. New York: Reinhold Book Corp.

Gerson, R. 1974. Karst processes of eastern upper Galilee, north Israel. *J. Hydrol.* 21:131-52.

Gibbs, R. J. 1967. The geochemistry of the Amazon River system, part I. *Geol. Soc. America Bull.* 78:1203-32.

Gilbert, G. K. 1877. *Geology of the Henry Mountains (Utah).* U.S. Geog. and Geol. Survey of the Rocky Mtn. Region. Washington, D.C.: U.S. Govt. Printing Office.

———. 1909. The convexity of hilltops. *Jour. Geology* 17: 344-50.

———. 1914. The transportation of debris by running water. U.S. Geol. Survey Prof. Paper 86.

———. 1917. Hydraulic-mining debris in the Sierra Nevada. U.S. Geol. Survey Prof. Paper 105.

Gile, L. H. 1966. Cambic and certain non-cambic horizons in desert soils of southern New Mexico. *Soil Sci. Soc. Am. Proc.* 30:773-81.

Gile, L. H.; Peterson, F.; and Grossman, R. 1965. The K horizon. A master soil horizon of carbonate accumulation. *Soil Sci.* 99:74-82.

———. 1966. Morphological and genetic sequences of carbonate accumulation in desert soils. *Soil Sci.* 101:347-60.

Gilkes, R.; Scholtz, G.; and Dimmock, G. 1973. Lateritic deep weathering of granite. *Jour. Soil Sci.* 24:523-36.

Gilluly, J. 1937. Physiography of the Ajo region, Arizona. *Geol. Soc. America Bull.* 43:323-48.

———. 1949. The distribution of mountain-building in geologic time. *Geol. Soc. America Bull.* 60:561-90.

———. 1955. *Geologic contrasts between continents and ocean basins.* Geol. Soc. America Spec. Paper 62:7-18.

———. 1964. Atlantic sediments, erosion rates, and the evolution of the Continental Shelf—some speculations. *Geol. Soc. America Bull.* 75:483-92.

———. 1969. Geological perspectives and the completeness of the geologic record. *Geol. Soc. America Bull.* 80:2303-12.

Gilluly, J.; Reed, J. C., Jr.; and Cady, W. M. 1970. Sedimentary volumes and their significance. *Geol. Soc. America Bull.* 80:353-76.

Gjessing, J. 1967. On plastic scouring and subglacial erosion. *Norsk. Geogr. Tidsskr.* 20:1-37.

Glen, J. W. 1952. Experiments on the deformation of ice. *Jour. Glaciol.* 2:111-14.

———. 1955. The creep of polycrystalline ice. *Proc. Royal Soc. London,* ser. A: 228:519-38.

———. 1958. Mechanical properties of ice. I. The plastic properties of ice. *Philos. Mag.,* Suppl. 7:254-65.

Glen, J. W., and Lewis, W. V. 1961. Measurements of side-slip at Austerdalsbreen, 1959. *Jour. Glaciol.* 3:1121.

Glennie, K. W. 1970. *Desert sedimentary environments.* Amsterdam: Elsevier.

Goldich, S. 1938. A study of rock weathering. *Jour. Geology* 46:17-58.

Goldthwait, R. P. 1951. Development of end moraines in east central Baffin Island. *Jour. Geology* 59:567-77.

———. 1956. Formation of ice cliff. U.S. Army Corps Engrs., Snow, Ice, Permafrost Res. Est. Tech. Rept. 39:139-150.

———. 1960. Study of ice cliff in Nunatarssuaq, Greenland. U.S. Army Corps Engrs., Cold Regions Res. and Eng. Lab Tech. Rept. 39.

———. 1969. Patterned soils and permafrost on the Presidential Range (abs). Paris. 8th INQUA Cong. *Résumés des Communications.* 150.

———, ed. 1971. *Till: A symposium.* Columbus: Ohio State Univ. Press.

———. 1973. Jerky glacier motion and meltwater. *Int. Assoc. Sci. Hydrol. Bull.* 95:183-88.

———. 1976. Frost sorted patterned ground: A review. *Quat. Res.* 6:27-35.

Goodwin, A. M. 1973. Plate tectonics and evolution of Precambrian crust. In *Implications of continental drift to the earth sciences,* edited by D. Tarling and S. Runcorn. New York: Academic Press.

———. 1975. The most ancient continental margins. In *The geology of continental margins,* edited by C. Burk and C. Drake. New York: Springer-Verlag.

Gordon, M.; Tracey, J; and Ellis, M. 1958. Geology of the Arkansas bauxite region. U.S. Geol. Survey Prof. Paper 299.

Gorsline, D. S. 1966. Dynamic characteristics of west Florida Gulf Coast beaches. *Marine Geol.* 4:187-206.

Gorycki, M. A. 1973. Sheetflood structure: Mechanism of beach cusp formation and related phenomena. *Jour. Geology* 81:109-17.

Goudie, A. 1973. *Duricrusts.* Oxford: Clarendon Press.

Graf, W. L. 1970. The geomorphology of the glacial valley cross section. *Arc. Alp. Res.* 2:303-12.

————. 1976. Cirques as glacier locations. *Arc. Alp. Res.* 8:79-90.

Gravenor, C. P. 1953. The origin of drumlins. *Am. Jour. Sci.* 251:674-81.

Gravenor, C. P., and Kupsch, W. O. 1959. Ice disintegration features in western Canada. *Jour. Geology* 67:48-64.

Gravenor, C. P., and Meneley, W. A. 1958. Glacial flutings in central and northern Alberta. *Am. Jour. Sci.* 256:715-28.

Grawe, O. R. 1936. Ice as an agent of rock weathering: A discussion. *Jour. Geology* 44:173-82.

Green, J., and Short, N. 1971. *Volcanic landforms and surface features.* New York: Springer-Verlag.

Gregory, K. J., and Walling, D. E. 1973. *Drainage basin form and process.* New York: Halsted Press.

Griggs, D. 1936a. The factor of fatigue in rock exfoliation. *Jour. Geology* 44:783-96.

————. 1936b. Deformation of rocks under high confining pressures. *Jour. Geology* 44:541-77.

Grim, R. 1962. *Applied clay mineralogy.* New York: McGraw-Hill.

Groom, G. E., and Williams, V. 1965. The solution of limestone in South Wales. *Geogr. Jour.* 131:37-41.

Gross, D. L., and Moran, S. R. 1971. Grain-size and mineralogical gradations within tills of the Allegheny Plateau. In *Till: A symposium,* edited by R. P. Goldthwait. Columbus: Ohio State Univ. Press.

Grove, J. M. 1960. The bands and layers of Vesl-Skautbreen. In *Norwegian cirque glaciers,* edited by W. V. Lewis, pp. 11-23. Royal Geog. Soc. Res. Ser. 4.

Grund, A. 1903. Die Karsthydrographie Studien aus Westbosnien. *Geogr. Abh.* 7:3.

————. 1914. Der geographische Zyklus im Karst. *Z. Ges. Erdkunde,* 1914: 621-40.

Guilcher, A. 1958. *Coastal and submarine morphology.* London: Methuen and Co.

Gutenberg, B. 1941. Changes in sea level, postglacial uplift, and mobility of the earth's interior. *Geol. Soc. America Bull.* 52:721-72.

Guza, R. T., and Inman, D. L. 1975. Edge waves and beach cusps. *Jour. Geophys. Research* 80:21:2997-3012.

Haan, C. T., and Johnson, H. P. 1966. Rapid determination of hypsometric curves. *Geol. Soc. America Bull.* 77:123-25.

Hack, J. T. 1941. Dunes of the western Navajo country. *Geogr. Rev.* 31: 240-63.

————. 1957. Studies of longitudinal stream profiles in Virginia and Maryland. U.S. Geol. Survey Prof. Paper 294-B:45-97.

————. 1960a. Relation of solution features to chemical character of water in the Shenandoah Valley, Virginia. U.S. Geol. Survey Prof. Paper 400-B:387-90.

————. 1960b. Interpretation of erosional topography in humid temperate regions. *Am. Jour. Sci.* (Bradley Vol.) 258-A:80-97.

————. 1965. Postglacial drainage evolution in the Ontonagan area, Michigan. U.S. Geol. Survey Prof. Paper 504-B: 1-40.

————. 1966. Circular patterns and exfoliation in crystalline terrane, Grandfather Mountain area, North Carolina. *Geol. Soc. America Bull.* 77:975-86.

————. 1973. Stream-profile analysis and stream-gradient index. *U.S. Geol. Survey Jour. Research,* 1:421-29.

Hack, J. T., and Goodlett, J. C. 1960. Geomorphology and forest ecology of a mountain region in the central Appalachians. U.S. Geol. Survey Prof. Paper 347.

Hadley, R. F. 1961. Influence of riparian vegetation on channel shape, northeastern Arizona. U.S. Geol. Survey Prof. Paper 424-C: 30-31.

————. 1967. Pediments and pediment-forming processes. *Jour. Geol. Educ.* 15:83-89.

Hadley, R. F., and Schumm, S. A. 1961. Sediment sources and drainage basin characteristics in upper Cheyenne River basin. U.S. Geol. Survey Water Supply Paper 1531-B:137-96.

Haefeli, R. 1963. Observations and measurements on the cold ice sheet on Jungfraujoch. *Int. Assoc. Sci. Hydrol. Bull.* 8:122.

Haefeli, R., and Brentani, F. 1955. Observations in a cold ice cap. *Jour. Glaciol.* 2:571-80.

Hall, R. D. 1976. Investigation of sinkhole stratigraphy and hydrogeology, south-central Indiana. *Natl. Speleol. Soc. Bull.* 38:88-93.

Hallam, A. 1973. *A revolution in the earth sciences.* London: Oxford Univ. Press.

Hanna, S. R. 1969. The formation of longitudinal sand dunes by large helical eddies in the atmosphere. *Jour. Applied Meteorology* 8: 874-83.

Happ, S. C.; Rittenhouse, G.; and Dobson, G. C. 1940. Some principles of accelerated stream and valley sedimentation. U.S. Dept. Agri. Tech. Bull. 695.

Harmon, R. S.; White, W. B.; Drake, J. J.; and Hess, J. W. 1975. Regional hydrochemistry of North American carbonate terrains. *Water Resour. Res.* 11:963-67.

Harris, C. 1973. Some factors affecting the rates and processes of periglacial mass movement. *Geogr. Annlr.* 55A: 24-58.

Harris, S. E. 1943. Friction cracks and the direction of glacial movement. *Jour. Geology* 51:244-58.

Harrison, A. E. 1964. Ice surges on the Muldrow Glacier, Alaska. *Jour. Glaciol.* 5:365-68.

Harrison, W. 1958. Marginal zones of vanished glaciers reconstructed from the preconsolidation-pressure values of overridden silts. *Jour. Geology* 66:72-95.

———. 1960. Original bedrock composition of Wisconsin till in central Indiana. *Jour. Sed. Petrology* 30:432-46.

———. 1969. Empirical equations for foreshore changes over a tidal cycle. *Marine Geol.* 7:529-51.

Harrison, W. D. 1972. Temperature of a temperate glacier. *Jour. Glaciol.* 11:15-29.

———. 1975. Temperature measurements in a temperate glacier. *Jour. Glaciol.* 14:23-30.

Hartshorn, J. H. 1958. Flowtill in southeastern Massachusetts. *Geol. Soc. America Bull.* 69:477-82.

Harwood, T. A. 1969. Some possible problems with pipelines in permafrost regions. *Proc. 3rd Canadian Conf. on Permafrost,* pp. 79-84. Natl. Res. Council of Canada, Tech. Memo 96.

Hastenrath, S. L. 1967. The barchans of the Arequipa region, southern Peru. *Zeit. f. Geomorph.* 11:300-331.

Haynes, C. V., Jr. 1968. Geochronology of late-Quaternary alluvium. In *Means of correlation of Quaternary successions,* edited by H. Wright and R. Morrison, pp. 591-631. Salt Lake City: Univ. Utah Press.

Heezen, B. C. 1975. Atlantic-type continental margins. In *The geology of continental margins,* edited by C. Burk and C. Drake. New York: Springer-Verlag.

Heezen, B. C.; Tharp, M.; and Ewing, M. 1959. *The floor of the ocean, I. The North Atlantic.* Geol. Soc. America Spec. Paper 65.

Heim, A. 1932. *Bergsturz und Menschenleben.* Zurich: Fretz and Wasmuth Verlag.

Heirtzler, J.; Dickson, G. O.; Herron, E.; Pitman, W.; and Le Pichon, X. 1968. Marine magnetic anomalies, geomagnetic field reversals, and motions of the ocean floor and continents. *Jour. Geophys. Research* 73:2119-36.

Hellaakoski, Aaro. 1931. On the transportation of materials in the esker of Laitila. *Fennia* 52:7.

Hembree, C., and Rainwater, F. 1961. Chemical degradation on opposite flanks of the Wind River Range, Wyoming. U.S. Geol. Survey Water Supply Paper 1535-E.

Hennion, F. B., and Lobacz, E. F. 1973. Corps of engineers technology related to design of pavements in areas of permafrost. In *Proc. Permafrost 2nd Internat. Conf.,* pp. 426-29. Natl. Acad. Sci.-Natl. Res. Council, Yakutsk, U.S.S.R., 1973.

Herak, M., and Stringfield, V. T. 1972. *Karst. Important karst regions of the northern hemisphere.* Amsterdam: Elsevier.

Hess, H. H. 1955. Serpentines, orogeny and epeirogeny. In *Crust of the Earth,* edited by A. Poldervaart. Geol. Soc. America Spec. Paper 62:391-408.

———. 1962. History of ocean basins. In *Petrologic studies. A volume to honor A. F. Buddington,* edited by A. E. J. Engel, H. L. James, and B. F. Leonard, pp. 599-620. Boulder, Colo.: Geol. Soc. America.

Hickin, E. J. 1974. The development of meanders in natural river channels. *Am. Jour. Sci.* 274:414-42.

Hickin, E. J., and Nanson, G. C. 1975. The character of channel migration on the Beatton River, northeast British Columbia, Canada. *Geol. Soc. America Bull.* 86:487-94.

Higgins, C. G. 1953. Miniature "pediments" near Calistoga, California. *Jour. Geology* 61:461-65.

Hill, A. R. 1971. The internal composition and structure of drumlins in north Down and south Antrim, northern Ireland. *Geogr. Annlr.* 53:14-31.

Hino, M. 1975. Theory on formation of rip current and cuspidal coast. *Proc. 14th Conf. on Coast. Eng.,* pp. 901-19.

Hjulström, F. 1939. Transportation of detritus by moving water. In *Recent marine sediments: A symposium,* edited by P. Trask. Tulsa, Okla.: Am. Assoc. Petroleum Geologists.

Hobbs, P. V., and Mason, B. J. 1964. The sintering and adhesion of ice. *Philos. Mag.* 9:181-97.

Hodge, S. M. 1974. Variations in the sliding of a temperate glacier. *Jour. Glaciol.* 13:349-69.

———. 1976. Direct measurement of basal water pressures: A pilot study. *Jour. Glaciol.* 16:205-17.

Holdsworth, G. 1973. Ice calving into the proglacial Generator Lake Baffin Island, N.W.T., Canada. *Jour. Glaciol.* 12:235-50.

Holeman, J. N. 1968. The sediment yield of major rivers of the world. *Water Resour. Res.* 4:737-47.

Holland, H. D.; Kirsipu, T. V.; Huebner, J. S.; and Oxburgh, V. M. 1964. On some aspects of the chemical evolution of cave waters. *Jour. Geology* 72:36-67.

Holmes, A. 1929. Radioactivity and earth movements. *Geol. Soc. Glasgow Trans.* 18:559-606.

Holmes, C. D. 1947. Kames. *Am. Jour. Sci.* 245:240-49.

————. 1952. Drift dispersion in west-central New York. *Geol. Soc. America Bull.* 63:993-1010.

————. 1960. Evolution of till-stone shapes, central New York. *Geol. Soc. America Bull.* 71:1645-60.

Holmes, G. W.; Hopkins, D. M.; and Foster, H. L. 1968. Pingos in central Alaska. *U.S. Geol. Survey Bull.* 1241-H.

Holmes, J. W., and Colville, J. S. 1970a. Grassland hydrology in a karstic region of southern Australia. *J. Hydrol.* 10:38-58.

————. 1970b. Forest hydrology in a karstic region of southern Australia. *J. Hydrol.* 10:59-74.

Holtedahl, H. 1967. Notes on the formation of fjords and fjord valleys. *Geogr. Annlr.* 49:188-203.

Hooke, R. LeB. 1965. Alluvial fans. Ph.D. dissertation, California Inst. of Technology.

————. 1967. Processes on arid-region alluvial fans. *Jour. Geology* 75:438-60.

————. 1968. Steady-state relationships on arid-region alluvial fans in closed basins. *Am. Jour. Sci.* 266:609-29.

————. 1972. Geomorphic evidence for Late Wisconsin and Holocene tectonic deformation, Death Valley, California. *Geol. Soc. America Bull.* 83:2073-97.

Hopkins, D. M., and Sigafoos, R. S. 1951. Frost action and vegetation patterns on Seward Peninsula, Alaska. *U.S. Geol. Survey Bull.* 974-C:51-100.

Hopkins, D. M.; Karlstrom, T. N. V.; and others. 1955. Permafrost and groundwater in Alaska. U.S. Geol. Survey Prof. Paper 264-F.

Hoppe, G. 1959. Glacial morphology and the inland ice recession in northern Sweden. *Geogr. Annlr.* 41:193-212.

Hoppe, G., and Schytt, V. 1953. Some observations on fluted moraine surfaces. *Geogr. Annlr.* 35:105-15.

Horton, R. E. 1945. Erosional development of streams and their drainage basins: Hydrophysical approach to quantitative morphology. *Geol. Soc. America Bull.* 56:275-370.

Howard, Arthur D. 1942. Pediment passes and the pediment problem. *Jour. Geomorph.* 5:3-31, 95-136.

————. 1959. Numerical systems of terrace nomenclature: A critique. *Jour. Geology* 67:239-43.

————. 1967. Drainage analysis in geologic interpretation: A summation. *Am. Assoc. Petroleum Geologists Bull.* 51:2246-59.

Howard, A. D.; Fairbridge, R. W.; and Quinn, J. H. 1968. Terraces, fluvial—Introduction. In *Encyclopedia of geomorphology,* edited by R. W. Fairbridge, pp. 1117-23. New York: Reinhold Book Corp.

Howard, A. D., and Spock, L. 1940. A classification of landforms. *Jour. Geomorph.* 3:332-45.

Howard, Alan D. 1968. General systems theory in geomorphology. In *Encyclopedia of geomorphology,* edited by R. W. Fairbridge, pp. 382-84. New York: Reinhold Book Corp.

————. 1971a. Problems of interpretation of simulation models of geologic processes. In *Quantitative geomorphology,* edited by M. Morisawa, pp. 61-82. S.U.N.Y., Binghamton, Pubs. in Geomorphology, Proc. 2nd Symposium.

————. 1971b. Simulation model of stream capture. *Geol. Soc. America Bull.* 82:1355-76.

Hoyt, J. H. 1966. Air and sand movements in the lee of dunes. *Sedimentology* 7:137-44.

————. 1967. Barrier island formation. *Geol. Soc. America Bull.* 78:1125-36.

————. 1968. Barrier island formation: Reply. *Geol. Soc. America Bull.* 79(7):947; 79(10):1427-32.

Hoyt, J. H., and Henry, V. J. 1967. Influence on island migration on barrier island sedimentation. *Geol. Soc. America Bull.* 78:77-86.

————. 1971. Origin of capes and shoals along the southeastern coast of the United States. *Geol. Soc. America Bull.* 82:59-66.

Hsu, K. J. 1965. Isostasy, crustal thinning, mantle changes and the disappearance of ancient land masses. *Am. Jour. Sci.* 263:97-109.

————. 1975. Catastrophic debris streams (sturzstroms) generated by rockfalls. *Geol. Soc. America Bull.* 86:129-40.

Hubbert, M. K. 1940. The theory of groundwater motion. *Jour. Geology* 48:785-944.

Humphreys, B., and Hughes, D. J. 1974. Development of alluvial stream channels. A five-stage model. Discussion. *Geol. Soc. America Bull.* 85:149.

Hunt, C. B. 1954. Pleistocene and Recent deposits in the Denver area, Colorado. *U.S. Geol. Survey Bull.* 966-C: 91-140.

————. 1972. *Geology of soils.* San Francisco: W. H. Freeman.

Hunt, C. B.; Averitt, P.; and Miller, R. L. 1953. Geology and geography of the Henry Mountains region, Utah. U.S. Geol. Survey Prof. Paper 228.

Hunt, C. B., and Mabey, D. R. 1966. Stratigraphy and structure. Death Valley, California. U.S. Geol. Survey Prof. Paper 494-A.

Huntley, D. A., and Bowen, A. J. 1973. Field observations of edge waves. *Nature* 243:160-61.

Huntoon, P. W. 1974. The karstic groundwater basins of the Kaibab Plateau, Arizona. *Water Resour. Res.* 10:579-90.

Hutchinson, J. N. 1968. Mass movement. In *Encyclopedia of geomorphology,* edited by R. W. Fairbridge, pp. 688-96. New York: Reinhold Book Corp.

Hutton, C. E. 1947. Studies of loess-derived soils in southwestern Iowa. *Soil Sci. Soc. Am. Proc.* 12:424-31.

Hyndman, D. W. 1972. *Petrology of igneous and metamorphic rocks.* New York: McGraw-Hill.

Inglis, C. C. 1949. The behavior and control of rivers and canals. *Research Pub.,* Poona (India), no. 13, 2 vols.

Inglis, D. R. 1965. Particle sorting and stone migration by freezing and thawing. *Science* 148:1616-17.

Inman, D. L., and Bagnold, R. A. 1963. Littoral processes. In *The sea,* edited by M. N. Hill, 3:529-53. New York: Interscience.

Inman, D. L., and Brush, B. M. 1973. The coastal challenge. *Science* 181:20-32.

Inman, D. L.; Ewing, G. C.; and Corliss, J. B. 1966. Coastal sand dunes of Guerrero Negro; Baja, California, Mexico. *Geol. Soc. America Bull.* 77:787-802.

Inman, D. L., and Filloux, J. 1960. Beach cycles related to tide and local wind wave regime. *Jour. Geology* 68:225-31.

Inman, D. L., and Frautschy, J. D. 1966. Littoral processes and the development of shoreline. *Proc. Coast. Eng. Speciality Conf.,* Am. Soc. Civil Engineers (Santa Barbara, Calif.), pp. 511-36.

Inman, D. L. and Nordstrom, C. E. 1971. On the tectonic and morphologic classification of coasts. *Jour. Geology* 79:1-21.

Ippen, A. T., ed. 1966. *Estuary and coastline hydrodynamics.* New York: McGraw-Hill.

Isacks, B.; Oliver, J.; and Sykes, L. 1968. Seismology and the new global tectonics. *Jour. Geophys. Research* 73:5855-99.

Ives, J. D., and Fahey, B. D. 1971. Permafrost occurrence in the Front Range, Colorado Rocky Mountains, U.S.A. *Jour. Glaciol.* 10:105-11.

Jackson, M.; Hseung, Y.; Corey, R.; Evans, E.; and Heuval, R. 1952. Weathering sequence of clay size minerals in soils and sediments. *Soil Sci. Soc. Am. Proc.* 16:3-6.

Jackson, R. G. 1975. Velocity-bed form-texture patterns of meander bends in the lower Wabash River of Illinois and Indiana. *Geol. Soc. America Bull.* 86:1511-22.

Jackson, T., and Keller, W. 1970. A comparative study of the role of lichens and "inorganic" processes in the chemical weathering of recent Hawaiian lava flows. *Am. Jour. Sci.* 269:446-66.

Jahn, A. 1960. Some remarks on evolution of slopes on Spitsbergen. *Zeit. f. Geomorph.,* Suppl. 1:49-58.

Jansen, J. M. L., and Painter, R. B. 1974. Predicting sediment yield from climate and topography. *J. Hydrol.* 21:371-80.

Jarvis, G. T., and Clarke, G. K. C. 1975. The thermal regime of Trapridge Glacier and its relevance to glacier surging. *Jour. Glaciol.* 14:235-49.

Jennings, J. N. 1967. Some karst areas of Australia. In *Landform studies from Australia and New Guinea,* edited by J. N. Jennings and J. A. Mabbutt, pp. 256-92. Canberra.

———. 1969. Karst of the seasonally humid tropics in Australia. In *Problems of the karst denudation,* edited by O. Štelcl, pp. 149-58. Brno.

———. 1971. *Karst.* Cambridge, Mass.: M.I.T. Press.

Jennings, J. N., and Bik, M. J. 1962. Karst morphology in Australian New Guinea. *Nature* 194:1036-38.

Jenny, H. 1941. *Factors of soil formation.* New York: McGraw-Hill.

———. 1950. Origin of soils. In *Applied sedimentation,* edited by P. Trask, pp. 41-61. New York: John Wiley and Sons.

Jenny, H., and Leonard, C. 1939. Functional relationships between soil properties and rainfall. *Soil Sci.* 38:363-81.

Johnson, A. M., and Rahn, P. H. 1970. Mobilization of debris flows. *Zeit. f. Geomorph.,* Suppl. 9:168-86.

Johnson, D. W. 1910. Beach cusps. *Geol. Soc. America Bull.* 21:604-21.

———. 1919. *Shore processes and shoreline development.* New York: John Wiley and Sons. Facsimile edition: Hafner, New York, 1965.

———. 1925. *New England-Acadian shoreline.* New York: John Wiley and Sons.

———. 1932. Rock fans of arid regions. *Am. Jour. Sci.* 23 (5th ser.):389-416.

———. 1944. Problems of terrace correlation. *Geol. Soc. America Bull.* 55:793-818.

Johnson, J. W. 1956. Dynamics of nearshore sediment movement. *Am. Assoc. Petroleum Geologists Bull.* 40:2211-32.

————. 1966. Ocean currents. Introduction. In *Encyclopedia of oceanography,* edited by R. W. Fairbridge. New York: Reinhold Book Corp.

Johnson, R. B. 1968. Geology of the igneous rocks of the Spanish Peaks region, Colorado. U.S. Geol. Survey Prof. Paper 594-C.

Johnson, W. D. 1904. The profile of maturity in alpine glacial erosion. *Jour. Geology* 12:7:569-78.

Johnston, G. H. 1963. Pile construction in permafrost. *Proc. Permafrost Internat. Conf.,* (Lafayette, Ind., 1963). Natl. Acad. Sci.-Natl. Res. Council Pub. 1287, pp. 477-81.

Jopling, A. V. 1966. Some application of theory and experiment to the study of bedding genesis. *Sedimentology* 7:71-102.

Judson, S. 1960. William Morris Davis, an appraisal. *Zeit. f. Geomorph.* 4:3/4:193-201.

————. 1968a. Erosion rates near Rome, Italy. *Science* 160:1444-46.

————. 1968b. Erosion of the land. *Am. Scientist* 56:356-74.

Judson, S., and Ritter, D. F. 1964. Rates of regional denudation in the United States. *Jour. Geophys. Research* 69:3395-401.

Kachadoorian, R., and Ferrians, O. J., Jr. 1973. Permafrost-related engineering problems posed by the Trans-Alaskan Pipeline. *Permafrost 2nd Internat. Conf.,* pp. 684-87. Natl. Acad. Sci.-Natl. Res. Council, Yakutsk, U.S.S.R., 1973.

Kamb, B. 1964. Glacier mechanics. *Science* 146:353-65.

————. 1970. Sliding motion of glaciers: Theory and observation *Rev. Geophys. and Space Phys.* 8:673-728.

Kamb, B., and LaChapelle, E. 1964. Direct observation of the mechanism of glacier sliding over bedrock. *Jour. Glaciol.* 5:159-72.

Kaye, C. A. 1957. The effect of solvent motion on limestone solutions. *Jour. Geology* 65:34-47.

————. 1964a. Outline of Pleistocene geology of Martha's Vineyard, Massachusetts. U.S. Geol. Survey Prof. Paper 501-C:134-39.

————. 1964b. Illinoian and early Wisconsin moraines of Martha's Vineyard, Massachusetts. U.S. Geol. Survey Prof. Paper 501-C: 140-43.

Kazmann, R. 1972. *Modern hydrology.* New York: Harper and Row.

Keller, E. A. 1971a. Pools, riffles and meanders. Discussion. *Geol. Soc. America Bull.* 82:279-80.

————. 1971b. Areal sorting of bedload material. *Geol. Soc. America Bull.* 82:753-56.

————. 1972. Development of alluvial stream channels. A five-stage model. *Geol. Soc. America Bull.* 83:1531-36.

————. 1974. Development of alluvial stream channels. A five-stage model. Reply. *Geol. Soc. America Bull.* 85:150-52.

Keller, W. 1954. Bonding energies of some silicate minerals. *Am. Mineralogist* 39:783-93.

Kellerhals, R. 1967. Stable channels with gravel-paved beds. Am Soc. Civil Engineers Proc., *Jour. Waterways and Harbors* 93:63-84.

Kemmerly, P. R. 1976. Definitive doline characteristics in the Clarksville quadrangle, Tennessee. *Geol. Soc. America Bull.* 87:42-46.

Kennedy, G. C. 1959. The origin of continents, mountain ranges, and ocean basins. *Am. Scientist* 47:491-504.

Kennedy, W. 1933. Trends of differentiation in basaltic magmas. *Am. Jour. Sci.* 25:239-56.

Kenney, T. 1964. Sea-level movements and the geological histories of the post-glacial marine soils at Boston, Nicolet, Ottawa, and Oslo. *Geotechnique* 14:203-30.

Kent, P. E. 1975. Continental margin of East Africa—a region of vertical movement. In *The geology of continental margins,* edited by C. Burk and C. Drake. New York: Springer-Verlag.

Kesel, R. H.; Dunne, K.; McDonald, R.; Allison, K.; and Spicer, B. 1974. Lateral erosion and overbank deposition on the Mississippi River in Louisiana caused by 1973 flooding. *Geology* 2:461-64.

Kesseli, J. E. 1941. Rock streams in the Sierra Nevada, California. *Geogr. Rev.* 31:203-27.

Keulegan, G. H. 1948. An experimental study of submarine sand bars. U.S. Army Corps Engrs., Beach Erosion Board Tech. Rept. 3.

Keyes, C. R. 1910. Deflation and the relative efficiencies of erosional processes under conditions of aridity. *Geol. Soc. America Bull.* 21:565-98.

————. 1912. Deflative scheme of the geographic cycle in an arid climate. *Geol. Soc. America Bull.* 23:537-62.

Kiersch, G. A. 1964. Vaiont reservoir disaster. *Civil Engineering* 34:32-39.

King, C. A. M. 1972. *Beaches and coasts.* New York: St. Martin's Press.

King, C. A. M., and Buckley, J. T. 1968. The analysis of stone size and shape in Arctic environments. *Jour. Sed. Petrology* 38:200-214.

King, C. A. M., and Lewis, W. V. 1961. A tentative theory of ogive formation. *Jour. Glaciol.* 3:913-39.

King, C. A. M., and McCullagh, M. J. 1971. A simulation model of a complex recurved spit. *Jour. Geology* 79:22-37.

King, C. A. M., and Williams, W. W. 1949. The formation and movement of sand bars by wave action. *Geogr. Jour.* 107:70-84.

King, D. 1956. The Quaternary stratigraphic record at Lake Eyre North and the evolution of existing topographic forms. *Trans. Royal Soc. Australia* 79:93-103.

King, L. C. 1953. Canons of landscape evolution. *Geol. Soc. America Bull.* 64:751-52.

Kirkby, M. J. 1967. Measurement and theory of soil creep. *Jour. Geology* 75:359-78.

————. 1969. Infiltration, throughflow, and overland flow; and erosion by water on hillslopes. In *Water, earth, and man,* edited by R. J. Chorley, pp. 215-38. London: Methuen and Co.

Kirkby, M. J., and Chorley, R. J. 1967. Throughflow, overland flow and erosion. *Int. Assoc. Sci. Hydrol. Bull.* 12:5-21.

Kirkby, M. J., and Kirkby, A. V. 1969. Erosion and deposition on a beach raised by the 1964 earthquake, Montague Island, Alaska. U.S. Geol. Survey Prof. Paper 543-H:1-41.

Kirkby, R. P. 1969. Variation in glacial deposition in a subglacial environment: An example from Midlothian. *Scott. J. Geol.* 5:49-53.

Knighton, A. D. 1974. Variation in width-discharge relation and some implications for hydraulic geometry. *Geol. Soc. America Bull.* 85:1069-76.

Knox, J. C. 1972. Valley alluviation in southwestern Wisconsin. *Ann. Assoc. Am. Geog.* 62:401-10.

Kolb, C. R., and Van Lopik, J. R. 1958. Geology of the Mississippi River deltaic plain, southeastern Louisiana. U.S. Army Corps Engrs., Waterway Exp. Sta. Rept. 3-483, Vicksburg.

Komar, P. D. 1970. The competence of turbidity current flow. *Geol. Soc. America Bull.* 81:1555-62.

————. 1971a. The mechanics of sand transport on beaches. *Jour. Geophys. Research* 76:3:713-21.

————. 1971b. Nearshore cell circulation and the formation of giant cusps. *Geol. Soc. America Bull.* 82:2643-50.

————. 1973. Observations of beach cusps at Mono Lake, California. *Geol. Soc. America Bull.* 84:3593-3600.

————. 1975. Nearshore currents: Generation by obliquely incident waves and longshore variations in breaker height. In *Proc. symposium on nearshore sediment dynamics,* edited by J. R. Hails and A. Carr. pp. 17-45. London: John Wiley & Sons.

————. 1976. *Beach processes and sedimentation.* Englewood Cliffs, N.J.: Prentice-Hall.

Komar, P. D., and Inman, D. L. 1970. Longshore sand transport on beaches. *Jour. Geophys. Research* 75:30:5914-27.

Kottlowski, F., Cooley, M. and Ruhe, R. 1965. Quaternary geology of the southwest. In *The Quaternary of the United States,* edited by H. Wright and D. Frey. Princeton, N.J.: Princeton Univ. Press.

Krigstrom, A. 1962. Geomorphological studies of sandar plains and their braided rivers in Iceland. *Geogr. Annlr.* 44:328-46.

Krinsley, D. H., and Donahue, J. 1968. Environmental interpretation of sand grain surface textures of electron microscopy. *Geol. Soc. America Bull.* 79:743-48.

Krumbein, W. C. 1944. Shore currents and sand movement on a model beach. U.S. Army Corps Engrs., Beach Erosion Board Tech. Memo 7.

Krumbein, W. C., and Oshiek, L. E. 1950. Pulsation transport of sand by shore agents. *Am. Geophys. Union Trans.* 31:216-20.

Kuenen, Ph. H. 1928. Experiments on the formation of wind-worn pebbles. *Leidsche Geol. Meded.* 3:17-38.

————. 1948. The formation of beach cusps. *Jour. Geology* 56:34-40.

————. 1960. Experiment abrasion, 4. Eolian action. *Jour. Geology* 68:427-49.

Kuno, H. 1969. Plateau basalts. In *The Earth's crust and upper mantle,* edited by P. Hart, pp. 495-500. Am. Geophys. Union, Geophys. Monograph 13.

Kupsch, W. O. 1955. Drumlins with jointed boulders near Dollard, Saskatchewan. *Geol. Soc. America Bull.* 66:327-38.

Lachenbruch, A. H. 1966. Contraction theory of ice wedge polygons: A qualitative discussion. *Proc. Permafrost Internat. Conf.* (Lafayette, Ind. 1963). Natl. Acad. Sci.-Natl. Res. Council Pub. 1287, pp. 63-71.

————. 1970. Some estimates of the thermal effects of a heated pipeline in permafrost. U.S. Geol. Survey Circ. 632.

Lagally, M. 1934. *Mechanik und Thermodynamik des stationären Gletschers.* Leipzig.

Lambe, T. 1953. The structure of inorganic soils. *Am. Soc. Civil Engineers Proc.* 79: sep. 315.

Lane, E. W. 1955. Design of stable channels. *Am. Soc. Civil Engineers Trans.* 120:1234-79.

————. 1957. A study of the shape of channels formed by natural streams in erodible material. *M.R.D. Sediments Series no. 9,* U.S. Army Corps Engrs., Eng. Div., Missouri River, Omaha, Neb.

Langbein, W. B., and Leopold, L. B. 1964. Quasi-equilibrium states in channel morphology. *Am. Jour. Sci.* 262:782-94.

————. 1966. River meanders: Theory of minimum variance. U.S. Geol. Survey Prof. Paper 422-H.

Langbein, W. B., and Schumm, S. A. 1958. Yield of sediment in relation to mean annual precipitation. *Am. Geophys. Union Trans.* 39:1076-84.

Langford-Smith, T., and Dury, G. H. 1964. A pediment at Middle Pinnacle, near Broken Hill, New South Wales. *Jour. Geol. Soc. Australia* 11:79-88.

Lattman, L. H. 1960. Cross section of a flood plain in a moist region of moderate relief. *Jour. Sed. Petrology* 30:275-82.

————. 1968. Structural control in geomorphology. In *Encyclopedia of geomorphology,* edited by R. W. Fairbridge, pp. 1074-79. New York: Reinhold Book Corp.

————. 1973. Calcium carbonate cementation of alluvial fans in Southern Nevada. *Geol. Soc. America Bull.* 84:3013-28.

Lattman, L. H., and Simonberg, E. 1971. Case-hardening of carbonate alluvium and colluvium, Spring Mountains, Nevada. *Jour. Sed. Petrology* 41:274-81.

Lau, J., and Travis, B. 1973. Slowly varying stokes waves and submarine longshore bars. *Jour. Geophys. Research* 78:4489-97.

Laury, R. L. 1971. Stream bank failure and rotational slumping. Preservation and significance in the geologic record. *Geol. Soc. America Bull.* 82:1251-66.

LaValle, P. 1967. Some aspects of linear karst depression development in south central Kentucky. *Ann. Assoc. Am. Geog.* 57:49-71.

————. 1968. Karst depression morphology in south-central Kentucky. *Geogr. Annlr.* 50A:94-108.

Lawson, A. C. 1915. The epigene profiles of the desert. *Univ. of Calif. Dept. Geol. Bull.* 9:23-48.

Lee, H. A. 1965. Investigation of eskers for mineral exploration. Canada Geol. Survey Paper 65-14:1-17.

Lee, W., and Uyeda, S. 1965. Review of heat flow data. In *Terrestrial heat flow,* edited by W. Lee. Am. Geophys. Union, Geophys. Monograph 8.

Legget, R. 1967. Soil: Its geology and use. *Geol. Soc. America Bull.* 78:1433-60.

Lehman, D. 1963. Some principles of chelation chemistry. *Soil Sci. Soc. Am. Proc.* 27:167-70.

Lehmann, H. 1936. Morphologische Studien auf Java. Stuttgart: *Geogr. Abh.* 3:9.

Lehmann, O. 1932. *Die Hydrographie des Karstes.* Leipzig.

Leighton, M. W., and Pendexter, C. 1962. Carbonate rock types. In W. E. Ham, editor, *Classification of carbonate rocks. Am. Assoc. Petroleum Geologists Mem.* 1:33-61.

Leliavsky, S. L. 1966. An introduction to fluvial hydraulics. New York: Dover Publications.

Lemke, R. W. 1958. Narrow linear drumlins near Velva, North Dakota. *Am. Jour. Sci.* 256:270-83.

Leopold, L. B. 1953. Downstream change of velocity in rivers. *Am. Jour. Sci.* 251:606-24.

Leopold, L. B., and Langbein, W. B. 1962. The concept of entropy in landscape evolution. U.S. Geol. Survey Prof. Paper 500-A:20.

————. 1963. Association and indeterminancy in geomorphology. In *The fabric of geology,* edited by C. C. Albritton, pp. 184-92. Reading, Mass.: Addison-Wesley.

Leopold, L. B. and Maddock, T., Jr. 1953. The hydraulic geometry of stream channels and some physiographic implications. U.S. Geol. Survey Prof. Paper 252.

Leopold, L. B., and Miller, J. P. 1954. A post-glacial chronology for some alluvial valleys in Wyoming, U.S. Geol. Survey Water Supply Paper 1261.

————. 1956. Ephemeral streams—hydraulic factors and their relation to the drainage net. U.S. Geol. Survey Prof. Paper 282-A.

Leopold, L. B., and Wolman, M. G. 1957. River channel patterns; braided, meandering and straight. U.S. Geol. Survey Prof. Paper 282-B.

————. 1960. River meanders. *Geol. Soc. America Bull.* 71:769-794.

Leopold, L. B.; Wolman, M. G.; and Miller, J. P. 1964. *Fluvial processes in geomorphology.* San Francisco: W. H. Freeman.

LePichon, X. 1968. Sea-floor spreading and continental drift. *Jour. Geophys. Research* 73:3611-97.

Lewin, J. 1976. Initiation of bed forms and meanders in coarse-grained sediment. *Geol. Soc. America Bull.* 87:281-85.

Lewis, B., and Dorman, L. 1970. Experimental isostasy, 2. An isostatic model for the USA derived from gravity and topographic data. *Jour. Geophys. Research* 75:3367-86.

Lewis, W. V. 1932. The formation of Dungeness foreland. *Geogr. Annlr.* 80:309-24.

————. 1947. Valley steps and glacial valley erosion. *Inst. Brit. Geog. Trans.* 14:19-44.

————. 1954. Pressure release and glacial erosion. *Jour. Glaciol.* 2:417-22.

————, ed. 1960. *Norwegian cirque glaciers.* Royal Geog. Soc. Res. Ser. 4.

Li, Y. H. 1976. Denudation of Taiwan Island since the Pliocene Epoch. *Geology* 4:105-07.

Lindsay, J. F. 1970. Clast fabric of till and its development. *Jour. Sed. Petrology* 40:629-41.

Linell, K. A., and Johnston, G. H. 1973. Engineering design and construction in permafrost regions. *Proc. Permafrost 2nd Internat. Conf.,* pp. 553-75. Natl. Acad. Sci.-Natl. Res. Council, Yakutsk, U.S.S.R., 1973.

Linton, D. L. 1963. The forms of glacial erosion. *Inst. Brit. Geog. Trans.* 33:1-28.

Livingstone, D. A. 1963. Chemical composition of rivers and lakes. U.S. Geol. Survey Prof. Paper 440-G.

Lliboutry, L. 1964. Subglacial ''supercavitation'' as a cause of the rapid advances of glaciers. *Nature* 202:77.

————. 1968. General theory of subglacial cavitation and sliding of temperate glaciers. *Jour. Glaciol.* 7:21-58.

————. 1971. Permeability, brine content and temperature of temperate ice. *Jour. Glaciol.* 10:15-29.

Lobacz, E. F., and Quinn, W. F. 1963. Thermal regime beneath buildings constructed on permafrost. *Proc. Permafrost Internat. Conf.,* (Lafayette, Ind., 1963). Natl. Acad. Sci.-Natl. Res. Council Pub. 1287, pp. 159-64.

Loewe, F. 1966. The temperature of the Sukkertoppen ice cap. *Jour. Glaciol.* 6:179.

Lohnes, R. A., and Handy, R. L. 1968. Slope angles in friable loess. *Jour. Geology* 76:247-58.

Long, J. T., and Sharp, R. P. 1964. Barchan-dune movement in the Imperial Valley, California. *Geol. Soc. America Bull.* 75:149-56.

Longuet-Higgins, M. S. 1970. Longshore currents generated by obliquely incident sea waves. *Jour. Geophys. Research* 75:6778-6801.

Longwell, C. R. 1960. Interpretation of the leveling data. U.S. Geol. Survey Prof. Paper 295:33-38.

Loughnan, F. 1969. *Chemical weathering of the silicate minerals.* New York: American Elsevier.

Loughnan, F., and Bayliss, P. 1961. The mineralogy of the bauxite deposits near Weipa, Queensland. *Am. Mineralogist* 46:209-17.

Loziński, W. 1912. Die periglaziale Fazies der mechanischen Verwitterung. Internat. Geol. Cong., 11th, Stockholm, 1910, *Compte rendu,* pp. 1039-53.

Lugn, A. L. 1962. *The origin and sources of loess.* Lincoln, Neb.: Univ. Nebraska Studies, new series, no. 26.

————. 1968. The origin of loesses and their relation to the Great Plains in North America. In *Loess and related eolian deposits of the world,* edited by C. B. Schultz and J. C. Frye, p. 139. Lincoln, Neb.: Univ. Nebraska Press.

Lumley, J. L., and Panofski, H. A. 1964. *The structure of atmospheric turbulence.* New York: John Wiley and Sons.

Lustig, L. K. 1965. Clastic sedimentation in Deep Springs Valley, California. U.S. Geol. Survey Prof. Paper 352-F.

————. 1969. Trend surface analysis of the Basin and Range province and some geomorphic implications. U.S. Geol. Survey Prof. Paper 500-D.

Mabbutt, J. A. 1965. The weathered land surface in central Australia. *Zeit. f. Geomorph.* 9:82-114.

————. 1966. Mantle-controlled planation of pediments. *Am. Jour. Sci.* 264:78-91.

————. 1971. The Australian arid zone as a prehistoric environment. In *Aboriginal man and environment in Australia,* edited by D. Mulvaney and J. Golson, pp. 66-79. Canberra: Australian Natl. Univ. Press.

MacClintock, P., and Dreimanis, A. 1964. Reorientation of till fabric by overriding glacier in the St. Lawrence Valley. *Am. Jour. Sci.* 262:1:133-42.

Macdonald, G. 1972. *Volcanoes.* Englewood Cliffs, N.J.: Prentice-Hall.

Mackay, J. R. 1970. Disturbances to the tundra and forest tundra environment of the western Arctic. *Can. Geotechnical Jour.* 7:420-32.

————. 1973. The growth of pingos, western Arctic coast, Canada. *Can. J. Earth Sci.* 10:979-1004.

————. 1974. Ice-wedge cracks, Garry Island, Northwest Territories. *Can. J. Earth Sci.* 11:1366-83.

————. 1975. The closing of ice-wedge cracks in permafrost, Garry Island, Northwest Territories. *Can. J. Earth Sci.* 12:1668-74.

Mackay, J. R., and Matthews, W. H. 1974. Movement of sorted stripes, the Cinder Cone, Garibaldi Park, B.C., Canada. *Arc. Alp. Res.* 6:347-59.

Mackin, J. H. 1936. The capture of the Greybull River. *Am. Jour. Sci.* 31:373-85.

————. 1937. Erosional history of the Big Horn Basin, Wyoming. *Geol. Soc. America Bull.* 48:813-93.

————. 1948. Concept of the graded river. *Geol. Soc. America Bull.* 59:463-512.

————. 1963. Rational and empirical methods of investigation in geology. In *The fabric of geology,* edited by C. Albritton. Reading, Mass.: Addison-Wesley.

Maddock, T., Jr. 1969. The behavior of straight open channels with movable beds. U.S. Geol. Survey Prof. Paper 622-A:70.

Malott, C. A. 1921. A subterranean cut-off and other subterranean phenomena along Indian Creek, Laurence Co., Indiana. *Indiana Acad. Sci. Proc.* 31:203-10.

———. 1938. Invasion theory of cavern development (abs.). *Geol. Soc. America Proc. 1937*, p. 323.

———. 1939. Karst valleys. *Geol. Soc. America Bull.* 50:1984.

Mammerickx, J. 1964. Quantitative observations on pediments in the Mojave and Sonoran deserts (southwestern United States). *Am. Jour. Sci.* 262:417-35.

Manley, S., and Manley, R. 1968. *Beaches; their lives, legends and lore.* Philadelphia: Chilton.

Mark, D. M. 1974. Line intersection method for estimating drainage density. *Geology* 2:235-36.

Martel, E. A. 1921. *Nouveau traité des eaux souterraines.* Paris: Delagrave.

Matschinski, M. 1968. Alignment of dolines northwest of Lake Constance, Germany. *Geol. Mag.* 105(1):56-61.

Matsumota, T. 1967. Fundamental problems in the circum-Pacific orogenesis. *Tectonophysics* 4:595-613.

Matthes, F. E. 1900. Glacial sculpture of the Bighorn Mountains, Wyoming. *U.S. Geol. Survey 21st Ann. Rept., 1899-1900,* pt. 2, pp. 167-90.

———. 1930. Geologic history of the Yosemite Valley. U.S. Geol. Survey Prof. Paper 160.

Maxwell, A.; Von Herzen, R.; Hsu, K.; Andrews, J.; Saito, T.; Percival, S.; Milow, E.; and Boyce, R. 1970. Deep-sea drilling the south Atlantic. *Science* 168:1047-59.

Maxwell, J. C. 1975. Early western margin of the United States. In *The geology of continental margins,* edited by C. Burk and C. Drake. New York: Springer-Verlag.

May, J. P., and Tanner, W. F. 1973. The littoral power gradient and shoreline changes. In *Coastal geomorphology,* edited by D. R. Coates. pp. 43-60. S.U.N.Y., Binghamton, 3rd Ann. Geomorph. Symposium.

McCall, J. G. 1952. The internal structure of a cirque glacier: Report on studies of the englacial movements and temperatures. *Jour. Glaciol.* 2:122-31.

———. 1960. The flow characteristics of a cirque glacier and their effect on glacial structure and cirque formation. In *Norwegian cirque glaciers,* edited by W. V. Lewis, pp. 39-62. Royal Geog. Soc. Res. Ser. 4.

McComas, M.; Hinkley, K.; and Kempton, J. 1969. Coordinated mapping of geology and soils for land-use planning. Illinois Geol. Survey Environ. Geol. Note 29.

McCoy, R. M. 1971. Rapid measurement of drainage density. *Geol. Soc. America Bull.* 82:757-62.

McDonald, R. C. 1975. Observations on hillslope erosion in tower karst topography of Belize. *Geol. Soc. America Bull.* 86:255-56.

McDowall, I. C. 1960. Particle size reduction of clay minerals by freezing and thawing. *New Zealand J. Geol. and Geophys.* 3:337-43.

McGee, W. J. 1897. Sheetflood erosion. *Geol. Soc. America Bull.* 8:87-112.

McGinnis, L. D. 1966. Crustal tectonics and Precambrian basement in northeastern Illinois. Illinois Geol. Survey Rept. of Inv. 219.

McGowen, J. H., and Garner, J. H. 1970. Physiographic features and stratification types of coarse-grained point bars. Modern and ancient examples. *Sedimentology* 14:77-111.

McKee, E. D. 1966. Structures of dunes at White Sands National Monument, New Mexico, and a comparison with structures of dunes from other selected areas. *Sedimentology* 7:1-69.

McKee, E. D., and Tibbitts, G. C., Jr. 1964. Primary structures of a seif dune and associated deposits in Libya. *Jour. Sed. Petrology* 34:5-17.

McKenzie, D. P., and Parker, R. L. 1967. The north Pacific: An example of tectonics on a sphere. *Nature* 216:1276-80.

McKenzie, G. D. 1969. Observations on a collapsing kame terrace in Glacier Bay National Monument, S.E. Alaska. *Jour. Glaciol.* 8:413-25.

McLean, R. F., and Kirk, R. M. 1969. Relationship between grain size, size-sorting and foreshore slope on mixed sand-shingle beaches. *New Zealand J. Geol. and Geophys.* 12:138-55.

McLeish, W. 1968. On the mechanics of wind-slick generation. *Deep Sea Res.* 15:461-89.

McPherson, H. J., and Rannie, W. F. 1967. Geomorphic effects of the May, 1967, flood in Graburn watershed, Cypress Hills, Alberta, Canada. *J. Hydrol.* 9:307-21.

McPherson, M. B. 1974. *Hydrological effects of urbanization.* Paris: UNESCO Press.

Meade, R. H. 1969. Errors in using modern stream-load data to estimate natural rates of denudation. *Geol. Soc. America Bull.* 80:1265-74.

Meier, M. F. 1960. Mode of flow of Saskatchewan glacier, Alberta, Canada. U.S. Geol. Survey Prof. Paper 351.

————. 1962. Proposed definitions for glacier mass budget terms. *Jour. Glaciol.* 4:252-65.

Meier, M. F., and Johnson, A. 1962. The kinematic wave on Nisqually Glacier, Washington. *Jour. Geophys. Research* 67:886.

Meier, M. F.; Kamb, W. B.; Allen, C. R.; and Sharp, R. P. 1974. Flow of Blue Glacier, Olympic Mountains, Washington, U.S.A. *Jour. Glaciol.* 13:187-212.

Meier, M. F., and Post, A. S. 1969. What are glacier surges? *Can. J. Earth Sci.* 6:807-17.

Meier, M. F., and Tangborn, W. V. 1965. Net budget and flow of South Cascade Glacier, Washington, *Jour. Glaciol.* 5:547-66.

Meinzer, O. E. 1923. Outline of ground-water hydrology, with definitions. U.S Geol. Survey Water Supply Paper 494:1-71.

Mellor, M. 1970. Phase composition of pore water in cold rocks. U.S. Army Corps Engrs., Cold Regions Res. and Eng. Lab. Research Rept. 292.

Melton, F. A. 1940. A tentative classification of sand dunes. *Jour. Geology* 48:113-73.

Melton, M. A. 1957. Analysis of the relations among elements of climate, surface properties and geomorphology. Proj. NR 389-042, Tech. Rept. 11, Columbia Univ. Dept. of Geol.; ONR, Geog. Branch, New York.

————. 1958. Correlation structure of morphometric properties of drainage systems and their controlling agents. *Jour. Geology* 66:442-60.

————. 1965a. The geomorphic and paleoclimatic significance of alluvial deposits in southern Arizona. *Jour. Geology* 73:1-38.

————. 1965b. Debris-covered hillslopes of the southern Arizona desert—consideration of their stability and sediment contribution. *Jour. Geology* 73:715-29.

Menard, H. W. 1961. Some rates of regional erosion. *Jour. Geology* 69:155-61.

Menard, H. W., and Smith, S. M. 1966. Hypsometry of ocean basin provinces. *Jour. Geophys. Research* 71:4305-25.

Merriam, R. 1960. Portuguese Bend landslide, Palos Verdes Hills, California. *Jour. Geology* 68:140-53.

Meyerhoff, A., and Meyerhoff, H. 1972. The new global tectonics: Major inconsistencies. *Am. Assoc. Petroleum Geologists Bull.* 56:269-336.

Meyer-Peter, E., and Muller, R. 1948. Formulas for bed-load transport. Intnal. Assoc. for Hydr. Structures Res. Proc., 2nd Meeting, Stockholm: 39-65.

Mickelson, D. M. 1973. Nature and rate of basal till deposition in a stagnating ice mass, Burroughs Glacier, Alaska. *Arc. Alp. Res.* 5:17-27.

Mickelson, D. M., and Berkson, J. M. 1974. Till ridges presently forming above and below sea level in Wachusett Inlet, Glacier Bay, Alaska. *Geogr. Annlr.* 56A;111-19.

Mielenz, R., and King. M. 1955. Physical-chemical properties and engineering performance of clays. *California Div. of Mines Bull.* 169:196-254.

Millar, C.; Turk, L.; and Foth, H. 1958. *Fundamentals of soil science.* New York: John Wiley and Sons.

Miller, H. 1883. Methods and results of river terracing. *Royal Phys. Soc. Edinburgh* 7.

Miller, J. P. 1958. High mountain streams; effects of geology on channel characteristics and bed material. New Mexico State Bur. Mines and Min. Res. Memo 4.

Miller, R. D. 1966. Phase equilibria and soil freezing. *Proc. Permafrost Internat. Conf.* (Lafayette, Ind., 1963). Natl. Acad. Sci.-Natl. Res. Council Pub. 1287, pp. 193-97.

Millette, J. F. G., and Higbee, H. W. 1958. Periglacial loess, I. Morphological properties. *Am. Jour. Sci.* 256:284-93.

Mills, H. C., and Wells, P. D. 1974. Ice-shove deformation and glacial stratigraphy of Port Washington, Long Island, New York. *Geol. Soc. America Bull.* 85:357-64.

Miotke, F. D. 1973. The subsidence of the surfaces between mogotes in Puerto Rico east of 'Arecibo. *Caves and Karst* 15:1-12.

Mitchell, A., and Reading, H. 1969. Continental margins, geosynclines, and ocean floor spreading. *Jour. Geology* 77:629-46.

Miyashiro, A. 1961. Evolution of metamorphic belts. *Jour. Petrology* 2:277-311.

Monroe, W. H. 1969. Evidence of subterranean sheet solution under weathered detrital cover in Puerto Rico. In *Problems of karst denudation.* Internat. Speleol. Cong., 5th, Stuttgart.

————. 1970. A glossary of karst terminology. U.S. Geol. Survey Water Supply Paper 1899 K.

————. 1976. The karst landforms of Puerto Rico. U.S. Geol. Survey Prof. Paper 899.

Moore, J. 1970. Relationship between subsidence and volcanic load, Hawaii. *Bull. Volcanol.* 34:562-76.

Morehouse, D. F. 1968. Cave development via the sulfuric acid reactions. *Natl. Speleol. Soc. Bull.* 30:1-10.

Morey, G.; Fournier, R.; and Rowe, J. 1962. The solubility of quartz in water in the temperature interval from 25°C to 300°C. *Geochim. et Cosmochim. Acta* 26:1029-43.

————. 1964. The solubility of amorphous silica at 25°C. *Jour. Geophys. Research* 69:1995-2002.

Morgan, J. P. 1970. Deltas—a résumé. *Jour. Geol. Educ.* 18:107-17.

Morgan, W. J. 1968. Rises, trenches, great faults, and crustal blocks. *Jour. Geophys. Research* 73:1959-82.

————. 1972. Deep mantle convection plumes and plate motion. *Am. Assoc. Petroleum Geologists Bull.* 56:203-13.

Morisawa, M. E. 1962. Quantitative geomorphology of some watersheds in the Appalachian Plateau. *Geol. Soc. America Bull.* 73:1025-46.

————. 1964. Development of drainage systems on an upraised lake floor. *Am. Jour. Sci.* 262:340-54.

————. 1968. *Streams, their dynamics and morphology.* New York: McGraw-Hill.

Morris, E. M. 1976. An experimental study of the motion of ice past obstacles by the process of regelation. *Jour. Glaciol.* 17:79-98.

Morrison, R. 1967. Principles of Quaternary soil stratigraphy. In *Quaternary soils,* edited by R. Morrison and H. Wright, pp. 1-69. Univ. Nevada, Reno, Desert Research Inst.

Moss, J. H. 1974. The relation of river terrace formation to glaciation in the Shoshone River basin, western Wyoming. In *Glacial geomorphology,* edited by D. R. Coates, pp. 294-314. S.U.N.Y., Binghamton, Pubs. in Geomorphology, 5th Ann. Symposium.

Moss, J. H., and Bonini, W. 1961. Seismic evidence supporting a new interpretation of the Cody terrace near Cody, Wyo. *Geol. Soc. America Bull.* 72:547-56.

Moss, J. H., and Whitney, J. H. 1971. Diversity of origin of the Cody and Powell terraces along the Shoshone River, Bighorn Basin, Wyoming. *Geol. Soc. America, Abs. with Programs* 3:652-53.

Muehlberger, W.; Denison, R.; and Lidiak, E. 1967. Basement rocks in the continental interior of the United States. *Am. Assoc. Petroleum Geologists Bull.* 51:2351-80.

Mueller, J. E. 1972. Re-evaluation of the relationship of master streams and drainage basins. *Geol. Soc. America Bull.* 83:3471-74.

Muguruma, J.; Mae, S.; and Higashi, R. 1966. Void formation by non-basal glide in ice single crystals. *Philos. Mag.* 13:625-30.

Muir Wood, A. M. 1969. *Coastal hydraulics.* London: MacMillan.

Mullenders, W., and Gullentops, F. 1969. The age of the pingos of Belgium. In *The periglacial environment,* edited by T. Péwé. Montreal: McGill-Queens Univ. Press.

Muller, E. H. 1974. Origin of drumlins. In *Glacial geomorphology,* edited by D. R. Coates, pp. 187-204. S.U.N.Y., Binghamton. Pubs. in Geomorphology, 5th Ann. Symposium.

Müller, F. 1962. Zonation in the accumulation areas of the glaciers of Axel Heiberg Island, N.W.T., Canada. *Jour. Glaciol.* 4:302-11.

————. 1963. *Observations on pingos (Beobachtungen über Pingos).* Can. Natl. Res. Council Tech. Translation 1073.

Muller, S. W. 1947. *Permafrost or permanently frozen ground and related engineering problems.* Ann Arbor, Mich.: J. W. Edwards.

Munk, W. H. 1949. Surf beats. *Am. Geophys. Union Trans.* 30:849-54.

Myrick, R. M., and Leopold, L. B. 1963. Hydraulic geometry of a small tidal estuary. U.S. Geol. survey Prof. Paper 422-B.

Nelson, R. 1954. Glacial geology of the Frying Pan River drainage, Colorado. *Jour. Geology* 62:325-43.

Nobles, L. H., and Weertman, J. 1971. Influence of irregularities of the bed of an ice sheet on deposition rate of till. In *Till: A symposium,* edited by R. P. Goldthwait. Columbus: Ohio State Univ. Press.

Norris, R. M. 1966. Barchan dunes of Imperial Valley, California. *Jour. Geology* 74:292-306.

Nunn, K. R., and Rowell, D. M. 1967. Regelation experiments with wires. *Philos. Mag.* 16:1281-83.

Nye, J. F. 1952a. The mechanics of glacier flow. *Jour. Glaciol.* 2:82-93.

————. 1952b. A comparison between the theoretical and the measured long profiles of the Unteraar glacier. *Jour. Glaciol.* 2:103-07.

————. 1957. The distribution of stress and velocity in glaciers and ice-sheets. *Proc. Royal Soc. London,* ser. A:239:113-33.

————. 1960. The response of glaciers and ice-sheets to seasonal and climatic changes. *Proc. Royal Soc. London,* ser. A:256:559-84.

————. 1965. The flow of a glacier in a channel of rectangular, elliptic, or parabolic cross-section. *Jour. Glaciol.* 5:661-90.

————. 1969. A calculation on the sliding of ice over a wavy surface using a Newtonian viscous approximation. *Proc. Royal Soc. London,* Ser. A:311:445-67.

————. 1970. Glacier sliding without cavitation in a linear viscous approximation. *Proc. Royal Soc. London,* ser. A:315:381-403.

Nye, J. F., and Martin, P. C. S. 1967. *Glacial erosion,* pp. 78-83. Int. Assoc. Sci. Hydrol., Comm. Snow and Ice, Bern.

Oberlander, T. M. 1972. Morphogenesis of granitic boulder slopes in the Mojave Desert, California. *Jour. Geology* 80:1-20.

———. 1974. Landscape inheritance and the pediment problem in the Mojave Desert of southern California. *Am. Jour. Sci.* 274:849-75.

Okko, V. 1955. Glacial drift in Iceland, its origin and morphology. *Bull. Comm. Geol. Finland* 170:1-133.

Ollier, C. D. 1963. Insolation weathering: Examples from central Australia. *Am. Jour. Sci.* 261:376-81.

———. 1969. *Weathering.* Edinburgh: Oliver and Boyd.

Olson, J. S. 1958. Lake Michigan dune development. *Jour. Geology* 66:254-63; 345-51; 437-83.

Ongley, E. D. 1968. An analysis of the meandering tendency of Serpentine Cave, N.S.W. *J. Hydrol.* 6:15-32.

Orme, A. R. 1973. Barrier and lagoon systems along the Zululand coast, South Africa. In *Coastal geomorphology,* edited by D. R. Coates, pp. 181-217. S.U.N.Y., Binghamton, 3rd Ann. Geomorph. Symposium.

———. 1974. Quaternary deformation of marine terraces between Ensenada and El Rosario, Baja California, Mexico. In *Geology of peninsular California, Pacific sections,* pp. 67-79. AAPG, SEPM and SEG.

Osborn, G. D. 1975. Advancing rock glaciers in the Lake Louise area, Banff National Park, Alberta. *Can. J. Earth Sci.* 12:1060-62.

Østrem, G. 1964. Ice-cored moraines in Scandinavia. *Geogr. Annlr.* 46:282-337.

———. 1971. Rock glaciers and ice-cored moraines, a reply to D. Barsch. *Geogr. Annlr.* 53A:207-13.

———. 1975. ERTS data in glaciology—an effort to monitor glacier mass balance from satellite imagery. *Jour. Glaciol.* 15:403-14.

Otvos, E. G. 1970. Development and migration of barrier islands, northern Gulf of Mexico. *Geol. Soc. America Bull.* 81:241-46.

Outcalt, S. I., and Benedict, J. B. 1965. Photo-interpretation of two types of rock glaciers in the Colorado Front Range. U.S.A. *Jour. Glaciol.* 5: 849-56.

Paige, S. 1912. Rock-cut surfaces in the desert ranges. *Jour. Geology* 20:442-50.

Pakiser, L. C., and Robinson, R. 1966. Composition of the continental crust as estimated from seismic observations. In *The Earth beneath the continents,* edited by J. Steinhart and T. Smith, pp. 620-26. Am. Geophys. Union, Geophys. Monograph 10.

Palmer, A. C. 1972. A kinematic wave model of glacier surges. *Jour. Glaciol.* 11:65-72.

Palmer, A. N. 1975. Origin of maze caves. *Natl. Speleol. Soc. Bull.* 37:57-76.

Palmer, V. E., and Palmer, A. N. 1975. Landform development of the Mitchell Plain of southern Indiana: Origin of a partially karstic plain. *Zeit. f. Geomorph.* 19:1-39.

Panoś, V., and Štelcl, O. 1968. Physiographic and geologic control in development of Cuban mogotes. *Zeit. f. Geomorph.* 12:117-65.

Parizek, R. 1969. *Glacial ice-contact rings and ridges.* Geol. Soc. America Spec. Paper 123:49-102.

Paterson, W. S. B. 1964. Variations in velocity of Athabasca Glacier with time. *Jour. Glaciol.* 5:277-85.

———. 1969. *The physics of glaciers.* Oxford: Pergamon Press.

———. 1971. Temperature measurements in Athabasca Glacier, Alberta, Canada. *Jour. Glaciol.* 10:339-49.

———. 1972. Temperature distribution in the upper layers of the ablation area of Athabasca Glacier, Alberta, Canada. *Jour. Glaciol.* 11:31-41.

Paton, T., and Williams, M. 1972. The concept of laterite. *Ann. Assoc. Am. Geog.* 62:42-56.

Patton, H. B. 1910. Rockstreams of Veta Park, Colorado. *Geol. Soc. America Bull.* 22:663-76.

Patton, P. C., and Baker, V. R. 1976. Morphometry and floods in small drainage basins subject to diverse hydrogeomorphic controls. *Water Resour. Res.* 12:941-52.

Patton, P. C., and Schumm, S. A. 1975. Gulley erosion, northwestern Colorado: A threshold phenomenon. *Geology* 3:88-90.

Peltier, L. 1950. The geographical cycle in periglacial regions as it is related to climatic geomorphology. *Ann. Assoc. Am. Geog.* 40:214-36.

Perutz, M. F. 1940. Mechanism of glacier flow. *Proc. Royal Soc. London,* ser. A:52:132-35.

Pesci, M. 1968. Loess. In *Encyclopedia of geomorphology,* edited by R. W. Fairbridge. New York: Reinhold Book Corp.

Petrie, G., and Price, R. J. 1966. Photogrammetric measurements of the ice wastage and morphological changes near the Casement Glacier, Alaska. *Can. J. Earth Sci.* 3:827-40.

Petterssen, S. 1964. Meteorology. In *Handbook of applied hydrology,* edited by V. T. Chow, pp. 3-1–3-39. New York: McGraw-Hill.

Pettijohn, F. 1941. Persistence of heavy minerals and geologic age. *Jour. Geology* 49:610-25.

Peyronnin, C. A., Jr. 1962. Erosion of Isles Dernieres and Timbalier Islands. Am. Soc. Civil Engineers, *Jour. Waterways and Harbors,* 1:57-69.

Péwé, T. L. 1955. Origin of the upland silt near Fairbanks, Alaska. *Geol. Soc. America Bull.* 66:699-724.

————. 1966. Ice-wedges in Alaska—Classification, distribution and climatic significance. In *Proc. Permafrost Internat. Conf.* (Lafayette, Ind., 1963). Natl. Acad. Sci.-Natl. Res. Council Pub. 1287, pp. 76-81.

————. 1969. *The periglacial environment.* Montreal: McGill-Queens Univ. Press.

————. 1973. Ice wedge casts and past permafrost distribution in North America. *Geoform* 15:15-26.

Phillip, H. 1920. Geologische Untersuchungen über den Mechanismus der Gletscher Bewegung und die Entstehung der gletschertextur. *Neuer Jb. Miner. Geol. Paläont.* 43:439-556.

Pierce, J. W. 1970. Tidal inlets and washover fans. *Jour. Geology* 78(2): 230-34.

Pike, R. J., and Wilson, S. E. 1971. Elevation-relief ratio, hypsometric integral, and geomorphic area-altitude analysis. *Geol. Soc. America Bull.* 82:1079-84.

Piper, A. M. 1932. Ground water in north-central Tennessee. U.S. Geol. Survey Water Supply Paper 640:69-89.

Pissart, A. 1970. *The pingos of Prince Patrick Island (76°N–120°W).* Natl. Res. Council of Canada, Tech. Translation 1401.

Pluhar, A., and Ford, D. C. 1970. Dolomite karren of the Niagara Escarpment. Ontario, Canada. *Zeit. f. Geomorph.* 14:392-410.

Pohl, E. R. 1955. *Vertical shafts in limestone caves.* Natl. Speleol. Soc. Occasional Paper 2.

Pohl, E. R., and White, W. B. 1965. Sulfate minerals: Their origin in the central Kentucky karst. *Am. Mineralogist* 50:1461-65.

Poldervaart, A., ed. 1955. Chemistry of the Earth's crust. In *Crust of Earth,* Geol. Soc. America Spec. Paper 62: 119-44.

Porslid, A. E. 1938. Earth mounds in unglaciated Arctic northwestern America. *Geogr. Rev.* 28:46-58.

Porter, S. C. 1975. Weathering rinds as a relative-age criterion. Application to subdivision of glacial deposits in the Cascade Range. *Geology* 3:101-04.

————. 1977. Present and past glaciation threshold in the Cascade Range, Washington, U.S.A. Topographic and climatic controls, and paleoclimatic implications. *Jour. Glaciol.* 18:101-16.

Post, A. S. 1960. The exceptional advances of the Muldrow, Black Rapids and Sustina glaciers. *Jour. Geophys. Research* 65:3703-12.

————. 1966. The recent surge of Walsh glacier, Yukon and Alaska. *Jour. Glaciol.* 6:375-81.

————. 1967. Effects of the March 1964 Alaskan earthquake on glaciers. U.S. Geol. Survey Prof. Paper 544-D.

————. 1969. Distribution of surging glaciers in western North America. *Jour. Glaciol.* 8:229-40.

Potter, N., Jr. 1972. Ice-cored rock glacier, Galena Creek, northern Absaroka Mountains, Wyoming. *Geol. Soc. America Bull.* 83:3025-57

Potter, N., Jr., and Moss, J. H. 1968. Origin of the Blue Rocks block field deposits, Berks County, Pennsylvania. *Geol. Soc. America Bull.* 79:255-62.

Potter, W. D. 1953. Rainfall and topographic factors that affect runoff. *Am. Geophys. Union Trans.* 34:67-73.

Potts. A. S. 1970. Frost action in rocks: Some experimental data. *Inst. Brit. Geog. Trans.* 49:109-24.

Powell, J. W. 1875. *Exploration of the Colorado River of the West (1869-72).* Washington, D.C.: U.S. Govt. Printing Office.

————. 1876. *Report on the geology of the eastern portion of the Uinta Mountains.* Washington, D.C.: U.S. Govt. Printing Office.

Powers, R. W. 1962. Arabian Upper Turassic carbonate reservoir rocks. In *Classification of carbonate rocks,* edited by W. E. Ham. Am. Assoc. Petroleum Geologists Mem 1:122-92.

Price, L. W. 1972a. *The periglacial environment, permafrost, and man.* Assoc. Am. Geog., Comm. on College Geog. Resource Paper 14.

————. 1972b. Up-heaved blocks: A curious feature of instability in the tundra. *Icefield Ranges Res. Proj.* 3:177-82.

Price, R. J. 1966. Eskers near the Casement glacier, Alaska. *Geogr. Annlr.* 48:111-25.

————. 1969. Moraines, sandar, kames, and eskers near Breidamerkurjökull, Iceland. *Inst. Brit. Geog. Trans.* 46:17-43.

————. 1970. Moraines at Fjallsjökull, Iceland. *Arc. Alp. Res.* 2:27-42.

————. 1973. *Glacial and fluvioglacial landforms.* New York: Hafner.

Price, W. A. 1968. Barriers—beaches and islands. In *Encyclopedia of geomorphology,* edited by R. W. Fairbridge, pp. 51-54. New York: Reinhold Book Corp.

Price, W. A., and Kornicker, L. S. 1961. Marine and lagoonal deposits in clay dunes, Gulf Coast, Texas. *Jour. Sed. Petrology* 31:245-55.

Pritchard, G. B. 1962. Inuvik, Canada's new Arctic town. *Polar Record* 11:71:145-54.

Qureshy, M.; Venkatachalam, S.; and Subrahmanyam, C. 1974. Vertical tectonics in the middle Himalayas: An appraisal from recent gravity data. *Geol. Soc. America Bull.* 85:921-26.

Rahn, P. H. 1966. Inselbergs and nickpoints in southwestern Arizona. *Zeit. f. Geomorph.* 10:217-25.

———. 1967. Sheetfloods, streamfloods, and the formation of pediments. *Ann. Assoc. Am. Geog.* 57: 593-604.

Ramsden, J., and Westgate, J. A. 1971. Evidence for reorientation of a till fabric in Alberta. In *Till: A symposium,* edited by R. P. Goldthwait. Columbus: Ohio State Univ. Press.

Rankin, J. K. 1952. Development of the New Jersey shore. In *3rd Coastal Engr. Conf. Proc.,* edited by J. W. Johnson, pp. 306-17. Cambridge, Mass.: Council of Wave Research.

Rapp, A. 1960. Recent development of mountain slopes in Karkevagge and surroundings, northern Scandinavia. *Geogr. Annlr.* 42:65-206.

Raraty, L. E., and Tabor, D. 1958. The adhesion and strength properties of ice. *Proc. Royal Soc. London,* ser. A:245:184.

Raudkivi, A. J. 1967. *Loose boundary hydraulics.* Oxford: Pergamon Press.

Ray, L. L. 1951. Permafrost. *Arctic* 4:196-203.

Rector, R. L. 1954. Laboratory study of the equilibrium profiles of beaches. U.S Army Corps Engrs., Beach Erosion Board Tech. Memo 41.

Reed, B.; Galvin, C. J.; and Miller, J. P. 1962. Some aspects of drumlin geometry. *Am. Jour. Sci.* 260:200-210.

Reiche, P. 1943. Graphic representation of chemical weathering. *Jour. Sed. Petrology* 13:58-68.

Revue de géomorphologie dynamique. 1967. Field methods for the study of slope and fluvial processes. *Rev. géomorph. dynamique* 17:145-88.

Rhoades, R., and Sinacori, M. N. 1941. Pattern of groundwater flow and solution. *Jour. Geology* 49:785-94.

Rich, J. L. 1935. Origin and evolution of rock fans and pediments. *Geol. Soc. America Bull.* 46:999-1024.

Richardson, H. W. 1942. Alcan-America's glory road, parts I and II. *Engr. News Record* 129:25:81-96 and 27: 35-42.

———. 1943. Alcan-America's glory road, part III. *Engr. News Record* 130:1:131-38.

———. 1944. Controversial Canol. *Engr. News Record* 132:2:78-84.

Richmond, G. 1962. Quaternary stratigraphy of the LaSal Mountains, Utah. U.S. Geol. Survey Prof. Paper 324.

Rigsby, G. P. 1960. Crystal orientation in glacier and in experimentally deformed ice. *Jour. Glaciol.* 3:589-606.

Ritter, D. F. 1967. Rates of denudation. *Jour. Geol. Educ.* 15, C.E.G.S. short rev. #6:154-59.

———. 1972. The significance of stream capture in the evolution of a piedmont region, southern Montana. *Zeit. f. Geomorph.* 16:83-92.

———. 1974. Holocene climate change and fluvial systems, abs. *Am. Quat. Assoc., Abs. 3rd Biennial Mtg.* (Madison, Wis.), pp. 58-62.

———. 1975. Stratigraphic implications of coarse-grained gravel deposited as overbank sediment, southern Illinois. *Jour. Geology* 83:645-50.

Ritter, D. F.; Kinsey, W. F.; and Kauffman, M. E. 1973. Overbank sedimentation in the Delaware River valley during the last 6,000 years. *Science* 179:374-75.

Robin, G. deQ. 1955. Ice movement and temperature distribution in glaciers and ice sheets. *Jour. Glaciol.* 2:523-32.

———. 1976. Is the basal ice of a temperate glacier at the pressure-melting point? *Jour. Glaciol.* 16:183-95.

Robin, G. deQ., and Weertman, J. 1973. Cyclic surging of glaciers. *Jour. Glaciol.* 12:3-18.

Robinson, G. 1966. Some residual hillslopes in the Great Fish River Basin, South Africa. *Geogr. Jour.* 132:386-90.

Roglić, J. 1972. Historical review of morphologic concepts. In *Karst,* edited by M. Herak and V. T. Stringfield, pp. 1-18. New York: American Elsevier.

Ronov, A., and Yaroshevsky, A. 1969. Chemical composition of the Earth's crust. In *The Earth's crust and upper mantle,* edited by P. Hart, pp. 37-57. Am. Geophys. Union, Geophys. Monograph 13.

Rossby, C. G. 1941. The scientific basis of modern meteorology. In *Climate and man,* U.S. Dept. Agri. Yearbook. pp. 599-655.

Röthlisberger, H. 1972. Water pressure in intra- and sub-glacial channels. *Jour. Glaciol.* 11:177-203.

Rubey, W. W. 1938. The force required to move particles on a stream bed. U.S. Geol. Survey Prof. Paper 189-E.

———. 1952. Geology and mineral resources of the Hardin and Brussels quadrangles (in Illinois). U.S. Geol. Survey Prof. Paper 218.

Ruhe, R. V. 1952. Topographic discontinuities of the Des Moines lobe. *Am. Jour. Sci.* 250:46-56.

———. 1965. Quaternary paleopedology. In *The Quaternary of the United States,* edited by H. Wright and D. Frey, pp. 735-64. Princeton, N.J.: Princeton Univ. Press.

———. 1969. *Quaternary landscapes in Iowa*. Ames, Iowa: Iowa State Univ. Press.

———. 1975. *Geomorphology*. Boston: Houghton Mifflin Co.

Russell, I. C. 1892. Mt. St. Elias and its glaciers. *Am. Jour. Sci.* 43:169-82.

———. 1893. Malaspina Glacier. *Jour. Geology* 1:219-45.

Russell, R. 1943. Freeze-thaw frequencies in the United States. *Am. Geophys. Union Trans.* 24:125-33.

Russell, R. J. 1944. Lower Mississippi valley loess. *Geol. Soc. America Bull.* 55:1-40.

Russell, R. J. 1967a. Aspects of coastal morphology. *Geogr. Annlr.* 49A:299-309.

———. 1967b. *River and delta morphology*. Louisiana State Univ., Coastal Studies Inst. Tech. Rept. 52.

Russell, R. J., and McIntire, W. G. 1965. Beach cusps. *Geol. Soc. America Bull.* 76:307-20.

Rust, B. R. 1972. Structure and process in a braided river. *Sedimentology* 18:221-45.

Ruxton, B. P. 1968. Measures of the degree of chemical weathering of rocks. *Jour. Geology* 76:518-27.

Ruxton, B. P., and Berry, L. 1961. Weathering profiles and geomorphic position on granite in two tropical regions. *Rev. géomorph. dynamique* 12:16-31.

Ruxton, B. P., and McDougall, I. 1967. Denudation rates in northeast Papua from potassium-argon dating of lavas. *Am. Jour. Sci.* 265:545-61.

Ryckborst, H. 1975. On the origin of pingos. *J. Hydrol.* 26:303-14.

Ryder, J. M. 1971a. The stratigraphy and morphology of para-glacial alluvial fans in south-central British Columbia. *Can. J. Earth Sci.* 8:279-98.

———. 1971b. Some aspects of the morphometry of paraglacial alluvial fans in south-central British Columbia. *Can. J. Earth Sci.* 8:1252-64.

St. Amand, P. 1957. Geological and geophysical synthesis of the tectonics of portions of British Columbia, the Yukon Territory, and Alaska. *Geol. Soc. America Bull.* 68:1343-70.

Salter, P., and Williams, J. 1965. The influence of texture on the moisture characteristics of soils, I. A critical comparison of techniques for determining the available-water capacity and moisture characteristic curve of a soil. *Jour. Soil Sci.* 16:1-5.

Samuels, S. 1950. The effect of base exchange on the engineering properties of soils. Bldg. Res. Sta. Gr. Britain, Note C176.

Saucier, R. T. 1968. A new chronology for braided stream surface formation in the lower Mississippi Valley. *Southeastern Geol.* 9:65-76.

Saucier, R. T., and Fleetwood, A. R. 1970. Origin and chronologic significance of Late Quaternary terraces, Ouachita River, Arkansas and Louisiana. *Geol. Soc. America Bull.* 81:869-90.

Savage, C. N. 1968. Mass wasting. In *Encyclopedia of geomorphology*, edited by R. W. Fairbridge, pp. 696-700. New York: Reinhold Book Corp.

Savage, J. C., and Paterson, W. S. B. 1963. Borehole measurements in the Athabasca Glacier. *Jour. Geophys. Research* 68:4521-36.

Sawkins, J. 1869. Report on the geology of America. *Mem. Geol. Survey*.

Schalscha, E.; Appelt, H.; and Schatz, A. 1967. Chelation as a weathering mechanism, I. Effect of complexing agents on the solubilization of iron from minerals and granodiorite. *Geochim. et Cosmochim. Acta* 31:587-96.

Schatz, A. 1963. Chelation in nutrition, soil microorganisms and soil chelation. The pedogenic action of lichens and lichen acids. *Jour. Agri. and Food Chem.* 11:112-18.

Schatz, A.; Cheronis, N.; Schatz, V.; and Trelawney, G. 1954. Chelation (sequestration) as a biological weathering factor in pedogenesis. *Pennsylvania Acad. Sci. Proc.* 28:44-57.

Scheidegger, A. E. 1965. The algebra of stream-order numbers. U.S. Geol. Survey Prof. Paper 525-B:187-89.

Scheidegger, A. E., and Potter, P. E. 1968. Textural studies of grading: Volcanic ash falls. *Sedimentology* 11:163-70.

Schoewe, W. M. 1932. Experiments on the formation of wind-faceted pebbles. *Am. Jour. Sci.* 24:111-34.

Schultz, C. B., and Frye, J. C. 1968. *Loess and related eolian deposits of the world*. Lincoln, Neb.: Univ. Nebraska Press.

Schumm, S. A. 1956. Evolution of drainage systems and slopes in badlands at Perth Amboy, New Jersey. *Geol. Soc. America Bull.* 67:597-646.

———. 1960. The shape of alluvial channels in relation to sediment type. U.S. Geol. Survey Prof. Paper 352-B.

———. 1962. Erosion of miniature pediments in Badlands National Monument, South Dakota. *Geol. Soc. America Bull.* 73:719-24.

———. 1963a. A tentative classification of alluvial river channels. U.S. Geol. Survey Circ. 477.

———. 1963b. Sinuosity of alluvial channels on the Great Plains. *Geol. Soc. America Bull.* 74:1089-1100.

———. 1963c. Disparity between present rates of denudation and orogeny. U.S. Geol. Survey Prof. Paper 454-H.

———. 1965. Quaternary paleohydrology. In *The Quaternary of the United States,* edited by H. E. Wright and D. G. Frey, pp. 783-94. Princeton, N.J.: Princeton Univ. Press.

———. 1967a. Rates of surficial rock creep on hillslopes in western Colorado. *Science* 155:560-61.

———. 1967b. Meander wavelength of alluvial rivers. *Science* 157:1549-50.

———. 1968. River adjustment to altered hydrologic regimen, Murrumbidgee River and paleochannels, Australia. U.S. Geol. Survey Prof. Paper 598.

———. 1969. River metamorphosis. Am. Soc. Civil Engineers. *Jour. Hydraulics Div., HY1*:255-73.

———. 1971. Fluvial geomorphology. In *River mechanics,* edited by H. W. Shen, chs. 4 and 5. Ft. Collins, Colo.: Colorado State Univ.

———, ed. 1972. *River morphology.* Stroudsburg, Pa.: Dowden, Hutchinson, and Ross.

———. 1973. Geomorphic thresholds and complex response of drainage systems. In *Fluvial geomorphology,* edited by M. Morisawa, pp. 299-310. S.U.N.Y., Binghamton, Pubs. in Geomorphology, 4th Ann. Mtg.

Schumm, S. A., and Chorley, R. J. 1964. The fall of Threatening Rock. *Am. Jour. Sci.* 262:1041-54.

———. 1966. Talus weathering and scarp recession in the Colorado Plateaus. *Zeit. f. Geomorph.* 10:11-36.

Schumm, S. A., and Hadley, R. F. 1957. Arroyos and the semiarid cycle of erosion. *Am. Jour. sci.* 255:161-74.

Schumm, S. A., and Khan, H. R. 1972. Experimental study of channel patterns. *Geol. Soc. America Bull.* 83:1755-70.

Schumm, S. A., and Lichty, R. W. 1963. Channel widening and flood-plain construction along Cimarron River in southwestern Kansas. U.S. Geol. Survey Prof. Paper 352-D.

———. 1965. Time, space and causality in geomorphology. *Am. Jour. Sci.* 263:110-19.

Schwartz, M. L. 1967. Littoral zone tidal cycle sedimentation. *Jour. Sed. Petrology* 37:677-83.

Schwartz, M. P. 1970. Multiple causality of barrier islands. *Jour. Geology* 79:91-94.

Schytt, V. 1964. Scientific results of the Swedish expedition to Nordaustlandet, Spitzbergen, 1957 and 1958. *Geogr. Annlr.* 46:243-81.

———. 1968. Notes on glaciological activities in Kebnekaise, Sweden during 1966 and 1967. *Geogr. Annlr.,* ser. A:50:111-20.

———. 1969. Some comments on glacier surges in eastern Svalbard. *Can. J. Earth Sci.* 6:867-73.

Scott, A. J., and Fisher, W. L. 1969. Delta systems and deltaic deposition. In *Delta systems in the exploration for oil and gas,* edited by W. Fisher, L. Brown, A. Scott, and J. McGowen. Univ. Texas, Austin, Bureau Econ. Geology.

Scott, G. 1963. Quaternary geology and geomorphic history of the Kassler Quadrangle, Colorado. U.S. Geol. Survey Prof. Paper 421-A.

Scott, K. M., and Gravlee, G. C., Jr., 1968. Flood surge on the Rubicon River, California—Hydrology, hydraulics, and boulder transport. U.S. Geol. Survey Prof. Paper 422-M.

Seed, H.; Woodward, R.; and Lundgren, R. 1964. Clay mineralogical aspects of the Atterberg limits. *Am. Soc. Civil Engineers Proc.* 90:SM4:107-31.

Selby, M. J. 1966. Methods of measuring soil creep. *J. Hydrol.* 5:54-63.

———. 1967. Aspects of the geomorphology of the Greywacke ranges bordering the lower and middle Waikato basins. *Earth Sci. Jour.* 1:1-22.

Senstius, M. 1958. Climax forms of chemical rock-weathering. *Am. Scientist* 46:355-67.

Sevon, W. D. 1969. Sedimentology of some Mississippian and Pleistocene deposits of northeastern Pennsylvania. In *Geology of selected areas in New Jersey and eastern Pennsylvania.* New Brunswick, N.J.: Rutgers Univ. Press.

Sharon, D. 1962. On the nature of hamadas in Israel. *Zeit. f. Geomorph.* 6:129-47.

Sharp, R. P. 1940. Geomorphology of the Ruby-East Humboldt Range, Nevada. *Geol. Soc. America Bull.* 51:337-72.

———. 1951. Features of the firn on upper Seward Glacier, St. Elias Mountains, Canada. *Jour. Geology* 59:599-621.

———. 1953. Deformation of a vertical bore hole in a piedmont glacier. *Jour. Glaciol.* 2:182-84.

———. 1960. *Glaciers.* Eugene, Ore.: Univ. Oregon Press.

———. 1963. Wind ripples. *Jour. Geology* 71:617-36.

———. 1964. Wind-driven sand in Coachella Valley, California. *Geol. Soc. America Bull.* 75:785-804.

———. 1966. Kelso Dunes, Mohave Desert, California. *Geol. Soc. America Bull.* 77:1045-74.

Sharp, R. P., and Noble, L. H. 1953. Mudflow of 1941 at Wrightwood, southern California. *Geol. Soc. America Bull.* 64:547-60.

Sharpe, C. F. S. 1938. *Landslides and related phenomena.* New York: Columbia Univ. Press.

Shaw, J. 1972. Sedimentation in the ice-contact environment, with examples from Shropshire (England). *Sedimentology* 18:23-62.

Shawe, D. R. 1963. Possible wind-erosion origin of linear scarps on the Saga Plain, south-western Colorado. U.S. Geol. Survey Prof. Paper 475-C:138-42.

Shepard, F. P. 1950. Beach cycles in Southern California. U.S. Army Corps Engrs., Beach Erosion Board Tech. Memo 20.

————. 1952. Revised nomenclature for depositional coastal features. *Am. Assoc. Petroleum Geologists Bull.* 36:1902-12.

————. 1960. Gulf Coast barriers. In *Recent sediments, northwest Gulf of Mexico,* edited by F. P. Shepard, F. B. Phleger, and T. H. van Andel, pp. 197-220. Tulsa, Okla.: Am. Assoc. Petroleum Geologists.

————. 1963. *Submarine geology.* New York: Harper and Row.

Shepard, F. P.; Emery, K. O.; and LaFond, E. C. 1941. Rip currents: A process of geological importance. *Jour. Geology* 49:337-69.

Shepard F. P., and Grant, U.S., IV. 1947. Wave erosion along the southern California coast. *Geol. Soc. America Bull.* 58:919-26.

Shepard, F. P., and Inman, D. L. 1950. Nearshore circulation related to bottom topography and wave refraction. *Am. Geophys. Union Trans.* 31:2:196-212.

Shepard, F. P., and Wanless, H. R. 1971. *Our changing coastlines.* New York: McGraw-Hill.

Shields, A. 1936. Arwendung der Ahnlichkeits mechanik und der Turbulenzforschung auf die Geschiebebewegung. *Mitteilungen der Preuss. Versuchganst fur Wasser, Erd und Schiffsbau,* no. 26. Berlin.

Shlemon, R. J. 1974. Holocene climate change and fluvial systems: Discussion. *Am. Quat. Assoc. Abs. 3rd Biennial Mtg.* (Madison, Wis.), p. 67.

Shreve, R. L. 1966a. Statistical law of stream numbers. *Jour. Geology* 74:17-37.

————. 1966b. Sherman landslide, Alaska. *Science* 154:1639-43.

————. 1967. Infinite topologically random channel networks. *Jour. Geology* 75:178-86.

————. 1968. *The Blackhawk landslide.* Geol. Soc. America Spec. Paper 108.

————. 1972. Movement of water in glaciers. *Jour. Glaciol.* 11:205-14.

————. 1974. Variation of mainstream length with basin area in river networks. *Water Resour. Res.* 10:1167-77.

————. 1975. The probabilistic-topologic approach to drainage-basin geomorphology. *Geology* 3:527-29.

Shreve, R. L., and Sharp, R. P. 1970. Internal deformation and thermal anomalies in lower Blue Glacier, Mount Olympus, Washington, U.S.A. *Jour. Glaciol.* 9:65-86.

Shumskii, P. A. 1964. *Principles of structural glaciology.* New York: Dover Publications.

Shuster, E. T., and White, W. B. 1971. Seasonal fluctuations in the chemistry of limestone springs: A possible means for characterizing carbonate aquifers. *J. Hydrol.* 14:93-128.

Silar, J. 1965. Development of tower karst of China and North Vietnam. *Natl. Speleol. Soc. Bull.* 27(2):35-46.

Silvester, R. 1966. Wave refraction. In *Encyclopedia of oceanography,* edited by R. W. Fairbridge, pp. 975-76. New York: Reinhold Book Corp.

Simonen, A. 1969. Batholiths and their orogenic setting. In *The Earth's crust and upper mantle,* edited by P. Hart, pp. 483-88. Am. Geophys. Union, Geophys. Monograph 13.

Simons, D. B., and Richardson, E. V. 1962. Resistance to flow in alluvial channels. *Am. Soc. Civil Engineers Trans.* 127:927-52.

————. 1963. Forms of bed roughness in alluvial channels. *Am. Soc. Civil Engineers Trans.* 128:284-302.

————. 1966. Resistance to flow in alluvial channels. U.S. Geol. Survey Prof. Paper 422-J.

Simpson, D. 1964. Exfoliation in the upper Pocahontas Sandstone, Mercer County, West Virginia. *Am. Jour. Sci.* 262:545-51.

Skempton, A. W. 1953. Soils mechanics in relation to geology. *Yorkshire Geol. Soc. Proc.* 29:33-62.

————. 1964. The long-term stability of clay slopes. *Geotechnique* 2:75-102.

Smalley, I. J. 1966. Drumlin formation. A rheological model. *Science* 151:1379.

————. 1970. Cohesion of soil particles and the intrinsic resistance of simple soil systems to wind erosion. *Jour. Soil Sci.* 21:154-61.

Smalley, I. J., and Unwin, D. J. 1968. The formation and shape of drumlins and their distribution and orientation in drumlin fields. *Jour. Glaciol.* 7:377-90.

Smart, J. S. 1969. Topological properties of channel networks. *Geol. Soc. America Bull.* 80:1757-73.

————. 1974. The random model in fluvial geomorphology. In *Fluvial geomorphology,* edited by M. Morisawa, pp. 27-49. S.U.N.Y., Binghamton, Pubs. in Geomorphology, 4th Ann. Mtg.

Smart, J. S., and Surkan, A. J. 1967. The relation between mainstream length and area in drainage basins. *Water Resour. Res.* 3:963-74.

Smith, D. D., and Dolan, R. G. 1960. Erosional development of beach cusps along the Outer Banks of North Carolina (abs). *Geol. Soc. America Bull.* 71, n.12:p1979.

Smith, D. D., and Wischmeier, W. H. 1962. Rainfall erosion. *Advances in Agron.* 14:109-48.

Smith, D. G. 1976. Effect of vegetation on lateral migration of anastomosed channels of a glacier meltwater river. *Geol. Soc. America Bull.* 87:857-60.

Smith, D. I. 1969. The solution erosion of limestone in an arctic morphogenetic region. In *Problems of the karst denudation,* edited by O. Štelcl, pp. 99-110. Brno.

Smith, G. D. 1942. Illinois loess-variations in its properties and distribution; a pedologic interpretation. *Univ. Illinois Agr. Exp. Sta. Bull.* 490:137-84.

Smith, H. T. U. 1948. Giant glacial grooves in northwest Canada. *Am. Jour. Sci.* 246:503-14.

———. 1953. The Hickory Run boulder field, Carbon County, Pennsylvania. *Am. Jour. Sci.* 251:625-42.

Smith, K. G. 1958. Erosional processes and landforms in Badlands National Monument, South Dakota. *Geol. Soc. America Bull.* 69:975-1008.

Smith, N. D. 1970. The braided stream depositional environment: Comparison of the Platte River with some Silurian clastic rocks, north-central Appalachians. *Geol. Soc. America Bull.* 81:2993-3014.

———. 1971. Transverse bars and braiding in the Lower Platte River, Nebraska. *Geol. Soc. America Bull.* 82:3407-20.

———. 1974. Sedimentology and bar formation in the upper Kicking Horse River, a braided outwash stream. *Jour. Geology* 82:205-23.

Snodgrass, D.; Groves, G.; Hasselmann, K.; Miller, G.; Munk, W.; and Powers, W. 1966. Propagation of ocean swell across the Pacific. *Phil. Trans. Royal Soc. London,* ser. A: 259:431-497.

Soil Survey Staff. 1951. *Soil survey manual.* U.S. Dept. Agri. Handbook 18, Soil Conserv. Serv.

———. 1960. *Soil classification, a comprehensive system—7th approximation.* U.S. Dept. Agri., Soil Conserv. Serv.

Sonu, C. J. 1972. Field observation of nearshore circulation and meandering currents. *Jour. Geophys. Research* 77:18:3232-47.

Soucie, G. 1973. Where beaches have been going: Into the ocean. *Smithsonian* 4:3:55-61.

Springer, M. E. 1958. Desert pavement and vesicular layer of some desert soils in the desert of the Lahontan Basin, Nevada. *Soil Sci. Soc. Am. Proc.* 22:63-66.

Stalker, A. MacS. 1960. Ice-pressed drift forms and associated deposits in Alberta. *Canada Geol. Survey Bull.* 57.

Stanley, D. J.; Krinitzsky, E. L.; and Compton, J. R. 1966. Mississippi River bank failure, Fort Jackson, Louisiana. *Geol. Soc. America Bull.* 77:850-66.

Stearns, S. R. 1966. Permafrost (perennially frozen ground). U.S. Army Corps Engrs., Cold Regions Res. and Eng. Lab, Cold Regions Sci. and Eng. 1 (A2).

Steers, J. A. 1962. *The sea coast.* London: Collins.

Steinemann, S. 1954. Results of preliminary experiments on the plasticity of ice crystals. *Jour. Glaciol.* 2:404-12.

———. 1958. Flow and recrystallization of ice. *Intl. Assoc. Sci. Hydrol. Pub.* 39:449-62.

Steinhart, J., and Woollard, G. 1961. Seismic evidence concerning continental crustal structure. In *Explosion studies of continental structure,* edited by J. Steinhart and R. Meyer. Washington, D.C.: Carnegie Institute Pub. 622.

Stephens, C. G. 1964. Silcretes of central Australia. *Nature* 203:1407.

Stevens, R., and Carron, M. 1948. Simple field test for distinguishing minerals by abrasion pH. *Am. Mineralogist* 33:31-49.

Stewart, J. H., and LaMarche, V. C., Jr. 1967. Erosion and deposition produced by the flood of Dec. 1964 on Coffee Creek, Trinity County, California. U.S. Geol. Survey Prof. Paper 422-K:1-22.

Stokes, W. L. 1964. Incised, wind-aligned stream patterns of the Colorado Plateau. *Am. Jour. Sci.* 262:808-16.

Stone, R. 1968. Deserts and desert landforms. In *Encyclopedia of geomorphology,* edited by R. W. Fairbridge, pp. 271-79. New York: Reinhold Book Corp.

Strahler, A. N. 1946. Geomorphic terminology and classification of land masses. *Jour. Geology* 54:32-42.

———. 1950. Equilibrium theory of slopes approached by frequency distribution analysis. *Am. Jour. Sci.* 248:800-814.

———. 1952a. Dynamic basis of geomorphology. *Geol. Soc. America Bull.* 63:923-38.

————. 1952b. Hypsometric (area-altitude) analysis of erosional topography. *Geol. Soc. America Bull.* 63:1117-42.

————. 1957. Quantitative analysis of watershed geomorphology, *Am. Geophys. Union Trans.* 38:913-20.

————. 1958. Dimensional analysis applied to fluvially eroded landforms. *Geol. Soc. America Bull.* 69:279-99.

————. 1964. Quantitative geomorphology of drainage basins and channel networks. In *Handbook of applied hydrology,* edited by V. T. Chow, pp. 4-39–4-76. New York: McGraw-Hill.

————. 1965. *Introduction to physical geography.* New York: John Wiley and Sons.

————. 1966. Tidal cycle of changes on an equilibrium beach. *Jour. Geology* 74:247-68.

————. 1968. Quantitative geomorphology. In *Encyclopedia of geomorphology,* edited by R. W. Fairbridge, pp. 898-912. New York: Reinhold Book Corp.

Strakhov, N. 1967. *Principles of lithogenesis.* London: Oliver and Boyd.

Stringfield, V. T., and LeGrand, H. E. 1969a. Hydrology of carbonate rock terranes—a review. *J. Hydrol* 8:349-413.

————. 1969b. Relation of sea water to fresh water in carbonate rocks in coastal areas, with special reference to Florida, U.S.A., and Cephalonia (Kephallinia), Greece. *J. Hydrol.* 9:387-404.

Sugden, D. E. 1968. The selectivity of glacial erosion in the Cairngorm Mountains, Scotland. *Inst. Brit. Geog. Trans.* 45:79-92.

————. 1970. Landforms of deglaciation in the Cairngorm Mountains, Scotland, *Inst. Brit. Geog. Trans.* 51:201-19.

Sundborg, A. 1956. The river Klarälven, a study of fluvial processes. *Geogr. Annlr.* 38:280-91.

Svasek, J. N., and Terwindt, J. H. J. 1974. Measurements of sand transport by wind on a natural beach. *Sedimentology* 21:311-22.

Svensson, H. 1959. Is the cross-section of a glacial valley a parabola? *Jour. Glaciol.* 3:362-63.

Swanson, D. 1972. Magma supply rate at Kilauea Volcano, 1925-1971. *Science* 175:169-70.

Sweeting, M. M. 1950. Erosion cycles and limestone caverns in the Ingleborough District of Yorkshire. *Geogr. Jour.* 115:63-78.

————. 1958. The karstlands of Jamaica. *Geogr. Jour.* 124:184-99.

————. 1968. Karstic morphology. In The University of Edinburgh British Honduras-Yucatan Expedition. *Geogr. Jour.* 134:49-54.

————. 1973. *Karst landforms.* New York: Columbia Univ. Press.

Swineford, A., and Frye, J. C. 1951. Petrography of the Peorian loess in Kansas. *Jour. Geology* 59:306-22.

Swinnerton, A. C. 1932. Origin of limestone caverns. *Geol. Soc. America Bull.* 43:662-93.

Swinzow, G. K. 1969. Certain aspects of engineering geology in permafrost. *Eng. Geol.* 3:177-215.

Symposium. 1940. Walter Penck's contribution to geomorphology. *Ann. Assoc. Am. Geog.* 30:219-80.

Taber, S. 1929. Frost heaving. *Jour. Geology* 37:428-61.

————. 1930. The mechanics of frost heaving. *Jour. Geology* 38:303-17.

————. 1943. Perennially frozen ground in Alaska: Its origin and history. *Geol. Soc. America Bull.* 54:1433-1548.

————. 1953. Origin of Alaska silts. *Am. Jour. Sci.* 251:321-36.

Tanner, W. F. 1958. The equilibrium beach. *Am. Geophys. Union Trans.* 39:889-91.

————. 1974. The incomplete flood plain. *Geology* 2:105-06.

Tarr, R. S., and Butler, B. S. 1909. The Yakutat Bay region, Alaska. U.S. Geol. Survey Prof. Paper 64:1-178.

Tator, B. A. 1952. Pediment characteristics and terminology. *Ann. Assoc. Am. Geog.* 42:295-317.

————. 1953. Pediment characteristics and terminology. *Ann. Assoc. Am. Geog.* 43:47-53.

Taylor, A. B., and Schwarz, H. E. 1952. Unit-hydrograph lag and peak flow related to basin characteristics. *Am. Geophys. Union Trans.* 33:235-46.

Terzaghi, K. 1936. The shearing resistance of saturated soils. *Proc. 1st Internat. Conf. on Soils Mech. and Foundation Eng.* 1:54-66.

————. 1943. *Theoretical soils mechanics.* New York: John Wiley and Sons.

————. 1950. Mechanism of landslides. In *Application of geology to engineering practice,* edited by S. Paige. Geol. Soc. America Berkey Vol., pp. 83-123.

————. 1955. Influence of geological factors on the engineering properties of sediments. *Econ. Geology* (50th Anniv. Vol.): 557-618.

————. 1962. Stability of steep slopes on hard unweathered rock. *Geotechnique* 12:251-70.

Thie, J. 1974. Distribution and thawing of permafrost in the southern part of the discontinuous permafrost zone in Manitoba. *Arctic* 27:189-200.

Thomas, T. M. 1973. Solution subsidence mechanisms and end-products in south-east Breconshire. *Inst. Brit. Geog. Trans.* 60:69-86.

Thornes, J. B. 1970. Hydraulic geometry of stream channels in the Xingu-Araguaia headwaters. *Geogr. Jour.* 136:376-82.

Thornthwaite, C. W. 1931. The climates of North America according to a new classification. *Geogr. Rev.* 21:633-55.

Thornton, E. B. 1973. Distribution of sediment transport across the surf zone. *Proc. 13th Conf. on Coast. Eng.*: 1049-68.

Thorp, J. 1934. The asymmetry of the Pepino Hills of Puerto Rico in relation to the Trade Winds. *Jour. Geology* 42:537-45.

Thorp, J., and Smith, G. 1949. Higher categories of soil classifications. Order, Suborder, and Great Soil Groups. *Soil Sci.* 67:117-26.

Thorp, J., and Smith, H. T. U. 1952. Pleistocene eolian deposits of the United States, Alaska, and parts of Canada. Geol. Soc. America Spec. Map.

Thrailkill, J. 1968. Chemical and hydrologic factors in the excavation of limestone caves. *Geol. Soc. America Bull.* 79:19-45.

———. 1972. Carbonate chemistry of aquifer and stream water in Kentucky. *J. Hydrol.* 16:93-104.

Tinkler, K. J. 1970. Pools, riffles, and meanders. *Geol. Soc. America Bull.* 81:547-52.

———. 1971. Pools, riffles, and meanders. Reply. *Geol. Soc. America Bull.* 82:282-83.

Todd, D. K. 1959. *Ground water hydrology.* New York: John Wiley and Sons.

———, ed. 1970. *The water encyclopedia.* Port Washington, N.Y.: Water Information Center.

Toth, S. 1964. The physical chemistry of soils. In *Chemistry of the soil,* edited by F. Bear, pp. 142-205. New York: Reinhold Book Corp.

Townsend, D. W., and Vickery, R. P. 1967. An experiment in regelation. *Philos. Mag.* 16:1275-80.

Trenhaile, A. S. 1971. Drumlins: Their distribution, orientation and morphology. *Can. Geog.* 15:113-26.

Tricart, J. 1967. Le modelé des régions périglaciaires. In J. Tricart and A. Cailleux, eds., *Traité de géomorphologie 2.* Paris: SEDES.

———. 1968. Notes géomorphologiques sur la karstification en Barbade (Antilles). *Mém. docums. cent. docum. cartogr. géogr.* 4:329-34.

———. 1969. *Geomorphology of cold environments,* trans. by Edward Watson. New York: St. Martin's Press.

Troeh, F. R. 1965. Landform equation fitted to contour maps. *Am. Jour. Sci.* 263:616-27.

Troll, C. 1958. Structure soils, solifluction, and frost climates of the Earth. U.S. Army Corps Engrs., Snow, Ice, Permafrost Res. Est. Trans. 43.

Tuan, Ti-Fu. 1959 *Pediments in southeastern Arizona.* Berkeley, Calif.: Univ. Calif. Pubs. in Geog. 13.

———. 1962. Structure, climate and basin land forms in Arizona and New Mexico. *Ann. Assoc. Am. Geog.* 52:51-68.

Tucker, M. J. 1950. Surf beats: Sea waves of 1 to 5 minute period. *Proc. Royal Soc. London,* ser. A, 202:565-73.

Turekian, K. K., ed. 1971. *The Late Cenozoic glacial ages.* New Haven, Conn.: Yale Univ. Press.

Tuttle, O., and Bowen, N. 1958. Origin of granite in light of experimental studies. *Geol. Soc. America Mem.* 74.

Twidale, C. R. 1967. Origin of the piedmont angle as evidence in South Australia. *Jour. Geology* 75:393-411.

———. 1968a. Yardang. In *Encyclopedia of geomorphology,* edited by R. W. Fairbridge, pp. 1237-38. New York: Reinhold Book Corp.

———. 1968b. Weathering. In *Encyclopedia of geomorphology,* edited by R. W. Fairbridge, pp. 1228-32. New York: Reinhold Book Corp.

———. 1972. Evolution of sand dunes in the Simpson Desert, central Australia. *Inst. Brit. Geog. Trans.* 56:77-110.

Ueta, H. T., and Garfield, D. E. 1968. Deep core drilling program at Byrd Station 1967-68. *U.S. Arctic Journal* 3:111-12.

Ursic, S. J., and Dendy, F. E. 1965. Sediment yields from small watersheds under various land uses and forest covers. *Proc. Fed. Inter-Agency Sedimentation Conf.* (1963). U.S. Dept. Agri. Misc. Publ. 970:47-52.

Vagners, V. J. 1966. Lithologic relationship of till to carbonate bedrock in southern Ontario. M.Sc. thesis, Geology Dept., Univ. Western Ontario.

Van Dorn, W. G. 1965. Tsunamis. In *Hydroscience Advances 2.* New York: Academic Press.

———. 1966. Tsunamis. In *Encyclopedia of oceanography,* edited by R. W. Fairbridge, pp. 941-43. New York: Reinhold Book Corp.

Vanoni, V. A. 1941. Some experiments on the transportation of suspended load. *Am. Geophys. Union Trans.,* 22nd Ann. Mtg., pt. 3:608-20.

———. 1946. Transportation of suspended sediment by water. *Am. Soc. Civil Engineers Trans.* 3:67-133.

Vanoni, V. A., and others. 1966. Sediment transportation mechanics. Initiation of motion. Am. Soc. Civil Engineers, *Jour. Hydraulics Div.* 92:HY2:291-313.

Varnes, D. J. 1958. Landslide types and processes. In *Landslides and engineering practice,* edited by E. Eckel, pp. 20-47. Washington, D.C.: Highway Research Board Spec. Rept. 29.

Verhoogen, J.; Turner, F.; Weiss, L.; Wahrhaftig, C.; and Fyfe, W. 1970. *The Earth.* New York: Holt, Rinehart and Winston.

Vernon, P. 1966. Drumlins and Pleistocene ice flow over the Ards Peninsula. *Jour. Glaciol.* 6:401-09.

Verstappen, H. Th. 1964. Karst morphology of the Star Mountains (central New Guinea) and its relation to lithology and climate. *Zeit. f. Geomorph.* 8:40-49.

Vine, F. J. 1966. Spreading of the ocean floor: New evidence. *Science* 154:1405-15.

Vine, F. J., and Matthews, D. H. 1963. Magnetic anomalies over oceanic ridges. *Nature* 199:947-49.

Vivian, R., and Bouquet, G. 1973. Subglacial cavitation phenomena under Glacier d' Argentiere, Mont Blanc, France. *Jour. Glaciol.* 12:439-52.

Wahrhaftig, C. 1965. Stepped topography of the southern Sierra Nevada. *Geol. Soc. America Bull.* 76:1165-90.

Wahrhaftig, C., and Cox, A. 1959. Rock glaciers in the Alaska Range. *Geol. Soc. America Bull.* 70:383-436.

Walcott, R. I. 1972. Late Quaternary vertical movements in eastern North America. Quantitative evidence of glacio-isostatic rebound. *Rev. Geophys. and Space Physics* 10:849-84.

Wall, J. R. D., and Wilford, G. E. 1966. Two small-scale solution features of limestone outcrops in Sarawak, Malaysia. *Zeit. f. Geomorph.* 10:90-94.

Ward, W. H. 1945. The stability of natural slopes. *Geogr. Jour.* 105: 170-97.

Warnke, D. A. 1969. Pediment evolution in the Halloran Hills, central Mojave Desert, California. *Zeit. f. Geomorph.* 13:357-89.

Warren, A. 1970. Dune trends and their implications in the central Sudan. *Zeit. f. Geomorph.,* Suppl. 10:154-80.

———. 1971. Dunes in the Ténéré Desert. *Geogr. Jour.* 137:458-61.

———. 1972. Observations on dunes and bi-modal sand in the Ténéré Desert. *Sedimentology* 19:37-44.

Washburn, A. L. 1956. Classification of patterned ground and review of suggested origins. *Geol. Soc. America Bull.* 67:823-66.

———. 1967. Instrumental observations of mass-wasting in the Mesters Vig district, Northeast Greenland. *Medd. om Grønland* 166:4.

———. 1970. An approach to a genetic classification of patterned ground. *Acta Geogr. Lodz.* 24:437-46.

———. 1973. *Periglacial processes and environments.* London. Edward Arnold Ltd.

Wayne, W. J. 1967. Periglacial features and climatic gradient in Illinois, Indiana and western Ohio, east-central United States. In *Quaternary paleoecology,* edited by E. Cushing and H. Wright, pp. 393-414. New Haven, Conn.: Yale Univ. Press.

Wear, J., and White, J. 1951. Potassium fixation in clay mineral studies as related to crystal structure. *Soil Sci.* 71:1-14.

Weertman, J. 1957. On the sliding of glaciers. *Jour. Glaciol.* 3:33-38.

———. 1961a. Stability of Ice Age ice sheets. *Jour. Geophys. Research* 66:3783-92.

———. 1961b. Mechanism for the formation of inner moraines found near the edge of cold ice caps and ice sheets. *Jour. Glaciol.* 3:965-78.

———. 1964. The theory of glacier sliding. *Jour. Glaciol.* 5:287-303.

———. 1967. An examination of the Lliboutry theory of glacier sliding. *Jour. Glaciol.* 6:489-94.

Weertman, J., and Weertman, J. B. 1964. *Elementary dislocation theory.* New York: Macmillan.

Weinert, H. 1961. Climate and weathered karroo dolerites. *Nature* 191: 325-29.

———. 1965. Climatic factors affecting the weathering of igneous rocks. *Agri. Meterol.* 2:27-42.

Welder, F. A. 1959. *Processes of deltaic sedimentation in the lower Mississippi River.* Louisiana State Univ., Coastal Studies Inst. Tech. Rept. 12.

Wendler, G., and Ishikawa, N. 1973. Experimental study of the amount of ice melt using three different methods: A contribution to the International Hydrological Decade. *Jour. Glaciol.* 12:399-410.

———. 1974. The combined heat, ice and water balance of McCall Glacier, Alaska: A contribution to the International Hydrological Decade. *Jour. Glaciol.* 13:227-41.

Wendler, G., and Streten, N. A. 1969. A short term heat balance study on a coast range glacier. *Pure Appl. Geophys.* 77:68-77.

Wentworth, C. K. 1938. Marine beach formation: Water level weathering. *Jour. Geomorphology* 1:6-32.

Westgate, J. A. 1968. Linear sole markings in Pleistocene till. *Geol. Mag.* 105:501-05.

Weyl, P. K. 1958. The solution kinetics of calcite. *Jour. Geology* 66:163-76.

Wheeler, S. S. 1968. *The desert lake.* Caldwell, Idaho: Caxton Printers, Ltd.

Whitaker, R. H.; Buol, S. W.; Niering, W. A.; and Havens, Y. H. 1968. A soil and vegetation pattern in the Santa Catalina Mountains, Arizona. *Soil Sci.* 105:440-51.

White, E. L., and Reich, B. M. 1970. Behaviour of annual floods in limestone basins in Pennsylvania. *J. Hydrol.* 10:193-98.

White, S. E., 1976a. Is frost action really only hydration shattering? A review. *Arc. Alp. Res.* 8:1-6.

———. 1976b. Rock glaciers and block fields, review and new data. *Quat. Res.* 6:77-98.

White, W. A. 1955. Water sorption properties of homoionic clay minerals. Ph.D. thesis, Univ. Illinois.

White, W. A. 1966. Drainage asymmetry and the Carolina capes. *Geol. Soc. America Bull.* 77:223-40.

White, W. B. 1960. Termination of passages in Appalachian caves as evidence for a shallow phreatic origin. *Natl. Speleol. Soc. Bull.* 22:43-53.

———. 1969. Conceptual models for carbonate aquifers. *Ground water* 7:15-21.

———. 1976. Geology and biology of Pennsylvania caves. Pennsylvania Geol. Survey Gen. Rept. 66:1-71.

White, W. B., and Schmidt, V. A. 1966. Hydrology of a karst area in east central West Virginia. *Water Resour. Res.* 2:549-60.

White, W. B., and Stellmack, J. A. 1968. Seasonal fluctuations in the chemistry of karst groundwater. *Proc. 4th Intl. Cong. Speleol.,* Ljubljana, 1965, 3:261-67.

Whitney, M. I., and Dietrich, R. V. 1973. Ventifact sculpture by wind-blown dust. *Geol. Soc. America Bull.* 84:2561-82.

Whittecar, G. R., and Mickelson, D. M. 1976. A theory of the formation of drumlins based upon their internal structures, Waukesha, Wisconsin: Balance between erosion and growth. *Geol. Soc. America Abs. with Programs* 8:6:1168-69.

Wiegel, R. L. 1964. *Oceanographical engineering.* Englewood Cliffs, N.J.: Prentice-Hall.

———. 1966. Wave theory. In *Encyclopedia of oceanography,* edited by R. W. Fairbridge, pp. 977-86. New York: Reinhold Book Corp.

Wilcock, D. N. 1971. Investigation into the relations between bedload transport and channel shape. *Geol. Soc. America Bull.* 82:2159-76.

———. 1975. Relations between planimetric and hypsometric variables in third- and fourth-order drainage basins. *Geol. Soc. America Bull.* 86:47-50.

Wilford, G. E., and Wall, J. R. D. 1965. Karst topography in Sarawak. *J. Trop. Geogr.* 21:44-70.

Williams, A. T. 1973. The problem of beach cusp development. *Jour. Sed. Petrology* 43:857-66.

Williams, G. 1964. Some aspects of the eolian saltation load. *Sedimentology* 3:257-87.

Williams, J. R. 1970. Ground water in the permafrost region of Alaska. U.S. Geol. Survey Prof. Paper 696.

Williams, L. 1974. Regionalization of freeze-thaw activity. *Ann. Assoc. Am. Geog.* 54:597-611.

Williams, P. J. 1966. Downslope soil movement at a Sub-Arctic location with regard to variations with depth. *Canadian Geotech. Jour.* 3:191-203.

Williams, P. W. 1963. An initial estimate of the speed of limestone solution in County Clare. *Ir. Geogr.* 4:432-41.

———. 1966a. Limestone pavements with special reference to Western Ireland. *Inst. Brit. Geog. Trans.* 40: 155-72.

———. 1966b. Morphometric analysis of temperate karst landforms. *Ir. Speleol.* 1:23-31.

———. 1971. Illustrating morphometric analyses of karst with examples from New Guinea. *Zeit. f. Geomorph.* 15(1):40-61.

———. 1972a. The analysis of spatial characteristics of karst terrains. In *Spatial analysis in geomorphology,* edited by R. J. Chorley. London: Methuen and Co.

———. 1972b. Morphometric analysis of polygonal karst in New Guinea. *Geol. Soc. America Bull.* 83:761-96.

Willman, H. B., and Frye, J. C. 1970. Pleistocene stratigraphy of Illinois. *Illinois Geol. Survey Bull.* 94.

Willman, H. B.; Glass, H.; and Frye, J. 1963. Mineralogy of glacial tills and their weathering profiles in Illinois, I. Glacial tills. *Illinois Geol. Survey Circ.* 347.

———. 1966. Mineralogy of glacial tills and their weathering profiles in Illinois, II. Weathering profiles. *Illinois Geol. Survey Circ.* 400.

Wilson, B. W. 1966. Seiche. In *Encyclopedia of oceanography,* edited by R. W. Fairbridge, pp. 804-11. New York: Reinhold Book Corp.

Wilson, I. G. 1972. Aeolian bedforms—their development and origins. *Sedimentology* 19:173-210.

———. 1973. Ergs. *Sed. Geol.* 10:77-106.

Wilson, J. T. 1963. A possible origin of the Hawaiian Islands. *Can. Jour. Physics* 41:863-70.

———. 1965. Transform faults, oceanic ridges, and magnetic anomalies southwest of Vancouver Island. *Science* 150:482-85.

Wilson, L. 1968. Morphogenetic classification. In *Encyclopedia of geomorphology*, edited by R. W. Fairbridge, pp. 717-28. New York: Reinhold Book Corp.

————. 1969. Les relations entre les processus géomorphologiques et le climat moderne comme méthode de paleoclimatologie. *Rev. géog. phys. et de géologie dyn.* 11:303-14.

————. 1972. Seasonal sediment yield patterns of United States rivers. *Water Resour. Res.* 8:1470-79.

————. 1973. Variations in mean annual sediment yield as a function of mean annual precipitation. *Am. Jour. Sci.* 273:335-49.

Winkler, E. M. 1965. Weathering rates as exemplified by Cleopatra's Needle in New York City. *Jour. Geol. Educ.* 13:50-52.

Winkler, E. M., and Wilhelm, E. J. 1970. Salt bursts by hydration pressures in architectural stone in urban atmosphere. *Geol. Soc. America Bull.* 81:567-72.

Winkler, H. 1965. *Petrogenesis of metamorphic rocks*. New York: Springer-Verlag.

Wise, D. U. 1972. Freeboard of continents through time. *Geol. Soc. America Mem.* (Hess Vol.) 132:87-100.

————. 1975. Continental margins, freeboard and volumes of continents and oceans through time. In *The geology of continental margins*, edited by C. Burk and C. Drake. New York: Springer-Verlag.

Woldenberg, M. J. 1966. Horton's laws justified in terms of allometric growth and steady state in open systems. *Geol. Soc. America Bull.* 77:431-34.

————. 1969. Spatial order in fluvial systems. Horton's laws derived from mixed hexagonal hierarchies of drainage basin areas. *Geol. Soc. America Bull.* 80:97-112.

————. 1971. The two dimensional spatial organization of Clear Creek and Old Man Creek, Iowa. In *Quantitative geomorphology,* edited by M. Morisawa. pp. 83-106. S.U.N.Y., Binghamton, Pubs. in Geomorphology, Proc. 2nd Symposium.

Wolfe, T. E. 1964. Cavern development in the Greenbrier Series, West Virginia. *Natl. Speleol. Soc. Bull.* 26:37-60.

Wollast, R. 1967. Kinetics of the alteration of K-feldspar in buffered solutions at low temperature. *Geochim. et Cosmochim. Acta* 31:635-48.

Wolman, M. G. 1955. The natural channel of Brandywine Creek, Pennsylvania. U.S. Geol. Survey Prof. Paper 271.

————. 1967. A cycle of sedimentation and erosion in urban river channels. *Geogr. Annlr.* 49-A:385-95.

Wolman, M. G., and Brush, L. M., Jr. 1961. Factors controlling the size and shape of stream channels in coarse noncohesive sands. U.S. Geol. Survey Prof. Paper 282-G.

Wolman, M. G., and Leopold, L. B. 1957. River flood plains; some observations on their formation. U.S. Geol. Survey Prof. Paper 282-C.

Wolman, M. G., and Miller, J. P. 1960. Magnitude and frequency of forces in geomorphic processes. *Jour. Geology* 68:54-74.

Wood, A. 1942. The development of hillside slopes. *Proc. Geol. Assoc.* 53:128-40.

Woodcock, A. H. 1974. Permafrost and climatology of a Hawaii volcano crater. *Arc. Alp. Res.* 6:49-62.

Woodcock, A. H.; Furumoto, A. S.; and Woollard, G. P. 1970. Fossil ice in Hawaii. *Nature* 226:873.

Woodruff, N. P., and Siddoway, F. H. 1965. A wind erosion equation. *Soil Sci. Soc. Am. Proc.* 29:602-08.

Woollard, G. P. 1962. *The determination of gravity from elevation and geologic data*. Univ. Wisconsin Geophys. and Polar Res. Center, Res. Rept. Ser. 62-9.

————. 1966. Regional isostatic relations in the United States. In *The Earth beneath the continents,* edited by J. Steinhart and T. Smith, pp. 557-94. Am. Geophys. Union, Geophys. Monograph 10.

————. 1972. Regional variations in gravity. In *The nature of the solid Earth,* edited by E. Robertson. New York: McGraw-Hill.

Woolnough, W. G. 1927. The duricrust of Australia. *Jour. Royal Soc. New South Wales* 61:24-53.

Wright, H. E. 1957. Stone orientation in the Wadena drumlin field, Minnesota. *Geogr. Annlr.* 39:19-31.

Wright, H. E., and Frey, D. G., eds. 1965. *The Quaternary of the United States.* Princeton, N.J.: Princeton Univ. Press.

Wright, J., and Schnitzer, M. 1963. Metallo-organic interactions associated with podsolisation. *Soil Sci. Soc. Am. Proc.* 27:171-76.

Wyllie, P. 1971. *The dynamic Earth.* New York: John Wiley and Sons.

Yaalon, D. H. 1975. Conceptual models in pedogenesis. Can soil-forming functions be solved? *Geoderma* 14:189-205.

Yaalon, D. H., and Ganor, E. 1966. The climatic factor of wind erodibility and dust blowing in Israel. *Israel Jour. Earth Sci.* 15:27-32.

Yang, C. T. 1971. On river meanders. *J. Hydrol.* 13:251-53.

————. 1973. Incipient motion and sediment transport. Am. Soc. Civil Engineers, *Jour. Hydraulics Div.* 99:HY10:1679-1704.

Young, A. 1960. Soil movement by denudational processes on slopes. *Nature* 188:120-22.

———. 1961. Characteristic and limiting slope angles. *Zeit. f. Geomorph.* 5:126-31.

———. 1972. *Slopes.* Edinburgh: Oliver and Boyd.

Zeigler, J. M.; Hayes, C. R.; and Tuttle, S. D. 1959. Beach changes during storms on outer Cape Cod, Massachusetts. *Jour. Geology* 67:318-36.

Zeigler, J. M., and Tuttle, S. D. 1961. Beach changes based on daily measurements of four Cape Cod beaches. *Jour. Geology* 69:583-99.

Zenkovich, V. P. 1967. *Processes of coastal development,* translated by D. G. Fry, edited by J. A. Steers. Edinburgh: Oliver and Boyd.

Zeuner, F. E. 1959. *The Pleistocene period. Its climate, chronology and faunal successions.* London: Hutchinson.

Zingg, A. W. 1940. Degree and length of land slope as it affects soil loss in runoff. *Agri. Engineering* 21:59-64.